THE HARMONY DEBATES

Exploring a practical philosophy for a sustainable future

Edited by Nicholas Campion

SOPHIA CENTRE PRESS

In partnership with
The University of Wales Trinity Saint David
and
The Sustainable Food Trust

The Harmony Debates
edited by Nicholas Campion

© Sophia Centre Press 2020

First published in 2020.

Sophia Centre Press
University of Wales, Trinity St David
Ceredigion, Wales SA48 7ED, United Kingdom.
www.sophiacentrepress.com

Cover Design: Jenn Zahrt

ISBN: 978-1-907767-22-7

British Library Cataloguing in Publication Data.
A catalogue card for this book is available from the British Library.

Typeset by Daniela Puia.
Printed by Lightning Source.

CONTENTS

Acknowledgments

We wish to acknowledge the generous assistance of the Harmony Institute at the University of Wales Trinity Saint David, Professor David Cadman, Patrick Holden, the Sustainable Food Trust, Bonnie Welch and the Harmony Project. Considerable support was also received from the University's Sophia Centre for the Study of Cosmology in Culture. We are also indebted to Jenn Zahrt, Kate White and Daniela Puia for their editorial and production skills.

The Harmony Institute
www.uwtsd.ac.uk/harmony-institute

The idea that the universe exists as a single whole is fundamental to many ancient philosophies and indigenous worldviews, as well as to modern science. This gives rise to the concept that the cosmos exists in a state of Harmony (with a capital 'H'), underpins the belief that human beings can live in harmony with the wider environment, and resonates with many traditions from both the East and West and Old and New Worlds. In practice, such ideas have implications in health studies, education, business, architecture, agriculture, conflict resolution and a range of other activities and disciplines. If all things are connected then they are also related, and the well-being of one depends on the well-being of another. Such notions can be contextualised within a framework of social justice and equality and underpin wider notions of sustainability. The Harmony Institute critically explores the theory and implementation of such ideas and practices across periods and cultures.

FOREWORD

HRH *The Prince of Wales*

From a speech to the Harmony in Food and Farming Conference, Llandovery College, 10 July 2017.

I FEEL TREMENDOUSLY HUMBLED by your creating an entire conference out of a desire to understand what I have been trying to point out for so long. For what has seemed rather an eternity, I found people have tended to think – or have been encouraged to think – that I was just concerned to pursue some sort of pet, 'New Age', niche farming project for food fads in this country, or just concerned about efforts to preserve the heritage of the past for some sort of ridiculous nostalgic reason, or that I wanted to see a kind of housing development that harked back to some long lost, golden age of building with everything covered in classical columns. What never seemed to be reported was that my concern has always been focused not on the past, but on the future and how best to address the critical environmental, economic and social issues of our day. In the end, I felt I simply had to produce a book that explained my proposition in a bit more detail, and that book was of course *Harmony*.[1] And it was my attempt to set out how we might approach the way we do things by looking at how nature herself operates, and it endeavoured to explain the simple tenets of the ancient philosophical standpoint that lay behind all of my efforts to put its tireless, perennial wisdom into action, not least when it comes to food and farming.

Now I must say it is particularly appropriate that here in Wales (which, of course, is renowned as the land of song), you should be exploring why I chose that important word as the title of my book. What you may not know is that the concept of harmony also lies at the very heart of traditional Welsh poetry. Some years ago, I had the pleasure of meeting the poet and former Archdruid of Wales, Dic Jones, who, as well as being a farmer in Ceredigion, was a master of *cynghanedd*, which is the ancient system of poetic meters in Welsh poetry. Dic's poetry followed the same system – with its meters, precise syllable counts and rhymes – so brilliantly that his work was compared with that of Dafydd ap Gwilym who was a contemporary of Chaucer and one of the chief glories of Welsh literature. The system is actually far older than Chaucer's day. It goes back over a thousand years and, thankfully, is still thriving today.

Crucially, the word cynghanedd may be translated as 'harmony' and embodies an approach that seeks to embody the principles of symmetry, proportion and beauty, not just in every poem, but, literally, within every line. Dic Jones actually wrote one of his *englyns* about cynghanedd itself:

Yn enaid yr awenydd – ei geiriau
Fel dau gariad newydd
Drwy ei sain a'u hystyr sydd
Yn galw ar ei gilydd.

In the soul of the author – harmony's words,[2]
like lover and lover,
through music and meaning are
calling to one another.

As I say, Dic was a farmer as well as a poet, and that is a rare combination, so I am very pleased to see that the arts also have a place at your conference. It is more commonly the view that things like beauty and harmony, a reverence for the sacred, putting nature at the heart of our thinking and so on, have no place at all in agricultural matters, in the design process, the way we do business, our approach to engineering and, certainly, to the way we might gear our entire economy. It is argued that in a world where resources are scarce, where populations are ballooning, where all that matters is the bottom line and where computers and digital technology can supposedly do the job much more efficiently and dispassionately, to consider a notion like beauty or harmony is to divert attention away from what matters most. Well, I would just say, be very careful. It is worth taking a step back and considering the consequences; what happens when we separate what we *are* from what we do.

That is what I believe has happened. We are struggling with the deep-rooted consequences of an immense separation. As I try to explain in the book, it has a long history that goes back beyond the dawning of the Scientific Revolution in the seventeenth century.

The first hint of a shift occurs during the course of the twelfth century when the very notion of the divine started to change. For all sorts of reasons, 'God' began to be seen as a separate entity – 'out there,' beyond creation, separate from nature. And with that came the idea that nature was an unpredictable force without inherent order. Humanity was seen as the instrument of the will of God, rather than a 'participant' in creation. And so, as God became separate from His Creation, so humanity became separate from nature, and thus, what I might call 'the organic unity of reality' began to fragment. It put paid to thousands of years of understanding of our place in the world and so put the teachings of all the great sacred traditions at odds with the way Western thinking was starting to go.

Now it is important to note that the ancient, but perennial philosophical principles lay at the root of every one of the world's great traditions, including the Western tradition founded by the ancient Greeks. To put Plato very simply – it was the philosophy of wholeness. It was a perception of the world that lasted right up to the thirteenth century in Christian philosophy too, and it taught that everything is interconnected and therefore interdependent, so that we inhabit a world where no one part of the whole can grow well or true without it serving the well being of the whole. What is more, there is an underlying geometry at work, a constantly moving pattern of life that is proportioned and remarkably balanced.

Sadly, as I charted in the book, that idea of humanity existing within 'a living whole' was abandoned by those who led the mechanistic revolution that found its feet in the seventeenth century. We kept the words but tended to forget their meaning. What has happened is that the sense of an animate nature in which we live and move and have our being has been replaced more or less wholesale by a rather more artificial idea that nature is some kind of autonomous machine with no purpose and no self-organising principles. And for me, that is a very damaging consequence of separating what we *are* from what we do. You only have to look at the precise and detailed scientific observations we now have to realise how uncomfortably close to the brink it has taken us, particularly when it comes to the appalling risks we are running with climate change.

Nowhere is this separation more starkly apparent than in agriculture. Food production in its rich variety of forms effectively covers some seventy per cent of the land in the United Kingdom, yet in my lifetime I have watched the industrialisation of food production turn the living organism of an individual farm into little more than a factory, where finite raw materials are fed in at one end, and food of varying quality comes out the other.

My great hope is that your conference might strengthen the common understanding of why this approach has to change – why we have to find ways of bringing about a widespread transition to farming, where farms become more balanced and harmonious entities – within nature, within their communities, and certainly within the capacity of the planet.

If you think about it, there is no technical reason why farms cannot become more diverse, nor why they cannot care more for the soil they depend upon; nor why farm animals can't be treated more humanely. Restoring harmony to farming means having to put back as much as you take out and thus working with the grain of nature – there is no reason why food cannot be produced in ways that enhance biodiversity rather than destroy it, and why, ultimately, the vital connection between the food producer and the food consumer can't also be restored. Re-forging that critical relationship would, I suspect, improve the chances of us making progress in all these other areas I've just mentioned.

What is encouraging, though, is that attitudes do seem to be changing. When once there would have been a discordant chorus of outraged abuse for talking about there being a comprehensive systemic relationship between all things, now eminent bodies in science and learning acknowledge there is truth in this. In many scientific fields, for instance, there is a growing realisation that we are, indeed, utterly embedded within nature's self-organising living web. To the extent that we are not simply a part of that web; we *are* the web ourselves. We *are* nature – *her* patterns are *our* patterns. We live and move and have our being within Nature's benevolent complexity and it is this living system that makes us – and which, incidentally, we are doing out utmost to test to destruction.

This is why, ladies and gentlemen, I find it so unbelievable when people ask why should we bother with the conservation and protection of the Earth's dwindling biodiversity, or why we should strive to make the terrifying environmental issues we now face such a priority. It is, of course, the diversity of life on Earth which actually enables us 'to have our being'. Deplete it, reduce it, erode and destroy it and we will succeed in causing such disorder that we risk de-railing humanity's place on Earth for good.

This is why I have been trying to say for so long that we have to look urgently at what will restore nature's balance before it is finally too late – and that moment, I hate to say, is upon us. We have to restore that perception of the world as a joined up, integrated unity. We have to reconcile the voices of both sides of our being, the intuitive and the rational; between, if you like, the East and the West in our consciousness.

So I am immensely encouraged by what is going on here in Wales, particularly at the University of Wales Trinity Saint David. And I cannot thank enough the Venerable Master Chin Kung for his decision to invest so much of his time and resources in supporting the university's Harmony Programme which is striving to teach the importance and process of interfaith dialogue and peace, but moreover

– and this, for me, is immensely significant – to explore ways in which Harmony itself can be developed as a proper discipline; one which takes a much more integrative view of things, in that farming is as related to the way we build things, as are the ways we approach, say, healthcare or business.

Work is already being done, on the ground, in education, and I gather you will be hearing in a little while how this all works from a head teacher, Richard Dunne who, for some unaccountable reason, as he's explained, was seemingly inspired by my explanation of the principles of harmony and went on to apply them in the classrooms of his own state primary school in Surrey. Now, as you will hear, enquiries of learning are carried out across the entire curriculum from the viewpoint of the principles of harmony. Which is to say that, rather than separating out the different subjects, as others have preferred, individually studying maths or chemistry, geography or economics, a subject like climate change becomes the subject of an enquiry of learning, which involves the application of all of those key disciplines, and others too.

This then is one very good example of how we might change our view of the world. And perhaps it might be a good start to this conference as you take a look at what can be gained from a study of the systemic web of life we call nature and how so many processes and patterns work so coherently to keep the whole of nature going. If we can apply ourselves to this, my hope is that we might begin to mimic that approach in so many fields of our endeavour. This leads me to my final point, which is to put this question to you – how might these patterns of behaviour, this 'grammar of harmony', better serve a more sustainable approach to food production and farming? And to that end, what can we learn from things like traditional architecture, traditional crafts, music, education and engineering, that might enable us to establish a much more sure-footed response to the enormous problems we face by forging a more circular form of economy, as Dame Ellen MacArthur has articulated so brilliantly.

Ladies and gentlemen, I began my own efforts to understand such questions with self-doubt. Now I have no doubt. We simply cannot solve the problems we have caused by responding with a 'business as usual' approach, trying to bounce back from every knock we take using the conventional approach, which only compounds the problem. What we have to do is bounce forward by learning from the past. We have to look again very seriously at the philosophy of wholeness that held sway for so long in all of the world's great sacred traditions. The clues are to be found in the arts of the past, in the music of the past, in the methods and approach of the traditional crafts, in the way we once revered the Earth and spoke openly of our inherent sense of the sacred, but above all in the inherent genius of nature herself. There lie the seeds of the answers, I promise you. This is not backward-looking and anti-science; it is reinstating the discarded baby that was rashly removed with the bathwater. So, the fact that you are about to do just that over these next two very full days is more encouraging to me than you can ever imagine and I much look forward to hearing if you can resuscitate the baby – harmoniously!

Notes

1.   HRH The Prince of Wales, Tony Juniper and Ian Skelly, *Harmony: A New Way of Looking at our World* (London: Harper Collins, 2010).
2.   Literally 'its words'.

INTRODUCTION

Nicholas Campion

This is a call to revolution. The Earth is under threat. It cannot cope with all that we demand of it. It is losing its balance and we humans are causing this to happen.

HRH The Prince of Wales.[1]

The solar system reminds us that, just as the Earth is not at the centre of the Universe, neither are we humans the centre of the Earth. We, along with the rest of the natural world, are all interconnected within the larger web of life.

The United Nations.[2]

WE ALL LIVE ON ONE PLANET. And we share its resources, its water, land and air. Currents in the oceans and atmosphere travel the Earth, sometimes in a matter of days. As we have found out to our cost, climate change, nuclear leaks and virus pandemics do not respect national boundaries, social class, or ethnic divisions. To guard against, or recover from, any such calamity requires a global effort and new ways of thinking, as we have seen.

This current volume is inspired by the publication in 2010 of *Harmony: A New Way of Looking at our World*. The three authors were His Royal Highness The Prince of Wales; Tony Juniper, the environmental campaigner who had recently ended a term as Executive Director of Friends of the Earth and is currently Chair of Natural England; and Ian Skelly, the broadcaster and trustee of the Temenos Academy.[3] The Prince of Wales, of course, is well-known both for his interest in traditional philosophies and his engagement with practical solutions to the natural and social problems of our time. He is also Patron of the University of Wales Trinity Saint David (UWTSD), which accounts for the University's current involvement with the practice and philosophy of harmony.

The concept of harmony assumes that everything in the universe is interconnected, interrelated, and existing in a state of balance. It's a radical idea which draws no distinction between the physical, emotional and intellectual – bodies, feelings and thoughts. It's also a notion that is found in many cultures, ancient and modern.

The Prince of Wales' book is distinctive within the English speaking world in that it is the first detailed exploration of the implications of the concept of cosmic harmony for social, political and economic life since the 1619 publication of *The Harmony of the World* by Johannes Kepler. There is a long lineage behind such ideas, and the model which Kepler adopted had previously been set out in around 400 BCE by the Greek philosopher Plato, who described a system of education, law and politics designed to preserve stability in a Harmonious cosmos. Kepler himself hinted at the political consequences of Harmony, but largely focused on its mathematical, astronomical and geometrical features. In his own book, the Prince of Wales includes material on planetary cycles and geometry, but also devotes a great deal of attention to exploring harmony as a foundational philosophy for sustainability. Beginning with a foundation in classical notions of geometry, pattern, proportion and beauty, *Harmony* explores the natural environment and our own place within it. The book is notable for the range of material it includes, from the global to the local and from the philosophical to the practical, crossing agriculture, architecture, business and education, and always with a view as to how we as individuals can act in pursuit of a more balanced, sustainable and equal world. It raises questions such as how we manage or place ourselves in, and engage with, the natural world, develop the built environment, and nurture communities. The book is also embedded in current conversations concerning cultural politics; it engages in a serious critique of the downside of the metaphysics of modernism, with its uncritical worship of technological progress at all costs, its obsession with profit as the measure of economic well-being, and its model of life as a kind of machine. There is a distinct urgency underlying the book's message:

> We are at an historic moment – because we face a future where there is a real prospect that if we fail the Earth, we fail humanity. To avoid such an outcome, which will comprehensively destroy our children's future, we must urgently confront and then make choices which carry monumental implications.[4]

This makes the need for practical action even more important:

> If we want to hand on to our children a much more durable way of operating in the world, then we have to embark on what I can only describe as a 'Sustainability Revolution' – and with some urgency. This will involve taking all sorts of dramatic steps to change the way we consider the world and act in it, but I believe we have the capacity to take the steps. All we have to see is that the solutions are close at hand.[5]

This, it is argued, means putting nature at the heart of everything we do. The Prince and his fellow authors then enumerate seven key areas for action: sustainable agriculture, especially organic farming; sustainable urbanism based on social and environmental value; mixed-use development; putting pedestrians at the centre of the design process; emphasising local identity and using ecological building techniques; action to preserve natural ecosystems; a balanced approach to healthcare and medicine; more rounded educational systems; and a more holistic approach to science and technology.[6] Success in implementing these goals then depends both on effective partnerships between private and public sectors and non-governmental organisations, in order to extend lines of communication and take into account the widest range of issues. There is, of course, a huge, valuable and growing literature on such questions, stretching back well over half-a-century, adopting philosophical positions and seeking practical solutions to global problems. . This book aims to add a new dimension to such work, exploring an approach to environmental questions which is both modern and rooted in antiquity and a variety of cultures.

The notion of harmony as a desirable state of existence attracted the attention of the United Nations in 2009 and, since then, the UN's General Assembly has adopted no less than nine resolutions on 'Harmony with Nature'.[7] The culmination of these discussions was the drafting in 2015 of a major document, the 2030 Agenda for Sustainable Development. As the UN states,

> This Agenda is a plan of action for people, planet and prosperity. It also seeks to strengthen universal peace in larger freedom. We recognize that eradicating poverty in all its forms and dimensions, including extreme poverty, is the greatest global challenge and an indispensable requirement for sustainable development.[8]

All UN member states then adopted the now well-known seventeen Development Goals, together with a fifteen-year implementation plan. Harmony appears first in the 2030 Agenda in the Preamble under the heading 'Prosperity', which established the document's global aspirations:

> We are determined to ensure that all human beings can enjoy prosperous and fulfilling lives and that economic, social and technological progress occurs in harmony with nature.[9]

Taken together, the various sections in the UN's documents offer a fairly comprehensive definition of what it means by harmony: these are sustainability, care for the natural environment, economic equality, government based on democracy and the rule of law, and justice and rights for all.[10] However, there are clear tensions between the need to promote economic development on the one hand, and maintain the health of the natural environment on the other. This is hardly surprising considering the compromises which are necessary if a multitude of national representatives are to reach a consensus. Still, it is remarkable that the document was even agreed.

Harmony and sustainability, though, are not exactly the same. The term sustainability is itself a problematic one. Nobody can disagree with the objective of being sustainable. However, like other words in the environmental lexicon it has been appropriated by some industries and businesses in ways that many may find uncomfortable. For example, it is possible to talk about mining industries as sustainable in a narrow sense, as being profitable, even when they cause huge damage to the environment. However, the word is used by the United Nations, as it is by most environmentalists and ecologists, in the sense of making the whole planet sustainable, with all its natural systems. And this is the sense in which harmony provides a framework for thinking about sustainability.

The UN suggests that harmony will be the outcome of the successful implementation of sustainable policies, which may well be true, but there is an opposite view which sees sustainability as being based in harmony. In other words, harmony precedes sustainability rather than follows it. In this perspective, sustainability by itself is only part of the picture, and comes with its own limitations. This point was made by David Cadman, a Harmony Professor of Practice, in a talk at the University of Wales Trinity Saint David:

> Sustainability, in the way in which it has come to be used, describes the relationships between environment, society and economy that can be sustained and nourished over long periods of time for mutual benefit. It tends towards being instrumental: 'We do this so that...' The problem is this: that for many the broad matter of sustainability has been reduced to the more narrow matter of environmental management, and within this to the even more limited matters of energy and waste. Turning off the lights and recycling. This is necessary but not sufficient.[11]

The pursuit of sustainability, from this perspective, is open to criticism on the grounds that it can be too managerial, and concerned primarily with quick fixes to environmental problems. This is David Cadman again:

So, is harmony a tool of sustainability or does sustainability sit within principles of harmony? My view is that sustainability, in either its wider or more narrow form, cannot be explored other than in the context of harmony. It is a circle within a circle... perhaps this is something we should explore together.[12]

Harmony is therefore an overarching philosophy which seeks to provide a broad guide for action. The notion of a harmonious world in which all things operate together in a system which is ultimately balanced and benign, is a particularly appealing one. The idea of harmony as balance and order can be traced back to the classical Greek world, where *harmonia* meant 'union' or 'fitting together'. And it was the Greek philosophers who articulated the concept, widespread in the ancient world, that the entire universe is a single integrated whole. The movements of the stars and planets, they believed, make sounds as they travel and, if we could hear these, they would make a beautiful melody. This is the foundation of the 'Harmony of the Spheres', a notion which was popular amongst Renaissance thinkers and inspired a belief that the purpose of culture, politics and religion should be to avoid conflict and manage collective affairs for the benefit of all. The same ideas about universal balance and the integration of all things occur in many cultures in multiple forms.

The worldviews which maintain this notion are well established. They include Stoicism from the classical Greek and Roman world, Buddhism from India, and Taoism and Confucianism from China. The belief that all things are related also pervades traditional and indigenous cultures. All these ways of thinking and living are alive and influential in the modern world and have much to say about our relationship with the environment and politics. For example, Stoicism, which sees humanity as wholly part of nature rather than separate from it, is fundamental to Deep Ecology. Debates on the nature of Confucianism are central to arguments about the modern Chinese State on both sides of the debate, both pro- and anti-Chinese government. In parts of Africa, notions of the relationship between spirits of the ancestors and the natural world have been used to encourage communities to maintain rather than degrade their environments and resources.

This does not mean that there are necessarily simple parallels between different cultures and languages, which may not have a precise equivalent of the Greek term *harmonia*. In this volume M. A. Rashed identifies the modern Arabic synonym for 'harmony' or 'to be in harmony' as *tanāghum*, and David Rubin talks of three individual Hebrew terms – *sholom* (or *shalom*; 'peace'), *yachad* ('together'), and *tif'ereth* ('beauty' or 'splendour') – as representing different aspects of harmony.[13] And in China, the Mandarin word *he* (和), which is now often taken as a synonym for harmony, may not have had the same meaning historically. We should therefore be aware of our tendency to project the Greek word *harmonia* onto other cultures and, instead, we should acknowledge a range of meanings – and recognise that we are also currently redefining the word. As has become clear in the editing of this book, we are now using the word in two distinct yet often overlapping senses: Harmony (with a capital 'H') implies that there is an essential cosmic order written into the fabric of the universe, while harmony (with a small 'h') is a loose synonym for balance, peace and reconciliation. Both spellings are used throughout this book – and this chapter – depending on context.

With this in mind, David Cadman has provided a definition of harmony which has caught a mood and is widely quoted through this volume. In his words,

> Harmony is an expression of wholeness, a way of looking at ourselves and the world of which we are part. It's about connections and relationships. The emotional, intellectual and physical are all connected. We are connected to our environments, both built and natural; and all the parts of our communities and their environments are connected, too. Harmony asks questions about relationship, justice, fairness and respect in economic, social and political relationships. As an integrative discipline it can be expressed in ideas and practice.[14]

The crucial concept here is that Harmony asks questions. It is not a fixed entity, but a way of looking at the world, of exploring the manner in which land, sea, sky, plants, animals, agriculture, industry, society, education and politics are interconnected; it is also a framework for developing policies and practices which recognise such interconnections. The notion of interconnectedness leads to the idea of correct, or right, action, which requires respect for all people, life, and the Earth itself. And this, in turn, means that some forms of political and economic activity are inherently inharmonious, including political structures which depend on repression and conflict, as well as economic systems which are based on exploitation and pollution. In this sense, harmony takes a position which is radical and revolutionary. A softer word, which many prefer, is transformational. For example, in politics, the priority must be dialogue, debate and respect for individual freedoms and our fellow humans rather than conflict, confrontation and authoritarianism: we see, then, that the harmony perspective prioritises non-violent action, social justice and diversity.

Interconnectedness has another consequence: if all things are linked then there are no linear causes. Object A does not cause object B to move. Instead the world is complex. It is chaotic, although not in the sense of random confusion but, as modern chaos theory argues, in terms of complex, related and interlocking patterns. And it is cyclical. Life follows the patterns of day and night (as do the circadian rhythms which regulate both our body-clocks) and the seasons.

There are many – probably thousands – of examples of harmonious activity within political organisations, private and public institutions, communities and businesses, but few relate what they do specifically to the theory of Harmony. One who does is Richard Dunne, the former head teacher of a primary school in south-east England. Dunne has explored the relationship between harmony and sustainability through the curriculum which he designed. He is careful to talk about 'principles' of Harmony rather than '*the* principles', in order to stress that there is no fixed list – and that there *can be no* fixed list. In order to explore the application of Harmony in education, he has delineated seven key principles of Harmony in the natural world which could be used as a framework for developing more enlightened practice in education and in many other sectors. The principles which Dunne has identified are: Geometry (the patterns which underpin many natural phenomena); the Cycle (as in natural cycles); Interdependence; Diversity; Health and Wellbeing; Adaptation (as in the need to respond to change); and Oneness (the dynamic and interrelated quality of all things). He then gives a series of practical examples, taken from his experience with Year 4 children (aged 8 to 9). In each of Dunne's examples a Harmony principle provides the underlying idea, while the resulting practical project is a means of promoting sustainability. Harmony and sustainability therefore exist in a symbiotic relationship, in a kind of ecology of mind and action.

Autumn (Fall) term

Harmony Principle: Adaptation. Children pursue a project combining an appreciation of local history and an understanding of food at that time, looking at how they might prepare a sixteenth century banquet and gaining an idea of how the world has changed.
Sustainability: Children look for ways to source food locally for their banquet.

Harmony Principle: Interdependence. Children learn about their local community and organise a community partnership project.
Sustainability: Children work together with local partners to organise a

community project.

Spring term

Harmony Principle: The Cycle. Children learn about the order of the solar system and its cycles.
Sustainability: Children develop a project to understand and promote solar energy.

Harmony Principle: Diversity. Children observe the night sky and the constellations of the stars, learning about their stories.
Sustainability: Children learn to appreciate the awe and wonder of the universe.

Summer term

Harmony Principle: Oneness. Children learn about notions of balance and limit in relation to life in Ancient Egypt.
Sustainability: Children consider how they can learn to live within the carrying capacity of the planet.

Harmony Principle: Health. Children learn where their food comes from and how it is produced.
Sustainability: Children explore options for local, seasonal, free range and organic food.[15]

Dunne's analytical model could easily be applied to other areas, business being a notable example. We can imagine a class full of executives being asked to think about why they wrap their products in packaging which serves no purpose, is difficult to remove, and impossible to recycle, and to consider the total cost of manufacturing, including sourcing and transporting the original materials, not to mention disposing of (hopefully recycling) the final article at the end of its useful life. It is from Richard Dunne and David Cadman that we derive this book's subtitle, 'Exploring a practical philosophy for a sustainable future': Harmony is both theoretical and practical.

     In terms of practical politics one way forward is to look at every policy decision in terms of the widest range of consequences that can be anticipated, and we must act on the basis of the widest available evidence. In an absolute sense this may be impossible and could be a recipe for complete inaction. Ultimately, action is necessary and so too are fine judgments on which path to take and which policies to implement. Pragmatism therefore requires that choices are made and policies implemented, even if compromises are necessary. In difficult situations there may well be no obvious solution to any single problem. Indeed, even to think of single problems may be a mistake, for if all things are interconnected and interrelated, nothing can be understood separately from its social, economic, political, environmental, psychological and spiritual consequences. The world is complex and the essence of the Harmony perspective is therefore pragmatic rather than dogmatic. It means that we must consider the consequences of our actions; it requires transparency in policy making and open consideration of all options. If we follow this path this then we may better avoid the problems that can result when we implement simple solutions to complex problems. Some single solutions have actually made a bad situation worse. In business, examples include the introduction of the plastic bag to replace paper in order to save trees, a worthy policy which resulted in a plague of plastic. Another was the promotion of diesel cars in the 1990s as an alternative to petrol, in spite of the fact that, as was well-known at the time, they produce their own form of poison. And in politics, military action which might have solved one problem has generated many others: we may think of Iraq. In ideal circumstances, the harmony approach requires working with people rather than against them, enabling, encouraging, and facilitating improvements in the ways in which we behave. As an example, the belief that wild-life conservation could be pursued by expelling farmers or herders from their land has been replaced by the understanding that engaging communities is far more effective.

For individuals, the questions include asking what difference harmony makes, and are there personal benefits? One of the most important implications relates to how we manage our health. While many of us experience medical emergencies in which surgery or drug treatment are the only solutions, the maintenance of a healthy lifestyle is of much wider concern and underpins the notion of well-being, or wellness, including mental well-being. Well-being in turn depends on our understanding of the wider environment, and a wealth of current research has shown that enjoyment of the natural world – even a daily walk – is vital for a balanced state of mind. Such advice is obvious, but often ignored. The question of personal gain can then be answered in terms of health, well-being and participation in a peaceful world.

The wider environment includes our relationships, domestic contexts and communities, all of which are necessary for mental well-being. Maintaining physical health requires a healthy and balanced diet; as nutritionists know, what we eat can have a dramatic effect on some diseases. And from there we move to the global level, to which all individuals are inextricably linked. Where does our food come from? How was it produced? And how does it travel? What about cars, planes and mass tourism? How do we manage our own movement across the planet with a light touch?

In education, children need a rounded curriculum consisting of a balance of academic subjects, creative arts, crafts and physical activities, which takes into account their personal needs. Some children are better with words, others with numbers or painting. For children who are bad at sport, exercise systems such as dance, yoga or tai chi could be ideal. Children should also understand where their food comes from and, where possible, engage with community farms or help grow their own food.

In architecture and design, urban planning must take into account the needs of the natural environment, the individual and the community, and property developers must act with regard to natural, historic and social sensibilities. In business, the circular economy, in which there is no waste, is a priority, as is care by employers for their employees. Manufacturers should take into account the need to recycle their products (the built-in obsolescence in so many modern high-tech goods runs counter to harmony principles). In business, total cost accounting, in which the costs of producing and transporting goods across an entire supply chain, including pollution and costs to communities, should be taken into account.[16]

One more point needs to be made. While we ourselves may strive for harmony, models of cosmic Harmony assume that the universe moves between harmonious and inharmonious states. As John Eliot Gardiner says in this volume, 'Harmony by its very essence predicates and generates a balance between opposing forces, or in musical terms, "the combination of simultaneously sounded musical notes to produce a pleasing or agreeable effect"'. In some of the earliest writings on the topic, the Greek philosopher Empedocles saw the world alternating between periods of Love and Strife, not unlike the alternation between Yin and Yang in Chinese thought. Paradoxically, being out of balance may be as much a form of balance as being in balance. For, as the various philosophical schools that deal with the topic insist, it is this that makes our own pursuit of peace and balance so crucial.

For we in the University of Wales Trinity Saint David, the study of Harmony and its practice is an essential part of the implementation of the Well-Being of Future Generations Act. This visionary piece of legislation embodies the Welsh

Government's legal commitment to making sure that all policy decisions take into account the welfare and well-being of those who come after us. While the Act has legal force only in Wales, the University is keen to develop international links, recognizing that we all live on one planet. Within the academic world we can take a variety of approaches. In vocational subjects (such as education and business), we may explore the application of harmony principles, but all academic perspectives require that we also ask what we mean by harmony. In this book, then, we are not saying what harmony – or Harmony – is or is not, or how it can or cannot, or should or should not, be applied. Individual authors may well suggest such things, but the totality of all the contributions is a conversation.

Thus this book explores harmony, and Harmony, through a variety of voices. Many of contributions are based on presentations at two conferences at the University of Wales Trinity Saint David in Lampeter in 2017 and 2018, and the Harmony in Food and Farming conference organised by the Sustainable Food Trust and the Harmony Project at Llandovery College in 2018. Other chapters have been solicited from thinkers, practitioners and activists who are engaged in areas ranging from religion and philosophy to the arts and agriculture.

The chapters include critiques of the concept of Harmony in classical thought and ecology, research into its application in education, exploration of its role as a form of perennial philosophy, its study in anthropology, its expression in the arts, and the ways it can underpin sustainable development and business. We also include personal statements on what harmony is or ought to be and how we can work with it. We ask whether Harmony is a state to be attained, or a process, and whether it involves protest as much as it does reconciliation and resolution: being out of balance may be as much a form of balance as being in balance as long as the process continues. To extend John Eliot Gardiner's musical analogy, we might think of the contrast between dissonance and consonance.

The contributors come from a wide range of backgrounds. Between them they all address the seven key areas set out in *Harmony*, along with the philosophical discussions which run through that book. We are very proud that the foreword has been contributed by the Prince of Wales, from his opening talk at the 2018 Food and Farming in Harmony conference. We also include contributions from two people who have been instrumental in developing the enquiry into, and application of, harmony principles in the University of Wales Trinity Saint David: David Cadman, one of the University's Harmony Professors of Practice, and Patrick Holden, Director of the Sustainable Food Trust and the Harmony Project. Also from the world of food and farming we include Helen Browning, organic farmer and director of the Soil Association; Gunhild Stordalen, founder and president of EAT Foundation; and Angie Polkey of the Lampeter Permaculture Group. Luci Attala extends the discussion into the project in which she is engaged in Kenya concerning water. Louise Emanuel, a Commissioner for the Royal Commission on the Ancient and Historical Monuments, explores harmony and a sense of place, while Ilaria Cristofaro looks at the interplay of land, sea and sky. From the arts we have composer John Eliot Gardiner, musician Kayleen Asbo, sculptor Sophie Howard and dancer and writer Wendy Buonaventura. From the world of environmental activism and campaigning, we have another two of the University's Harmony Professors: John Sauven, head of Greenpeace UK, and Tony Juniper, Chair of Natural England and co-author of the Prince of Wales' book on harmony, also both Harmony Professors. On business we include Mark Goyder and Dame Ellen MacArthur. A number of contributors engage with community issues, including Rachel Parker, Mike Durke and, on the basis of her work in South Africa, Eve Annecke. Richard Dunne and Emilie Martin, Caroline Lohmann-Hancock and Nichola Welton, and Glenda Tinney all examine harmony in education, while Tania Davies discusses her research within the University. Rupert Sheldrake, Marc Andrus, Bishop of the Episcopal Diocese of California, and Stephan Harding, Head of Holistic Science at Schumacher College, explore issues arising from science. Rhodri Thomas, M. A. Rashed and David Rubin deal with harmony in Christianity, Islam and Judaism respectively. Joseph Milne and Crystal Addey examine harmony in Plato and Platonic philosophy, a narrative which Angela Voss extends into the Renaissance. Angus Slater, Jack Hunter and Jeremy Naydler each take a wide philosophical brief, and Toto Gill explores recent scholarly literature on harmony in various philosophical traditions. Alan Ereira, the celebrated-documentary maker and honorary Professor, explores harmony amongst the Kogi of Colombia, while Trevor Leaman examines land and sky in the Australian Aboriginal dreaming. Scherto Gill and Sneha Roy examine peace and reconciliation, which is both a prerequisite for, and consequence of, harmony. And last, my own contribution looks at cosmology, politics and ecology.

And finally a word about the book's title, *The Harmony Debates*. This was the title of the 2017 Harmony conference at the University of Wales Trinity Saint David (subtitle: 'What is the Relationship between Harmony and Sustainability?'), and was inspired by the Putney Debates, one of the great events of the English radical tradition and so in an entire strand of thought and action in Western politics. The Debates were held in 1647 when a group of activists assembled at Putney, then just outside London, in order to debate the country's political future. Their demands included votes for all men (even to revolutionaries it was inconceivable that women should vote), together with a series of natural rights including freedom of conscience and equality before the law. Even though they were suppressed at the time, these demands underpin Harmony's political project, as evident in the United Nations statements on the subject; and, as David Cadman writes, 'Harmony asks questions about relationship, justice, fairness and respect in economic, social and political relationships'. It is to these areas, along with business, education, faith and philosophy, sustainability, conflict resolution and the promotion of peace, social justice and the environment, that we turn in the following contributions.

NOTES

1.   HRH The Prince of Wales in HRH the Prince of Wales, Tony Juniper and Ian Skelly, *Harmony: A New Way of Looking at our World* (London: Harper Collins, 2010), p. 3.

2.   United Nations, 'Harmony with Nature', http://www.harmonywithnatureun.org/ [Accessed 28 March 2020].

3.   HRH the Prince of Wales, Tony Juniper and Ian Skelly, *Harmony: A New Way of Looking at our World* (London: Harper Collins, 2010).

4.   HRH The Prince of Wales, Tony Juniper and Ian Skelly, *Harmony*, p. 325.

5.   HRH The Prince of Wales, Tony Juniper and Ian Skelly, *Harmony*, p. 3.

6.   HRH The Prince of Wales, Tony Juniper and Ian Skelly, *Harmony*, pp. 2–3.

7.   United Nations, 'Harmony with Nature', http://www.harmonywithnatureun.org/ [Accessed 28 March 2020].

8.   United Nations, 'Transforming our World: The 2030 Agenda for Sustainable Development', 2015, https://sustainabledevelopment.un.org/post2015/ transformingourworld/publication [Accessed 28 March 2020].

9.   United Nations, 'Transforming our world: the 2030 Agenda for Sustainable Development' Preamble.

10.   United Nations, 'Transforming our world: the 2030 Agenda for Sustainable Development', 'Declaration', para. 8.

11.   David Cadman, 'The Relationship between Harmony and Sustainability', The University of Wales Trinity Saint David Harmony Conference, Lampeter on 2nd and 3rd March 2017.

12.   Cadman, 'The Relationship between Harmony and Sustainability'.

13.   M. A. Rashed, 'Harmony in Islamic Cosmology: Subjugation, Sujūd and Oneness in Islamic Philosophical Thought', in *The Harmony Debates*, ed. Nicholas Campion (Lampeter: Sophia Centre Press, 2020); David Rubin, 'Harmony and Judaism', in *The Harmony Debates*, ed. Nicholas Campion (Lampeter: Sophia Centre Press, 2020).

14.   David Cadman, 23 May 2017.

15.   Richard Dunne, *Harmony: A new way of looking at and learning about our world. A teacher's guide to purposeful learning* (London and Bristol: The Harmony Project/ Sustainable Food Trust, 2019), p. 89.

16.   For an excellent series of papers see Sustainable Food Trust, *The True Cost of American Food*, proceedings of the conference held in San Francisco, April 14th–17th 2016.

Towards a Definition of Harmony
in Music and Agriculture
An Open Letter to HRH The Prince of Wales

John Eliot Gardiner

SIR, YOU KNOW ME IN MY CAPACITY both as a musician and as a farmer. To some people these two disciplines seem odd bedfellows; yet to me they complement each other naturally. Ever since I was a boy growing up on a Dorset farm I loved being involved in all its activities – from lambing and calving, to feeding the livestock, harvesting the crops and threshing the wheat. As luck would have it my parents were pioneer organic farmers as well as keen amateur musicians. Thanks to them it never occurred to me that there was anything odd in the way they linked farming with music, or in the plays they devised to mark the changing seasons of the farming year with folk song and dance, polyphony and chamber music. I grew up to appreciate how vital both agriculture and music are to the health of our species – one to feed the body, the other to feed the soul. I came to realise that I wouldn't be satisfied in life with the one without the other. That was reason enough for me to attempt to combine and balance them in my professional life. In practise, allocating time and giving sufficient attention to both roles has proved a complicated juggling act, for both professions are all-consuming and carry immense responsibilities and challenges. Depressingly, both are often officially undervalued, misunderstood or pushed to the margins of public consciousness through other pressures. One can gauge the degree of importance attached to them by UK governments over the past thirty years by the fact that Agriculture and the Arts are usually the last two ministerial portfolios to be handed out by successive Prime Ministers. How can that be seeing that the one carries responsibility for the safety of the food we eat, the water we drink and the air we breathe, and the other for the cultural and spiritual sustenance of the whole community and, crucially, for the imaginative development of our children?

This downgrading of two such fundamental disciplines is symptomatic, it seems to me, of a prevalent loss of wonder at the earth's delicate, fragile life-supporting systems – the very things that nourish our existence and the wellsprings of our creativity. Over time we have lost much, if not all, of our sense of harmony – or of what it's like to live harmoniously within our physical environment. Harmony by its very essence predicates and generates a balance between opposing forces, or in musical terms, 'the combination of simultaneously sounded musical notes to produce a pleasing or agreeable effect'. That sounds all very fine in theory, some people will say; but in practice, what benefit is it to them in their daily struggle to earn a living and to keep pace with bewildering change, they might ask. What has it got to do with our reconnecting with nature or with saving the planet? Doesn't music along with the other arts just cloud reality? Escapism is not what the natural world in crisis needs at this point, it could be argued.

I have to admit that classical music is particularly bad at presenting its own case. On the one hand it is widely recognised as one of the greatest achievements of the Western mind, but on the other it betrays its origins in social privilege and exclusivity. Most of the musicians I work with regularly are acutely aware of this paradox. Coming from all over Europe, Asia, Australasia and the Americas they have learnt the hard way: how to juggle cultures and from direct experience how opinions and prejudices divide men and women, knowing that the activity of music-making unites people regardless of social class, gender, age and ethnicity. Perhaps the most urgent task for us musicians now is to convince our audiences that, for all its mysterious, indefinable attributes, music is rooted in the soil and earth of our – and their – everyday lives. Music exists in an imperfect world in which social injustice and inequality threaten to constrain the potential of individuals; but it has the power to refresh, give hope and purpose, and even to transform people's lives.

A few years ago I came across a note that J.S. Bach wrote to himself in the margin of his copy of the Holy Bible: 'NB. Where there is devotional music, God with His grace is always present'.[i] This was a covenant as binding on Bach (perhaps the greatest musician the world has ever seen) as God's promise of the rainbow was to Noah. Bach is referring to a powerful phenomenon known to (but not always admitted by) musicians whenever they meet to make music together: the potential access we are given to a world of grace and truthful experience that no other art form can offer to the same degree. Engagement in the spirit of devotion with music by Bach or any of the great composers of the past four or five hundred years can help us to understand that vibrant creativity and the rooting of music in a given order need not work against each other, but can be combined harmoniously.

Harmony is a beautiful word and concept. Yet I'm bothered by the way it is used in common parlance. It doesn't seem to have much inherent substance beyond a vague, feel-good notion of 'being in a state of agreement or concord'.

As a definition this seems to me to fall well short of the mark. Slightly better might be 'a means of connecting unrelated or loosely related concepts to concur and meld in a benign way', but even that sidesteps the crucial mathematical and philosophical roots of musical harmony.

So for a moment may I ask you, Sir, to return with me to the very origins of harmony – to the discovery at some point in the fifth century B.C. of the harmonic structure of music? Pythagoras experimented with a stretched piece of string. When plucked, the string sounded a certain note. When halved in length and plucked again, the string sounded a higher note – completely consonant with the first. In fact it was the identical note but at a higher pitch. Pythagoras had discovered the ratio 2:1 – that of the octave, in other words. Plucking the string at two-thirds of its original length produced a perfect fifth in the ratio of 3:2. When a three-quarters length of string was plucked, a perfect fourth was sounded in the ratio of 4:3, and so on and so forth. The pitch of a musical note is thus in inverse proportion to the length of the string that produces it. These sounds being all consonant are generally accepted as pleasing to the human ear – and they have stood the test of time. Yet the significance that Pythagoras attributed to this discovery went far beyond mere euphony. Finding that harmonic music is expressed in exact numerical ratios of whole numbers demonstrated to him the intelligibility of reality and the existence of a reasoning intelligence behind it. He considered that the harmonious sounds that men make, either when singing or playing their instruments, were an approximation to a larger harmony that exists in the universe, one which is also expressed by numbers. This, he proposed, was 'the music (or harmony) of the spheres'.[ii] If the Sun, Moon and planets all emit their own unique hum based on their orbital revolution, he concluded, then music must be the ordering principle of the universe. In this way music came to be understood as being based on natural laws. Its value derived from its capacity to frame and elaborate these laws in musical form. Aristotle tells us that the Pythagoreans 'supposed the elements of numbers to be the elements of all things, and the whole heaven to be a musical scale and a number'. In other words, music is number made audible. The heavenly spheres and their rotations through the sky produce tones at various levels and pitches. When brought together in concert these tones create harmonious sounds that resonate with man's sense of participation in the harmony of the universe.[1]

To many people the appeal to planetary motion as the foundation of musical harmony might seem faintly ludicrous; yet in today's noise-polluted world it retains traces of the elemental force that our ancestors acknowledged and valued when they celebrated music's magical, mythical origins. Claude Lévi-Strauss, France's leading 20th-century social anthropologist, described the function of myths in primitive society as 'resolving contradictions'. He went on to define music as 'the supreme mystery of the science of man'.[iii] However hard we try to explain the beauty of music in the end it can only deprive it of its mystery. What audiences hear in a concert performance is just the finished article – like a polished stone or a diamond – but they have no yardstick with which to assess its true worth. It is part of our job as musicians to convince listeners that they can participate not just as spectators watching and listening while others of us perform, but by being inspired through direct exposure to its beauties in the very process of its being put together. Unless we can show people how harmony is formed music could become a dying art. We can do more, perhaps by inviting them into the fulcrum of rehearsal – into our work-place. For just as the priest needs to recognise that harmony can be achieved not just in church but in a secular context as well, so the musician needs to acknowledge that harmonic rhythms can be replicated in something as seemingly unconnected as agriculture, while even the experienced farmer can learn that there is a harmonic component to the right and responsible way to treat his livestock as well as his children.

The idea of 'the music of the spheres' runs through the history of Western civilisation like a golden thread and with tenacious consistency. At first it was treated quite literally as the sounds of a cosmos regulated by the fixed and constant motions of the planets, audible to God but not, since the Fall, to man. Music was seen more as a discovery than a creation, because it relied on pre-existing principles of order in nature for its operation. During the early sixth century CE a book called *The Principles of Music* appeared which was to have enormous influence throughout the Middle Ages and beyond.[iv] Its author, Boethius, believed that arithmetic and music were intertwined. He proposed that music is related not only to speculation but to morality as well, for nothing is more consistent with human nature than to be soothed by sweet modes and disturbed by their opposites.[2]

By the early 17th-century astronomers and mathematicians were struggling to discover a set of rules common to both music and the newly observed movements within the solar system. In *The Harmony of the Universe* Johannes Kepler proposed a new theory of musical harmony and a cosmology of the heavens and the earth. Much of this fascinating book is devoted to proving how harmonious consonances correspond to the geometrical archetypes in the mind of God and the soul of man. Kepler set out to prove the relationship between the musical ratios and the motions of the planets. He discovered that the solar system's planets follow elliptical, not circular, paths. He based his version of music of the spheres on the relative maximum and minimum angular velocities of the planets measured from the sun. 'The movements of the heavens are nothing except a certain everlasting polyphony', he wrote.[v]

At around the time as astronomers like Kepler and Galileo were transforming the way people understood planet Earth's place in the cosmos, a radical shift was taking place in the world of music. Suddenly there was a challenge to the old Pythagorean view that the rules of musical practice could simply be deduced from nature itself through a mathematical study of harmonic proportions. Throughout the Renaissance composers had struggled to achieve perfection in linear counterpoint as a way of reflecting divine order. Then along came Claudio Monteverdi. The exact contemporary of Galileo, Kepler and Shakespeare, he bucked this trend by looking at music's role from a more human perspective: how we stand in relation to God and to our neighbour and, above all, how our emotional life can be expressed through music. Monteverdi's primary concern as a composer was not so much structural perfection (though he excelled at that, too) as expressive force. He was in effect the first composer to harness the power of music to express and encompass the full gamut of human emotions. His whole life was spent probing and extending music's capacity to reflect the relationships of men to women and women to men and their collective response to the existence of God. Replacing the old church modes with something far more assertive and even abrasive at times, Monteverdi uncovered a new richness in diatonic harmony (as we would now call it). Harshly criticised by conservative theorists of the day, he was a pioneer in using dissonant chords to generate moments of acute emotional and sexual tension via a series of harmonic crunch points which always resolve back to a consonance. As Ben Jonson said around the same time, 'All concord's born of contraries'.[vi]

By the early 1600s we arrive at a way of describing harmony in music based on the Christian symbol of the Cross – as the fruitful intersection of a horizontal plane made up of melody and rhythm and a vertical plane comprising chords and the spacing of musical intervals. Now the thing is this: could we not say something similar when defining the harmonious basis of agriculture? Isn't it possible to see soil structure and biodynamic vitality as forming the 'horizontal' dimension standing in a symbiotic relationship to crop- and animal-husbandry which form the 'vertical' dimension? Problems, all-too-familiar in today's industrialised agriculture, pile up when these two dimensions cease to operate harmoniously and start to collide. Just as some people find discordant or atonal music disturbing, many others feel that chemically-based agriculture is out of kilter with the principles of harmony: that it disrespects and ultimately degrades the very source of our nourishment, fracturing the two-way relationship whereby the soil yields its fruits in return for the honour and sanctity we accord to it. In the past fifty years this veneration for the process has become diluted and has now almost disappeared from farming. Yet in the same period it has increased significantly in the sphere of music.[3] The central core of both farming, forestry and music is not ownership, but stewardship. As stewards we have a duty to be fully accountable, as well as grateful, for all that has been entrusted to us. It is the essential 'give' and 'take' which is such a vital ingredient in the fruitful interaction between musicians.

Music and farming share this, too: they come to life and unfold in real time – in the duration it takes to perform a piece of music or for the growth cycle of a plant to complete its life-span. Agriculture like music is part interpretation (in its contingency upon the weather, the water table, and on soil analysis), and part performance (in the ways we sow, cultivate and harvest our crops). Craftsmanship is common to both activities. Inspiration, on the other hand, is a much less dependable commodity: for while it is possible to create favourable circumstances which allows inspiration to strike (and it can be humbling if and when it does), it needs to be treated more like a bonus or even a blessing and not as a dead certainty. Our forebears treated farming as a devotional activity culminating in a ceremony of thanks at the conclusion of each of its cycles. These celebrations were marked through sacred forms of song and dance. Yet although many of these rituals are disappearing I am convinced that children can still respond to them – especially harvest festivals – just as I did when I was growing up. The Ashley C. of E. primary school at Walton on Thames was a model of its kind during the time Bryony, my youngest daughter, taught there. Richard Dunne was headmaster at that time, and gave children a wonderful introduction to the principles of harmony, showing them, for instance, how to grow their own food and encouraging them to participate in the cooking of it. Miraculous moments occur in the preparation of food just as they do in rehearsing music, no matter if it's slanted towards a school concert or a grand symphonic affair. As in the rehearsal room, the kitchen becomes a laboratory where children respond to the enthralling alchemy when ingredients suddenly fuse together. I have never forgotten the thrilling moment as a child when my mother, after stirring thick cream in a jar for what seemed like hours on end, suddenly caused it to turn into butter.

My great-uncle Balfour Gardiner was a composer of lush romantic music inspired by the English countryside, all of it well-crafted and some of it strikingly beautiful.[4] But after hearing the experimental music of the second Viennese school in which chromatic expressionism without a firm tonal centre began to take hold, he decided that his time was up. During the bleak years that followed the Great War he gave up composing music altogether. He decided to burn the bulk of his compositions and to plant trees instead. In 1924 he bought seventy-five derelict acres on an outcrop of the North Dorset Downs. He appealed to his nephew Rolf (my father), who was then in his early twenties, to join forces with him in a venture of rural regeneration centred on Gore Farm in the parish of Ashmore. It was to provide much needed employment at a time of economic depression. My father accepted the challenge and went off to Dartington Hall to train as a forester. Soon afterwards uncle and nephew embarked on restoring this stretch of chalk downland on this western edge of Cranborne Chase, that once huge tract of royal forest which had become denuded of trees and badly eroded through rabbit scrapes and screes. They had big ambitions to create 'a harmonious matrix of hedgerows, orchards and well-managed woodland for shelter, to prevent erosion, to provide trace elements from the subsoil, to maintain the water table, to moderate extremes of heat and cold, to provide humus from the forest floor, to harbour wildlife, an efficacious drainage system, a varied diet for livestock, to provide continuous stability, to avoid having to import wood products from abroad'. Quite some list! Over a period of nearly forty years they planted three million deciduous and non-deciduous trees on the impoverished soils on the fringes of the Chase. The results of their efforts are still there today for all to see.

Unlike his uncle, my father never gave up music. He sang on his horse or on a tractor and was a great believer in the value of collective music-making in binding communities together. To him the real issue of the time was the conflict between what he described as 'the divine order of music and the diabolical disorder of noise'. As a farmer and tree-planter (and a founder member of the Soil Association) he saw the reassertion of the principle of music, of rhythmic form, both as a grave duty and as a source of heart-giving joy. He set a premium on the sanctity of life, and adopted unfashionably what he called a 'sacramental approach' to agriculture. He strove all his life to restore a balance between the new and the old, to create new contexts for ancient structures and sacred mysteries that had been lost over time. When I was aged 20, I told him that I had set my heart on becoming a professional musician and a conductor, but he was vehemently opposed to the idea. Music should never be an end in itself, he insisted. The way he saw it I was about to sign a Faustian pact which would fracture the vital practice of communal music-making linked to the seasonal rituals of farming and the turning year – the very things he had taught me and my siblings to respect all through our childhood. And so we argued. Then, just months before his untimely death in 1971, he gamely came round to the idea that there might after all be merits in my dream of combining professional music-making with the tending of the land. We were finally reconciled. It is to him that I owe my awareness of the cycles of nature and the rooting of music in the soil and its connection to the rhythms of sustainable farming and forestry. And there was another valuable lesson I learnt from him (and one so easy to forget): how to take a nourishing intake of breath before any strenuous creative effort, and how to gather oneself in humility on the edge of performance through a conscious surrender of self-will. It is a pathway towards achieving that X factor – the one thing which can transfigure a routine musical performance into an illuminating and irradiating miracle.

I am greatly indebted to you, Sir, for appreciating that music-making at all levels fosters human interaction: that as a shared activity it is collectively empowering and that it speaks directly to the individual in ways inaccessible to rational argument and dispute. Beyond that I believe that music in its widest sense provides a paradigm of balanced, harmonious coexistence and that it engenders a sense of reverence and mystery which has largely leached out of our approach to nature and to the soil and to the way we grow and treat our food. Many people who respond eagerly to classical music feel alienated by the jargon and hype that surrounds it. Yet who can blame them? We now have proof that the music of Bach and Mozart opens up neural pathways to the brain which can otherwise become starved of nutrients. If our children are not given regular access to these nutrients [5] they remain deprived, maimed even, in some fundamental way. It used to be thought that we process art and music in the right hemisphere of our brains, with language and mathematics located in the left. No longer. Recent research in the neuroscience of music now shows that the activities of composing, performing and listening to music spread to almost every area of the brain and involve nearly every neural subsystem. I find that a wonderful thought.

You, Sir, have consistently and courageously shown us the paramount need for the principles of harmony to be respected, emulated and extended throughout our society and body politic. It is easy to be overawed by the size of the task and the constant fight with gremlins at every stage of the journey. Your generous patronage and skill in connecting people from different fields of expertise continues to have a palpable impact across the globe, as was so clear to me when talking to delegates at the Dumfries House conference on the Future of Food and Farming in 2017. It drummed home to me how our sense of shared culture, our irrefutable knowledge of our planetary environmental interdependence, has so far not registered widely enough to prevent us from the ecological imperative to cooperate at a national and international level in order to halt escalating, irreversible environmental damage. As human beings we are both a part and product of nature, yet we have the capacity to influence its processes – for good or evil. We need constant reminders that our actions have far-reaching effects on other living organisms and creatures. Like them, we need space and repose, music for our souls, food for our minds as well as our bodies.

At Dumfries House I sensed a general consensus about the aims of sustainable farming: to produce healthy food that gives nurture and pleasure to our taste buds while being harmonically consonant with nature. Speaker after speaker extolled the need for responsible animal and plant husbandry and to account for the true costs of production. These need to be reconciled harmoniously, yet in our century this rarely happens. Too often dissonant or what we musicians call 'bum' notes are struck through the promiscuous use of agro-chemicals. If our children are not encouraged to explore the responsible ways and practices of how food can be grown, planted, tended and harvested, they risk being cut off from the vital chain of nature and sustenance. Thirty years ago, Leonard Bernstein had something important to say on this topic.[6] Towards the end of his life, he wrote:

> We destroy our children's songs of existence by giving them inhibitions, teaching them to be cynical, manipulative, and all the rest of it… [vii]

> You become hardened; but you can find that playfulness again. We've got to find a way to get music and kids together, as well as to teach teachers how to discover their own love of learning. Then the infectious process begins… Think of what we can do with all that energy and all that spirit instead of eroding and degrading our planet on which we live, and disgracing ourselves as a race. I will spend my dying breath and my last blood and erg of energy to try to correct this impossible situation.[viii]

And so he did. That was in 1989. Surely the task is even more urgent today.

It is through the example of a Lenny Bernstein and the inspirational leadership of yourself, Sir, together with men of vision like Patrick Holden, Richard Dunne and a few others, that our best hopes for the future lie. Thank you for showing us the way and for supporting those of us who firmly believe that we need to bring back the principles of harmony – *musical* harmony – to their rightful place in the foreground of our discussions about, food, farming, forestry and the health of the environment as well as of ourselves. Meanwhile we musical pilgrims sing on.

Notes

i.   Howard H. Cox (ed.), *The Calov Bible of J. S. Bach* (Ann; Arbor: UMI Research Press, 1985), p. 419; Hans T. David, Arthur Mendel and Christoph Wolff, *The New Bach Reader* (New York : W.W. Norton & Company, 1998), p. 161.

ii.   Piero Weiss and Richard Taruskin, *Music in the Western World* (noplace: Cengage, 2008) p.3.

iii.   Claude Lévi-Strauss, *Mythologiques I: Le cru et le cuit* (1964), quoted by George Steiner, Errata: An Examined Life (London : Weidenfeld & Nicolson, 1997), p.63.

iv.   Boethius, *De institutione musica*, trans. Calvin M. Bower, Claude V. Palisca (ed.), (New Haven & London: Yale University Press, 1989).

v.   Johannes Kepler, 'Epitome of Copernican Astronomy: IV and V. The Harmonies of the World: V' trans. Charles Glenn Wallis, in Claudius Ptolemy, The Almagest, trans. R. Catesby Taliaferro (Chicago: Encyclopaedia Britannica, 1952), pp. 839-1085 (p.1048), https://www.sacred-texts.com/astro/how/how08.htm.

vi.   Ben Jonson, *Cynthia's Revels*, edited by C.H. Herford and Percy Simpson (Oxford: Clarendon Press, 1986), Act V scene ii.

vii.   Jonathan Cott, *Dinner with Lenny: The Last Long Interview with Leonard Bernstein* (Oxford: Oxford University Press, 2013), p. 74.

viii.   Cott, *Dinner with Lenny*, p. 76.

1.   In the *Republic* Plato taught that 'rhythm and harmony find their way into the inward places of the soul on which they mightily fasten, imparting grace, and making the soul of him who is rightly educated graceful' (Plato, *Republic*, 2 Vols., trans. Paul Shorey, Cambridge Mass., London: Harvard University Press, 1935, 401d). He pointed to Damon of Athens to demonstrate the political significance of music's power. Damon claimed that he would rather control the modes of music in a city than its laws, because the former have a more decisive impact than the latter on the formation of the character of its citizens. At the same time these ancient Greeks were wary of music's powers because they understood that, just as there was harmony, there was also disharmony. Just as musical concord can nourish the spirit, so musical discord can assault or damage it. It's an age-old theme.

2.   What is it, then, in and about music, that gives us an experience so outside of ourselves that we feel we can see reality anew, as if newborn in a strange but wonderful world? The composer, the late John Tavener proposed an answer to this mystery in his artistic credo: 'My goal is to recover one simple memory from which all art derives. The constant memory of the paradise from which we have fallen leads to the paradise which was promised to the repentant thief. The gentleness of our sleepy recollections promises something else. That which was once perceived as in a glass darkly, we shall see face to face. We shall not only see; we shall hear.'. (see Robert R. Reilly, *Surprised by Beauty: A Listener's Guide to the Recovery of Modern Music* (San Francisco: Ignatius Press, 2002) p. 438.

3.   This is shown in the respect and faithful attention that musicians now regularly give to the exact notation and implied intentions of composers of the past. It has been the springboard for what is now known as HIP (historically informed performance) which has given intellectual ballast to the Early Music Movement of the past 75 years.

4.   In 1912/13 Balfour Gardiner put together and conducted a trail-blazing series of eight symphonic and choral concerts at the Queen's Hall, London, with each programme devoted to living British composers. Amongst them was Hubert Parry whom Balfour

invited to conduct his own Fifth (and final) Symphony in February 1913.

5.   You will undoubtedly be aware, Sir, of the several green shoots of innovative music-making taking place within a social context today. So, for example, the Irene Taylor Trust not only brings music into prisons, but it also helps released prisoners to get their lives back on track through live performances and training placements. A few years ago the journal Psychology of Music published a study involving fifty children aged eight to eleven, from which the researchers concluded that interacting with music made them noticeably more emotionally attuned to others: just an hour a week of musical activity during term time increased their emotional empathy.

6.   People forget that Bernstein was not only a famous conductor and composer, he was one of the greatest musical educationalists of the last century. A selection of his Young People's Concerts, for example, which delighted and educated millions of children for 14 years (1958-1972), has been re-issued on video.

Principles of Harmony

David Cadman

From a talk at 'The Harmony Debates', a conference at the University of Wales Trinity Saint David, Lampeter 2 March 2017

I SPEAK TO YOU TODAY AS one of your Harmony Professors of Practice. My proposition is this: it would seem that all human societies seek to describe their relationships with each other and with that of which they feel themselves to be a part in terms of some kind of order, a set of relationships which govern them, and one attempt to do this is to speak of harmony, and within this task one attempt is to look at harmony as it is expressed in nature to try and discern nature's principles of harmony.

Within nature there are a number of characteristics qualities or principles that speak of harmony. I say a number, but by that I do not mean to presume that the number is limited by that which I have observed. That is why I speak of principles of harmony and not *the* principles of harmony. I do not presume to have discovered the defining set of principles, only to have observed what I take to be some of those principles. Nature is more wonderful than we can possibly imagine. The principles I have observed include wholeness, connection, interdependence, diversity within wholeness, cycles of time and season, patterns, rhythms, reciprocity and mutuality and justice and lawfulness, and I accept that what I have found may have been what I was looking for and others would find something else.

Nevertheless, what I have seen suggests a systemic order of intertwined and entangled patterns and rhythms that might constitute a form of governance that, if followed, would align us with that which is good for us and good for the Earth; and I propose this not simply by way of detached intellectual enquiry but also by way of experience, by practice and participation. For I find that when I live as if my life is ordered by harmony, harmonious relationships are inclined to manifest, and this must surely be so, for, as the Buddha made clear in the opening stanzas of the Dhammapada, it is with our thoughts that we make the world.[1] And in my own work I claim that the governing principle of this order is Love. Not love as virtue or romantic sentiment, although there's nothing wrong with either virtue or romance, but love as being of the essence, of shaping all that is. And in this I am encouraged by the work of the social biologist, Humberto Maturana, and the psychologist, Gerda Verden-Zöller, who come from South America. For in their book, *The Origins of Humanness in the Biology of Love,* which I recommend to you, they claim that we are by ancient nature a loving species, *homo sapiens amans.*[2]

The idea that we make the world according to our thoughts seems to me to be correct when I look at the world that we have already created, for there I find that it too has been shaped by a particular set of beliefs, in this case a set of principles framed within the lexicon of economy and most especially within the frame of what is termed the neo-liberal economy governed by assumptions of free markets, unbridled consumption, unlimited wealth, the requirement for unlimited growth and the notion of no such thing as society. Is it possible, do you think, that this dominant creed is wrong? And, if it is, is it then possible that a proposition of governance according to principles of harmony, including diversity within wholeness, might be more appropriate or at least worth exploring.

As I come to an end I've been asked to do one other thing, and that is to say something about the relationship between harmony and sustainability. I'll do my best, but again this is only my view. Harmony embodies broad and all-embracing principles about the nature of the cosmos and our relationship with it or, rather, our relationships within it, for we are a part of and not apart from, and this enquiry is timeless and is not instrumental. It asks questions about ways of true being, the way things are, not the way we would like them to be or might seek to manage them to be for our own benefit.

Sustainability, in the way in which the term has now come to be used, describes relationships between environment, society and economy that can be sustained and nourished over long periods of time for mutual benefit. It tends towards being instrumental and the problem is that, for many, the broad matter of sustainability has been reduced to the more narrow matter of environmental management, and within this to the even more limited matters of energy and waste; turning off the lights and recycling. These are necessary but not sufficient. An example of the broad definition of sustainability relevant here, not least because it is used in the 2015 Wellbeing of Future Generations Wales Act, is this: sustainable development means the process of improving the economic, social, environmental and cultural wellbeing of Wales by taking action in accordance with the sustainable development principle aimed at achieving wellbeing goals.[3] Again, you can see the instrumental intent of this definition, achieving and improving the wellbeing of Wales defined in terms of equality, communal cohesion, health, resilience, prosperity, responsibility and culture and language.

So, is harmony a tool of sustainability or does sustainability sit within principles of harmony? My view is that sustainability, in either its wider or more narrow form, cannot be explored other than in the context of harmony. It is a circle within a circle. I was hoping to bring clarity, but I fear I may have only brought more confusion and, if this is so, perhaps this is something we should explore together. Perhaps this is what universities are for. Perhaps this is what Harmony Professors of Practice are for.

NOTES

1.   Eknath Easwaran, *The Dhammapada* (Nigiri Press, 2007), I.1-2.

2.   Humberto Maturana and Gerda Verden-Zöller, *The Origins of Humanness in the Biology of Love* (Exeter: Imprint Academic, 2008).

3.   Wellbeing of Future Generations Wales Act 'The Sustainable Development principle' http://futuregenerations.wales/about-us/future-generations-act/ [accessed 2 March 2016].

Dance and Harmony

Wendy Buonaventura

THE ARTS ARE AN EXPRESSION OF THE HUMAN SPIRIT and are vital to our well-being; to our understanding of the human condition and the part we play in the drama that is life on earth. Every day of our lives we take part in one or other of the arts – listening to music, looking at films or graphic images, and subliminally responding to the architecture all around us. We may not be able to play a musical instrument, nor become a skilful painter or actor, but there is one arts activity that is uniquely accessible to all of us. For, despite what some may fear, we can all learn to move our bodies in harmony to music. In many cultures this is taken for granted. Yet in some, at any rate in the UK, there are those who don't believe this is so.

I have taught dance for many years and I cannot count the number of times someone has said to me, 'I don't go to see dance,' or 'I can't dance'. How many people have told me they have two left feet; that they are embarrassed to get up on the dance floor until they've been fortified with a stiff drink? Yet I have not heard this on my travels in Spain, Argentina, Morocco, or in any other country which has not lost its connection to this vital social activity.

In 2007 Sir Ken Robinson gave a TED talk on education entitled 'Do Schools Kill Creativity?'[1] It has become the most-watched of all TED talks: by September 2019 it had been viewed over 62 million times. In it he spoke about how Western educational systems set up in the 19th century were designed to meet the needs of industrialism and were hierarchically driven. Even today, he says, all over the world this situation has not changed. At the top of education's hierarchy of important subjects are maths and science; and at the bottom are the arts. As for the arts: they have their own hierarchy, and way down this hierarchy is dance.

Why, asks Robinson, Why, when we have daily maths lessons, don't we also have daily dance classes? After all, we all have bodies, we aren't just brains on sticks. Our bodies aren't simply a means of transporting our heads to meetings and conferences. This is education from the waist up, leaving the development of our physical well-being and talents way down the list of educational priorities. And without a type of education that includes both mind and body it is more difficult for us to become balanced, healthy human beings with access to all our natural talents.

Robinson recounted a lunch-time conversation with the distinguished British choreographer Gillian Lynn. Lynn, a child in the 1930s, recalled never being able to sit still and was disruptive in class. Her school decided Gillian had a learning disorder. According to Robinson,

> Gillian and I had lunch one day. I said, "How did you get to be a dancer?" It was interesting. When she was at school, she was really hopeless. And the school, in the '30s, wrote to her parents and said, "We think Gillian has a learning disorder." She couldn't concentrate; she was fidgeting. I think now they'd say she had ADHD. Wouldn't you? But this was the 1930s, and ADHD hadn't been invented at this point. It wasn't an available condition.[2]

This diagnosis of a mental problem happened to Gillian when she was just eight years old. Her mother took her along to a specialist and, after talking for a while, the doctor reached a view. Robinson picks up the story.

> In the end, the doctor went and sat next to Gillian and said, "I've listened to all these things your mother's told me. I need to speak to her privately. Wait here. We'll be back. We won't be very long," and they went and left her.

> But as they went out of the room, he turned on the radio that was sitting on his desk. And when they got out of the room, he said to her mother, "Just stand and watch her." And the minute they left the room, she was on her feet, moving to the music. And they watched for a few minutes, and he turned to her mother and said, "Mrs. Lynne, Gillian isn't sick. She's a dancer. Take her to a dance school."[3]

Gillian Lynn went on to become a member of the Royal Ballet and the choreographer of highly successful musical shows, including *Cats* and *Phantom of the Opera.*

In the Western world dance isn't considered a 'useful' educational subject. It is neither a health-promoting activity nor an art which is capable, like other arts, of exploring the great themes of existence, and remains a specialist activity, and it has a long history of attempted state suppression, as we shall see. In the UK dance is a component in physical education for school children aged 11–14 (so-called 'Key Stage 3'), although not for any other age group and, even then, it is completely overshadowed by sport.[4] In fact, evidence shows that education in dance in England is collapsing. The figures speak for themselves: entries to the

'A Level' exam in dance, which students normally take at age eighteen, declined from 2,261 in 2010 to 1455 in 2017 and 1316 in 2018, which translates from a decline of 42% from 2010 to 2018, and 10% from 2017 to 18.[5] We see from this that, in Britain, rejection of dance as a valuable activity from which we can all benefit begins at school and is part of a shutting down of the senses, a rejection of a fundamental part of human creative experience. The closing of the senses that begins at school involves music as well as dance. A recent report sponsored by the Musicians' Union concluded that even though the ability to play a musical instrument is central to academic achievement and general wellbeing, 'Music education in the United Kingdom is in a perilous state', due to what appears to be the government's 'chaotic education policies'.[6]

The irony of this is that music is ubiquitous in modern life. It's all around us, in the background as we go about our daily tasks, a soundtrack to our lives. What used to be derided as 'musak' or 'wallpaper' is now unquestioned in restaurants, shopping malls and lifts, to name only some of the places that bombard us with background music. Much of the time this music is rhythmic and demands that we move our bodies to it. In restaurants up-tempo music is played to get us to eat faster and make way for fresh customers, and if it's too loud it can kill conversation. This kind of rhythmic music does nothing so much as make us want to get up and go dancing between the tables! But we can't do this. Instead we shut it out, shut down our hearing, just as we have learned to shut down our physical impulse to dance along with it.

There are some forms of dance designed to create harmony between human beings and the wider universe. An example is the *zhikr*, the dance of the whirling dervishes in which the turning of circles of men around a central group of dervishes corresponds to the spinning of the planetary spheres. That this much is commonly understood is evident from the course material in at least one American university:

> Additionally, the way in which the dervishes are arranged is symbolic of the celestial bodies which orbit circularly around the sun. The turning of the dervishes is symbolic of the way the planets and celestial bodies spin on their axes as they orbit the sun.[7]

At a recent Sufi festival in Scotland, one of the organisers stated that, 'We aim to promote the core message of Sufism, to live life on this earth with love, beauty, harmony and worship'.[8] And as one modern Sufi group, the Threshold Society states,

In the Mevlevi *zhikr* it is important that we are in harmony and unity. The goal is to be as one and that takes subtlety, nuance, and attention. To achieve this we need to follow the lead of the person leading the *zhikr* by being conscious of their volume, speed, and tone. We also need to be aware that our own voice does not rise above that of those who are near to us or the general level of the group.[9]

In Renaissance Europe courtly dance rested on similar theoretical premises; at least according to poets such as John Davies, who saw the function of dances such as the Galliard as mirroring the harmonious movements of the cosmos:

> Behold the World how it is whirled round,
> And for it is so whirl'd, is named so;
> In whose large volume many rules are found
> Of this new Art, which it doth fairely show:
> ...
> Under that spangled skye, five wandring flames
> Besides the King of Day, and Queene of Night,
> Are wheel'd around, all in their sundry frames,
> And all in sundry measures doe delight:
> Yet altogether keepe no measure right.
> For by it selfe, each doth it selfe advance,
> And by it selfe each doth a Galliard daunce.[10]

The purpose of such dance, Davies continued, was to reinforce the love and harmony essential to the perfectly ordered society:

> Loe this is Dauncings true nobilitie.
> Dauncing, the child of Musick and of Love,
> Dauncing it selfe, both love and harmony,
> Where all agree, and all in order move;
> ...
> For that true Love which Dauncing did invent,
> Is he that tun'd the World's whole harmony,
> And linkt all men in sweet societie.[11]

Today we know that physical exercise releases endorphins that trigger positive feelings and lead to a higher degree of self-esteem.[12] And in our sedentary lives, how rewarding is it to use our physical energy in a pleasurable, creative way, rather than to exercise by rote? What Barbara Ehrenreich calls in her book *Dancing in the Streets* the 'collective joy' of dance – releases people from the prison of the self and restores balance and harmony to the spirit.[13]

In the long and complex history of attempts to suppress participatory public dancing, there is one particular type of dance that has, despite such attempts, largely escaped a general clamp-down. This is dance as therapy, which often takes the form of ecstatic social dance rituals. Ehrenreich recounts how, back in the 18th century, dance was proposed as a cure for what has been described as an epidemic of melancholy throughout Europe (an affliction that became, in England, something of a fashionable pose for the Romantic poets).[14] In Italy the tarantella, danced for hours on end on successive days, was thought to cure this melancholy state which had been brought about, it was said, by the poisonous bite of the tarantula spider.[15] Christina Pluhar, the creative director of the early music group, L'Arpeggiata, who have performed and recorded a version of Athanasius Kircher's 17th-century 'Tarantella Napolitana', describes the 'healing' version of the tarantella:

> In order to overcome the poison, he must overcome the broken equilibrium in himself. On his journey the sick person identifies himself with Nature, whose harmony he perceives through sounds and colours and whose vibrations he absorbs into his body. The sick person becomes the black sun (or black spider) in the centre, surrounded by the planets, which are symbolised by the people and the musicians who accompany him in his quest for healing.[16]

Writing about the source of this kind of profound depression, Ehrenreich gives the opinion that modernization was at the root of the problem; that 'urbanization and the rise of a competitive market-based economy favour a more anxious and isolated sort of person – potentially both prone to depression and distrustful of communal pleasures'.[17] She goes on to argue that government and church, by phasing out communal dance and music festivities destroyed the traditional cure for mental problems. Europeans, she says, 'could congratulate themselves for brilliant achievements in the areas of science, exploration, and industry... But with the suppression of festivities that accompanied modern European "progress" 'they had done something perhaps far more damaging'; they had, she says, rejected one of the most ancient sources of help – the mind-preserving, lifesaving techniques of ecstasy'.[18]

In the 19th century, attempts by Christian missionaries to take charge of indigenous cultures around the world often succeeded in repressing these dance rituals. Among the Namaquas of South Africa it was said of someone who converted to Christianity that 'he has given up dancing'.[19] By 1820 Protestant missionaries in Tahiti could claim proudly that they had 'restrained the natural levity of the natives'.[20] When even the weaving of garlands of flowers was forbidden, the Tahitians abandoned their normal dress, shaved their heads and gave up their healing music rituals. It was observed that, after this, no-one danced any longer, and that their chief social pleasure lay in drinking alcohol.

In other countries, folk rituals weren't so easily destroyed, and those involving music and dance carried on being used to heal the troubled spirit. African trance rituals continue down to this day as a form of healing for people suffering from mental and emotional problems. In North Africa sufferers are described as being 'possessed' by an evil spirit which has to be driven out. We would most likely describe such mental affliction as depression.[21]

In trance rituals which I have witnessed in Morocco, the afflicted person has a ceremony arranged for them, and musicians arrive with friends and family in a supportive role. The musicians, principal among them drummers, proceed to play successive rhythms, seeking out the particular one belonging to the troubled person. When he or she hears their own spirit rhythm, they will dance and dance in simple, repetitive movements. This may go on for many hours, until they drop, when they will be taken to lie down and will be looked after while they recover. Anyone who submits to this repetitive movement and driving rhythm may, after a while, enter a state of trance, characterised by a sense of ecstasy and transcendence. At the very least this dance experience leaves participants feeling nerve-tinglingly alive and at peace, with a deep sense of well-being.

Over the years I have taught a form of dance largely based on the solo female dance of the Middle East, which is performed by women and girls of all ages for each other at private celebrations. It is a dance both lyrical and humorous and is expressed primarily via movements of the torso and hips. I have taught dance not as therapy, but as a source of creativity and a key to tapping our hidden energies; yet many of my students have told me they have found a therapeutic benefit, both mental and physical, from simply moving their bodies to music and reawakening stiff muscles. A woman in her 50s who once came along had suffered for years from back problems, and consulted several specialists with no positive results. After coming regularly to my class she told me her back was improving daily and she was no longer in pain.

Some of my older students, I recall, were initially nervous of joining a class for all kinds of reasons: perhaps they had long harboured the desire to dance but felt they were not the right size or the right age. Others said they hated being looked at and were nervous of being watched by others in the class (of course, everyone was so busy trying to master the movements that they had no interest in looking around at what everyone else was doing.) I used to end the class by inviting one or two volunteers to come into the middle of the floor and improvise for the pleasure of the others There were certainly some who were too shy to do this at the beginning; one who went on to become a good friend, held back from volunteering until one week when she mock-grumbled, 'Alright, I'll do it'. Then she went and did her improvisation behind a semi-transparent wicker screen standing in the corner!

Initial nervousness generally gives way to physical confidence; training the body to master technique leads to pride in what one can do. With confidence comes the desire to show off one's skills, to be seen and appreciated. I have many times known students get together and form small performance groups, then go out and present small shows for charity or community events.

Learning dance from another culture stimulates interest in that culture, whether it's the flamenco of Spain, the dances of India or the women's dance of Egypt. Students take holidays in those countries, become interested in the cuisine, the customs, and I have even known some start to learn Spanish and Arabic, in order to understand the words of the songs they find themselves dancing to. In this way, the journey into other cultures becomes one of discovery and friendship. In Sweden I learnt of a group who organised a weekly afternoon dance social in a café owned by one of them, and promoted it to their local Middle Eastern community. There was one proviso though: no man was allowed to go to this social gathering unless he brought a woman with him. In this way those Middle Eastern women were being encouraged to get out in public, meet other women and mix with the local Swedish community in a protected environment. As one Egyptian dancer once said to me, Western women learning her native dance in Europe and the USA 'are building a bridge of understanding between East and West'.

When governments wake up to the idea that, for the sake of our health, we need to exercise more, it tends to be sports that they recommend rather than dance. This is despite the fact that, for women at any rate, dance is one of the most popular leisure activities.[22] Yet the UK government, for one, has decided to throw its weight behind promoting not dance but sport.[23] Throughout the media there is an overwhelming emphasis on traditional male fighting sports like football and rugby, sports which reinforce hierarchy, learning our place in the pecking order, the giving and taking of orders and, above all, winning. And when dance does feature in the media it is with an equally biased focus.

In the UK millions of people have become interested in dance after watching the TV programme *Strictly Come Dancing*, which has now been exported around the world.[24] The high emotion aroused by all that physical exertion is manifested during the programme in tearful, huggy interviews with contestants and their families. One feature of the programme is that many of the contestants have never really danced before, and the older ones often confess rather sadly that they have always longed to. They are astonished at what they achieve by taking part in the programme, and often say they haven't realised just how long they have been working to perfect their steps; for when an activity excites our spirit, time spent doing it takes on a fresh quality: an hour can pass in a flash, and ten minutes can seem like an hour. The programme leaves contestants with a new confidence in what the body can do, and watching someone who has never in their life experienced the rigorous training required for dance as performance, it is truly impressive to see their mastery of steps and choreography. Yet the programme demonstrates complex attitudes to the female body, with female participants encouraged to pile on the make-up and wear skimpy costumes that reinforce banal ideas about glamour and beauty. And its basic message is far from the truth: the message that social and partner dancing is all about winning; that it's about set routines and competition, with the worst performers suffering the humiliation of being disqualified. In fact, this type of social dance is a conversation, not a lecture where everything that is said has been fixed beforehand, and which produces winners and losers.

Social dancing has sometimes even been suppressed by those in power, notably the Calvinists in 16th- and 17th-century Europe, for whom dance was associated with drunkenness, gluttony and general depravity.[25] It isn't hard to understand how celebratory behaviour involving sensual movement has been considered threatening to sexual morality down the years. Political movements keen to preserve the status quo know how dangerous is the unleashing of a powerful, unpredictable, crowd energy. Ironically, both extreme right wing as well as revolutionary left-wing movements have clamped down on popular dancing in an attempt to control the anarchic nature of public passions.

For Robespierre after the French Revolution, and Lenin after the Russian Revolution, the central social ritual was, in Ehrenreich's words

> the meeting – experienced in a sitting position, requiring no form of participation other than an occasional speech and conducted according to strict rules of procedure. Dancing, singing, trances, these could only be a distraction from the weighty business in hand.[26]

Under Argentina's military junta from 1976 to 1983 the tango was suppressed.[27] And in Iran dance and music were outlawed, following the 1979 Islamic Revolution'.[28] All of which bear witness to the fact that when extreme political and religious movements seize power, the first thing they do is ban those aspects of culture that are to do with human joy and creativity.

In past times, social dancing in private and in public was a principal way in which people met to enjoy themselves, refresh their spirits and regain their equilibrium. In ballrooms, dancehalls and, in the open air, people met to let off steam. There, the classes mingled and people came to know each other in a visceral way, through the body, through touch and rhythm, moving together in harmony. Certain social dances, such as the cancan and the waltz, were considered wild, revolutionary even, and were condemned from the pulpit and in the press. In general, it was the dances of the masses that caused most controversy. Ironically, it was also these which were taken up most eagerly in the ballrooms of the wealthy: but only after they had been tamed and fitted for 'polite' society. The waltz, with its continual spinning and turning, was initially considered scandalous.[29] It allowed men and women who barely knew each other to indulge in unheard-of physical intimacy as they set off across the floor, locked together in a close embrace. This hold, which prevented them losing their balance as they spun, gave them the chance to press their bodies even closer and they were soon lost in their own private world. The waltz was considered so likely to cause dancers to lose their moral compass that it attracted opprobrium from all directions. A number of books of 19th century etiquette pointed the finger at the questionable nature of this dance, including *The Gentleman and Lady's Book* by Mme Celnart. In her opinion this dance was: 'of quite too loose a character, and unmarried ladies should refrain from it altogether'.[30] The waltz went on, though, to become the most popular social dance in Europe during the 19th century, and today it comes as a surprise to people to understand the furore it once caused. It was denounced for causing illness and being a source of bodily and mental weakness, especially in women, and it wasn't the only social dance that was discouraged.

During the 1920s' ballroom dance craze an anti-dance movement sprang up in the USA. Social commentators thundered about the sinful, sexual nature of popular dances emerging from newly liberated Black Americans; dances which imitated animal behaviour and led inevitably to hell. In Judeo-Christian culture, where mind and body have been separated for thousands of years, the first considered superior, the second something to be denied, the reminder of our 'lower' animal selves in dances such as The Bunny Hug and The Turkey Trot were disturbing to some. Yet certain 1920s commentators maintained a sense of proportion. One doctor contributed to the debate, saying, ten thousand people injure themselves by the abuse of eating, for one who does so by dancing. Meanwhile the well-known, highly respected ballroom dancer Irene Castle commented that, if dancing were a sin, then half the population of American cities was in danger.[31]

Dance lovers know that dancing is a source of pleasure and an expression of creative energy for its own sake, and in hard times it's one of life's least expensive ways of keeping up one's spirits. After Argentina's financial crash in 2002 tango became more popular than ever in its birthplace, Buenos Aires.[32] Schools offering courses in the dance sprang up all over the city to cater for visiting Europeans and Americans intent on honing their skills in a dance that had become hugely popular outside the land of its birth. Improvised partner dancing depends on mutual give and take between couples, and nowhere more so than in the complex partner dance that is the Argentinean tango. Among all partner dances with which I'm familiar, tango is the one that most requires couples to move together in harmony. It reflects a complex male-female relationship and is the best example of an improvised dance which depends on co-operation between partners. Traditionally it is the man who leads, or 'proposes' a movement and the woman who 'responds' (though this rule is not so rigid now as it once was, and women are freer to initiate the shape of the dance than they once were). The tango is a conversation, with all the hesitations, interruptions and negotiations of any verbal exchange. Sometimes the conversation turns into an argument, even a battle, and unfortunately, when two partners are not in harmony it may become a monologue! But when it flows, a tango can be mesmerising to watch and sublime to perform, and at such a moment there can occur one of those magical instances of total connection between people that I have seen happen most strikingly on the dance floor.

Like tango, flamenco is a social dance that has moved away from its early days at parties, in bars and in the streets, and become a performance art. When the French critic Theophile Gautier first saw flamenco dancers in Andalusia in the late 19th century he fell in love with them and – despite his passion for a French ballerina – declared them preferable to the gaslit dolls of the Paris Opera.[33] The word 'duende' is often used to describe a memorable flamenco performer. One dancer described it as a state in which the spirit moves through you, a power 'that climbs up inside you from the soles of the feet': a description, if ever there was one, of harmony in body and spirit.

Reviewing a flamenco company performing at London's pantheon of professional dance, Sadler's Wells Theatre, the critic Nicholas Dromgoole dismissed them with the comment that the dancers were 'past their prime and God's gift to the corset industry. Middle aged spread, or too willing an addiction to paella, is much in evidence'.[34] In his opinion the allure of flamenco lies in a slim body and a figure-hugging costume. The truth is, however, that audiences who turn out in their thousands to watch flamenco don't care if a dancer is past her youth and displays an ample waistline. They go to watch flamenco for its passion, its power of expression and defiant energy. However, in the rarified world of performance the glorious diversity of the body is regarded as an irrelevance. In the hierarchy of Western dance classical ballet remains dominant, with the confident underlying assertion that no other dance form is so beautiful, a view confirmed by the authority of the *Encyclopaedia Britannica*:

> Ballet has been the dominant genre in Western theatre dance since its development as an independent form in the 17th century, and its characteristic style of movement is still based on the positions and steps developed in the court dances of the 16th and 17th centuries.[35]

Yet classical ballet has codified the movements of courtly dance into rigid and damaging postures. And perhaps it is fitting that an art which rejoices in the artificial and celebrates human triumph over the body has come to represent a Western ideal based on a thousand-year-old distrust, both religious and philosophical, of the human body. How many women have been discouraged from learning any kind of dance because of the assumption that they have to be as young and as thin as ballerinas in order to look good on the dance floor? Yet ballerinas starve themselves sometimes to the point of anorexia in order to fulfil the demands of their ballet masters; and (unlike their male counterparts) wear crippling pointe shoes that cause pain and bunions and can make the feet bleed in the normal course of performance?[36] The obsession with youth and a particular kind of body is a rod which the dance profession has made to beat its own back. Its result is rejection of those who cannot conform to a narrow aesthetic.

At a festival where I was once performing in Spain I met a flamenco dancer in her seventies. She smoked like a chimney and was a woman who clearly enjoyed her food. But she was a demon onstage! In the dressing room after the show I told her I thought it was amazing that she was still performing at her age. Her eyes glinted dangerously as she looked at me. 'Let me tell you a secret, my dear' she said, leaning across the table. 'When you dance, there is no such thing as old age'.

NOTES

1.   Ken Robinson, 'Do Schools Kill Creativity' 6 January 2007, https://www.youtube.com/watch?v=iG9CE55wbtY [Accessed 6 September 2019].

2.   Ken Robinson, 'Do Schools Kill Creativity', https://www.ted.com/talks/ken_robinson_says_schools_kill_creativity/transcript, [Accessed 6 September 2019].

3.   Robinson, 'Creativity', transcript.

4.   'National Curriculum in England: physical education programme of study', Statutory Guidance, 11 September 2013, https://www.gov.uk/government/publications/national-curriculum-in-england-physical-education-programmes-of-study/national-curriculum-in-england-physical-education-programmes-of-study [Accessed 5 September 2019].

5.   Cultural Learning Alliance, 'Further decline in arts GCSE and A Level entries', 23 August 2018 https://culturallearningalliance.org.uk/further-decline-in-arts-gcse-and-a-level-entries/ [Accessed 6 September 2019].

6.   Jonathan Savage and David Barnard, *The State of Play: A Review of Music Education in England 2019*, Musicians' Union, Music Industries Association, Making Music Changes Lives, UK Music, 2019, p. 4, https://www.musiciansunion.org.uk/StateOfPlay [Accessed 6 September 2019].

7.   Miami University in Oxford Ohio, 'Peoples of the World Web Project Homepage: Interpetation',http://www.units.miamioh.edu/ath175/student/greenka2/interpretation.html [Accessed 5 September 2019].

8.   Gareth Vile, 'Sufi Festival aims to 'promote the core message of Sufism, to live life on this earth with love, beauty, harmony and worship', *The List*, 15 July 2019, https://www.list.co.uk/article/109710-sufi-festival-aims-to-promote-the-core-message-of-sufism-to-live-life-on-this-earth-with-love-beauty-harmony-and-worship/ [Accessed 5 September 2019].

9.   'Mevlevi *zhikr*', The Threshold Society, https://sufism.org/mevlevi-zhikr [Accessed 6 September 2019].

10.   John Davies, 'Orchestra or a Poeme of Dauncing. Judicially prooving the true observation of Time and Measure, in the authenticall and laudable use of Dauncing', 1596, lines 232-5, 252-8, in *Spenser and the Tradition: English poetry 1579–1830, Gathering of Texts, Biography, and Criticism compiled by David Hill Radcliffe, Virginia Tech.* http://spenserians.cath.vt.edu/TextRecord.php?textsid=32843 [Accessed 6 September 2019].

11.   Davies, 'Orchestra', lines 666-9, 712-4.

12.   See for example, University of Turku, 'HIIT releases endorphins in the brain', *Science Daily*, 24 August 2017, www.sciencedaily.com/releases/2017/08/170824101759.htm (Accessed 6 September 2019, citing Tiina Saanijoki, Lauri Tuominen, Jetro J Tuulari, Lauri Nummenmaa, Eveliina Arponen, Kari Kalliokoski, Jussi Hirvonen. **Opioid Release after High-Intensity Interval Training in Healthy Human Subjects,** *Neuropsychopharmacology*, 2017; Anxiety and Depression Association of America, 'Physical Activity Reduces Stress', https://adaa.org/understanding-anxiety/related-illnesses/other-related-conditions/stress/physical-activity-reduces-st [Accessed 6 September 2019].

13.   Barbara Ehrenreich *Dancing In The Streets: A History of Collective Joy* (2007 Metropolitan Books, New York), p.137.

14.   Ehrenreich, *Dancing*, p.131.

15.   Ehrenreich, *Dancing*,p.151.

16.   Quoted in Angela Last, 'The cosmology of the tarantella', Mutable Matter, 12 May 2017   https://mutablematter.wordpress.com/2017/05/12/the-cosmology-of-the-tarantella/ [Accessed 6 September 2019].

17. Ehrenreich, *Dancing*, p.147.

18. Ehrenreich *Dancing,* p.153.

19. Ehrenreich *Dancing,* p.160.

20. Ehrenreich *Dancing,*p.179.

21. Ava L Vinesett, Muriel Price & Kenneth H Wilson, 'Therapeutic Potential of a Drum & Dance Ceremony based on the African Ngoma Tradition'. Journal of Alternative and Complementary Medicine, 2015 Aug 1; 21(8): 460-465. For the kind of discussion becoming increasingly familiar see Kimerer LaMothe, '"Way of the Bushmen". Dance, Love and God in Africa: The Bushmen share the spiritual wisdom of their healing dance', *Psychology Today*, 31 May 2016.

22. Dance in the United Kingdom (UK) – Statistics & Facts, Statista: The Statistics Portal https://www.statista.com/topics/4015/dance-in-the-united-kingdom-uk/ [Accessed 16 April 2019].

23. The Government's sport strategy 'Sporting Future: A New Strategy for an Active Nation', Dept. for Digital, Culture, Media & Sport, 17 December 2015, https://www.gov.uk/government /organisations/department-for-digital-culture-media-sport [Accessed 16 April 2019].

24. BBC Worldwide celebrates 50th country sale of Strictly Come Dancing'. The Media Centre, 8 April 2014. https:www.bbc.co.uk/mediacentre/worldwide/50th-country-strictly [Accessed 16 April 2019].

25. H. P. Clive, 'Calvinists and the Question of Dancing', Bibliothèque d'Humanisme et Renaissance T. 23, No. 2 (1961), pp. 296-323.

26. Ehrenreich *Dancing*, p.176.

27 . Mariana Zapata. 'Rock 'n' Roll and Military Dictatorships Almost Destroyed Argentine Tango', *Atlas Obscura*, 7 December 2016, https://www.atlasobscura.com/articles/rock-n-roll-and-military-dictatorships-almost-destroyed-argentine-tango [Accessed 16 April 2019].

28. 'Iran: Women Arrested for Dancing', Human Rights Watch. https://www.hrw.org/news/2018/07/11/iran-women-arrested-dancing, 11 July 2018 [Accessed 16 April 2019].

29. Wendy Buonaventura *I Put A Spell On You: Dancing Women from Salome to Madonna* (London: Saqi Books, 2003), p. 71.

30. Mme. Celnart, *The Gentleman and Lady's Book* (Boston 1833), p. 187, cited in Elizabeth Aldrich *From The Ballroom To Hell* (Illinois: Northwestern University Press 1991), p. 154.

31. Wendy Buonaventura *I Put A Spell On You: Dancing Women from Salome to Madonna* (London: Saqi Books, 2003*)*, p.186.

32. For a good introduction see Ana C. Cara, 'Entangled Tangos: Passionate Displays, Intimate Dialogues', *The Journal of American Folklore*, vol. 122, no. 486, 2009, pp. 438–465.

33. Cited in Walter Sorell, *Dance In Its Time* (New York, Anchor Press/Doubleday, 1981), p.280.

34. Cited in Buonaventura *Spell*, p.287.

35. 'Theatre Dance: Dance', *Encyclopaedia Britannica*, https://www.britannica.com/art/dance/Theatre-dance [Accessed 6 September 2019].

36. Emma John, 'I was doing a solo and I heard my foot crack', *The Guardian*, 5 September 2006, https://www.theguardian.com/stage/2006/sep/05/dance [Accessed 16 April 2019]; Jennifer Dunning, 'Eating Disorders Haunt Ballerinas', *New York Times*, 16 July 1997, https://www.nytimes.com/1997/07/16/arts/eating-disorders-haunt-ballerinas.html [Accessed 16 April 2019].

What is the Meaning and Role of Harmony in Sculpture?

Sophie Howard

One definition of harmony in art is 'unity in variety'.[1] 'The quality of forming a pleasing and consistent whole' says the Oxford English Dictionary.[2] Thus, in the visual arts harmony is created through combining and balancing differing visual and conceptual elements. Artwork that achieves harmony brings together phenomena to communicate a unique experience.

'Sculpture' describes a wide range of artworks that use three-dimensional space. Along with the wider category of 'art', definitions of sculpture have undergone great change. 'Even the idea of sculpture being quintessentially a three-dimensional art has been challenged, as its unique feature has become the way space is structured and organised.'[3] Sculptures may be experienced directly, or through images and descriptions to be recreated in the mind's eye. As part of modulating its effect, the sculptor harmonises the elements that go to make a sculpture. The impact on viewers will be affected by the sculptor's work and how it harmonises with the context in which it is seen, whether a gallery or the screen of a phone. The OED definition insists on the whole being 'pleasing and consistent' if it is to have harmony. The object may not in itself be pleasing. It may be challenging, or shocking or disturbing, but nonetheless have gone through a process of harmonisation as part of its making to allow its message to have an impact on the audience.

In my view an excellent contemporary example is provided by the sculptures of Cornelia Parker, which have been through a process of harmonisation in order to reach their ultimate forms. The materials, imagery and spaces are arranged so as to press home complex and meaningful experiences about vulnerability and time through extensive refinement. This refining process, or harmonisation, making the piece pleasing and consistent, is what makes it legible and accessible. This is evident in her piece *Cold Dark Matter: An Exploded View*, in which Parker reassembled the pieces of an exploded garden shed plus its contents by suspending them in space.[4]

Harmony takes a specific role in my own, more traditional, sculptural practice: I make domestic scale objects. When I make a sculpture I want people to look at it, and see something that engages their attention and keeps them looking, so as to receive the visual narrative that is the sculpture: I intend my work to be pleasing and consistent. The language of sculpture is one of volume and plane, surface and imagery. If the piece is hard to look at, confusing, ugly, disjointed, without visual logic, the viewer will not generally dwell on it. Harmonising the elements that make up the sculpture creates a connection to the viewer's feelings, an engagement with the body of the sculpture and its meaning. The accents and dynamics will occur as planned, whilst the other parts will flow.

Definitions of harmony and beauty have changed over time. For a dip into some of the deeper analyses, one can do worse than read Jared S. Moore on the subject.[5] He writes of six, sometimes overlapping, kinds of harmony:

- Formal, or 'the pattern or arrangement of sensuous materials – colours, shapes, tones';[6]
- Ideal, as beauty and 'defined entirely in terms of [Platonic] Idea';[7]
- Expressive, meaning 'having significance, or as expressive of certain emotions, ideas and images';[8]
- Spiritual, as in responding to spiritual needs;
- Psychophysical, which may be expressed through empathy, or a 'feeling of oneness with the object';[9]
- Psychological, or a 'purely subjective harmony within the mind of the observer'.[10]

Although my experience is intuitive and practical, I find that it does align with Moore's own ideas in certain key respects. In his theory of 'expressive' beauty, Moore brings together three elements: the form of the artwork, the idea of the artwork and the observer of the artwork. He writes of 'the ease with which the stimuli from the object is assimilated into the neuronic patterns that already exist in the brain'.[11] To put it another way, the experience of seeing a sculpture is affected not only by its appearance or form, and the ideas that the artist has suggested, but also the mind of the viewer.

In conceiving of a piece and wanting to make it, I experience the form and the meaning *as one thing,* and it is that experience which I aim to share with the viewer. Every piece has its origins in my experience of the world, as well as a trajectory. That direction is not fully set when I begin. There is trial and error, and the unexpected is discovered, but my intention is that, as in Moore's words, 'the material object exactly embodies the idea'.[12] As Moore implies, the artist has to consider the mind of the viewer in the making process, if the sculpture is to carry its message to that mind. Seeing through the eyes of others is part of the job. The effort of the artist to understand and manage the perceptions of viewers is indeed a key part of their role. An artist needs to understand at least in part, how people see, and their frame of reference, if they are to affect that vision. Does the artwork harmonise with existing 'truths' within the mind of its audience? If so it may be that this is the key to harmony being associated with the classical view, that harmony and beauty are the 'natural' result of truth, whilst falseness is 'naturally' ugly.

To be able to share my conception of the *one thing*, the thought/form that I want to make, I have to break it down, and become aware of the form and thought, the visible physical object it will become and the meaning that is intrinsic to the planned piece. It is akin to a translation, a disassembling of the concept that I will then rebuild. Curves and angles, textures and colours, gestures and imagery are all elements that have to be harmonised. This is a process and does not happen all at once, and sometimes not at all.

As I conceive an idea for a new sculpture and begin to make it, I trust the piece will become a successfully completed unified whole, as it is gradually built. In the end, I intend, to quote Moore again, that 'Form and significance are not separable parts of an aesthetic object but inseparable aspects'.[13] The precise outcome is not certain, but grows, guided by the desire to share a certain vision, a partly hidden truth. The making process includes playing with options of form, scale and surface. Do I seek harmony? Not as an end in itself but it is a means, a tool to modulate the diverse elements and make the sculpture do its work.

As the practice of sculpture has changed, so have ideas of beauty and harmony. Arthur Woods writes succinctly 'In a good work of art, all of its elements are interrelated for a coherent unity-content and form fused into its overall message or meaning. Such images, exploiting the expressive qualities of form, go beyond mere representation of fact and can communicate a wide range of subtle and powerful feelings'.[14]

Dance is an aesthetic experience and one that can lead to exalted states of delight. Like other physical and performance arts, dance is an expressive activity. It can be seen as Vasudha Narayanan said as 'an articulation of the deeper nature of things and as connected with the other streams of life'.[15] My sculpture is an embodiment of the whole experience of dance, including the feeling, mood and music, reinterpreted in ceramic form.

I have made a series of sculptures of tango dancers, evoking in abstract form various aspects of the dance. I initially named them after cities where I have danced or evoke something about dance. One example is 'Sienna' (Figure 1), another is 'Utrecht' (Figure 2). My aim in making 'Utrecht' was to express

Figure 1: 'Sienna'. Original sculpture by Sophie Howard.

Figure 2: 'Utrecht'. Original sculpture by Sophie Howard.

an experience; seen, heard and danced, through the medium of the sculpture. Dancers move, and this kind of sculpture, at least, does not. 'Utrecht' is a hand built ceramic form, inspired by a particular step in Argentine tango, which I have been learning and dancing for many years. A couple are engaged in a swift but controlled action: this is a move I have seen and experienced as a dancer, so I can physically imagine this tiny moment which is one of a flow of postures. Knowing the way it feels helps to guide me in the making of the form.

I make tango sculptures in clay and fire them, often glazing. The results are three-dimensional forms that communicate the experience of the dance. I build in the moods and rhythms of the music, the connection and motion of the bodies, and include too the dancers' intention. To make the sculpture communicate all the things I want, I must consider many parts of the experience, and bring them together in a form that makes sense to the viewer. Elements to be included are the tempo and mood of the music, the interpretation that the two dancers might add in that particular moment and the weight, balance and co-ordination of the pair as they move together in the moment of movement, as well as any other narrative that I want to include.

Tango requires deep communication between the dancers, demanding harmonisation in the sense that they must align themselves, finding common, shifting ground within the music. Julie Taylor sees tango arising from a deeply disharmonious environment, becoming a therapy that is itself harmonious. She writes that 'dancers cling together in the uniquely intimate tango embrace', reconciling dramatic 'tense disharmonies', and citing one maestro who said 'it's the bodies talking together on their own', reconciling dramatic 'tense disharmonies'.[16] For me, tango is a strong and expressive improvisation à deux, a changing physical conversation between two people, even though the coordination between the partners is often impossible to see. The couple make shapes that move and change apparently without planning, and although one person leads and one follows, each can be fully expressive in responding to the tenderness and power in the complex layers of the music. I seek to embody moments of tango music and dance in my sculpture: dance and sculpture are both addictive.

If all the physical detail of the dancers was included in 'Utrecht', there would be too much to look at. Those details, of shoes, of hair, of clothing, would distract from the essential forms that the sculpture aims to manifest. 'Utrecht' shows just a moment but includes a sense of the moment before and the ones to come. The aspects I want the viewer to see and feel are specific: how the bodies press together but move around each other. How the dancers are both unified and

Figure 3: 'Before'. Original sculpture by Sophie Howard.

separate. How their dynamic movement and weight has turned them into one form that holds together as it responds to the music.

As well as the abstract dance sculptures of which 'Utrecht' is one, I also make more 'realistic' ones, which include details such as hair, clothing, shoes and faces. One, is just called 'Before' (Figure 3) and requires some effort by the viewer to think 'before what?' My answer, if needed, is that the whole thing is a memory: the dancers remember a time 'before' when they danced, in a nostalgic way. Perhaps ' before' they got together as a couple, implying that this moment was significant in their attraction. Another example is 'Heloise' (Figure 4). My description of the pieces as 'realistic' and 'abstract' is not precise, but more of a shorthand. The 'realistic' ones are more accurately 'sketched three-dimensional studies'. The

Figure 4: 'Heloise'. Original sculpture by Sophie Howard.

'abstract' ones are more accurately 'semi-abstract'. Through making the studies, I explore the subject thoroughly. In the studies, I learn a lot about the interlocking shapes of the bodies, the emphasis and flow of the forms and what is essential. This process allows me to discover what is most exciting, dynamic and dramatic in the particular moment of tango. To make a more abstract design based on a study, I start a new piece in which I define, refine, combine and edit forms and angles, weight and surface, which I describe as a process of harmonisation.

How does this harmonisation process work in practice? In the study I see repeating angles, surfaces and shapes that seem similar, curves and joins and textures that recur. The shapes that fit with the thrust and dynamism I want are brought to the fore in the new piece. In building abstract pieces I allow the shapes I have chosen to take over, making a curve repeat on different parts of the sculpture. I minimise some variations in the forms and accentuate angles so that only the essence of the figures' posture is visible after this process of refinement. The working study from which I develop the abstract sculpture may have separate colours for clothing and hair. In the more abstract piece of dance sculpture, colour and texture can be exploited across the whole form to be expressive of music and mood. Colour may, as part of the process take over with a swirl of blue and white or fragments of gold. The elements that have been harmonised allow the complex form to carry as much information as possible to the viewer's eye and mind, without interruption. If there is too much information to take in, unity is lost and the eye tires, the viewer loses track of the experience. Harmony is a tool to streamline the visual information embodied in the sculpture so the viewer can experience its impact.

A part of the making process can include a journey through the uninspiring sinking sands of over-harmonisation in which every dynamic is smoothed away. There is usually a way out. But can there be too much harmony? Does the boiling down of the visual soup ultimately result in a burnt pan? To resolve and quieten every difference ends with dull sculpture, a simple geometric form, or empty space. Raw unresolvedness also has a place in sculpture. The boiling down should leave me with perfect crystals, ideally, rather than sludge. In the case of tango sculptures, harmonising of body forms, music and movement should not render them so smooth that they disappear, but allow the impression to be received, the dynamic to be felt.

Some ideas for sculptures seem to leap to mind fully formed, and it is the satisfactory crafting of the object that is my only task, but usually the conversation with the clay is more complex. I do visual research, looking at other work, photos, videos, and reading around ideas. However, the most complex conversation and examination of ideas so often leads one back to the basics: What will this thing be? How should it look? What will it make a person feel?

If all the elements are working together, a sculpture can be achieved in which the appearance, feeling and ideas become one. To achieve this, there is a role for harmonising. The aim will be to expose a form/idea in a way that allows it to be better experienced by its viewers. Harmony is an issue when it comes to locations for sculpture, as there is always a context. When making a piece I consider how it looks from all angles, what it means and how viewers can engage with it. I make house trained sculpture, intended to be spied from all angles, and adaptable to setting. Ideally, mine don't fall over, won't damage their surroundings or viewers, and will blend into diverse settings.

Baudelaire had strong opinions about how sculpture affected architecture and natural settings. He was not a fan and complained in his 'Salon de 1846 ' entitled 'Why Sculpture is Boring' that sculpture was 'crude and direct like nature' and 'vague and elusive owing to the many sides that can be seen at the same time'.[17] This startling objection has been taken as a starting point for detailed discussions about exactly what sculpture can 'do' in various contexts. Theory and discussion on how sculpture can and should be viewed have implications for architects, designers and sculptors themselves. Arnold Cusmariu offers a highly detailed analysis of the aesthetics of sculpture that 'gives the third dimension a new aesthetic paradigm ... and creates extraordinary artworks'.[18]

In the physical world, sculptures are seen in a myriad of temporary and permanent settings. Pieces are often designed for the site they occupy. There are works that are both architectural and sculptural at the same time, such as the pyramids of Egypt. There is, as Art Brenner says 'a strong historical affinity between sculptor and architect stemming from their common origins as craftsmen of wood and stone".[19] He goes on to point out that 'Records from Romanesque and Gothic periods tend to confirm that they were often one and the same'.[20] Exploring the relationship that a sculpture has with its setting, we soon find that the two need to harmonise if they are to enhance a place, with the dynamics of one complementing the characteristics of the other. If a sculpture is made for a particular setting or type of place, both place and sculpture will be simultaneously enlivened. As Brenner points out, art plays a crucial, almost shamanic role of connecting people to their environment. Even a sculpture designed for temporary settings has to fit well into its surroundings to be seen at its best. Context is not everything, but even a movable sculpture benefits through proper placement.

Harmony has a role as an integral element of the experience of sculpture as it smooths the way, making space for the eye and mind to engage with a piece, be the form simple or the idea complex. As a concept, harmony has meaning independently of its practical role. There is a view, previously mentioned, that harmony and beauty are the 'natural' result of truth, whilst falseness is 'naturally' ugly.

Certainly, when things 'look wrong' or 'don't work' in the making of a sculpture it is a sign that something is amiss. The sculptor has to search for the truth of their intention and find a better way to share it.

Discussions on the making, viewing and placing of sculpture raise questions to consider when making new work. In the practical space of making, it is always useful to ask questions. Theoretical questions do help as a way to interrogate my own practice when the way ahead is foggy. My questions as a maker are mostly practical ones that seek to solve problems about why a work in progress does not yet do what I want it to. My answers are physical and involve changes to forms, surfaces and imagery. Fresh eyes and the interrogation of my own motivations and abilities lead to clarity. Making sculpture is not a purely intellectual exercise, but a practice without which I would be unable to engage with a world which is at once dissonant, harmonious and dynamic. Harmony quietens the babble of conflicting energies and allows clear communication.

NOTES

1.   Jared S. Moore,. 'Beauty as Harmony', *The Journal of Aesthetics and Art Criticism*, vol. 2, no. 7 (1942): pp. 40, 42.

2.   'Harmony', *Oxford English Dictionary*, https://en.oxforddictionaries.com/ definition/harmony [Accessed 25 May 2019].

3.   Anna Moszynska, *Sculpture Now* (London: Thames and Hudson, 2013), p. 7.

4.   The Tate, 'The Story of Cold Dark Matter', https://www.tate.org.uk/art/artworks/ parker-cold-dark-matter-an-exploded-view-to6949/story-cold-dark-matter [Accessed 25 May 2019].

5.   Moore, 'Beauty', p. 50.

6.   Moore, 'Beauty', p.41.

7.   Moore, 'Beauty', p. 42.

8.   Moore, 'Beauty', p. 41.

9.   Moore, 'Beauty', p. 46.

10.   Moore, 'Beauty, p. 43.

11.   Moore, 'Beauty', p. 46.

12.   Moore, 'Beauty', p. 45.

13.   Moore, 'Beauty', p. 44.

14.   Arthur Woods, 'The "Cosmic Dancer": Sculpture and the Absence of Gravity', *Leonardo*, vol. 26, no. 4 (1993): pp. 297-301.

15.   Vasudha Narayanan,. '2002 Presidential Address: Embodied Cosmologies: Sights of Piety, Sites of Power', *Journal of the American Academy of Religion*, vol. 71, no. 3 (2003): pp. 495-520.

16.   Julie Taylor, 'Death Dressed As a Dancer: The Grotesque, Violence, and the Argentine Tango', *TDR (1988-)*, vol. 57, no. 3 (2013): pp. 117-31, (pp. 117, 121, 124).

17.   Arnold Cusmariu, 'Baudelaire's Critique of Sculpture', *The Journal of Aesthetic Education*, vol. 49, no. 3 (2015): pp. 96, 119.

18.   Cusmariu, 'Baudelaire's Critique of Sculpture', p. 119.

19.   Art Brenner, 'Concerning Sculpture and Architecture', *Leonardo*, vol. 4, no. 2 (1971): pp. 99-100.

20.   Brenner, 'Concerning Sculpture and Architecture', p.100.

THE CONCEPT OF HARMONY: A LITERATURE REVIEW

Toto Gill

ALTHOUGH THE NOTION OF HARMONY IS CONCEPTUALLY RICH and varied in its definitions and emergent qualities, it resides either implicitly or explicitly at the heart of a plethora of philosophical religious, and/or spiritual traditions as a metaphysical principle/framework and as a value with ethical, moral or spiritual significance. Due to the limited scope of the project,[1] research for this review was conducted using digital resources and online databases (such as JSTOR), focussing upon secondary scholarly and academic sources in the English language. For example, I have summarised statements on Plato and Buddhism from scholarly books, journals and papers, rather than from original sources such as Platonic dialogues or Buddhist scriptures. All the statements I have collected therefore represent academic interpretations and perspectives; primary sources await specific, detailed research and review. For each major tradition and sub-category within the area of study, I performed a variety of searches, each including relevant keywords. My first search usually entailed the name of the tradition followed by 'harmony' and various permutations of the two terms, such as 'Buddhist conception of harmony', 'harmonisation in Buddhism' and so on, and became more specific based upon the initial results returned. Notably, the Confucian, Daoist, Hellenic and Buddhist traditions were those which returned more results which were specifically relevant to harmony. Other traditions which are less explicit in the articulation of notions and practices of harmony, such as the Abrahamic religions, often did not return results which featured explicit mention of harmony in titles. As the project went on, it became increasingly clear that 'harmony as a conceptual area of study is, from an academic perspective, esoteric and under-explored. Outside of this, harmony has been subject to discussion within the sphere of sociology, such as in the Global Harmony Index.[2]

This review is organised around three common themes within diverse conceptions of harmony: (a) harmony as unity within diversity, (b) harmony as a dynamic process, and (c) harmony as an attribute of the divine. The philosophical/religious/spiritual traditions to be examined are Confucianism, Daoism, Mohism, ancient Greek thought (Plato, Aristotle, Heraclitus and Pythagoras), Buddhism, Christianity (including Aquinas and Leibniz), Judaism and Islam. In presenting these diverse conceptions, differences, similarities and nuanced distinctions are also sought and discussed. Harmony is then examined from the perspective of practice, illuminating various approaches towards how harmony can be integrated into and as a way of being.

HARMONY AS UNITY WITHIN DIVERSITY

In order to begin to elucidate a basic understanding of the notion of harmony and its conceptual fundamentals, I shall turn first to ancient Chinese thought. Chenyang Li, who can be credited with having written the first full length book on the 'understudied' concept of harmony within the Confucian tradition,[3] argues that harmony is 'the most cherished ideal in Chinese culture' and a central concept within ancient Chinese philosophy more broadly.[4] The earliest iterations of the notion of harmony appeared primarily within metaphorical frameworks of music and cuisine.[5] Li provides an etymological deconstruction detailing the range of semantic meanings generated by *he* 和 (translating to harmony), arguing that its original meaning is derived from music, wherein different sounds are related in a 'mutually promoting, mutually complementing, and mutually stabilizing' rhythmic interplay.[6] Stephen C. Angle, citing the pre-Confucian *Classic of Odes*, details an analogy of harmony with a broth, wherein diverse ingredients are harmonized by a cook into a balanced proportionality, simultaneously regulating and rectifying both excess and deficiency.[7] From both analyses, it can be understood that at its most fundamental, harmony can be interpreted broadly as the 'coming-together of different things', within which a certain 'favourable relationship' is implied, analogous to musical (or at the least, sonorous) concordance.[8] In both Li and Angle's depictions of harmony, harmony implies a virtuous quality of a totality wherein the existence of differing components is presupposed as in the interplay of diverse sounds and the fusions of different cooking ingredients yet are set into a balanced relationship.

However, harmony in Confucianism is not merely a state of equanimity between varying elements. Li makes an emphatic distinction between the undesirable notion of *tong* (sameness or uniformity) and *he* (harmony).[9] According to Chung-Ying Cheng, *he* admits 'disagreement and difference in unison' whereas *tong* admits 'no disagreement and no difference at all'.[10] Harmony differs from 'stagnant concordance' in that it is 'sustained by energy generated through the interaction of different elements in creative tension'.[11] Li suggests that *tong* 'without adequate differences' precludes the possibility of meaningful harmony, stating that 'a single item does not harmonize'.[12] Thus, diversity and difference can be understood as requisite foundations for harmony, such that harmony can be characterized to a certain extent by *unity in diversity* or *balance in opposition*.

As Li observes, there are significant similarities between Confucian and Hellenic conceptual models of harmony during the axial age (8th–3rd century BCE), particularly evident in the thought of Heraclitus, who along with Pythagoras was one of the first Greek philosophers to explore harmony.[13] The basis of the English word 'harmony' is found in the Latin word *harmonia* and the Greek word ἁρμονία – derived from ἁρμός ('*harmos*'), meaning 'joint' or 'means of joint' – roughly translating to 'concord of sounds'.[14] Heraclitus defines harmony at its most fundamental as 'the opposites in concert', stating 'that which is in opposition is in concert, and from things that differ comes the most beautiful harmony'.[15] Much like Confucianism, Heraclitus construes deeper harmony as constituted in a unity of multiple contrasting elements, as opposed to a lesser harmony exemplified by relations marked by similarity. In his explanation, Li cites W. K. C Guthrie's interpretation of harmony from a Heraclitan perspective, which can be summarised in three points.

1. As everything is composed in some defining respect by opposing elements, all things are 'subject to internal tension'.

2. All opposites are in some significant sense identical. This statement loosely delineates several conceptual aspects of opposites: mutual transformability or 'reciprocal succession' (for example, day and night); relativity to the subject's perspectives; and the presence of multiple and seemingly contradictory facets of the same conceptual object (for example, the properties 'up' and 'down' belonging to the same continuum).

3. Tension between opposites is 'the universal and creative force of harmony'. Where there is no tension and opposition, there is no harmony. This harmony exists in the equilibrium of opposing movements and elements.[16]

The tropes of *unity in diversity* and *balance in opposition* manifest most distinctly in the ancient Chinese paradigm of *yin-yang*. Broadly, *yin-yang* refers to the unity of two 'mutually opposed' yet 'correlative and complementary' forces considered to exist within all things.[17] In a narrower sense, it refers to two dynamic and mutually complementary 'force elements within *qi*' whose fusion and interaction 'determines the existence of all things in the universe'.[18] Represented by the dots in each respective half of the flowing circle of the *yin-yang* symbol, the *yin* and *yang* are 'interdependent, interpenetrating and inter-transformational'.[19]

The reciprocal relationship between hot and cold, day and night, and up and down exemplify these qualities of mutual transformability. Cheng interprets the condition of this mutual 'interpenetration of things in a whole' to be the 'unity of opposites', such that genuine opposites are only found in unity, and vice versa; unity and opposites therefore become the 'condition for existence of the other'.[20]

Heraclitus was not the only ancient Greek philosopher to conceive of harmony as constituted within relationships of difference. The earliest examples of Pythagorean investigations into the concept of harmony manifested in his significant discovery of intervallic musical consonances by means of mathematical ratios.[21] Pythagoras is credited with being the first to conceive of the 'harmony of the spheres', the notion that the celestial bodies (Sun, Moon and planets) move in harmonious, 'musical' proportion.[22] He was also the first to apply the word *kosmos*, which translates to 'order', in describing the world and more broadly, all phenomena, acknowledging what he perceived as its immanent logic.[23] The Pythagoreans viewed music as a 'prototype' of the concept of harmony, but equated harmony with 'numerical ratios and abstract mathematical formulas', illustrating a characteristically and contrastingly 'quantitative tendency' in their conception of harmony, compared with the predominantly qualitative approach of ancient Chinese philosophy.[24] For Pythagoreans, God is not the primary mover in the cosmos; mathematics constitute the 'ultimate principle of everything', innately representing the 'harmonious unification of opposites' in their alternation of even and odd.[25] Harmony is thus presented within the Pythagorean tradition as a metaphysical principle and a means of qualifying our perception of order which in itself is constituted by unified differing elements.

Buddhist philosophy also regards diversity to be crucial in the formation of harmony. Michio T. Shinozaki states that 'oneness' in Buddhism is not conducive to or synonymous with sameness, arguing that Buddhists perceive the ideal way of being as a relational and inclusive 'oneness of self and others', stating that 'violence exists when sameness dominates over difference'.[26] Shinozaki, in his article on the notion of peace from a Mahayana Buddhist perspective, focusing on the thought of Nikkyo Niwano (one of the founders of Buddhist organisation Rissho Kosei Kai and its first president), summarises Niwano's conception of the state of peace as 'a harmonious state of diversity'.[27] The role of the concept of harmony in regards to Buddhist ethics, practice and metaphysics shall be discussed in further detail later.

The concept of harmony also played a 'crucial role' in the philosophical thoughts of Christian theologian, philosopher Gottfried Wilhelm Leibniz.[28] Laurence Carlin states that harmony is incorporated into Leibniz's mathematics, metaphysics, ethics and social philosophy 'as a central descriptive and explanatory concept'.[29] Similarly to Heraclitan and Confucian thought, Leibniz defines harmony as 'a similitude in dissimilar things', 'unity in variety' and more frequently as 'diversity compensated by identity'.[30] Leibniz summarises the basic premise of his concept of harmony as follows:

> Harmony is unity in variety ... Harmony is when many things are reduced to some unity. For where there is no variety, there is no harmony. Conversely, where variety is without order, without proportion, there is no harmony. Hence, it is evident that the greater the variety and the unity in variety, this variety is harmonious to a higher degree.[31]

Similarly, Daoism regards 'creative tension' and 'contrast' between opposing qualities as fundamental attributes of harmony, implying a 'mutually completing and mutually compensating relationship'.[32] The 'ultimate harmony' is exemplified by nature, in which all things, composed of both *yin* and *yang*, are harmonised by *qi* (energy), illustrating the significance of the relationship between humanity and the effortlessly harmonious world for Daoists.[33] It is useful to clarify what is meant by 'nature' and 'natural' in the Daoist context. According to Jiyuan Yu, in Daoism, nature is not the natural world *per se*, but relates to the 'fundamental operational principle of the natural world', which is in itself a 'self-transformational and spontaneous state' itself constituted by the interrelated concepts of *ziran* (naturalness) and *dao*.[34]

In contrast, the Mohist conception of harmony is centred upon human relationships, connecting harmony with love, with emphasis on the 'aspect of accord' rather than tension, diversity and dynamism.[35] According to Li, Mozi maintained that the sharing the same idea was a necessary condition for harmonisation in relation to his notion of *shang tong* (promoting uniformity or sameness).[36] Thus, Li claims that Confucians are 'more willing to accommodate difference' and less willing to embrace *tong*, whereas Mohist thought is 'wary of difference' and inclined towards unity through uniformity.[37]

It is pertinent to briefly address the relationship between 'harmony' and its multitude of inherited meanings ranging from the ontological to the ethical, and harmony as a musical phenomenon. Cheng offers a basic 'analytic reconstruction' of musical harmony highlighting four central aspects: (1) musical harmony is a 'totality of parts'; (2) each part of the totality relates to other parts in the totality (he argues that this relation is one of 'support and recognition'); (3) each part contributes to 'the formation of the totality in the sense of wholeness'; (4) the relating of parts in musical harmony is a dynamic *process* realised necessarily within spatiotemporal parameters.[38] Hence, he renders musical harmony as a 'four-dimensional totality' comprised of interrelated parts engaged in 'mutual support', realised in an 'explicit process of time and implicit reference to space'.[39] It is important to note, however, the fallacy in constructing a single theory of musical harmony, given the immense diversity within musical systems and ontologies across the world, and the changing definitions and aesthetic conceptions of, not only 'harmony', but music over time. Hellenic and ancient Chinese conceptions of musical processes would have been limited geographically and historically given the relative youth of music as a cultural medium and aesthetic form, rendering a qualitative analysis less problematic. Therefore, the analogising of musical and extramusical harmony must be approached tentatively; I would suggest that, if attempted, the analogy should be approached from the perspective of social mediation and the relational affordances of musical participation, rather than through an exclusively "musical" analytical perspective.

Nevertheless, the analogising of music and harmony, or more specifically in this case their co-conception, is central to the emergence of the concept of harmony within Hellenic and ancient Chinese philosophy.[40] This relationship also underscores the aspect of the aesthetic within the concept of harmony, inherent in our experience of 'agreeable totality' and the 'feeling of harmony' within art, music, cuisine and the natural world.[41] Even within the Pythagorean tradition, our perception of harmonious order in numbers and astronomical phenomena is fundamentally aesthetic in nature, given that order is presented in these contexts as inherently pleasing. Aesthetic balance presents itself as the fruit of coordination of diverse elements without sacrificing their distinctness; within this framework, aesthetic beauty lies precisely in the possibility of the contrasting and conflictual elements being held harmoniously in an underlying whole. Thus, harmony is 'both internally and externally real', existing as both a 'real structure' and as an 'ideal projection', rooted in human experience.[42]

HARMONY AS A DYNAMIC PROCESS

Another significant trope which is shared across several conceptions of harmony is its characterisation as a dynamic and continuous *process* as opposed to a *state* which is achieved and maintained. We have already seen how Daoism construes harmony as a constant dynamism wherein *qi* continuously harmonises *yin* and *yang* in the natural world, while humanity should strive to fluidly adapt to and harmonise with the natural world and its processes. In Confucianism, Li argues that while the Chinese word *he* translates to 'harmony', it is characterised more often as a verb than as a noun, and so may be more appropriately articulated as *harmonisation*, emphasising its nature as a 'dynamic process' requiring agency and action.[43] Cheng stresses the importance of the evolution of conflicting elements into 'different states of their existence' in a 'totalistic system' by virtue of the unifying and integrating process of harmonisation, thus producing a '*whole* of differences'.[44] The concept of harmony as unity in diversity but most pertinently as a metaphysical principal of *transformative process* is described by Cheng as the 'the basic insight of the philosophy of the *Yijing*'.[45]

A significant element in the characterisation of harmony as dynamic *harmonisation* is that of 'strife', which can be understood at its most fundamental within this context as struggle or conflict. Heraclitus has been interpreted to perceive change and transformation to be possible exclusively by virtue of strife, without which things would 'cease to exist'.[46] Edward Hussey describes harmony as conceived by Heraclitus as implying a 'purposive mutual adjustment of components to produce a unity', again suggesting the presence of agency and action as requisite elements within the harmonisation process.[47] Li concurs with this interpretation, arguing that a particular kind of strife is 'inherent' in Confucian harmony, detailing two types of strife: 'tension and cooperative opposition', and the more severe 'antagonistic opposition'.[48] The first is illustrated using the example of people moving in opposite directions in a busy train station yet accommodating one another's movement and achieving cooperation despite difference.[49] The second is more complex, illustrated through the mutually exclusive relationship between wolves and sheep, which Li argues necessitates a harmonisation process which transforms antagonistic opposition into cooperative opposition (notably exemplified in a process occurring 'naturally' in the world).[50] This illuminates a characterization of harmony as a quasi-dialectical creative principle, wherein differing elements enter into mutually transformative, generative relationships to form novel totalities in a dynamic and ongoing process.

Cheng criticises the Heraclitan concept of strife on the grounds that it lacks an explicit explanation of the 'principle of unity', critiquing what he construes as too great a focus on the concept of strife and the transformation of opposites.[51] Informed by principles of the *Yijing*, he suggests instead that the 'very presence of change and transformation in existing things should presuppose a unity of unity and disunity', underscoring the unity of opposites as a uniquely creative principle 'in a metaphysical sense'.[52] In this way Cheng argues that it is *unity* rather than disunity in which transformation occurs.[53]

While a search for uses of the word 'harmony' in the *Oxford Online*

Dictionary of Buddhism rendered only seven results, none of which explicitly defined a concept of harmony on its own terms but rather presented harmony in reference to other Buddhist terminology, concepts aligned with the broad parameters of harmony theory are fundamental to Buddhist philosophy.[54] The notion of *pratītyasamutpāda*, which attempts to 'capture the Buddha's account of causality', is the 'basic idea of Buddhism as a system of thought'.[55] Translating roughly as 'dependent co-arising', this fundamental concept assumes that 'reality appears as an interdependent process wherein change and choice, doer and deed, person and community are mutually causative'.[56] The mutuality of causation, phenomena and being implicitly places a metaphysics of harmony at the centre of Buddhist philosophy.

According to Shinozaki, many Japanese Buddhists, including Nikkyo Niwano, regard harmony as one of the 'highest values', interpreting nirvana (the 'ideal Buddhist state') as a 'dynamism of creation and harmony', rather than merely a 'state of mental peace and quiet'.[57] Harmony is again analogised with music, specifically with the performance of an orchestra, emphasising implicitly the aspect of the interdependence and relationality within group performance.[58] John Brinkman states that within Buddhism, harmony as a dynamic process, inextricable from the notion of dependent co-arising, can be interpreted such that it becomes 'the comprehensive character of the universe and the fundamental quality of each particular existence'.[59]

Harmony as an Attribute of the Divine

Historical difference between Eastern and Western epistemologies and hermeneutics may help illuminate the diverse and contrasting conceptual frameworks for the notion of harmony across the globe. David Hall and Roger Ames suggest that the prototypical Western philosophical approach is rooted in 'Truth-seeking' wherein Truth can be interpreted as essence, eternal and absolute, exemplified in Plato's Forms.[60] In contrast, ancient Chinese thought is characterised by 'Way-seeking', wherein the Way, in Li's words, is 'generated through human activity',[61] and is not predominated by pre-set narratives.[62]

Classical Chinese worldviews and metaphysics can be interpreted as 'revolving around the articulation of ongoing processes' which 'have no external, Godlike power behind them'.[63] While Western thought follows logical patterns, relying on 'the application of an antecedent pattern of relatedness to a given situation', the classical Chinese approach is rooted in aesthetic order, relying on 'creation of novel patterns', requiring 'openness, disclosure and flexibility'.[64]

The Confucian, Daoist and Buddhist concept of harmony, or harmonisation, epitomises this distinction, interpretable as an 'integration of different forces and as an on-going process in a fluid yet dynamic world' without a 'given, fixed underlying structure', which Li terms 'deep harmony'.[65] This is contrasted with the 'conformist harmony' of Pythagoras, Plato and the Abrahamic religions, which complies with a predetermined, 'perfect order' in the world, either resulting from the imposition of order on the world by a divine Other such as God, or from the primordiality of number/mathematics as the ultimate principle in the case of Pythagoras.[66]

Imtiyaz Yusuf argues that humanity has experienced the Ultimate predominantly in three ways: first, perceived 'from the outside' such as in the cases of the Semitic prophets; second, perceived 'from within', such as in the cases of Hinduism, Jainism and Buddhism; third, through a medium such as a shaman, as is the case in shamanistic and indigenous African religions/spiritual traditions.[67] John Hick argues that monotheistic religions understand God on 'personal terms', defined by the relationship of the individual believer and the divine Other, whereas non-theistic religions interact with Ultimate Reality either via worship of many gods 'at a popular level' or by 'adhering to the notion of non-personal Ultimate Reality at the philosophical level' (as is the case in Buddhism).[68] It is this 'personal' relation between a divine Other and humanity which underscores much of the Western religions conceptualisations of harmony.

The Abrahamic religions share similarities in their understanding of harmony on a conceptual level, insofar as harmony is constituted both in and by God. In other words, harmony is understood in relation to God and as an attribute of the divine. Before examining the concept and role of harmony within the Abrahamic religions, it is pertinent to examine harmony as an emergent quality of the creation of a divine Other within the thought of Plato as a precursor to the theocentric metaphysics of the Abrahamic religions. In *Timaeus*, Plato presents a cosmogony (notably distinct from that of contemporaneous Greek mythology) in which, from a state of chaos, the 'Demiurge' imposes mathematical order, generating a 'soul for the world'.[69] A moral element is attached to the construction of the harmonious cosmos from chaos in that Plato considers order to be implicitly good and disorder to be bad. Plato is one of the first thinkers to introduce the concept that there is an intrinsically good Divine other, who necessarily produces only the good.[70] Plato distinguishes the ever-changing phenomenological world with the eternal, unchanging and perfect world of the Forms which can only be 'apprehended by intelligence and reason', wherein, according to Li, Plato's concept of harmony is rooted.[71] Plato's understanding of harmony can therefore be interpreted as constituted in the divine, pre-established, perfect order of the cosmos, generated by a divine Other based upon the unchanging Forms.

Brinkman argues that within the Christian tradition, the universe is permeated with the 'attribute of divine peace', by virtue of God being the 'pure and simple' source of every being's 'integrity', 'accord' and 'the sacred cause of the universal integration and peace of all phenomena'.[72] Based upon the thought of Thomas Aquinas, Brinkman interprets the 'central perfection of the universe' to be 'the act of existence itself', due to, in simple terms, the immanence of God; the quality of harmonious perfection exists only in God's 'transcendent simplicity'.[73] Similarly, Leibniz's notion of order makes the same appeal to God's quality of perfection, arguing that God is not capable of doing that which is not orderly.[74] For Leibniz, events that are not regular are not conceivable; irregularity is simply a perceptual error on part of the human subject when a rule of order is more complex than the human mind can grasp.[75] This bears similarities to Heraclitus' doctrine of hidden harmony, wherein the most profound harmony is construed as that which is not immediately visible.

Judaism also sees harmony as an attribute of the divine. God, as I. A. Ben Yosef writes, is the 'Ultimate Reality', with God's quality as *creator* being one of the most significant aspects of the Jewish understanding of God.[76] In contrast with Confucianism, which is inclined towards a unified conception of human and ultimate, Judaism presents the 'separation between human beings and the ultimate' as the primary motive in 'the explanation for relatedness in the world'.[77] Galia Patt-Shamir offers the example of the story of Creation (*Genesis* 1-2) as the 'divine act of separation' wherein God becomes understood by mankind as 'distinct, detached, and perfect'.[78] For Patt-Shamir, God is understood in Judaism as an aspiration – the only being who truly understands what is good and what is not; mankind is eternally distinct from and 'subdued' by God.[79] Hence, Yosef argues that 'the concept of unity of the universe was foreign to Jews; they saw unity in God alone'.[80] Patt-Shamir argues that *disharmony* is established in the human world as a 'necessary human condition' and as a 'motivation for action and progress', citing the Biblical story of Babel as an example.[81] Subsequently, separation and disharmony can be understood as 'primary motivations for action'; it is God's 'demand' that mankind improves.[82] It seems that harmony from a Jewish perspective can be understood as actualised as an aspect of God's perfection and divine unity, and as an ideal that should be strived for by humanity in order to come closer with the divine.

Similarly, Islam conceives of harmony as a cosmic principle emerging from God's immanence and perfection. Ibrahim Kalin cites the traditional Islamic philosophy of 'the great chain of being', which refers to the interpretation of the world as the 'best of all possible worlds', due to its actuality, implying 'completion and plenitude' rather than potentiality, and prerequisite order derived from God.[83] He goes on to describe the natural world as 'a constant state of peace' as according to the Qur'an, it is *Muslim* in that it 'surrenders... itself to the will of God' and therefore transcends 'tension and discord'.[84] This suggests that harmony in Islam is conceived of not only as an inherent attribute of God and an emergent quality of his creation but most importantly in the relationship between mankind, nature and the divine Other.

In its conceptualisation of harmony, Islam seems to place particular emphasis on the relationship between the believer and God. While harmony is not an explicitly defined principle or concept in Islamic doctrine *per se* or (seemingly) in the broad spectrum of Islamic studies, the concept of *Tawhid* may inform the construction of an Islamic notion of harmony. Defined in the *Oxford Dictionary of Islam* as the 'defining doctrine' of Islam, *Tawhid* can be interpreted as 'oneness with God', the declaration of 'the unity and uniqueness of God as creator and sustainer of the universe'.[85] *Tawhid* is used as the 'organising principle for human society' and the fundament of 'religious knowledge, history, metaphysics, aesthetics, and ethics' as well as 'social, economic and world order'.[86]

Sumanto Al Qurtuby offers an analysis of the concept of *Tawhid* from an anthropocentric perspective, arguing that *Tawhid* is interpretable as a 'human act', a departure point for 'true understanding' of ourselves and our place in the cosmos, and 'human-God relations'.[87] He describes *Tawhid* as a 'principle of unity' which offers a foundation for 'achieving political justice, social harmony, civil liberty, and peacebuilding', underscoring not only the 'oneness of God' but the 'unity of humanity'.[88] He states that as created, humans exist in 'dispersion, disarray, disharmony, dissonance, and discord', and that *Tawhid* presents itself as a pathway to resolve this disharmony by overcoming 'false realities' (defined as 'anything that distracts the heart from ... God') and achieving social justice.[89]

In a similar vein to Judaism, in Islam, disharmony is conceived of as a natural state resulting from human action and agency, against which humanity must strive through commitment to God. Mohammed Abu-Nimer argues that, given humanity's capacity for wrongdoing, the prophets of God will encounter opposition, highlighting the nature of conflict as a 'natural phenomenon', emphasising the importance of conflict transformation, the devotion to God and God's will for mankind to come into unity.[90] It is worth noting, however, that despite this characterisation of humanity as in a state of disharmony, Islam recognises that the 'primordial nature' of humanity as created by God is a 'moral and spiritual substance' attracted essentially to goodness and 'God-consciousness', wherein imperfections are 'accidental' qualities which result from free will and emerge naturally in the struggle of the soul to do good and transcend 'subliminal desires'.[91] This is arguably a conceptual trait shared across the Abrahamic traditions, insofar as they all share the notions of the fall of humanity from Eden, God-given free will, and the necessity of human transcendence. Consequently, in Islam, humanity's capacity to live in harmony both with one another and themselves is conceived of as a responsibility to God. According to Sohail Hashimi, human nature is characterised by the 'will to live on earth in a state of harmony and peace with other living things'; this is the 'ultimate import' of man's responsibility as God's 'viceregent'.[92] Thus, peace can be interpreted as God's 'true purpose of humanity'.[93]

Abu-Nimer maintains that there are three principle responsibilities according to which each person will be judged by God: responsibility to Allah through faithful performance of religious duties; responsibility to oneself through existing in 'harmony with oneself' and responsibility to live harmoniously and peacefully with others.[94] Kalin argues that the 'normative depiction of natural phenomenon' sees nature as 'prostrating before God', and the 'process' of man's participation in praising and 'acknowledging God's unity' underscores the 'essential link' between *anthropos* and *cosmos*.[95] He calls this inherent commonality between humanity as 'subject' and universe as 'object' the 'anthropocosmic vision', illuminating man's God-given responsibility to the world and the importance of harmonising with nature in recognition of its divine essence in Islamic thought.[96]

This responsibility towards harmony is further underscored by the ever-present doctrine of the sanctity of life. Farhan Mujahid Chak notes that within the Islamic tradition, all humanity (past, present and future inclusive) are understood to be children of Adam and are asked to 'make witness to God', thus 'imbuing every human being with intrinsic goodness and recognition of the Divine', creating 'an inherent kinship in the entire human family'.[97] Thus, it can be inferred that harmony is conceived as unity within and of the cosmos by virtue of God's immanence, and as per God's will, a state which is to be strived for on both an interpersonal and intrapersonal level by humanity through full commitment to God, which in itself constitutes the ultimate harmony. The primacy of the relationship between mankind and God in terms of both outward and inward intentionality lends itself to a characterisation of harmony as a relational dynamic between humanity and the Divine, and furthermore, a process in which humanity strives for transcendence of his own imperfection in order to come closer to his inherent 'God-consciousness'.[98] Thus, while harmony is not understood as a metaphysical principle which embodies the divine self-referentially as it is more commonly in the East, within Islam, it can be argued that harmony is still encountered as a dynamic process which prioritises the relational.

However, the relationship which is of primordial significance is that with the divine-Other, through which all other relationships are mediated. While harmony or *harmonisation* in Confucianism is a self-referential process constituting the greatest good in itself, harmonisation in Abrahamic religions is experienced and striven for as part of a totality of intentionality towards God. This embodies what Kalin conceives the purpose of all religious teachings: to 're-establish the primordial harmony between heaven and earth, between the Creator and the created'.[99]

INTEGRATING UNDERSTANDINGS OF HARMONY IN PRACTICE AND AS A WAY OF BEING

Harmony can be understood not only as a metaphysical principle, but also as a value, or in other words, as a framework which denotes ethical and moral values and practices. This raises the question as to how harmony is perceived as a concept applicable within or in itself as a way of being.

The role of harmony within Hellenic thought extends to practice, manifesting as a principal virtue in the process of self-actualisation. In Plato's *Republic*, justice within the state is qualified by the 'harmonious existence of the three classes', and on a personal level within the harmonisation of the 'three elements of the soul'.[100] In *Phaedrus*, Plato offers the an allegorical vision of the soul, consisting of three parts, one being rational (the charioteer) and the other two being the 'spirited' and the 'appetitive'.[101] The harmonisation of these three elements, achieved through the attunement of the spirited element and reason over appetite, is the condition in which the soul is 'healthy', 'beautiful' and in the 'ontologically correct, hierarchic, inner order'.[102] Li interprets the harmony of the 'three classes and the three elements of the soul' as representing the 'cardinal virtue of justice'.[103] Similarly, Sidney Zink writes that Aristotle praised the cultivation and maintenance of a 'virtuous mean preventing contrary human propensities from operating to excess'.[104] Zink also argues that Heraclitus regarded the end of human conduct to be the 'preservation of proper harmony among the elements of the soul'.[105] Plato's *Republic* conceptualises an ideal education which is 'blended in perfect proportion', suggesting not only the importance of the cultivation of intellect and reflection in the process of self-actualisation but also the significance of the cultivation of the harmonious, balanced and appropriate relationships between virtues and elements of the soul.[106]

Confucian thought also features the notion of 'perfectibility of humanity through self-cultivation'.[107] The concept of *li* (ritual propriety) underscores the manifestation of harmony within social frameworks, embodying the application of *ren* ('love based on humanity') and *yi* (righteousness) in society.[108] Self-actualisation in itself constitutes a process of harmonisation, requiring the balance of different virtues (*ren*, *yi* and *li*). The cultivation of equilibrium within music and food is intended to 'bring peace and harmony' into the mind of the 'superior man', conducing to harmonious conduct in his interpersonal relations, illustrating that Confucian harmony is not only a 'quality of things and... of perception', but also a 'quality of mind and... of judgment and conduct'.[109] Thus, 'equilibrium of mind' is understood as the 'requisite foundation' for harmony.[110] Li suggests that Confucians attach a moral quality to harmony in that the *junzi* (superior or good person) is one who harmonises rather than seeks sameness, while a lesser person will seek sameness rather than harmonise.[111]

Li expands the notion that friendliness and love are not necessary conditions for harmony; the 'requirement of harmony' places constraints on each party interacting harmoniously, while simultaneously providing a context within which each party has 'the optimal space to flourish'.[112] Cheng suggests that this is the 'genuine Confucian position', reflective of the acknowledgement of not only the 'deep and hidden harmony value in the virtue of man' but also of the relationship between the actualisation of harmony on a personal level and the subsequent transformation of the 'external chaotic world' into a 'harmonious order of *ren*, *yi*, and *li*'.[113]

The neo-Confucian reading of *he* 和 reviewed in this chapter regards harmony as a fundamentally relational framework on both metaphysical, ontological and social levels. Exemplified in the *yin-yang* paradigm and its articulation of the principle of interdependence as previously expounded, transformation and creative processes are understood in much ancient Chinese thought and particularly within Confucianism as necessarily underscored by interpenetrating relationships. From a social and human perspective, an individual's capacity to harmonize is realized within their relational capacity; intrapersonal harmony is mutually constitutive of realization of harmony more broadly within the world.

Thaddeus Metz notes similarities between the centrality of self-actualisation in Confucianism, Hellenic *eudaimonism* and sub-Saharan African thought.[114] Harmony is described by Metz as a 'concept central to... the sub-Saharan ethic of *ubuntu*', ubuntu being 'the famous southern African... word for human excellence'.[115] Despite there being immense ethnic and linguistic diversity and thus cultural and social diversity across sub-Saharan Africa, Metz maintains that scholarship in this area has demonstrated 'broad commonalities' amongst different groups in regards to ethical fundamentals.[116] According to Tim Murithi, the concept of ubuntu is presents itself in 'diverse forms' across societies throughout Africa, and defines ubuntu as a 'cultural worldview that tries to capture the essence of what it means to be human'.[117] Metz notes that in sub-Saharan African thought, maxims are often utilised to articulate moral and ethical codes.[118] Desmond Tutu, considering the moral values of sub-Saharan Africa, writes:

> We say, 'a person is a person through other people'. It is not 'I think therefore I am'. It says rather: 'I am human because I belong'. I participate, I share... Harmony, friendliness, community are great goods. Social harmony is for us the *summum bonum* – the greatest good. Anything that subverts or undermines this sought-after good is to be avoided like the plague.[119]

Tutu's words demonstrate the centrality of harmony, both social and conceptual, within self-actualisation and ubuntu within sub-Saharan African philosophy. Metz elaborates that in African thought, the ultimate *telos* of a person is to achieve total personhood – to become 'a genuine human being, one who has *ubuntu*', arguing that self-realisation lies in 'living harmoniously, or communally, with others'.[120] He notes that in both Confucianism and sub-Saharan African thought, there are 'distinctively human' and 'more animalistic sides' to human nature, with personhood being constituted in this human aspect, achieved through the process of self-actualisation via harmonious relation to others.[121] Murithi echoes this statement, stating that ubuntu societies prioritise communal life and the maintenance of positive relationships, a priority in which 'all members of a community' participate and share.[122] Metz also suggests that in African thought, approaches to self-actualisation are fundamentally conceived as relational in nature in contrast with Western approaches which are often preoccupied with individualist concepts such as 'freedom, autonomy, agreement, contract, pleasure and happiness'.[123]

In the Daoist text *Zhuangzi*, harmony or harmonisation is promoted as a 'positive value' and the 'guiding philosophy for the enlightened', synonymous with the experience of profound happiness.[124] Harmonisation conceived of as value is manifested in the notion of *wu wei*, which Li translates to 'effortless action', meaning 'to take a path that harmonises with the world', wherein 'the world' in this context relates to the aforementioned Daoist concepts surrounding nature/the natural.[125] He argues that while Confucians actively seek to 'harmonise the world', Daoists strive to 'harmonise *with* the world'.[126] While Daoism places emphasis on harmony from an intrapersonal perspective, with the outward relationship existing towards the natural world, Confucianism places greater emphasis on interpersonal harmony, seeing 'more consistency than distinction between the "private" and the "public"... between the political and non-political, and the human society and the natural world'.[127]

It has been previously argued that Islam conceives of harmony as constituted in the relationship between the believer and God. In Islam, achieving harmony with God is the necessary condition to 'achieving internal and external peace and harmony'.[128] To harmonise and find peace both with and within God through total 'submission' conduces to 'physical, mental, spiritual and social harmony', and represents the essence of 'real Islam'.[129] According to Chak, the aforementioned principle of *Tawhid* serves as the conceptual framework of social harmony and tolerance, defined as the 'inherent recognition and embracing of diversity and plurality'.[130] Following the ethical framework of the Qur'an and harmonising with oneself, others and nature can be interpreted as the necessary conditions to harmonising with God, elucidating how the implicit Islamic concept of harmony denotes a moral and ethical system. This understanding of harmony as a process of spiritual necessity is similar to the Islamic concept of *ihsān,* meaning 'doing what is beautiful' and 'virtue, beauty, goodness, comportment, proportion, comeliness', highlighting the aesthetic, Way-based nature of harmony/harmonisation.[131]

The concept of harmony within Buddhist thought can be interpreted to manifest itself in its ethical and moral system as a prescriptive value in practice, as well as in the act of meditation and the perception of the inner truths of Buddhist Dharma. According to Joanna Rogers Macy, Buddhist morality is 'grounded' in the aforementioned notion of interdependent causality or *pratītyasamutpāda*, as in the 'corollary' Buddhist concepts of *anattā* and *karma*.[132] The notion of *anattā* in Buddhism conceptualises the self as an 'interdependent, self-organising process shaped by the flow of experience and the choices that condition this flow'.[133] The self is not a subject which perceives; rather than having experience of phenomena, the self *is* itself the experience of phenomena.[134] This notably relational and non-atomic view of self, emerging from the principal fundament of Buddhist dharma (*pratītyasamutpāda*), necessitates an ethics rooted in the capacity to perceive, acknowledge and act according to the fundamental metaphysical harmony of all things.

For Thich Nhat Hanh, meditative insight into 'the interrelatedness or interconnectedness of all things' must be the basis of all compassionate action;

compassionate action is grounded in mindful awareness of dependent co-arising.[135] By experiencing the insight into the 'inter-being' and causal interdependence of all things, one is able to perceive that suffering does not belong to a single self, but is 'symptomatic of the broader suffering of all beings'; perception of this inner truth through meditation in itself constitutes compassion and conduces to its realisation in practice.[136]

Brinkman notes that Kukai (the founder of Shingon Buddhism) uses the word *yuga* (a transliteration of *yoga*) for harmony, denoting the meaning of 'meditation and concentration of the mind'.[137] He notes that the particular human mode of perception/being has the potentiality to 'bring its awareness into accord with the harmony that pervades the universe';[138] it is this coming-into-accordance which is constitutive of harmony as practice by virtue of acknowledgment of cosmic harmony and the interrelatedness of all things. In this sense, in the process of becoming aware, harmony is constituted self-referentially, facilitated by the practice of meditation. Buddhists then strive to perceive themselves, nature and the cosmos as non-distinct, harmonising and interrelated mutualities, illustrating the 'reciprocal dynamic' between personal, social and metaphysical transformation.[139] Harmony and dependent co-arising can only be perceived through a cultivation process synonymous with harmonisation itself; in other words, harmony in Buddhism is a mutuality between harmony in the metaphysical sense and harmonisation as practice in dynamic interplay. The notion of causal interrelatedness as the basic essence of the Buddhist system of thought underscores Buddhist principles of embracing diversity, harmonising with the natural world and displaying unconditional compassion. This aligns with the Confucian and Hellenic models of harmony which see the realisation of harmony in the world as mutually constitutive of realisation of harmony within the person.[140] Harmony in both Confucian and Buddhist thought is the reality, experience and perception itself of 'unity of a whole together with the interrelationship of its parts'.[141]

CONCLUSION

Significantly, searching for articles based upon the inclusion of keywords and key concepts did not necessarily aid in the formulation of a broad and thoroughly developed theory of harmony, both in terms of pre-existing conceptions and novel intellectual ground. This is primarily due to the fact that the notion of harmony (that is to say, in the sense(s) that are implicit and meaningful within the scope of this project), is often not classified principally in terms of the word 'harmony'. Instead, 'harmony' may manifest in distinct and subtle ways, such as in terms of "tolerance" and "oneness". This last point brings me to iterate that this research project focuses on critically reviewing academic literature available in digital format online in order to generate an up-to-date image depicting how notions of harmony, situated within various religious, philosophical and/or wisdom traditions across the world, have been examined, interpreted and understood, thus forming a preliminary conceptual map. It would be beyond the scope of this project and its resources to critically examine fundamental source material (i.e. religious texts), which is something to consider for the future.

It goes without saying that all traditions examined are socially, conceptually and in terms of practice, diverse and pluralistic in themselves. Thus, most articles cited in this chapter deal with a singular tradition or focus on a scholar's particular interpretations of harmony within a particular tradition as opposed to generalising a vast spectrum of understandings across diverse traditions. Having said that, these hermeneutic snapshots can indeed help us construct a broader vision of the religions/philosophies in question and their understandings, notions and practices of harmony.

During the research, I found that a notable trend within academic scholarship relating to peace and harmony under the branch of Islamic studies was that of defence. It became clear that discussions of peace, peace-building and harmony were often expounded in relation to violence (or more accurately, the avoidance of it), dealing to some degree with the trope of why Islam is *not* violent, as opposed to why it *is* peaceful. This extends beyond violence to ideas of global harmony, wherein scholars must address the tendency to equate Islam with 'global dissonance'.[142] This is no fault of the scholarship itself. On the contrary, this scholarly trend emerges from various levels: the perspectives of Western media and the damaging stereotypes and scapegoating which are generated in its wake; the hijacking/abuse/manipulation of Islam for justification of violent means, often arising on account of more complex socio-economic and geo-political dynamics of violence and unrest which are not commensurate with Islam as a religion or Muslims as a global community.[143] Abu-Nimer highlights this issue, offering the example of how few articles are available on Islam and non-violence compared to Islam and violence in the Library of Congress, describing this phenomenon as looking at Islam through a '*jihad* lens'.[144] This phenomenon is to be taken into consideration in construing this literature review. My analysis has therefore aimed to positively construct a hypothetical theory of harmony in Islam, rather than negatively define why Islam does *not* promote disharmony.

It is important to recognise that, while a culture or tradition may not have an explicit discourse of harmony developed at a conceptual level, or even have a direct translation of the word 'harmony' (at least which coincides with Western definitions), harmony may still be implicitly conceptualised, internalised or outwardly practiced in various ways. In order to bring forward a truly authentic and meaningful account of the many conceptions of harmony across the globe, it would be necessary to carry out field research in addition to literature-based research, so that theories of harmony can be constructed, as opposed to projected upon traditions. Chenyang Li is one of the few scholars I have encountered in this project who attempts to argue the case for the applicability of harmony as an ideal to the real world, linking the conceptual to the pragmatic within contemporary society, Western and non-Western.[145] This in itself represents an important step in the progression of harmony studies towards practical materiality, illuminating a crucial question: how can we connect research into the field of harmony with the world today? And how can an advanced and informed theory of harmony aid in the envisaging and construction of a better world? A project such as this review begins to answer some of these questions; it is hoped that it will stimulate further discourse within the nascent field of harmony studies.

NOTES

1.   This literature review was sponsored by the GHFP Research Institute as part of the Harmony Project. The original task was identifying, examining and comparing differing conceptions of harmony as situated within the world's major religions and spiritual traditions. However, as a web-based literature review, and due to a lack of published scholarship in this area of study, the scope of research was broadened. Initial research pointed towards ancient Chinese and ancient Greek thought as primary focal points for further review. In keeping with the original brief, the role of harmony within the Abrahamic religions and Buddhism was also considered. While these intellectual, philosophical and religious traditions fundamentally differ in nature, it is precisely their diversity which renders this review constructive, gesturing towards a conceptual map of harmony philosophies across the world.

2.   Daniel A. Bell and Mo Yingchuan, 'Harmony in the World 2013: The Ideal and the Reality', *Social Indicators Research* 118, no. 2 (2014): pp. 797–818.

3.   Chenyang Li, 'The Confucian Philosophy of Harmony', (New York: Routledge, 2015)

4.   Li, 'The Confucian Ideal of Harmony', p. 583.

5.   S. Angle, *Sagehood: The Contemporary Significance of Neo-Confucian Philosophy* (New York and Oxford: Oxford University Press, 2009), p. 61.

6.   Chenyang Li, 'The Confucian Ideal of Harmony', *Philosophy of East and West* 56, no. 4 (2006): pp. 584.

7.   Angle, 'Sagehood', p. 62.

8.   Li, 'The Confucian Ideal of Harmony', p. 589.

9.   Li, 'The Confucian Ideal of Harmony', p. 585/6.

10.   Chung-Ying Cheng, 'On Harmony as Transformation: Paradigms from the Yijing', *Journal of Chinese Philosophy* 36, (2009): p. 14.

11. Li, 'The Confucian Ideal of Harmony', p. 589.

12. Chenyang Li, 'Being as Process of Harmonization: A Chinese View of Dynamic Being', in Vesselin Petrov and Adam C. Scarfe (eds.) *Dynamic Being: Essays in Process-Relational Ontology*, (New Castle: Cambridge Scholars Publishing, 2015), 165

13. Chenyang Li, 'The Confucian Ideal of Harmony', *Philosophy of East and West* 56, no. 4 (2006): pp. 583-603; Chenyang Li, 'The Ideal of Harmony in Ancient Chinese and Greek Philosophy', Dao 7, no. 1 (2008): pp. 81-98.

14. Li, Chenyang. "The Ideal of Harmony in Ancient Chinese and Greek Philosophy." Dao 7.1 (2008): 90.

15. K. Freeman, *Ancilla to the Pre-Socratic Philosophers: A Complete Translation of the Fragments in Diels Fragmente der Vorsokratiker* (Cambridge, Mass.: Harvard University Press, 1983), p. 90.

16. Li, 'The Ideal of Harmony', p. 91.

17. Bo Mou, *Chinese Philosophy A-Z*, (Edinburgh University Press, 2009), p. 176

18. Bo Mou, *Chinese Philosophy A-Z*, (Edinburgh University Press, 2009), p. 176

19. Bo Mou, *Chinese Philosophy A-Z*, (Edinburgh University Press, 2009), p. 176 For more on the characterisation of opposites in unity within Confucianism and beyond, see Cheng, 'On Harmony', p. 31.

20. Cheng, 'On Harmony', p. 28

21. See J. W. Roberts, 'Pythagoras', *The Oxford Dictionary of the Classical World* (Oxford: Oxford University Press, 2005).

22. C. A. Huffman, 'The Pythagorean tradition,' in A. A. Long, ed., *The Cambridge Companion to Early Greek Philosophy* (Cambridge: Cambridge University Press 1999), p. 74.

23. Li, ' The Ideal of Harmony, p. 92.

24. Li, 'The Ideal of Harmony', p. 94.

25. Li, 'The Ideal of Harmony', p. 92.

26. Michio T. Shinozaki, 'Peace and Nonviolence from a Mahayana Buddhist Perspective: Nikkyo Niwano's Thought', *Buddhist-Christian Studies* 21, (2001): p. 24.

27. Shinozaki, 'Peace', p. 24.

28. Laurence Carlin, 'On the Very Concept of Harmony in Leibniz', *The Review of Metaphysics* 54, no. 1 (2000): p. 99.

29. Carlin, 'The Very Concept', p. 99.

30. Carlin, 'The Very Concept', p. 100.

31. G. W. Leibniz, *Textes Inédits*, ed. Gaston Grua (Paris: Presses Universitaires de France, 1948), p. 12, cited in Carlin, 'Harmony in Leibniz', p. 101.

32. Li, 'The Ideal of Harmony', p. 87.

33. Li, 'The Ideal of Harmony', p. 87.

34. Yu, Jiyuan. 'Living with Nature: Stoicism and Daoism.' *History of Philosophy Quarterly*, vol. 25, no. 1, (2008): p. 4

35. Li, 'The Ideal of Harmony', p. 90.

36. Li, 'The Ideal of Harmony', p. 90.

37. Li, 'The Ideal of Harmony', p. 90.

38. Cheng, 'On Harmony', p. 11.

39. Cheng, 'On Harmony', p.12.

40. See Li, 'The Ideal of Harmony'.

41. Cheng, 'On Harmony', p. 12.

42. Cheng, 'On Harmony', p. 12.

43. Li, 'The Confucian Ideal of Harmony', p. 583 and p. 593.

44. Cheng, 'On Harmony', p. 15.

45. Cheng, 'On Harmony', p. 15.

46. Cheng, 'On Harmony', p. 29.

47. E. Hussey, 'Heraclitus', in A. A. Long, ed., *The Cambridge Companion to Early Greek Philosophy* (Cambridge: Cambridge University Press. 1999), p. 110

48. Li, 'The Confucian Ideal of Harmony', p. 591.

49. Li, 'The Confucian Ideal of Harmony', p. 591.

50. See Li, 'The Confucian Ideal of Harmony', p. 591 for a more detailed account.

51. Cheng, 'On Harmony', p. 29.

52. Cheng, 'On Harmony', p. 29.

53. Cheng, 'On Harmony', p. 29.

54. Damien Keown, *A Dictionary of Buddhism* (Oxford: Oxford University Press, 2003).

55. Stephen J. Laumakis, *An Introduction to Buddhist Philosophy* (Cambridge and New York: Cambridge University Press, 2008), p. 268; B. C. Law, 'Formulation of Pratītyasamutpāda', *Journal of the Royal Asiatic Society of Great Britain and Ireland*, no. 2 (1937): p. 287

56. Joanna Rogers Macy, 'Dependent Co-Arising: The Distinctiveness of Buddhist Ethics', *The Journal of Religious Ethics* 7, no. 1 (1979): p. 38.

57. Shinozaki, 'Peace', p. 16.

58. Shinozaki, 'Peace', p. 16.

59. John Brinkman, 'Harmony, Attribute of the Sacred and Phenomenal in Aquinas and Kūkai', *Buddhist-Christian Studies* 15, (1995): p. 109.

60. David Hall and Roger Ames, *Thinking from the Han* (Albany: State University of New York Press, 1998), p. 18, cited in Li, 'The Confucian Ideal of Harmony', p. 593.

61. Li, The Confucian Philosophy of Harmony, p. 20

62. Hall and Ames. *Thinking*. While one should remain cautious when applying epistemic generalizations to a projected dichotomization of East and West, this suggested framework is useful in construing the distinct points of departure for comparative analysis of 'Western' and 'Eastern' traditions respectively.

63. S. Angle, *Sagehood: The Contemporary Significance of Neo-Confucian Philosophy* (New York and Oxford: Oxford University Press, 2009), p. 64.

64. Hall and Ames, *Thinking*, p. 16

65. Li, 'The Ideal of Harmony', p. 91.

66. Li, 'The Ideal of Harmony', p. 91.

67. Imtiyaz Yusuf, 'Dialogue Between Islam and Buddhism through the Concepts Ummatan Wasaṭan (The Middle Nation) and Majjhima-Patipada (The Middle Way)', *Islamic Studies* 48, no. 3 (2009): p. 372.

68. John Hick, *God Has Many Names* (Philadelphia: The Westminster Press, 1982), cited in Yusuf, 'Dialogue', p. 372.

69. Li, 'The Ideal of Harmony', p. 96.

70. Julia Annas, *Plato: A Very Short Introduction* (Oxford: Oxford University Press, 2003), p. 79.

71. Li, 'The Ideal of Harmony', p. 96.

72. Brinkman, 'Harmony', p. 106.

73. Brinkman, 'Harmony', p. 116.

74. G. W. Leibniz, 'Die Philosophischen Schriften', 4:431, *Philosophical Essays*, trans, and ed. Roger Ariew and Daniel Garber (Indianapolis: Hackett, 1989) p. 39, cited in Laurence Carlin, 'On the Very Concept of Harmony in Leibniz', *The Review of*

Metaphysics 54, no. 1 (2000).

75. G. W. Leibniz, *New Essays on Human Understanding Book II: Of Ideas*, Edited by Peter Remnant and Jonathan Bennett; (Cambridge: Cambridge University Press, 1996)

76. I. A. Ben Yosef, 'Action and Non-Action in Judaism and Taoism', *Religion in Southern Africa* 5, no. 1 (1984): p. 64.

77. Galia Patt-Shamir, 'Way as Dao: Way as Halakha: Confucianism, Judaism, and Way Metaphors', *Dao: A Journal of Comparative Philosophy* 5, no. 1 (2005): p. 138.

78. Patt-Shamir, 'Way as Dao', p. 138.

79. Patt-Shamir, 'Way as Dao', p. 138.

80. Yosef, 'Action and Non-Action', p. 64.

81. Patt-Shamir, 'Way as Dao', p. 140.

82. Patt-Shamir, 'Way as Dao', p. 157.

83. Ibrahim Kalin, 'Islam and Peace: A Survey of the Sources of Peace in the Islamic Tradition', *Islamic Studies* 44, no. 3 (2005): p. 335.

84. Kalin, 'Islam and Peace', p. 336.

85. The Oxford Dictionary of Islam, 'Tawhid', *Oxford Islamic Studies Online*, (2019), http://www.oxfordislamicstudies.com/article/opr/t125/e2356 [accessed 4 June 2019].

86. *Oxford Islamic Studies Online.*

87. Sumanto Al Qurtuby, 'The Islamic Roots of Liberation, Justice, and Peace: An Anthropocentric Analysis of the Concept of 'Tawhīd', *Islamic Studies* 52, no. 3/4 (2013): p. 314.

88. Al Qurtuby, 'Islamic Roots', p. 324.

89. Al Qurtuby, 'Islamic Roots', p. 324.

90. Mohammed. Abu-Nimer, 'A Framework for Nonviolence and Peacebuilding in Islam', *Journal of Law and Religion* 15, no. 1/2 (2000): p. 224.

91. Kalin, 'Islam and Peace', p. 337.

92. Abu-Nimer, 'A Framework', p. 217, summarising Sohail Hashmi, 'Interpreting the Islamic Ethics of War and Peace', in Terry Nardin, ed., *The Ethics of War and Peace: Religious and Secular Perspectives* (Princeton, N.J.: Princeton University Press, 1996), pp. 142 and 146.

93. Abu-Nimer, 'A Framework', p. 142 and p. 146.

94. Abu-Nimer, 'A Framework', p. 249.

95. Kalin, 'Islam and Peace', p. 336.

96. Kalin, 'Islam and Peace', p. 336.

97. Farhan Mujahid Chak, 'La Convivencia: The Spirit of Co-Existence in Islam', *Islamic Studies* 48, no. 4 (2009), p. 573.

98. Kalin, 'Islam and Peace', p. 337.

99. Kalin, 'Islam and Peace', p. 328.

100. Li, 'The Ideal of Harmony', p. 95; see Plato, *Republic*, 2 Vols., trans. Paul Shorey, Cambridge Mass., London: Harvard University Press, 1935. 441 A-D.

101. Plato, *Phaedrus*, trans H. N. Fowler (Cambridge, Mass, and London: Harvard University Press, 1914), 253 C-D; see also, Plato, *Timaeus*, trans. R.G. Bury (Cambridge Mass., London: Harvard University Press, 1931), 41D-E.

102. Gregory Vlastos, 'Justice and Psychic Harmony in the Republic', *The Journal of Philosophy* 66, no. 16 (1969): p. 506.

103. Li, 'The Ideal of Harmony', p. 95.

104. Sidney Zink, 'The Good as Harmony', *The Philosophical Review* 53, no. 6 (1944): p. 557.

105. Zink, 'The Good', p. 557.

106. F. M. Cornford, *The Republic of Plato, Translated with introduction and notes* (London: Oxford University Press, 1968), p. 102.

107. Cheng, 'On Harmony', p. 17.

108. Cheng, 'On Harmony', p. 16.

109. Cheng, 'On Harmony', p. 13.

110. Cheng, 'On Harmony', p. 18.

111. Li, 'The Confucian Ideal of Harmony', p. 586.

112. Li, 'The Confucian Ideal of Harmony', p. 589.

113. Cheng, 'On Harmony', p. 17.

114. Thaddeus Metz, 'Confucian Harmony from an African Perspective', *African and Asian Studies 15*, 1, (2016): p. 6.

115. Metz, 'Confucian Harmony', p. 2.

116. Metz, 'Confucian Harmony', p. 8.

117. Tim Murithi, 'An African Perspective on Peace Education: Ubuntu lessons in reconciliation', *International Review of Education 55*, (2009): p. 226.

118. Metz, 'Confucian Harmony', p. 8

119. Desmond Tutu, *No Future without Forgiveness* (New York: Random House, 1999), p. 35, cited in Metz, 'Confucian Harmony', p. 8.

120. Metz, 'Confucian Harmony', p. 8.

121. Metz, 'Confucian Harmony', p. 9.

122. Murithi, 'An African Perspective', p. 227.

123. Metz, 'Confucian Harmony', p. 9.

124. Li, 'The Ideal of Harmony', p. 89.

125. Li, 'The Ideal of Harmony', p. 89.

126. Li, 'The Ideal of Harmony', p. 89.

127. Li, 'The Confucian Ideal of Harmony', p. 588.

128. Abu-Nimer, 'A Framework', p. 224.

129. Abu-Nimer, 'A Framework', p. 243.

130. Chak, 'La Convivencia', p. 572.

131. Kalin, 'Islam and Peace', p. 333.

132. Macy, 'Dependent Co-Arising', p. 38.

133. Macy, 'Dependent Co-Arising', p. 42.

134. Macy, 'Dependent Co-Arising', p. 42.

135. Laumakis, *An Introduction,* p. 257.

136. Laumakis, *An Introduction,* p. 257.

137. Brinkman, 'Harmony', p. 109-110.

138. Brinkman, 'Harmony', p. 109-110.

139. Macy, 'Dependent Co-Arising', p. 38.

140. Cheng, 'On Harmony', p. 13.

141. Cheng, 'On Harmony, p. 25.

142. Chak, 'La Convivencia', p. 567.

143. Chak, 'La Convivencia', p. 570.

144. Abu-Nimer, 'A Framework', pp. 221-2.

145. See Chenyang Li, 'The Confucian Philosophy of Harmony', (New York: Routledge, 2015)

DIFFERENCE, BEAUTY AND THE DIVINE: HARMONY IN RELIGIOUS PLURALITY

Angus Slater

[12] *Put on then, as God's chosen ones, holy and beloved, compassionate hearts, kindness, humility, meekness, and patience,* [13] *bearing with one another and, if one has a complaint against another, forgiving each other; as the Lord has forgiven you, so you also must forgive.* [14] *And above all these put on love, which binds everything together in perfect harmony.*

—*Colossians 3: 12-14*

THERE IS UNDOUBTEDLY SOMETHING DEEPLY ENCHANTING about the notion of harmony. It has an alluring and attractive sense of possibility that evokes both desire from within us and a reflective sense of our inner harmony. Yet this sense of possibility is one that always seems to lie just beyond our full comprehension, and sadly, just beyond our capability to produce and sustain. In some sense, our appreciation for harmony exposes a desire within us to believe in the perfection, beauty and purpose of things. This lies beyond their immediate appearance or circumstances and beyond our immediate understanding of their relation. Of course, this presupposes that there is indeed a deeper relation of things which we fleetingly perceive in our glimpses of harmony but that remains largely hidden from our everyday perception. This notion is a kind of utopian sense – expressing the feeling that things should fit together well, that the whole is both more beautiful and more fitful as a sum than as its parts and in which things although currently disconnected, distraught or destroyed could be made, with only a little finesse, to be better. Harmony, however received or experienced, offers us a passing induction into this sense, a fleeting glimpse of the potential of our world and, perhaps, a starting point for the imagination required for its renewal.

This feeling evoked by the experience of harmony is apparent across the different circumstances in which we remain surprised by the perception of harmony, whether found in our enjoyment of musically, naturally or physically harmonious experiences. This concerns the aesthetic quality of harmony – the way in which harmony is perceived, felt or experienced.[1] Harmony, in this context, contains within it innumerable connections to the other qualities that form and structure our experience of reality. Whether in the form of experiencing beauty, truth, or goodness, the process and unfolding of the aesthetic reception remains the same. The rapture of Beauty, that experience of perfection aside from our desire to possess or own, an appreciation of the thing as is without the need to tinker, correct or add. As Eco notes, 'we talk of beauty when we enjoy something for what it is, immaterial of whether we possess it or not'.[2] The piercing revelation of Truth, that sudden epiphany of relation and not just correctness, but fittingness, a solidity of knowledge that expresses itself not only finitely, within a particular context, but expresses within its infinite relation to the universal.[3] The deep realisation of Good, either in an action or within some physical thing, that restores not only the balance and harmony of the context but also, in some sense, our hope for the wider goodness of things. As the medieval scholar Boethius notes in the context of divine human relations, goodness flows one from thing to another, stating:

> For the first good, because it is, is good in virtue of the fact that it is. But a second good, because it flowed from that whose being itself is good, is itself also good.[4]

Goodness connects goodness and in doing so, illuminates the connections between us all. Harmony is the hidden glimpse behind each of these that is unveiled through and with each of them but remains floating, somewhat elusive, between them.

There is a sense that Harmony is the connecting aspect of the Good, the True, the Beautiful – that each of these other principles exists in relation with the others through their innate harmony, that each principle extends and opens beyond itself to the others, reflecting within the disclosure of one a relation beyond that context and into a wider harmonious whole whereby the Good, True, and Beautiful are to be found through one another. This describes, perhaps, where our experiences of harmony can come to be found, between and amongst our experiences of other pinnacles of human emotion and experience. Think of those places in which we revel in the experience of harmony:

> The slow, delicate, haunting build-up of voices in the choral work of Thomas Tallis.[5] An aching longing silence, pregnant with possibility and enticement, that flowers, ever so slowly, into a glorious expression of purity, clarity and perfection. A swelling coherence of purpose begins, building within and through the difference found between voices, expressing not only the clarity

of vision but also the myriad themes to be found within it. A resounding returning of those differences to the central theme, a love-struck longing for belonging, an expression of wholeness and relation, a revelation of the truly good through harmonious song.

The quiet, purposeful, busyness of a flourishing wood in spring, the tiny and the gigantic all striving for their place. A sense of shared purpose that extends beyond the confines of a single organism, bridging the divide between species, between the natural and created orders, between the human and the world in which we live.[6] Harmony here is the feeling of correctness and connectedness, correct in burgeoning form and freedom of action, connected in producing together the crescendo of spring.

There is a feeling of experiencing a great work of art, that dwells in relation to both the viewer, and to the truth that the art seeks to express. Whether the art aims to be representational, expressing the truth of the thing directly, or aims to express the ineffable quality of relation between the thing and its potential.[7] Think of the tortured visions of saints, showing both the connection between the divine and the human, mirrored in ourselves, and highlighting the utopian possibility of our bodies.

These experiences, far removed from each other in time, place, or context but sharing that inescapable aesthetic quality, enchant our imaginations with the promise of purpose and meaning behind their apparent differences. Much as our lives are quests for meaning, so too our experiences of meaning elucidate a hidden harmonious whole. As David Cadman in attempting to define 'harmony' has noted the experience of harmony is independent of context being rather 'an expression of wholeness', a 'way of looking at ourselves and the world of which we are part' and fundamentally about 'connections and relationships'.[8] This notion of harmony draws out a number of characteristics of the circumstances in which harmony is to be found, including, but not limited to, aspects of purpose, relation, vulnerability and desire. Each of these aspects is to found in our appreciation of harmony with these characteristics themselves being integral to our aesthetic sense of harmony beyond the immediate context.

This can be seen most clearly by examining the relationship between these particular characteristics and the experience of harmony within specific contexts. The aspect of relation is perhaps the clearest. We experience harmony when we feel the logic of the ongoing difference within the context we experience when we are assured that the difference experienced within the harmonious circumstance is itself manifested through the relation of the specific to a broader universal. If we return to the notion of musical harmony, perhaps the easiest experience of harmony to grasp, we are aware that the different sounds produced within a harmonious piece of music rely, for their harmony, on their relationship to the other sounds produced within the same moment. Without this relation, either simultaneous relation in the context of technical musical harmony, for example, between different notes played at the same time, or relation between a particular note played at a particular time and the other notes that surround it within a piece of music, perhaps best thought of as holistic harmony, the existence of harmony collapses into the unrelated existence of the note in isolation. Without the relation of the specific (the note, the insect, the depiction held in art) to the general (the theme of the piece of music, the balanced ecology of the forest, or the glimpse of truth offered by the artistic representation) the notion of harmony being present in that particular context collapses. In this, relationality, the ongoing connection of one thing to another, is a requisite and necessary aspect of the experience of harmony. Without relationality, anything in particular fails to express its harmonious connection to things beyond itself, remaining locked within the prison of its own isolated, and therefore meaningless, existence.

This aspect, of relation between the specific and the general, highlights the necessity of purpose for the presentation of harmony within our experience of reality. A random confluence of things does not represent a harmonious whole, unless within the supposed random order some greater arrangement or purpose can be received by the self which is experiencing. This is not to say that the experience of harmony requires arrangement – only the perception of it by the self. While experiences of harmony may appear random at first, the bewildering busyness of a forest, for example, harmony emerges as an aesthetic experience in our uncovering of the pattern and purpose within that immediately confusing multiplicity. The relationship between the things perceived to be in harmony must be purposeful, whether deliberate in the case of created art or music or naturally occurring in the case of mutually interacting forms of ecology. In each case, it is the appreciation of purpose behind the relationship that reveals the experience of harmony to the viewer.

The notion of purpose, however, is distinct from the notion of arrangement, although obviously related. While a natural and unmeaning confluence of factors can evoke harmony through their relation to and revelation of a universal truth, this is not necessarily formed from the deliberate arrangement of things to suggest or incur that revelation. After all, the viewer of the instance of harmony carries with them the heavy hermeneutic weight of their lives, their other experiences, their expectations and their beliefs, the *wirkungsgeschichtliches Bewußtsein* (historically-effected consciousness) espoused by the German philosopher Hans-Georg Gadamer.[9] This hermeneutic filter often serves not to uncover harmony, but rather to obscure its presence, hiding the potential revelation of purpose and relation beyond ourselves within the constructed confines of our own selves.[10] By shaping our experience of the world not as it is, but rather through our expectations and personal beliefs, we shape our ability to perceive beyond the limits of our comprehension.[11] Indeed, so often it is the appreciation of purpose beyond ourselves that reveals the harmonious connection of the self to the other, the sacred to the profane and the world to worlds beyond.

Yet, the appreciation of true purpose, the ability to be surprised by its revelation, requires a sense of epistemological vulnerability within our hermeneutical framing.[12] Without a vulnerability to error within our matrix of expectations, everything is expected, pedestrian and already illuminated. There is no space for surprise within a system that perfectly matches our predictions, each occurrence following predictably from the last or within a system which constantly reframes the ideological filter of our reality.[13] The danger of our contemporary world undoubtedly lies within this aspect of its epistemology, a desire for a resolute certainty that isolates us from the sense of wonder, surprise, and illumination that is critical to developing our awareness of the power and value of things outside ourselves.[14] Within contemporary mechanistic, modernist and materialistic constructions of our reality, there is a lack of the vulnerability and humility, so necessary for the revelation of a deeper harmony beyond our immediate apprehension.

Without this sense of vulnerability in our assumptions and expectations, we become epistemological didacts, whereby our epistemological assumptions override our experience of reality – never surprised, never enchanted, never to be shaken from our preconceived notions and valuations. We become sad and detached, trapped within a cage of certainty which eternally disappoints our very human desire for wonder. Yet, our experiences of harmony in all sorts of contexts provide a glimpse between the bars of the iron rationality of our contemporary world, a sudden revelation of both purpose and relation that exceeds the confines of our expectations. Harmony when experienced is a surprise, a movement of ourselves outside of the commonplace or ordinary and the dislocation of our ordinary world by the glimpse of our dependence on something outside ourselves.[15] It is an allowance and welcoming of change in our life, a change that, through illuminating our unthinking expectations, temporarily disrupts our ordinary course in life.

This vulnerability to change on the part of our epistemological foundations, the ability to be surprised by something unexpected, the openness to the correction of our suppositions, connects to the appreciation for relation and purpose evident within our experience of harmony. Indeed, these three form necessary characteristics of the aesthetic quality of harmony, shaping our experiences of moments of transcendence. However, these three together, while sufficient for an experience of harmony, are not the only way in which our experience of harmony is in relation with our selves. Here, we move beyond the experience of harmony, to the experience of the experience of harmony. It is not enough merely to grasp that a piece of art gestures towards the relation of the particular to the universal, expresses that that relation has some form of purpose or connection behind it, and to be open to the possibility that this experience could in some sense illuminate one's own considerations of the nature of reality. There is also the critical matter of desire within our experiences of harmony, a desire not only to seek out those fleeting instances of connection but also a desire towards the continual uncovering of the meaning and importance of those moments.[16] Our experiences of harmony are not, therefore, discrete illuminations without intrinsic meanings that float free and loose, impinging on our selves only during the time of our immediate experience. Our feeling of harmony is rather a quest of desire, a continuous attempt not only to experience moments of harmony through particular experiences but also to expand this feeling both chronologically and kairologically within our lives.[17] We desire, on experiencing moments of harmony, to understand not only how this moment of revelation relates to other moments we experience, how our illumination of harmony within the forest relates to the seeming disharmony of our built environments but also how the revelation to us of harmony within a moment of experience relates to those beliefs we hold about the ultimate meaning of reality, how the specific awareness of relationality and purpose contained within our experience connects to the relationality and purpose of the whole of reality. This exposes two forms of desire, the first towards the other, and the second to a continuation of the flourishing of the first. Both forms of desire expressed here are forms of the central desire to know more about the reality in which we live, the purpose that seems to animate our own existences. Our desire is a continual uncovering of more, a greater understanding of things and a revelling in the difference that things outside ourselves make to our lives.

We therefore require in our aesthetic appreciation of harmony a combination of both relation and purpose. It is the unveiling of both of these features that elicits the grasping of harmony within us, the awareness that things are connected and in motion with each other, and that this relation is structured in such a way that a broader relation is connected to the specific relation apparent. The notion of harmony in song is exposed both technically, within the relation of purposefully struck notes to create a particular vibration, cohesively, between the location of that vibration, whether musical or vocal, and those around it, composing a discrete emotional reaction to their positions, and on a somewhat grander scale, as the meaning and tone of the harmonious music reveals within itself and to ourselves the connection between our personal experiences and the broader truths of the world. Harmony surprises us, uncovering and illuminating itself in an attempt to seduce our interest and our self.[18] The seduction of potential meaning is critical to the entrancing quality of our experiences of harmony, gesturing towards a correction of our current position by the potentiality of the other, and the grasping of true purpose through the revelation of a continual falling short in our current understanding.

These characteristics of our experiences of harmony connect through one particular, but often overlooked, necessity for harmony – the essential presence of difference. As Diogenes identified, this difference is not only a feature of our reality, but its maintenance and negotiation are critical aspects of our underlying reality. He noted:

All things come into being by a conflict of opposites, and the sum of things flows like a stream.[19]

Yet, it is Heraclitus that shows us that this conflict results not in the desolation of one aspect of difference, its destruction or erasure, but rather a harmonious attitude of mediation without reduction.[20] It is only through the relation between the difference that purpose emerges:

There is harmony in the bending back as in the case of the bow and the lyre.[21]

Without difference, however conceived, harmony does not and cannot exist. Without difference, between notes, between creatures, between the sacred and the profane, harmony is a notion that fails to function at all. Our recognition of difference is an intrinsic feature of our ability to partake in the harmonious; indeed, it is through our recognition of difference that harmony emerges. The characteristics of relation, purpose, vulnerability, and desire, each has within it the expression and awareness of difference.

This examination of the characteristics of our experience of harmony is not without a purpose of its own. If our experiences of harmony are structured according to a certain pattern (a recognition of difference, its exposing of relation between that difference, its revelation of purpose within that relation, and our continual desire to uncover more of that illumination of meaning) then so too, should we wish for the exploration of harmony on the level of society; our social and communal desire for harmony must come to be structured accordingly. In attempting to engage with this movement, we must take stock of the current situation before attempting to determine fruitful courses of action for our performance of social harmony. This practice is itself an individual, and social, modelling of the eternal quest for personal and public meaning that is central to our human experience. In attempting to discover the nature of the meaning of our existences, we commit also to the discovery of our shared social meanings, the way we, as individuals, fit within the relational whole of society. Our meanings and their production, the meanings of others and their production, and the differences expressed between them are critical factors in our social living; indeed, they form our very notion of what society is.[22]

Yet, within our contemporary societies, social atomisation and rampant individualism have long encouraged the denial of the role that we each play in the lives of others.[23] The denial of the way that our actions, whether directly or indirectly, have an impact beyond our immediate surroundings opens up a dangerous path. Think of our callous exportation of environmental disasters to poorer societies across the globe through our cultural fixation on accumulative materialism or the casual disregard for yawning chasms of inequality, of both outcome and opportunity, that infest our developed societies, or the ever declining sense of an ethical 'common good' on which our communities can rely for mutual comprehension. It is within this context that our notions of meaning and the negotiation of difference between them becomes a central task.

I wish to address this question – *How does the experience of harmony inform our performance of social harmony?* – within the context of religious difference for three reasons.

The first that religions are meaning producing systems par excellence in that they attempt to articulate the meaning behind the totality of our existence to provide a narrative that not only explains our individual place within society but also articulates the meta-level purpose of our existence.[24] Religions attempt to tell stories to explain not only how things are but also how things have been and how things will be. Beyond this, they also provide meaning to this description by offering up a vision of how things should be, how things could be changed and how the way things were, are, and will be are connected to one another within a broader explanatory framework. In this sense, religions, like our currently dominant form of secular liberalism, attempt to articulate the totality of our existence, giving meaning to every aspect of our lives, our understanding, and our societies. Religions form the meta-narrative structure which binds together the various stories within which we live.[25] Harmony, as a fleeting glimpse of relation and purpose, uncovers this meta-narrative reality, exposing the nature of the inter-relation of things posited by religious traditions to our gaze.

The second reason is that plurality, particularly religious plurality, is an ever-increasing feature of our societies. Our societies within the contemporary West are kaleidoscopic in their internal variety, with cultural, social and religious differences being prominent features of the way we experience our communities and of the way we conceive of our place within society. This plurality of culture, social understanding, self-identification and religious faith has become an inescapable feature of our worlds, impinging on our daily life in a way that represents a significant, and often under-considered, break with our imagined pasts. This process of pluralisation experienced by our communities is not a static event, but rather an ever-increasing trend towards greater and greater plurality.[26] This is the fractalisation of shared values and meanings, under the influence of post-modern hyper-individualisation producing within itself deeper and deeper accounts of internal variation. Plurality, the expression of difference, is the pre-eminent feature of our contemporary societies, an inescapable ordering of our world around the distinction between the self and the other, the one and the many, the we and the them.

These two considerations point towards the third way in which the understanding of the experience of harmony gestures towards the consideration of specifically religious harmony. Religions, in attempting to produce definitive meaning, often struggle to effectively consider the essentiality of difference. While cultural and social differences (increasingly tolerated within the framework of secular liberalism as expressions of individuality and preference) are relatively unchallenging to accounts of social similarity, or able to be bracketed within particular communities, religious difference poses a distinctly different issue.

Religions, in their meta-narrative function, attempt the explanation of the entirety of our reality, not only describing reality but by attempting to explain the ultimate meaning of it. In doing so, their ability to articulate meaning depends on the ability of that articulation to encompass the totality of our reality.[27] The persuasive power of the meta-narrative comes from its claim to total meaning and its expression of logical coherence in doing so and, given this, the ability to hold within itself space for difference, particularly different accounts of the meta-narrative nature of reality, declines.[28] In attempting to explain everything, meta-narrative structures must give meaning to the differences experienced by individuals. In doing so, meta-narrative structures construct and position the other, providing their meaning to us, sadly so often in ways that fail to reflect their true nature.

Religious difference therefore represents a significant challenge to our notions of harmony, never mind the aspiration for a sense of religious harmony within our societies.[29] Religions are noted for their disagreement, their difficulty conceiving of the place of the other, and their claims to matters of truth and correctness.

I wish to turn to exploring some of the ways in which our aesthetic experience of harmony offers guiding points for our practice of relation in the service of social and religious harmony. How might our experience of harmony, that revealing glimpse of something beyond, change the way in which we attempt to build harmony between and amongst ourselves in the here and now? If our experience of moments of harmony are formed from the illumination of relation between difference, purpose within that relation, an expression of vulnerability regarding our certainty of knowledge, and our desiring to know more about the connections between instances of harmony, then so too must these characteristics form part of our experience of harmony within society, and our commitment to the ongoing expression of that harmony in our individual and social actions. Ultimately, our experience of harmony must come to inform our performance of harmony.

So, how exactly might we mirror aspects of relation, purpose, vulnerability, and desire within our social interactions with those of other faiths? How might we act, and think, in such a way that these characteristics form significant parts of our attitude towards the presence of the other? These questions are complicated and context-specific, being related to the specific ways in which the difference being addressed manifests itself within our plural societies. However, we are able to consider at least the ways in which our own actions and ways of thinking may be broadened by a consideration of these characteristics of harmony.

The notion of relation is, I believe, a useful place to begin our consideration. As outlined before, it is tempting in our current social climate to declaim the idea that we live our lives in relation with others. We exist often in an isolated fashion both physically, socially, and mentally. Our physical isolation, even within bustling and ever more crowded global cities, exists in our solitary living, working, and playing – partitioned in small single occupancy accommodation, working within confining cubicles which deliberately stifle communal inefficiency, and often connecting socially only with those available online.[30] This physical solitude mirrors the increasing examples of social isolation, the breakdown of community feeling, the decline in civic and democratic participation and the prevailing sense of anomie that exists within much of the sub-cultural milieu.[31] Beyond this, and as a consequence of both of these factors, we come to exist psychologically as increasingly individual, whereby our lack of connection to the other becomes a normal facet of our existence, rather than an aberration of capitalist materialist control. Yet, this perception of isolation, of a lack of relation to the other, is illusory. While our current forms of engagement with anything or anyone outside of ourselves are often fleeting, pursued selfishly, or undertaken subconsciously, this is not to say that the fact of interaction is not meaningful. Our contemporary realities are distinctly formed through plurality and multiplicity, containing within them an often bewildering range of possibility and difference. Think of any town or city these days, and then seek to imagine that location without any of the religious, cultural, or social difference present in our shops, restaurants, places of worship, public spaces or even our own families. While our lives often seem to be devoid of interaction or relation, in actuality we live our lives surrounded by difference, constantly engaged in the negotiation of differences, from the minor aspects of learning to live within a community formed of racially, culturally and religious different people, to more major social questions of how to construct shared social values.

Perhaps is it that we have become too used to having to undergo this process of negotiation, and so it has become an invisible and unthinking practice of ours. With religious difference, we need to pay further attention if we are to understand the nature of our relation with the other. While within previously more homogenous societies religious communities could deny the extent of relation with the other, whether Protestant to our Catholic, or Methodist to our Anglican, or Muslim to our Christian, this denial of experience rings increasingly false in an ever more global, more varied and more connected world. It is impossible to deny the existence of the other within our societies, to deny their continued importance within our societies or to deny the way in which their presence inter-connects with our own.

These denials, while undoubtedly tempting on the part of the religious community, form a false consciousness that promotes a disconnect between the nature of reality as actually experienced on the one hand, and the meta-narrative offered within the religious group on the other.[32] It is this disconnect which, without strong shared values and mutual recognition, provides space for promotion of divisive, polarising, and ultimately dangerous narratives about the other.[33] Our experience of harmony, our grasping that the relation expressed within the instance of harmony mirrors in some way the relation between ourselves and that beyond us, is a critical feature in our grasping that our individual existence is related to the existence of others, and that those others, by virtue of their humanity, share in the broader relation of existence to the beyond.

The experience of harmony therefore demands a recognition of relation between ourselves and the other, an understanding that we are connected, whether consciously or unconsciously, whether wanted or unwanted, whether deeply or shallowly, with the continuing existence of the other in our lives. This cannot be elided and attempts to do so, either philosophically or socially, must be resisted from within our own religious traditions. However, it is not enough to merely accept this existence of relation between ourselves and the other. In connection with the facets of purpose, vulnerability, and desire, this relation must be correctly understood as offering up the possibility of change, the possibility of learning more about the other, and through that, learning more about ourselves. This is an attitude toward the other, which undergirds our recognition of their existence with a recognition of the purpose within their existence. As the Quran 49:13 notes: 'O mankind, We have created you male and female, and appointed you races and tribes, that you may know one another. Surely the noblest among you in the sight of God is the most godfearing of you. God is All-knowing, All-aware' and so our meeting of the other in plural societies requires from us an attitude within that meeting which mirrors the actions of vulnerability, desire, and a recognition of inner purpose.[34] While, as we have seen, the recognition of relation is a critical first step to developing a common ground between communities of difference, this second step, one of recognising the purpose in that expression of difference, is a somewhat greater ask. While religious traditions generally promote universal views of reality, seeking to explain not only the totality of reality but also to be relevant and explanatory for the totality of humanity, their conceptualisation of the existence of difference is often lacking.

This is made even more apparent when the question of continued difference is considered. Although much current soteriological and theological speculation considers the place of the Christian tradition relative to people of other faiths, this is rarely extended to considerations of what the lack of gradual conversion to Christian truth by others means for Christian beliefs.[35] These related questions, the existence of the other and the others continued existence, demand a response from within our religious traditions beyond a retreat from the recognition of relation. If we accept that our lives are fundamentally informed by our relation with those that are different from us, then this difference must be recognised as holding meaning, both for the other, and, crucially, for ourselves. While withdrawal from difference is perhaps comforting, it not only retreats from the nature of our reality, our everyday lived experiences but also excuses us from the ethical consideration of the meaning of the other.[36]

This is perhaps the most dangerous and demeaning step that we could undertake and fails to recognise the implicit push to universality within our religious traditions. By refusing to take the moral and ethical value of the other's presence in our lives seriously, we diminish the moral value of the other, degrading their presence either through their instrumentalisation or their marginalisation. We either deny the role that the other plays in our lives – insulating ourselves from difference, denying the possibility of change or growth, denying the other any voice with which to question our systems or beliefs – or we deny the moral weight of the other as a shared part of our universal reality; we instrumentalise them as tools of our politics, we misrepresent them for our own ends and we use the other in order to assuage our own moral failings.[37] While devastatingly prevalent in our contemporary political systems, this treatment of the other is a failure on the part of differing religious traditions and of secular modernity not only to take the other seriously but also to take their own traditions seriously as well.

As the Bible notes:

Then God said, 'Let us make man in our image, after our likeness. And let them have dominion over the fish of the sea and over the birds of the heavens and over the livestock and over all the earth and over every creeping thing that creeps on the earth'. So God created man in his own image, in the image of God he created him; male and female he created them. —Genesis 1:26-27.

As the Quran states:

He created the heavens and the earth in truth, wrapping night about the day, and wrapping the day about the night; and He has subjected the sun and the moon, each of them running to a stated term. Is not He the All-mighty, the All-forgiving? —Qur'an 39:5.

These scriptural resources are important not just for their commandatory power. They also ascribe moral weight to the universality of humankind, a validation of the existence of plurality within that universality and a requirement to take our shared humanity seriously. These statements teach us not only that all share in the relationship to the divine that characterises our humanity, but that this relation of the other to the divine matters, not just to them but also that it must come to matter seriously to us. How can we come to understand the divine, even fleetingly, without committing to understand the other, fully and without reservation, projection or instrumentalisation?

A commitment to our own faith therefore requires a commitment to a treatment of the other that expresses vulnerability in the certainty of our own conceptions and an awareness that the difference of the other contains within it purpose and meaning.[38] Our being in relation with the other therefore imputes unto us a series of duties; to desire to always understand the other and to express this desire in a way that is both non-appropriative of the other and non-final in our expression of understanding.[39] So how do our actions and attitudes towards the religious other open space for the social revelation of harmony, how does our aesthetic experience of harmony fold into our social practice of harmony, and how might we develop and promote the instantiation of harmony in our communities? Our experience of harmony, formed through a revelation of relation, purpose and desire, points the way toward a reassessment of the place of the other within our social circumstances, a movement from our glimpse of the divine other in the instance of harmony to the place of the social other in our communities. This place of the other is formed through a recognition of the relationality present between the other and ourselves within the kaleidoscopic plurality of contemporary societies, an inescapable pas-de-deux of social interaction and negotiation that, through its constant performance, inevitably shapes and constructs fundamental aspects of our reality. The recognition of this relation is recognition of the nature of difference, as present, meaningful and a continuous aspect of our own self-formation. The real question therefore becomes not just what difference is but how we choose to engage with that difference.

We have a choice to make. That choice is not about whether we wish to engage with the other, that option is far gone, unsustainable and ultimately damaging to others and to ourselves, but rather the choice of how we will engage with the other. How we make this choice has an impact not only on our relationship with others in our communities but also on the relationship between ourselves and that beyond us. Each level of relation is symbiotically and rhythmically related to the other, mirroring and reflecting our actions at one level in the other, expressing the relationship between our differing levels of relation.

This is therefore not a call toward homogeneity, the occlusion of difference through toleration, or the obliteration of difference through agreement. Rather, it is a call for the embrace of multiplicity, relationality, and plurality in our conception of our selves and our societies. A plurality that expresses an awareness and recognition of the internal relationality within it, that takes that relationality seriously, and that mirrors externally our internal notion of our relation with the divine. This is not a momentary instance, as in our appreciation of aesthetic forms of harmony, but rather an attitudinal commitment to the performance of the characteristics of harmony, the performance of deliberate relation, of an awareness of shared purpose, and of a deliberate vulnerability of our certainties. This is therefore also not a question of truth or belief, of rightness or certainty. Our experience of harmony shows us the necessary characteristics for a real motivated discussion of truth, belief and certainty. It illuminates the background on which this engagement with the other, if it is to be conducted in a way that embodies integrity and respect for difference, may be begun. It is a commitment to the exploration of the meaning of the plurality within which we live with others, an embrace of mutual vulnerability and uncertainty, and a recognition of the ethical importance of the other for these deliberations.

Our experiences of harmony, that magical slipping of the ordinary into the extraordinary through the uncovering of its purposeful relation with the rest of reality, gives a glimpse not only of the aesthetic, the beauty of our reality but also the ethical, the duties we have toward those others who live with us within and among difference. Without difference, between ourselves and the other, there can be no sense of relation. Relation, that bedrock of our lives, expressed individually, spiritually, or communally, depends on our recognition and relationship with things that are not us, a recognition that they differ and that this difference matters. Without difference, between what is and what could be, there is no purpose. We are animated by purpose, we recognise purpose in others, and in doing so we come into relation with a veritable cosmos of imagination. Without difference, an awareness of the ability of the difference to have purpose, we cannot be vulnerable to the correction of the other. We depend on difference for change, for correction, for growth. We cannot be vulnerable, let down our guard, without a valuation of the other as worth that vulnerability. Yet, without that vulnerability we deny ourselves any possibility of change or growth. Without difference we cannot learn to desire, to learn to be seduced by the difference of the other, to fall in love with the inescapable and ever-elusive mystery of that which we are not.

Notes

1. A. Gethmann-Siefert, *Einführung in die Ästhetik* (Munich: Wilhelm Fink, 1995), p. 7.

2. U. Eco, *On Beauty: A History of a Western Idea* (London: MacLehose Press, 2010), p. 9.

3. T. Aquinas, 'Quaestiones disputatae de veritate', vol XXII/1-3 in C. L. (ed.), *Opera omnia* (Rome. 1975).

4. Boethius, 'De Hebdomadibus' in S. C. MacDonald, *Being and Goodness: The Concept of the Good in Metaphysics and Philosophical Theology* (New York: Cornell University Press, 1991), pp. 299-304.

5. J. Harley, *Thomas Tallis* (London: Ashgate Publishing, 2016).

6. R. J. Putman, *Community Ecology* (New York: Springer, 1994).

7. S. O'Sullivan, 'The Aesthetics of Affect: Thinking art beyond representation', *Angelaki: Journal of the Theoretical Humanities*, 6(3), (2010): pp. 125-135.

8. David Cadman, 'Harmony', https://www.uwtsd.ac.uk/harmony-institute/ [accessed 9 April 2019].

9. H. G. Gadamer, *Truth and Method* (2nd revised edition), J. W. Marshall, trans. (New York: Crossroad,1989), pp.164-169.

10. R. Palmer, *Hermeneutics: Interpretation Theory in Schleiermacher, Dilthey, Heidegger, and Gadamer* (Evanston, IL: Northwestern University Press, 1969.)

11. M. Heidegger, *Being and Time* (Oxford: Blackwell Publishing, 1962), p. 24.

12. R. Williams, 'Trinity and Pluralism' in G. D'Costa, *Christian Uniqueness Reconsidered: The Myth of a Pluralistic Theology of Religions* (London: Orbis Books, 1990), p. 10.

13. T. Eagleton, 'Ideology and Enlightenment' in T. Eagleton, *Ideology* (London: Routledge, 1994), pp. 1-4.

14. T. Albertini, 'The Seductiveness of Certainty: The Destruction of Islam's Intellectual Legacy by the Fundamentalists', *Philosophy East and West*, 53(4), (2003): pp. 455-470.

15. F. Schleiermacher, *The Christian Faith* (London: Continuum Books, 1999), p. 132.

16. G. Ward, *Cities of God* (London: Routledge, 2000), p. 77.

17. R. Panikkar, 'The Jordan, the Tiber, and the Ganges: Three Kairological Moments of Christic Self-Consciousness' in P. F. John Hick, *In The Myth of Christian Uniqueness* (London: SCM Press, 1988), p. 92.

18. J. Baudrillard, *Seduction* (London: St Martin's Press, 1991), p. 165.

19. Diogenes Laërtius, ix. 8

20. Eco, *On Beauty*, p. 72.

21. P. Wheelwright (trans), *Heraclitus* (New York: Atheneum, 1964), p. 102.

22. G. Loughlin, 'Christianity at the End of the Story or the Return of the Master-

Narrative', *Modern Theology*, 8(4), (1992): pp. 365-384.

23. Z. Bauman, *Liquid Modernity* (Cambridge: Polity Press, 2000).

24. J. F. Lyotard, *The Postmodern Condition: A Report on Knowledge* (Manchester: University of Manchester Press, 1984).

25. Loughlin, 'Christianity at the End of the Story or the Return of the Master-Narrative', p. 375.

26. A. Dawson, 'Globally Modern, Dynamically Diverse: How Global Modernity Engenders Dynamic Diversity' in L. Kühle, J. Borup and W. Hoverd, *The Critical Analysis of Religious Diversity* (Leiden: Brill, 2018), pp. 83-106.

27. A. M. Slater, *Radical Orthodoxy in a Pluralistic World: Desire, Beauty, and the Divine* (London: Routledge, 2018), pp. 117-18.

28. J. Milbank, 'The End of Dialogue' in G. D'Costa, *Christian Uniqueness Reconsidered* (London: Orbis Books, 1990), p. 184.

29. Milbank, 'The End of Dialogue', p. 183.

30. Z. Bauman, 'Modernity and Ambivalence', *Theory, Culture & Society*, 7(2-3) (1990): pp. 143-69.

31. E. Durkheim, *On Suicide* (London: Penguin Classics, 2006), p. 277.

32. J-P. Sartre, *Essays in Existentialism* (New York: Citadel Press, 1993), pp. 160-64.

33. J. M. Miller, K. L. Saunders and C. E. Farhart, 'Conspiracy Endorsement as Motivated Reasoning: The Moderating Roles of Political Knowledge and Trust', *American Journal of Political Science* (2016): pp. 827-28.

34. All Quranic translations are according to A. J. Arberry, *The Koran Interpreted* (trans.) (New York: MacMillan, 1955); all Biblical references are according to the English Standard Version.

35. B. Sargent, 'Proceeding Beyond Isolation: Bringing Milbank, Habermas, and Ockham to the Interfaith Table', *Heythrop Journal*, 51, (2010): pp. 819-30.

36. I. Levinas, *Totality and Infinity: An Essay on Exteriority* (A. Lingis, trans.) (Pittburgh: Duquense UP, 1968).

37. M. Foucault, *Naissance de la biopolitique: Cours au College de France 1978-1979* (Paris: Le Seuil, 2004), pp. 27-34.

38. Slater, *Radical Orthodoxy in a Pluralistic World*.

39. Williams, 'Trinity and Pluralism'.

THE HARMONY OF THE COSMOS, THE SOUL, AND SOCIETY IN PLATO

Joseph Milne

And the love, more especially, which is concerned with the good, and which is perfected in company with temperance and justice, whether among gods or men, has the greatest power, and is the source of all our happiness and harmony, and makes us friends with the gods who are above us, and with one another.[1]

GREEK PHILOSOPHY EMERGED THROUGH SPECULATION on the cosmic myths that symbolically revealed the divine order of the universe. From these speculations on the cosmic order arose the various notions of the elements, the planetary motions and mathematics, and these notions were related to the question of the human order and the order of society.[2] It was understood that the human order was distinct from that of the immortal gods, yet also distinct from biological necessity. Human nature dwelled in a region between the immortal and the mortal, open to eternity yet projected into time, apprehending the unchanging yet compelled to adapt to the ever-changing. In the primordial myths the order of nature (*physis*) and human law (*nomos*) arose together and were bound together.[3] The order of nature and the order of the city resided in the rule of the gods, and this order could be observed in the harmony and proportion found throughout the Earth and the heavenly motions. The cosmos was filled with intelligence and with reason (*nous*), and every part and every motion attended the good of the whole.

In the myth of the Golden Age the human realm and the divine realm lived in perfect harmony. For example, in Hesiod we read:

The gods, who live on Mount Olympus, first
Fashioned a golden race of mortal men;
These lived in the reign of Kronos, king of heaven,
And like the gods they lived with happy hearts
Untouched by work or sorrow. Vile old age
Never appeared, but always lively-limbed,
Far from all ills, they feasted happily.
Death came to them as sleep, and all good things

Were theirs; ungrudgingly, the fertile land
Gave up her fruits unasked. Happy to be
At peace, they lived with every want supplied,
[Rich in their flocks, dear to the blessed gods.]
And then this race was hidden in the ground.
But still they live as spirits of the earth,
Holy and good, guardians who keep off harm,
Givers of wealth: this kingly right is theirs.[4]

In the *Laws* Plato alludes to that ancient age in which Kronos 'set up at that time
kings and rulers within our cities – not human beings but demons, members of a
more divine and better species... They provided peace and awe and good laws and
justice without stint'.[5] Yet this could not endure, and men began to devise their
own laws, forgetting the gods and breaking the bond between the eternal and the
temporal. The visible, temporal world may at best embody the divine pattern,
and be regulated by it, and laws ought to be made as like as possible to the age
of Kronos.[6] This distinction drawn between the eternal and the temporal realms
becomes the birthplace of philosophical enquiry because the distinction arouses a
part of the soul that seeks reconciliation between the eternal and temporal orders
of truth. The eternal realm beckons the soul, which finds itself dwelling between
the two orders, to enquire into the truth of things for its own sake, as an end in
itself. But once this yearning for truth is born, the mythological symbols of reality
no longer suffice. They were born from primordial intuition, a form of knowing
the essentially true at a single stroke but which is not yet reflective upon itself.
The desire to *understand* this truth, beyond simply assenting to it, is the birth of
philosophy.

In this way, Greek philosophy originated in meditation on cosmic myth, the
primordial apprehension of the whole, with a view to affirming its truth through
reason. And this meditation takes the form of the question: how may the human
being and society live in accord with the cosmic good? What is the appropriate
life of the human person or citizen? It is at once a rational and a religious
question. For the Greek philosophers, questions of the explanation of things are
secondary to this essential question that awakens questioning in the first place.
Philosophical enquiry is not a precursor to the scientific explanation of things,
because explanation is not a final end in itself, while the question of how should
life be lived is. And so Greek philosophy, even in its weaker or degenerate forms,
for example, with the sophists whom Plato frequently challenges in the dialogues,

always remains concerned with the relation of the divine cosmic order and the order of society or the *polis*.[7] The *polis* and the cosmos are bound together, just as the *polis* and the soul are bound together. Greek society drifted into political decline as it forsook these connections. Thus Voegelin writes:

> In their acts of resistance to the disorder of the age, Socrates, Plato, and Aristotle experienced and explored the movements of a force that structured the *psyche* of man and enabled it to resist disorder. To this force, its movements, and the resulting structure, they gave the name *nous*. As far as the ordering structure of his humanity is concerned, Aristotle characterized man as the *zoon noun echon*, as the living being that possesses nous.[8]

And it is with a view to restoring these connections that Plato and Aristotle enquired into the nature of the *polis* and the question of the relation between nature (*physis*) and law (*nomos*). Thus Heraclitus says 'Those who speak with understanding must hold fast to what is common to all as a city holds fast to its law (*nomos*), and even more strongly. For all human laws (*anthropeioi nomoi*) are fed by the one divine law (*theios nomos*). It prevails as much as it will, and suffices for all things with something to spare'.[9] Hence the nature of the *polis* and the divine law that sustains it cannot be separated without causing harm.

Neither Plato nor Aristotle abandon the gods nor the mythic symbols from which Greek philosophy was born.[10] This may be seen in their insistence that virtue and knowledge are bound together and that only the virtuous soul may contemplate the truth of things and live in accord with nature. In his *Nicomachean Ethics*, for example, Aristotle writes that the contemplative life is the happiest, as it is the life and activity of the gods:

> For the gods, the whole of life is blessed, and for human beings it is so to the extent that there is some likeness to such a way of being-at-work; but none of the other animals is happy since they do not share in contemplation at all. So happiness extends as far as contemplation does, and the more it belongs to any being to contemplate, the more it belongs to them to be happy, not incidentally but as a result of contemplating since this is worthwhile in itself.[11]

Coming at the close of the *Nicomachean Ethics* in Book 10, it is clear that it is only those who follow a noble life of virtue and excellence have the capacity to participate in the contemplative life. Contemplative knowledge is a kind of living alignment with truth, and this is possible only in the soul of the virtuous person who has self-mastery. In the human realm this relation between knowledge and virtue is the first harmony, the harmony where the soul comes into accord with itself and with the divine order of things. Plato elaborates on this in the *Timaeus*:

> God invented and gave us sight to the end that we might behold the courses of intelligence in the heaven, and apply them to the courses of our own intelligence which are akin to them, the unperturbed to the perturbed; and that we, learning them and partaking of the natural truth of reason, might imitate the absolutely unerring courses of God and regulate our own vagaries ... Moreover, so much of music as is adapted to the sound of the voice and to the sense of hearing is granted to us for the sake of harmony; and harmony, which has motions akin to the revolutions of our souls, is not regarded by the intelligent votary of the Muses as given by them with a view to irrational pleasure, which is deemed to be the purpose of it in our day, but as meant to correct any discord which may have arisen in the courses of the soul, and to be our ally in bringing her into harmony and agreement with herself; and rhythm too was given by them for the same reason, on account of the irregular and graceless ways which prevail among mankind generally, and to help us against them.[12]

Yet for the soul to come into this accord remains a calling, something to be worked towards and not something simply given, even though it is the proper end or *telos* of human nature.

In the cosmic myths the relationship between the cosmic order and the ethical is implicit, because truth and falsehood, and justice and injustice are bound together in action, just as in Greek drama.[13] But once the true and the just can be abstractly or metaphysically distinguished from one another, then their necessary unity comes into peril. It requires deep philosophical reflection to understand how they are ultimately bound together and originate from the Good. Thus Plato writes: 'Therefore, say that what provides the truth of the things known and gives the power to the one who knows, is the *idea* of the good'.[14] It is for this reason that Plato is always asking questions about the essence of things. For it is, according to Plato, only through knowing the essence of a thing that one can see how it originates in the good. Likewise, the convergence of the true and the just in the order of things is the ground of the original harmony that extends into the cosmic order and into every particular being.[15] Thus Apollo presides over law and the celestial song of the Muses and over healing through his son Asclepius. The bringer of order is also the bringer of law and healing. Plato often likens the art of the lawmaker to that of the physician.[16]

For example, while speaking of how people desire only laws that will please them the Athenian Stranger says 'Such a provision is in opposition to the common notion that the lawgiver should make only such laws as the people like; but we say that he should rather be like a physician, prepared to effect a cure even at the cost of considerable suffering'.[17] Also *Gorgias* draws a comparison between the physician and the judge in administering justice, one curing the body, the other the soul.[18] Again in the *Laws* Plato sees judicial penalty as aiming to restore the soul rather than merely causing it to suffer the consequences of injustice.[19] In *Laws* we also read: 'This, then – the knowledge of the natures and habits of souls – is one of the things that is of the greatest use for the art whose business it is to care for souls. And we assert (I think) that that art is politics. Or what?'[20] In the *Republic* Plato draws a direct analogy between health and sickness and justice and injustice, arguing that sickness and injustice are alike contrary to nature, while health and justice are according to nature.[21] Or as Brill observes: 'Plato's infamous employment of the language of medicine to characterize the work of the laws, language which we have seen play a critical role in the *Republic*, is in part a function of his focus upon the condition of the soul of the citizen. This is to say that Plato's therapeutic conception of law is inextricably linked to his psychology… Plato thus allots a dual educative/therapeutic function to the law'.[22]

Thus the step from the primordial mythic apprehension of the cosmos to the reflective philosophical understanding of the truth of things, which took place in classical Greece after the age of Homer, also brings reason into reflection upon itself. This raises the question of the capacity of human intelligence to know the truth of things, and so the soul is brought into self-reflection and self-examination. Self-knowledge, knowledge of nature (*physis*), and divine knowledge reveal themselves as distinct orders of knowledge and yet bound together. We can see this most clearly wherever Plato raises the question of justice. Those who cannot or will not truthfully observe themselves, such as Thrasymachus, who in the *Republic* argues that justice is rule by the strongest for their own benefit and gives up and leaves the discussion once his argument does not stand up to scrutiny, or Meno who likewise sees virtue as doing what is to one's advantage and harmful to one's enemies, cannot grasp the true nature of justice.[23] They conceive the just or the good only as what is advantageous to themselves. They cannot consider justice in itself as it belongs to the right relation of all things with one another, or as belonging to the harmony of the soul. Justice for Plato signifies more than anything else the great harmony that is of the essence of all things. So to conceive justice as privately advantageous is not only to mistake the nature of justice but also to divide the human individual off from the *polis*, and the *polis* from the cosmos. Civil fragmentation or factionalism is one of the perils of the step from the holistic mythic apprehension of the world and the human situation to the reflective philosophical apprehension of things.[24] There is an ever-present danger of losing the sense of the whole that belongs to the mythic and cosmic symbolism.[25] In philosophical reflection, reason must trace a path towards the whole from the particular, and from the immanent to the transcendent, as, for example, in the allegory of the cave in the *Republic*[26] or the ascent to the Beautiful in the *Symposium*.[27]

Plato draws upon the earlier philosophers, as well as the poets, for the themes that occupy his dialogues. With some he draws out further what they express only tersely, for example, Heraclitus and Parmenides while others he strongly disputes, such as the Sophists Gorgias and Meno.[28] The great question that distinguishes these different interlocutors lies in the understanding of the relation of language (*logos*) to the truth (*aletheia*) of things. Speech may be divinely uttered and inspired or deviously uttered for private advantage, giving birth either to order and friendship, or to chaos and tyranny. For Plato there is a correspondence between words uttered and the truth of things, expressed in the word *logos* itself which means at once language and reason or intelligence. To 'speak truthfully' is possible either by divine inspiration, as with poetic frenzy, or where the soul is in harmony with itself and perceives the true order of things and can speak their right names.[29] Such speech arises from reverence for truth or piety and is profitable to all. The Sophists have separated speech from the *logos* of things through false employment of rhetoric. For the Sophists the art of speech is nothing else than the art of persuasion. They taught this art to those seeking a successful political career. In this sense the Sophists are utilitarian and pragmatic. But to reduce rhetoric merely to the art of persuasion divorces the *logos* from truth and from virtue.[30]

There is, however, a middle place between true Platonic speech and sophistic speech, and this is *doxa* or opinion. *Doxa* is opinion held without proper enquiry or reflection, or views believed on hearsay.[31] Such opinions may indeed be true, or a mixture of true and false. For Plato and for Aristotle common opinion is insufficient to establish either the truth of things or the good life. This applies as much to opinions about the gods, nature or politics as it does about everyday things. The examination of common opinions plays a major role in the dialogues of Plato. Aristotle also will often begin his examination of a topic with a statement of what people generally believe. There is a strong dialectical element in his *Nicomachean Ethics* and *Politics* which, like Plato, seeks a path from what seems to be true to what is actually true.[32] Since this involves an examination of the premises or presuppositions of one's own thought it also demands self-examination and honesty, because the aim is not to establish a new opinion or *doxa* but to come into a transformed relationship with truth, to harmonise the soul with the nature and intelligent order of things.

For Plato the problem of the truth of things does not lie in correct or incorrect theories or doctrines, or in true or false statements, but in bringing the soul into a right disposition towards truth, ethically as well as intellectually. It may only be open towards it or closed towards it. Paradoxically, the soul is closer to truth by a recognition of its *not knowing*, as we see in Socrates' insistence in *Apology* that "I do not think I know what I do not know".[33] The urge for certainty can be an obstacle because it tends to reduce truth to mere propositions. What the philosopher seeks to accomplish is to bring the soul into harmony with itself and with the cosmic or heavenly order, as in *Timaeus* 47C.

That passage in the *Timaeus* alludes to the Pythagorean concept of the 'music of the spheres' in which the motions of the heavens form a choir of divine music, which only few can hear, and which regulates the universe in perfect proportion and symmetry, described in detail at *Timaeus* 35-37.[34] It is a symbolic idea, not to be taken literally, embodying an understanding of the universe governed entirely by intelligent order – not as imposed upon it from outside, but as its own living intelligence. Because it is intelligent harmony it is akin to the human soul when it is brought into its own proper order.[35] From this comes the tradition of the soul as a 'microcosm', containing within itself the same intelligence and beauty as the 'macrocosm'. The soul is not to be understood as a mere replica of the macrocosm, but as participating in the same intelligence, just as different living creatures participate in the same life. Intelligence and life are universals, just as the 'numbers' discerned in the music of the spheres are universals. Plato took 'number' very seriously and the study of mathematics played a major part in the Academy. The 'mathematical' is a type of learning or knowledge that shows itself as self-evident, as Plato demonstrates in the *Meno* with the slave who could solve a geometrical problem without any prior knowledge.[36] Nevertheless, the question of the real nature or essence of number remains a profound mystery, even though it manifests everywhere in the forms, symmetry, rhythms and cycles of nature.

But this passage from the *Timaeus* refers to an order of another kind, in which the soul is brought into accord with itself through the virtues. If the soul is to truly govern itself, as the cosmos does, then its various powers must be coordinated, and this is the work of the virtues. The virtue of temperance brings the *appetitive*, the *spirited*, and the *rational* parts of the soul into concord, with reason ruling.[37] But this virtue comes about through the cultivation of the virtues of *phronesis* (right judgement), courage and justice. For Plato, as for Aristotle also, ethics is not based on moral principles but on the virtues, which are states of the soul being at work or in action. The virtues are like skills. Temperance refers to the inner order of the soul, while the other virtues refer to its relationship with the world while maintaining its inner order. In the *Republic* and the *Laws* Plato assigns to the cultivation of the virtues the principle concern of education, not only for the sake of each individual but also because only virtuous souls can truly become citizens and live in harmony and friendship. For Plato citizenship and friendship are practically identical, since friendship and citizenship are sustained by a common love of excellence and justice.[38] Friendship is the proper proportion of the *polis*, resembling the cosmic order and the law of the gods. Plato speaks of this in the *Gorgias*:

> Now philosophers tell us Callicles, that communion and friendship and orderliness and temperance and justice bind together heaven and earth and gods and men, and that this universe is therefore called Cosmos or order.[39]

This passage demonstrates how for Plato the understanding of 'order' always bears an ethical meaning, and so the order of the cosmos is at once a proportional and virtuous order. Likewise, the proper relation between gods and men is at once proportionate, just and temperate. We find the same idea in Plato's *Laws* when discussing the ends the lawmaker must seek to attain: 'When we asserted one should look toward moderation, or prudence (*phronesis*), or friendship, these goals are not different but the same'.[40] This connection of the proportionate and the virtuous in the order of nature passes down through the Stoics, Neoplatonists, the Christian Fathers, and through to the High Middle Ages where it is given full expression by Aquinas in the *Summa Theologica*.[41] There is a particular word, *homonoia*, which bears this special sense. It is made up of Greek prefix *homo-*, which means 'alike' or 'same' and *nous* which means 'mind' or 'understanding' or 'insight'. So *homonoia* means to be 'like-minded' or of common understanding or agreement. Its opposite is *stasis*, 'internal division', which in the political sense means 'civil war' or 'factionalism'. For Plato these words are strongly connected with justice and injustice. Justice is a form of harmony and right proportion, while injustice is a form of discord and disproportion. This is clear in *Republic* Book I:

> Injustice, Thrasymachus, causes civil war [*stasis*], hatred and fighting among themselves, while justice brings friendship and a sense of common purpose [*homonoia*]. Isn't that so?
>
> Let it be so, in order not to disagree with you.
>
> You're still doing well on that front. So tell me this: If the effect of injustice is to produce hatred wherever it occurs, then, whenever it arises, whether among free men or slaves, won't it cause them to hate one another, engage in civil war [*stasis*], and prevent them from achieving a sense of common purpose [*homonoia*]?
>
> Certainly.
>
> What if it arises between two people? Won't they be at odds, hate each other, and be enemies to one another and to just people?
>
> They will.
>
> Does injustice lose its power to cause dissension when it arises within a single individual, or will it preserve it intact?
>
> Let it preserve it intact.
>
> Apparently, then, injustice has the power, first, to make whatever it arises in — whether in a city, a family, an army, or anything else — incapable of achieving anything as a unit, because of the civil wars [*stasiazonta*] and differences it creates, and, second, it makes that unit an enemy to itself and to what is in every way its opposite, namely, justice. Isn't that so?
>
> And even in a single individual, it has by its nature the very same effect. First, it makes him incapable of achieving anything, because he is in a state of civil war [*stasis*] and not of one mind [*homonoia*]; second, it makes him his own enemy, as well as the enemy of just people.
>
> Hasn't it that effect?
>
> Yes.[42]

This discussion is an attempt to refute the Sophist position that justice is not a universal principle and that injustice for the individual may be advantageous. If each individual seeks their own advantage, the Sophist holds, then somehow all will gain and justice is superfluous. The Sophist cannot see that strife between the different parts of the individual soul will follow from any form of injustice, internal or external. For Plato the individual soul cannot be broken off from the universal order without harming itself. Hence justice has the peculiar quality of being at once a principle (*arche*) ordering nature as a whole and an active state of being of the just person. For Plato only the just person really knows the nature of justice. Or the nature of justice is known only in its active performance.

As we observed at the outset, Plato is seeking to articulate in philosophy what was previously evident in myth where the gods presided over the cosmic order and in every particular down to the smallest detail. Thus 'cosmos' and 'law' were practically identical, as is clear in the passage from *Gorgias* above. But with the rise of early philosophy, which began to consider the cosmic order in rational rather than in mythic terms,[43] there also arose various forms of agnosticism, especially with the Sophists. Here two words already discussed become especially important: *physis* (nature) and *nomos* (law). Originally these two words formed a single concept, as may be seen in Heraclitus' fragment B 114:

> Thou who speak with the intellect [*xyn nooi*] must strengthen themselves with that which is common [*xynoi*] to all, as the polis does with the law [*nomos*], and more strongly so. For all human laws [*anthropeioi nomoi*] nourish themselves from the divine law [*theios nomos*] which governs as far as it will, and suffices for all things, and more than suffices.[44]

For Heraclitus to speak of the divine law (*theios nomos*) is to speak of law that 'suffices for all things', including the laws of the *polis* that 'nourish themselves from the divine law', and there is no appearance of *physis* as separate from *nomos*. Nature and law are bound together. And the *polis* likewise comes into being through *nomos*, since human laws take their existence from the same divine law that governs all things. The human citizen, by definition, is the being that reflects and deliberates on law, or on justice and injustice.[45] That is the original philosophical understanding. But later *physis* began to be conceived as separate from divinity and *nomos* and then the notion arose that human laws (*anthropeioi nomoi*) derived neither from divine law (*theios nomos*) nor from *physis*.[46] Rather, human law began to be conceived as merely conventional, differing from city to city, with no ground in *physis*. Thus arose the notion that individuals could follow their own nature (*physis*) and ignore the laws of the *polis*. And since the laws of the *polis* existed only by convention, the Sophists believed that no harm could come to them through disobeying them, at least in private if not in public. From this arose the further notion that the laws of the *polis* were made by the strongest and that justice was nothing else than the rule of the strong over the weak.[47]

While the Sophists could argue private advantage with this teaching, for Plato it indicated the decline of Athens and the destruction of citizenship.[48] But it also indicated, on a more profound level, the loss of the symbolic understanding of the order of the cosmos as revealed through the myths of the gods. The loss of knowledge of the *divine order* signified the fragmentation of the *human order*. Thus for Plato *homonoia*, *physis*, and *nomos* form a single complex, and it is the challenge of human reason to grasp this. We find the same insight articulated centuries later by Cicero: 'But those who have reason in common also have right reason in common. Since that is law, we men must also be reckoned to be associated with the gods in law. But further, those who have these things in common must be held to belong to the same state (*civitas*)'.[49] The citizen comes into being through *homonoia*, oneness of mind, agreement on a common purpose. Human reason is rooted in cosmic reason. The flourishing of the *polis* depends upon this grounding of the soul in the universal order and unity of *physis* and *nomos*.

One remarkable way in which Plato conceives the proportionate ordering of the *polis* in the *Laws*[50] is to limit the population to 5,040 households or extended families. The land for such a city should be large enough to support its population moderately, without excess, yet sufficient for defence against injustice from neighbouring cities, and strong enough to aid neighbours if they suffer injustices. The number 5,040 has exactly 60 divisors, counting itself and 1, and also is the sum of 42 consecutive primes. It therefore lends itself to complex proportionate divisions of functions of the population. This is not the place to elaborate on the special characteristics of this number. But the notion of a *natural size* of a self-sufficient *polis* which accords with the fertility of the land, the natural division of the human crafts and due administration of law, education and religious rites, and is of sufficient strength to have good relations with neighbouring cities, indicates that the human person naturally belongs to society. In his study of Plato's *Republic* Voegelin writes: 'Human nature is conceived as dispersed in variants over a multitude of human beings, so that only a group as a whole will embody the fullness of the nature. Order in society would then mean the harmonisation of the various types in correct super- and subordination'.[51] There is a final argument that gives the natural size of the *polis* strong support. Such a *polis* is of a size where all citizens may know one another and be friends, and this is conducive of virtue:

> There is no greater good for a city than that its inhabitants be well known to one another; for where men's characters are obscured from one another by the dark instead of being visible in the light, no one ever obtains in a correct way the honour he deserves, either in terms of office or justice. Above everything else, every man in every city must strive to avoid deceit on every occasion and to appear always in simple fashion, as he truly is – and, at the same time, to prevent other such men from deceiving him.[52]

Friendship emerges yet again as a principle of harmony promoting justice, openness and honesty. Human happiness is not attained through amassing wealth or by taking advantage of fellow citizens or of other cities or nations. For Plato the economic aspect of the city belongs to the realm of necessity and is therefore the least dignified of human concerns. The regulation of the population to 5,040 where each household has equal land to support itself, maintained by a prohibition on selling its land, removes the need for competition or opportunity for exploitation and frees all citizens to pursue the arts, learning and culture. Plato introduces another mathematical proportion, suggesting that the difference in wealth between citizens should never be more than ten times, and so the realm of necessity does not become an occasion for strife. Indeed, Plato says the earth must be acknowledged and honoured as the mother and sustainer of all living beings and must never be abused. There is a natural apportionment in which things ought to be honoured:

> We say, then, that the likelihood is that if a city is to be preserved and is to become happy within the limits of human power, it must necessarily apportion honours and dishonours correctly. The correct apportionment is one which honours most the good things pertaining to the soul (provided it has moderation), second, to the beautiful and good things pertaining to the body, and third, the things said to accrue from property and money. If some lawgiver or city steps outside this ranking either by promoting money to a position of honour or by raising one of the lesser things to a more honourable status, he will do a deed that is neither pious nor statesmanlike.[53]

This apportionment of honours corresponds with the cosmic hierarchy, where the divine intelligence descends through the orders of nature, ruling things justly and according to their proper ends. In Book X of the *Laws* Plato disputes the Sophist view that denies this divine hierarchy and holds that things come into being instead by nature, by art and by chance.[54] This view separates *physis* and *nomos* where the laws of cities are held to be arbitrary conventions devised by art. It conceives of intelligence coming into being last in the order of things, rather than first since nature (*physis*) here signifies only blind necessity. This view brings the gods into dispute, or at least their origin. For if the universe came into being through blind necessity, then the gods can be neither wise nor beneficent to the cosmos, the city or the soul, but will themselves be ruled by blind necessity.

From this state of affairs there arise various positions in relation to the gods: (a) that they do not exist, (b) they exist but care nothing for humanity, and (c) they exist and may be bribed into granting human desires. These positions derive from the belief that the gods came into being after the elements and the heavens, and that 'intelligence' is an incidental or chance product of nature (*physis*), and so all human laws and institutions have no ground in the cosmic order and exist only by human invention. It is a consequence of separating *physis* and *nomos*. It reduces *physis* to a mere *mechanism*, and *nomos* to arbitrary invention, and removes intelligence and justice from the cosmos – rendering it no longer a *cosmos*.

Plato devotes the whole of Book 10 of the *Laws* to this question, and how reasoned argument can overcome this false interpretation of the order of things. It is here where we can see most clearly how Plato is concerned to recover philosophically what has been lost or corrupted in the understanding of cosmic myth. The truth of the ancient myths is no longer intelligible to the Athenians. The symbols that once communicated the presence of the divine intelligence in all things, and in the art of law-making, no longer reveal their meaning. If indeed the universe is ordered by blind mechanism rather than divine intelligence, then there is no basis to cosmic justice or justice in civilisation. There is no ground for preferring a virtuous life to the opposite. And even if the mechanisms of nature may be discerned through empirical investigation and calculation, they will have no intelligible purpose or end. What emerges is a universe with no *telos*, where things exist without meaning. Enquiry into such a universe itself has no meaning.

There is, however, an alternative approach that Plato takes to the question of the order and harmony of things. This is through *kallos*, beauty. All that is truthful, harmonious, or virtuous appeals not only to the rational part of the soul but also to *eros*, the love of beauty. But just as reason can go astray with sophistry, so likewise *eros* can go astray by identifying beauty with particular objects. In the final speech in the *Symposium*, Socrates reports a discourse he had with Diotima on the ascent of *eros* from temporal things to eternal Beauty. In the *Phaedrus* Plato demonstrates that, whatever we behold here on earth as beautiful, moves the soul to a great passion because it is reminded of Absolute Beauty which it once beheld before coming into the human body.[55] This great passion is *eros*, which desires at once to unite with and to create beautiful things. In the earthly sense it is the desire for bodily generation, which is to attain a kind of immortality. But what *eros* truly desires is not particular instances of beauty in temporal things, but the Beautiful itself which is eternal and the source of all beauty. What the soul most desires is to give birth to the beautiful within itself, to become that divine Beauty. Thus whenever it beholds the beauty of goodness, it desires to become good, or in beholding the beauty of justice it desires to become just, or in beholding the beauty of wisdom it desires to become wise. It desires both to unite with these beautiful things and to give birth to them. For Plato truth is always associated with beauty, and beauty always associated with goodness, and so the true and the beautiful give birth to virtue in the soul.

One of the great questions of philosophy is: how is truth known to be truth? One of Plato's answers is that the soul *recognises* truth whenever it presents itself. It is an act of *anamnesis*, remembering. This is what occurs with the slave in the *Meno* discussed earlier. In myth this is the goddess *Mnemosyne*, mother of the Nine Muses. Since the intelligence of the soul corresponds with the universal intelligence, or with the 'rational motions of the heavens' as described in the *Timaeus*, it responds through kinship with the universal intelligence. Or, as Aristotle says in the opening of *Metaphysics*, 'All human beings by nature stretch themselves out towards knowing',[56] just as he says in the *Nicomachean Ethics* that the senses are oriented towards what is best or most beautiful: '. . . since every one of the senses is at work in relation to something perceptible, and is completely at work when it is in its best condition and directed towards the most beautiful of the things perceptible by that sense'.[57] This is the ground of reason, and why it is drawn towards truth, with the senses directed to the same end.

For Plato it is the same with *eros*, love. *Eros* is the ground of all the different kinds of love, and the root of all desire or yearning. It seeks to unite with divine Beauty, and this is the reason why the soul is moved whenever it is struck by anything beautiful. But in order for it to arrive where it desires to be, it must learn to distinguish between particular instances of beauty and Beauty itself. In the *Symposium* Diotima explains to *Socrates* how, upon seeing the beauty of one beautiful body, the soul must learn to see that it is the same beauty present in every beautiful body.[58] The same procedure must be followed with the virtues, and with institutions, and with laws, moving each time from the particular instances of beauty to the universal, until it finally arrives at Beauty itself which has no form, but which gives to all beautiful things their form.

It is clear that Plato understands that the truth of things, or goodness or beauty, can be known only through the ascent of the mind from the temporal realm to the eternal. The soul is by nature open to eternity. This is what defines the soul as dwelling between the mortal and the immortal. If it judges or measures only by the temporal or finite, then it will never arrive where it seeks to be and will have only relative or contingent knowledge, or at best what Plato calls 'right opinion'. In many ways, this principle may be demonstrated. For example, we only know the finite by an intuitive reference to the infinite. Yet the infinite is never visible. Or we recognise the imperfect because we have an intuitive knowledge of the perfect. Yet the perfect is never visible. Likewise with justice or goodness. But also there is a kind of 'poetic frenzy' that embraces the divine, where the mind goes out of itself in giving birth to beauty:

> If anyone comes to the gates of poetry and expects to become an adequate poet by acquiring expert knowledge of the subject without the Muses' madness, he will fail, and his self-controlled verses will be eclipsed by the poetry of men who have been driven out of their minds.[59]

By the constant reference to the eternal or the transcendent, Plato opens the door to a philosophical understanding of what was previously established through cosmic myths and the gods. Yet, as is clear in Book X of the *Laws*, piety towards the gods remains essential if the harmony of the *polis* is to be maintained. The proper life of the city, which brings harmony to the soul, is possible only so far as the civil laws derive from the harmonious order of the universe permeated by divine intelligence. It is this divine intelligence that manifests in number and proportion everywhere, and in the providential laws that nourish life and draw human intelligence, through awakening *eros*, towards the contemplation of truth.

NOTES

1. Plato, *Symposium*, 188a, in *The Dialogues of Plato*, translated by B. Jowett, Volume I (New York: Random House, 1937).

2. In Book 1 Chapter 3 of *Metaphysics* Aristotle gives a wide-ranging account of how the ancient thinkers conceived the origin and order of things, both of how the gods brought things into being and later how they conceived various elements, such as air, fire or water, as being the origin of things. In Chapters 4 and 5 he recounts how Hesiod, Parmenides, Empedocles and Anaxagoras conceived things coming into being in various ways. In Chapter 6 he gives an account of how Plato, following the lead of all these previous thinkers, sought to give more precise definitions of things, and that there was a distinction to be drawn between sensible changeable things and their forms and numbers which do not change, here drawing upon the Pythagoreans. For a penetrating study of the political and philosophical conditions of Athens that Plato confronts and seeks to remedy see Eric Voegelin, *Plato and Aristotle, Order and History*, Volume III, (Columbia: University of Missouri Press, 2000). For an excellent introduction to the emergence of philosophy in Greece see H. and H.A. Frankfort; John Wilson and Thorkild Jacobson, *Before Philosophy* (Harmondsworth, UK: Penguin Books, 1973). See also Alexander P.D. Mourelatos, *The Presocratics: A Collection of Critical Essays* (Princeton, NJ: Princeton University Press, 1974).

3. See Frankfort, *Before Philosophy*, Chapter VIII, pp. 248-62 for an account of how Greek poetry and myth were transformed into philosophy. See also Martin Heidegger *An Introduction to Metaphysics* (New Haven: Yale University Press, 1987), pp. 13-17 for an account of the original meaning of *physis* in Greek thought.

4. Hesiod, *Works and Days*, translated by Dorothea Wender (Harmondsworth, UK: Penguin Books, 1979), pp. 62-63, lines 108-130.

5. Plato, *The Laws of Plato*, translated by Thomas Pangle (New York: Basic Books, 1980), 713c.

6. Plato, *Laws*, 714a.

7. The word *polis* has no exact English equivalent and is often misleadingly translated as 'city' or 'city-state'. In Classical Greece it meant a self-ruling people, where every citizen took part in the political rule of the community, including the making of laws. In the opening of his *Politics* Aristotle describes the *polis* as the coming together of the family, the village and the agricultural community into a single 'natural' society, the kind of society that human nature is inherently inclined towards, embracing the common good through rational discourse, able to sustain itself without the need of external trade, and strong enough to defend itself. Its aim is to live virtuously and nobly. In the discussion of the founding of Magnesia in the *Laws* Plato likewise sees the *polis* as self-sufficient and even having a natural limit of 5,040 households.

8. Eric Voegelin, *Anamnesis* (Columbia: University of Missouri Press, 1990), p. 59. For an account of Socrates' and Plato's challenge to political corruption that prevailed in Athens, see Melissa Lane, *The Birth of Politics* (Princeton, NJ: Princeton University Press, 2014), Chapter 4.

9. Heraclitus, Fragment B 114, quoted from Max Hamburger, *The Awakening of Western Legal Thought*, translated by Bernard Mial (New York: Biblo and Tannen, 1969) p. 9, with Greek terms inserted as given by Voegelin in *Anamnesis*, p. 59.

10. Every society has its founding myths and associated symbols, even a modern secular society, such as symbols of justice, liberty or sovereignty. Such symbols are also part of the

social narrative or history through which a community identifies itself. A contemporary illustrative narrative is the materialist myth of progress, with its symbols of mastery over nature, an atheist narrative such as Plato critiques in *Laws* Book 10. In the dialogues, Plato often refers to or calls upon the presiding gods, even when speaking abstractly about justice, education or an art. Most dialogues begin with or imply a dedication to one of the gods or take place on a journey to a sacred shrine, as for example, in the *Laws* where the Athenian Stranger and his companions Kleinias and Megillus discourse on their way to the shrine of Zeus, or the opening of the *Republic* where Socrates goes with Glaucon to say a prayer to the goddess Bendis.

11. Aristotle, *Nicomachean Ethics* translated by Joe Sachs (Newbury MA: Focus Publishing, 2002), Book 10 1178b.

12. Plato, *Timaeus* 47d.

13. For a full study of Plato's understanding of the connection between cosmic order and virtue as exemplified in the *Timaeus* and *Critias* see T. K Johansen, *Plato's Natural Philosophy* (Cambridge: Cambridge University Press, 2004).

14. Plato, *Republic*, translated by Allan Bloom (New York: Basic Books, 1968), 508e.

15. See *Laws* Book 10.

16. For example, in *Laws* 684c while speaking of how people desire only laws that will please them, the Athenian Stranger says 'Such a provision is in opposition to the common notion that the lawgiver should make only such laws as the people like; but we say that he should rather be like a physician, prepared to effect a cure even at the cost of considerable suffering'. Also *Gorgias* 478-79 draws a comparison between the physician and the judge in administering justice, one curing the body, the other the soul. Again in the *Laws* 728b Plato sees judicial penalty as aiming to restore the soul rather than merely causing it to suffer the consequences of injustice. In *Laws* 650b we read: 'This, then – the knowledge of the natures and habits of souls – is one of the things that is of the greatest use for the art whose business it is to care for souls. And we assert (I think) that that art is politics. Or what?' In *Republic* 444c-e Plato draws a direct analogy between health and sickness and justice injustice, arguing that sickness and injustice are alike contrary to nature, while health and justice are according to nature.

17. Plato, *Laws*, 684c.

18. Plato, *Gorgias*, 478-79 in *The Dialogues of Plato*, translated by B. Jowett, Volume I (New York: Random House, 1937).

19. Plato, *Laws*, 728b.

20. Plato, *Laws*, 650b.

21. Plato, *Republic*, 444c-e.

22. Sara Brill, *Plato and the Limits of Human Life* (Bloomington, IN: Indiana University Press, 2013), pp. 168-69 and 173.

23. Plato, *Republic*, 337c; Plato, *Meno*, 71e in Jowett, *The Dialogues of Plato*.

24. For a detailed discussion of the break from myth and the transition to philosophy in Athens see Eric Voegelin *The World of the Polis, Order and History*, Volume II (Columbia: University of Missouri Press, 2000), Chapter 6, 'The Break with Myth'.

25. On the place of myth and symbol in any society Ricoeur observes 'The first function of the myths of evil is to embrace mankind as a whole in one ideal history. By means of a time that represents all times, 'man' is manifested as a concrete universal; Adam signifies man. 'In' Adam, says Saint Paul, we have all sinned. Thus experience escapes its singularity; it is transmuted in its own 'archetype'. Through the figure of the hero, the ancestor, the Titan, the first man, the demigod, experience is put on the track of existential

structures: one can now *say* man, existence, human being, because in the myth the human type is recapitulated, summed up. See Paul Ricoeur, *The Symbolism of Evil* (Boston: Beacon Press, 1969), p. 162. Also see Ford Russel, *Northrop Frye on Myth* (New York: Routledge, 1998), Chapter 14, 'Ricoeur and Fry on Myth'.

26. Plato, *Republic*, 514a–520a.

27. Plato, *Symposium* 201d-207a.

28. For a valuable historical and philosophical discussion of the relation of Parmenides and Heraclitus to Plato See Voegelin, *The World of the Polis*, Chapters 8 and 9, and for the Sophists see Chapter 11. See also Mourelatos *The Presocratics*.

29. For frenzy see, for example, Plato's *Phaedrus*.

30. For a valuable discussion of Plato's views on the Sophists see the 'Introduction' in Joe Sachs, *Socrates and the Sophists: Plato's Protagoras, Euthydemus, Hippias Major and Cratylus* (Indianapolis, IN: Focus Publishing, 2011).

31. See, for example, *Laws* 899d-902c on false or misguided opinions about the gods. For a detailed discussion of the special meaning of *doxa* in *Parmenides* and the shaping of Plato's philosophy of being see Voegelin *The World of the Polis* Chapter 8, especially p. 285ff.

32. In the *Politics* 1252A Aristotle argues that those who claim that skill in political rule is the same as household management or mastery of slaves, but on a larger scale, 'do not speak beautifully'. As Sachs remarks on this passage in note 39, 'the same assumption is made by the Eleatic Stranger at the beginning of Plato's *Statesman* (258E-258C)'. The classic example of the movement from appearance to the true is the allegory of the Cave in *Republic* 514a-520a.

33. Plato, *Apology* 21-22 in *Plato: Complete Works*, translated by G. M. A. Grubb (Indianapolis: Hackett Publishing, 1997).

34. For a valuable study of this tradition see S. K. Heninger, *Touches of Sweet Harmony* (Tacoma, WA: Angelico Press, 2013).

35. For a detailed study of Plato's understanding of how the soul is brought into harmony see Francesco Pelosi *Plato on Music, Soul and Body* (Cambridge: Cambridge University Press, 2010).

36. See Plato, *Meno* 84c-85d. The question the dialogue poses is whether virtue can be learned and if a distinction can be drawn between given or innate knowledge and acquired knowledge. That the slave who can solve a geometrical problem without prior study suggests, Socrates argues, that certain kinds of knowledge are already within the soul. Nevertheless, the dialogue comes to no conclusion as to whether virtue is innate or can be taught. The final suggestion is that virtue may be a gift from the gods. It is worth bearing in mind that for Plato it is the enquiry itself that matters, even if it leads to contradictory conclusions or no conclusion at all. Through the act of enquiring into the truth of things the soul already comes into a more harmonious relation with itself and with the greater order of things. It becomes temperate. The path any dialogue follows depends upon the condition of the souls of the interlocutors. This should make us particularly cautious about drawing fixed theories or doctrines from them.

37. Plato, *Republic*, 441e4-6.

38. For a wide-ranging study of Plato on friendship see Mary P. Nichols, *Socrates of Friendship and Community: Reflections on Plato's Symposium, Phaedrus, and Lysis* (Cambridge: Cambridge University Press, 2009).

39. Plato, *Gorgias*, 507d in *The Dialogues of Plato*, translated by B. Jowett,

40. Plato, *Laws*, 693c.

41. Thomas Aquinas, *Summa Theologica*, 4 Vols., (London, Second and Revised

Edition, Literally translated by Fathers of the English Dominican Province, London Burns, Oates and Washbourne, 1920) 1a. 110-19 and 1a2æ. 90-97. See also on justice, community and the common good 2a2 æ. 58.

42. Plato, *Republic,* 351d-352a in *Complete Works,* edited by John M. Cooper (Indianapolis/Cambridge: Hackett Publishing, 1997).

43. See 'The Emancipation of Thought from Myth' in Frankfort, *Before Philosophy.*

44. Quoted from Voegelin, *Order and History Volume II: The World of the Polis,* p. 380.

45. Aristotle, *Politics,* translated by Joe Sachs (Newbury MA: Focus Publishing, 2012), 1253a8.

46. For a penetrating and comprehensive study of the emergence of *physis* from earliest Greek thought and its meaning in Plato see Gerard Naddaf, *The Greek Concept of Nature* (New York: State University of New York Press, 2005). For an excellent history of the rise of law on Greece see Michael Gagarin, *Early Greek Law* (Berkeley: University of California Press, 1989).

47. In this regard it is worth noting that a unique characteristic of early Greek law was that it was governed by the citizens collectively and separately from political rule. Elsewhere laws were usually imposed upon citizens by a ruling class. See Gagarin, *Early Greek Law.* That by Plato's time it could be thought to be imposed by the strong for their own advantage, as maintained by some Sophists, shows how the understanding of *nomos* had changed since the time of Homer and Hesiod.

48. See Voegelin, *Order and History, Volume III,* p. 68ff for a discussion of how Plato saw the decline of Athenian politics and how this led him to enquire into the nature of justice and the order of the *polis.*

49. Quoted by Malcolm Schofield, *The Stoic Idea of the City* (Chicago: University of Chicago Press, 1999), p. 68 from Cicero's *De Legibus* 1 23. There is a clear resonance here with Heraclitus' fragment B 114 quoted earlier. Also see Katja Maria Vogt, *Law, Reason, and the Cosmic City: Political Philosophy in the Early Stoa* (Oxford: Oxford University Press, 2008) for an excellent study of the political and ethical philosophy of the Stoa.

50. Plato, *Laws,* 737c-738e.

51. Voegelin, *Order and History, Volume III,* p. 164. See also the discussion of the natural division of labour in the community in Brill, *Plato and the Limits of the Human,* Chapter 4, especially p. 98-99.

52. Plato, *Laws,* 738e.

53. Plato, *Laws of Plato, Laws* 697b.

54. See Gabriela Roxana Carone, *Plato's Cosmology and its Ethical Dimensions,* (Cambridge: Cambridge University Press, 2011), p. 164 ff. for a useful discussion of this argument and its importance in the Laws, 888e.

55. *Phaedrus,* 244-245, in Jowett, *The Dialogues of Plato.*

56. Aristotle *Metaphysics,* 980a translated by Joe Sachs (Santa Fe, New Mexico: Green Lion Press, 2002).

57. Aristotle *Metaphysics,* 1174b.

58. *Symposium,* 201d-207a.

59. Plato, *Phaedrus,* 245a, in *Plato: Complete Works* edited by John M. Cooper.

The Connected Cosmos:
Harmony, Cosmology and Theurgy in Neoplatonism

Crystal Addey

HARMONY (ἁρμονία) WAS A CENTRAL AND IMPORTANT CONCEPT within Neoplatonism, underpinning Neoplatonic metaphysics, cosmology, ethics and psychology, as well as approaches towards ritual and soteriology (the salvation of the soul). 'Neoplatonism' is a modern term used to describe an ancient philosophical movement which flourished in the Graeco-Roman world in the period known as late antiquity (second to seventh centuries CE); the philosopher Plotinus (205-70 CE) is considered to be the founder of this movement.[1] The term 'Neoplatonism' (and its cognates) is a modern term which originated in eighteenth-century German scholarship, coined pejoratively to denote the philosophers and movement inspired by Plotinus, as distinguished from Plato's own school and the so-called 'Middle Platonists'. These modern classifications are unsatisfactory, in part because they break the strong continuity linking 'Middle Platonists' and 'Neoplatonists'. Within this chapter, the terms 'Neoplatonism' and 'Neoplatonists' are used as purely chronological designations, without the pejorative and ideological assumptions which have often been attached to them.

Neoplatonist philosophers, including, most prominently, Plotinus, Porphyry (c.234-305 CE), Iamblichus (c.240-325 CE) and Proclus (c.410/12-485 CE), followed, endorsed and interpreted the philosophy of Plato, which included the philosophical doctrine that the human soul is immortal. Their notion of harmony, the 'joining together' of the parts of the universe and alignment of the parts with the whole, which formed a significant and central aspect of the Neoplatonic worldview, was influenced by the Pythagorean tradition as well as by Plato; Pythagoreans seemingly connected harmony with music and number as foundational principles in the cosmos.[2] Drawing on Plato's comments on harmony as set out in many of his works, particularly the *Timaeus* (30b-31a) and the Palinode speech (myth) of the *Phaedrus* (244a-257b), Neoplatonist philosophers from Plotinus onwards followed Plato in maintaining that the cosmos is a single living being, complete with soul, reason and intelligence, an interconnected entity containing all other entities, and a harmonious whole. They further developed the notion that the human being is a microcosm reflecting, expressing and containing (on a smaller scale) the macrocosm, the universe, which they considered to be both visible and non-visible or invisible.[3] They argued for a series of hypostases (principles and levels of reality or existence) arranged hierarchically – the One, considered to be the supreme source of all and identified with the Good (but also considered to be beyond and causally prior to 'being' or 'existence'), Intellect and Soul (as originally set out by Plotinus; later Neoplatonists elaborated further on this system).[4] Within Neoplatonism, 'the articulation of reality is the articulation of the relational patterns ordering being'.[5] Creation is seen as an eternal and spontaneous process – its operations relate to causes and their effects, rather than to any kind of production in time: from the One proceeds or 'emanates' Intellect, from Intellect, Soul, and from Soul – in its lowest phase, or nature – the visible or perceptible ('sensible') universe. Each hypostasis is undiminished by the giving of its power and a trace of each is immanent in every subsequent level of creation.[6] This cosmological and metaphysical structure reveals a worldview that is multilayered and based on a web of connections: all things (posterior to the One) are 'one' – as a unified, individual entity – and 'many' since their causes are inherent within them, although Intellect is described as 'one many' since it was considered to contain multiple noetic 'Forms' in one unity. Neoplatonist philosophers often use the language of 'participation' to describe the way in which entities relate to their causes and to the prior hypostases (or levels of reality). Proclus uses a specific terminology of abiding, proceeding and reverting, stating that every effect remains in its cause, proceeds from it and reverts upon it.[7] He also expresses the interconnected nature of everything by stating that all things are in all things, but in each thing according to its proper nature.[8]

Later Neoplatonist philosophers, particularly Iamblichus and Proclus, practised and endorsed theurgy, which literally means 'divine work': theurgy refers to a lifelong endeavour incorporating a set of ritual practices alongside the development of ethical and intellectual capacities which aimed to use symbols to reawaken the soul's pre-ontological, causal connection with the gods.[9] The goal of theurgy was the cumulative and progressive contact, assimilation and union with the divine and the consequent divinisation of the theurgist; that is to say, the ascent of the human soul to the divine, intelligible realm and the consequent manifestation of the divine in embodied, human life.[10] Theurgic ritual practice was based on and used traditional polytheistic religious practices, including divination, sacrifice, invocations and prayer. In his work *On the Mysteries* (*De mysteriis*), which sets out the philosophical and theological basis for theurgy, Iamblichus links theurgy with traditional Greek, Egyptian and Chaldaean (or Assyrian) religious practices.[11] Both Iamblichus and Proclus argued that theurgic ritual operates through or by means of the harmonious connections inherent within the visible and, even more importantly and crucially, the non-visible cosmos. Iamblichus claims that the human soul is inverted or 'upside down' and must restore itself to a harmonious whole through the lifelong endeavour of philosophical and theurgic praxis. Since harmony played such a central role within Neoplatonism, the scope of this topic is vast. Therefore, this chapter will necessarily be extremely selective and partial, providing: (1) a brief and selective summary of the role of harmony in Neoplatonic cosmology and (2) an exploration of the importance of harmony for the practice of theurgic ritual among the late Neoplatonists, especially Iamblichus and Proclus.

HARMONY, COSMOLOGY AND METAPHYSICS

Before examining Neoplatonic cosmology, it is important to note that the ancient Greeks had long conceived of the universe as inherently ordered; the Greek term for the 'universe', 'cosmos' (κόσμος) meant 'order' and 'good order', as well as referring to the world or universe itself, because of the latter's perfect arrangement. Following Plato's *Timaeus*, Neoplatonist philosophers consistently endorsed the idea that the universe, or cosmos, is one single, living being, that is to say, a harmonious, interconnected whole, most succinctly expressed by Plotinus, 'So the universe is one, single harmony' (οὕτω γὰρ ἓν καὶ μία ἁρμονία).[12] Iamblichus also maintains that the universe is one, single living being.[13] Plotinus quotes directly from Plato's *Timaeus* to support this view of the cosmos:

> First of all, we must posit that this All is a 'single living being which encompasses all the living beings that are within it': it has one soul which extends to all its parts, in so far as each individual thing is part of it; and each thing in the perceptible All is part of it, and completely a part of it as regards its body.[14]

Here, Plotinus follows Plato in seeing the universe (the 'All') as a single, living being, with a soul, the World Soul.[15] According to Plotinus, 'The universe lies in soul which bears it up, and nothing is without a share of soul'.[16] Human souls are not parts or products of the World Soul; rather, the World Soul, human souls and all other souls are part of the hypostasis (the principle of reality) 'Soul'.[17] The World Soul, or 'Soul of the All' has created the universe (this refers to creation in a causal rather than a temporal sense since Plotinus and later Neoplatonist philosophers held that the universe is everlasting) while individual souls each direct a part of the universe.[18] However, Plotinus maintains that the World Soul and individual souls are linked by a certain kind of 'sympathy' (συμπαθεία) or 'community of feeling', which will be discussed further below.[19] Both are related to the hypostasis or principle 'Soul' which has the rational formative principles (*logoi*) of the gods and of everything in existence – according to Plotinus, this is why the universe has everything.[20] Plotinus often uses the image of a pantomime dancer to illustrate the organic unity and harmony of the universe.[21] Although in English 'pantomime' usually refers to a theatrical show performed by a whole troupe of performers, late antique pantomime dancing refers to a dramatic performance by a solitary dancer who acted out a story or myth with specific gestures and movements without speaking or singing, although accompanied by music and choral song.[22] The Greek term *pantomimos* meant 'imitator of everything'; thus, as Anne Sheppard points out, 'For Plotinus the whole cosmos functions as a vast living organism in which all the parts ultimately cohere and serve the whole; the pantomime dancer too is a being of this kind, at a microcosmic level'.[23]

Moreover, the concept of harmony is crucial to Neoplatonic metaphysics, as well as cosmology. 'Metaphysics' refers to the branch of philosophy that deals with the first principles of things, including concepts such as being, knowing, time and space. The universe was considered to be caused by and based on the 'intelligible' (non-visible) realm or cosmos, which was considered to comprise the causes of everything within the perceptible universe. According to Plotinus, not only is the visible universe a harmonious whole but it exists in harmony with invisible intelligible realities which are prior (in a causal sense) to the visible universe; he states that the nature of the All made all things in imitation of the intelligible realities (the prior hypostases).[24] Thus, the harmony of the intelligible (invisible) world is linked closely to the harmony of the sensible world, as Plotinus makes clear:

> For how could there be a musician who sees the harmony in the intelligible world and will not be stirred when he hears the harmony in sensible sounds? Or how could there be anyone skilled in geometry and numbers who will not be pleased when he sees right relation, proportion and order with his eyes?[25]

Plotinus maintains that if one can 'see' or understand the harmony, order and proportion of the intelligible (i.e., non-visible) world in any way, this implies that one should be able to appreciate the visible manifestations of harmony, right relation, proportion and order in the sensible world, because the latter (the sensible realm) reflects and is intimately related to the intelligible world. Iamblichus defines the intelligible cosmos as real existence, the intelligible paradigms of the cosmos and as causes which pre-exist all things in nature; he further claims that the Demiurge god (identified with the Demiurge or 'craftsman' god of Plato's *Timaeus* who creates the sensible universe) gathers all of the intelligible paradigms into one and holds them within himself.[26] In doing so, he draws on Plato's *Timaeus*, where it is argued that the One Living Creature contains all of the intelligible living creatures and the cosmos resembles the One Living Creature because it contains all visible living creatures within itself.[27]

Iamblichus also describes the way in which everything in the universe 'participates' in the gods; the divine light or illumination of the gods permeates everything within the universe because it causes all things:

> The fact is that divinity illuminates everything from without, even as the sun lights everything from without with its rays. Even as the sunlight, then envelops what it illuminates, so also does the power of the gods embrace from outside that which participates in it ... so the light of the gods illuminates its subject transcendently, and is fixed steadfastly in itself even as it proceeds throughout the totality of existence.[28]

Iamblichus uses the terminology of divine light and illumination to describe how the gods permeate everything within the cosmos with no diminution of themselves, using the light of the Sun to explain this activity of the gods; the latter permeate everything and so are immanent throughout the universe while simultaneously remaining transcendent. He also maintains that it is in imitation of the light of the gods that the cosmos is harmonious and united:

> It is, indeed, in imitation of it [i.e., the indivisible light of the gods] that the whole heaven and cosmos performs its circular revolution, is united with itself, and leads the elements round in their cyclic dance, holds together all things as they rest within each other or are borne towards each other, defines by equal measures even the most far-flung objects, causes lasts to be joined to firsts, as for example earth to heaven, and produces a single continuity and harmony of all with all.[29]

Here, Iamblichus attests to the key Neoplatonic concept that there is no gap, break or vacuum in the order of reality: the gods and intelligible realities permeate everything in the universe and produce a single continuity and harmony between all, including causing Earth to be joined to heaven.[30] It is in this sense that Iamblichus then calls the perceptible universe 'the visible image (or statue) of the gods' (Τὸ δὴ τῶν θεῶν ἐμφανές τις ἄγαλμα).[31] Furthermore, the circular revolution of the heaven and cosmos alludes to several ideas: first, the idea that the planets and fixed stars, which were considered by Iamblichus and Proclus to be the visible bodies of the gods, move in a circular or spherical motion in harmony with one another, an idea related to the theory of the 'Music of the Spheres' (which will be examined further below).[32] The idea that the planets are gods was widespread in ancient Greek culture; in their accounts of the divinity of the planets, later Platonist philosophers were influenced particularly by Plato's *Timaeus*, where they were already described as 'visible gods' (τὰ περὶ θεῶν ὁρατῶν ... φύσεως).[33] It is also interesting that Iamblichus describes the heavens and cosmos leading the elements in a cyclic dance. Plotinus compares the circular movement of the heavenly bodies to a choral dance, an extremely common form of dance in antiquity performed by groups dancing and often singing rhythmically together in a circular space, accompanied by musicians.[34] One form of the choral dance involved the dancers singing and dancing around their lyre player, a point that may be significant in relation to Apollo's role as the god of harmony (also discussed below), since he was also considered to be the god of music and the lyre in particular.[35] Plotinus appears to have this image in mind (dancing around a lyre-player) since he discusses the heavenly bodies moving around the same centre and given that he describes the direction of the souls of the heavenly bodies towards one object as 'like strings on a lyre plucked harmoniously they sing a song which is naturally in tune' (ὥσπερ χορδαὶ ἐν λύρᾳ συμπαθῶς κινηθεῖσαι μέλος ἂν ᾄσειαν ἐν φυσικῇ τινι ἁρμονίᾳ).[36] As Stephen Clark notes, further support for the idea that we are to think of a choral dance around a lyre-player sitting in the centre, in place of Apollo, comes from Plato's *Republic* which describes Apollo, as the god of the oracle at Delphi and thus, in his role as the god of oracles and divination, as sitting at the centre of the earth on his sacred stone (referring to the omphalos stone at Delphi which the Greeks considered to mark the navel and centre of the earth) and guiding human beings through his advice given in the form of oracles.[37]

Furthermore, the circular revolution discussed by Iamblichus (in the passage above) alludes to the idea that the motion of the universe is carried around uniformly in the same place in imitation of the universal Soul which envelops heaven; the universal Soul imitates Intellect in this sense – by 'participating' in Intellect, the soul of the universe ascends to the intelligible.[38] Iamblichus relates that the Demiurge constructed the universe in the form of the sphere, to be an image of the Soul's self-motion, drawing on Plato's *Timaeus* which had asserted that the universe is constructed in the form of a sphere and revolves in a circular motion uniformly in the same spot.[39] Plotinus also discusses the way in which heaven moves in a circular fashion because it imitates Intellect.[40] Both philosophers draw on Plato's suggestion that human beings must correct the orbits in the head which were corrupted at birth and so bring ourselves into line with the heavens.[41] This statement points towards the necessity of human beings aligning themselves with the harmony of the heavens.

HARMONY AND THE SOUL

Although the universe is harmonious, the human soul is generally seen as having forgotten or lost its essential harmonious state (through its disorientation in its descent into a mortal body), which it must restore by assimilating itself to the harmonious reality of the universe.[42] Late Platonist philosophers draw on the account of the formation of the human soul set out in Plato's *Timaeus*, where the Demiurge makes human souls from the residue of the mixture used to make the World Soul (comprised of being, sameness and difference mixed by harmonic ratios divided into two parts which he then joined together and bent into a circular form comprised of two circles), although he mixes the mixture for human souls in a less uniform and more variable manner; the Demiurge then hands the mixture to the cosmic gods to make the human body and complete the soul; when the latter bind the human soul to the mortal body it disturbs the revolutions of the soul further and twists the harmonic ratios binding the mixture.[43] In Plato's *Timaeus*, we also see the idea that music was given to the universe for the sake of harmony; furthermore, the Muses give harmony to those who are inspired by creative pursuits in order to restore the soul to a harmonious order:

> Music too, in so far as it uses audible sound, was bestowed for the sake of harmony. And harmony, which has motions akin to the revolutions of the Soul within us, was given by the Muses to him who makes intelligent use of the Muses, not as an aid to irrational pleasure, as is now supposed, but as an auxiliary to the inner revolution of the Soul, when it has lost its harmony, to assist in restoring it to order and concord with itself.[44]

The Muses were seen in antiquity as semi-divine beings who cause and engender inspiration in human beings and were especially connected with music, poetry and dance. They were related to Apollo, who was considered to be the leader of the Muses. In Plato's *Republic*, musical education is seen as particularly important because rhythm and harmony find their way to the innermost depths of the soul and take the strongest hold upon it, leading one to detect beauty in other phenomena.[45] Music and beauty are characterised as leading to the philosophical path, based on love, set out in Plato's *Phaedrus* and *Symposium*.

Elaborating on Plato's point about the need for the soul to restore its inherent order, Proclus connects the perfection of the human soul with the need for the human soul to assimilate and align itself with the harmony of the universe:

> Furthermore, the whole celestial order and its motion demonstrates the harmonic work of the god [i.e., Apollo]. This is why individual souls as well, once they have removed the disharmony that results from generation, are perfected in no other way than by harmonic assimilation to the universe. For then they achieve the best life that is offered by the god [i.e., Apollo].[46]

Proclus links the harmony of the universe with the god Apollo, following Plato's *Cratylus* which states that Apollo directs the harmony of the heavens and the harmony inherent in music.[47] Proclus notes that human souls have to remove the disharmony which results from 'generation', which refers to the whole process of birth, life and death, 'coming-to-be' and 'passing away'; the descent of the soul into generation was thought to leave the soul disorientated and out of alignment with its true nature and purpose. After this purification from disharmony, Proclus claims that human souls are perfected by aligning and assimilating themselves with the harmony of the cosmos. Proclus explicitly relates Apollo's role as the cause of all harmony, both visible and invisible, with his role as the god of music; furthermore, according to Proclus, Apollo, together with Mnemosyne (the goddess of memory) and Zeus, engenders the Muses and co-operates in organising the perceptible universe. The musical functions of Apollo are intimately linked with Apollo's 'harmonising' of the whole universe into a single unity:

> His [i.e., Apollo's] musical activity is more prevalent at the leading and principal order. For it is this god who harmonizes even the whole cosmos into a single unity, establishing the chorus of Muses around himself, 'Taking pride in the harmony of light', as one of the theurgists says (*Or. Chald.* 71).[48]

Apollo's traditional role as *Mousêgetês*, the leader of the Muses, is related by Proclus to his role as the god who establishes the harmony of the universe. We are reminded of Plotinus' discussion of the movement of the heaven around a stable centre, which he compares to a circular choral dance with a lyre-player at the centre (discussed above), since Apollo and the Muses were seen as the original, 'divine' chorus who engender all *mousikê* (poetry, drama, dance and, according to Plato and the Neoplatonists, philosophy as well) through the giving of divine inspiration to mortals.

HARMONY, THEURGY AND COSMOLOGY

Proclus' linking of Apollo with the harmony underlying the universe is also relevant to the later Neoplatonist approval and use of theurgy, a kind of ritual which included as a central element the use of divination; Apollo was traditionally the god of divination in the ancient Greek world. In this sense, harmony is particularly important for understanding why late Platonist philosophers (such as Iamblichus and Proclus) used theurgy since they considered that the harmony of the universe and the sympathy of its parts for one another are both causal explanatory factors that partially explain the workings of theurgy. As we have seen, Neoplatonist cosmology and metaphysics is rooted in the notion that the universe is a harmonious, multi-layered whole and an interconnected entity containing all other entities, which is partially explained by the idea of cosmic sympathy (συμπάθεια), according to which all things within the universe are connected by a certain affinity, sympathy and likeness.[49] Thus, sympathy was considered to work on a 'horizontal' level – connecting things in the sensible universe which are spatially distant but similar in some way with one another. For theurgists, sympathy was also held to work on a 'vertical' level – connecting things in the sensible universe with the gods, their divine causes. For example, Proclus explains that theurgy aims to link the theurgic practitioner to the gods on the basis of this sympathy: 'For the powers of the gods descend from above all the way down to the lowest realm, being appropriately manifested at each level, all of which, indeed, theurgy undertakes to link to the gods by way of a sympathetic relationship'.[50] Since everything in the universe was considered to be permeated by divine light and thus connected with the gods, theurgist practitioners held that every natural entity and object in the cosmos was a 'symbol' (σύμβολον) or 'token' (συνθήμα) of the god or goddess to which it was intrinsically connected and thus harmoniously aligned with. The theurgic practitioner seeks to develop right relationship and alignment with the gods by using symbols in an efficacious manner. It is crucial to understand that theurgic symbols were considered to have an ontological link with the deity from which they were derived. Within a theurgic context, a symbol could be a physical object such as a plant, gemstone, herb or type of incense – the theurgist assembled the specific symbols adapted to and harmoniously aligned with each deity as a way of creating a pure and integrated receptacle for the manifestation of that deity, and for the consequent contact with and assimilation to that deity.[51] For example, Proclus tells us that the lion, cockerel, sunflower (heliotrope) and the lotus are all symbols of the Sun god.[52] Significantly, for theurgists, the gods work *through* nature and *through* human souls.[53] Iamblichus explains that the theurgist used the symbols found in nature within their ritual practices and by virtue of these symbols was able to contact the gods and to direct him or herself harmoniously in accordance with the gods' disposition, thereby assuming the sacred role of the gods [54]

Iamblichus claims that an even greater cause than sympathy (συμπάθεια) underlies theurgy: theurgic practices have the potential to connect the human soul with the gods by virtue of divine love or friendship (θεία φιλία), which causes cosmic sympathy to arise and from which sympathy is derived.[55] A fragment from the *Chaldean Oracles* (a mystical collection of oracles in hexameter verse which set out ethical and ritual instructions for the ascent of the soul in relation to a complex cosmological and metaphysical background and which were used by theurgists and attained the status of a sacred text within Neoplatonism) also refers to the bond of love which the Paternal Intellect sowed into all things in the cosmos, so that the 'All' might continue to love for an infinite time and that things woven by the intellectual light of the Intellect might not collapse; it also adds that the elements of the cosmos remain on course because of this love.[56] The role of divine love and friendship in theurgy is critical since it linked theurgic practice to the divine providence and goodness of the gods and the One. Furthermore, the role of divine love within theurgy is used by Iamblichus to argue against the idea that theurgy is just a manipulation of horizontal cosmic sympathy and thus an automatic or 'mechanical' process unrelated to the ethical or intellectual condition of the theurgic practitioner. On the contrary, theurgy was considered to involve the practitioner assimilating herself or himself to the harmonious nature of the cosmos and thus to the divine harmony from which the cosmos derives its unity.[57]

For Iamblichus, harmony with divine, oracular power is a key feature of divine possession or inspiration, the foundation of oracles and other forms of inspired divination.[58] Harmony is a particularly important feature in his account of the operations of the rituals associated with the cults of the Korybantes, Sabazios (sometimes identified with Dionysos) and the Great Mother, Cybele.[59] A significant feature of these cults was divine possession or inspiration, whereby the human initiate was considered to become possessed by the relevant god or goddess; music, including listening to pipes, cymbals or tambourines was used within these cults to engender or trigger the possessed or inspired state of consciousness.[60] In explaining the phenomenon of divine possession within these cults, Iamblichus dismisses as irrelevant the possible explanations raised by his fellow-philosopher Porphyry, such as the fact that the music displaces the temperaments or dispositions of the body and that irregular tunes are appropriate for ecstatic trance; according to Iamblichus, these features of the cult are physical and human, accomplishments of human skill.[61] Therefore, Iamblichus argues that they do not explain or account for the ultimate origins, or causes, of divine possession, which must be divine:

> What we would rather say, then, is this: that those things such as sounds and tunes are properly consecrated to each of the gods, and kinship is probably assigned to them in accord with their proper orders and powers, the motions in the universe itself and the harmonious sounds rushing from its motions. It is then, in virtue of such connections of the tunes with the gods that their presence occurs (for nothing intervenes to stop them) so that whatever has a fortuitous likeness with them, immediately participates in them, and a total possession and filling with superior being and power takes place at once. It is not that the body and soul interact with one another or with the tones, but since the inspiration of the gods is not separated from the divine harmony, having been allied with it from the beginning, it is shared by it in suitable measures.[62]

Iamblichus argues that the sounds and tunes involved in the music used in these rituals already have an intrinsic and inherent connection with the gods, which means that they also have an inherent connection with the motions of the cosmos and with what Iamblichus refers to as 'the harmonious sounds rushing from these motions', a reference to the 'harmony' or 'music of the spheres'. Iamblichus' use of the term ῥοῖζος here points in this direction since this was a Chaldean and Pythagorean term for the sound caused by the planetary revolutions, the so-called 'Harmony' or 'Music of the Spheres'.[63] This doctrine, which unites music and astronomy, is based on the idea of cosmic sounds made by the circular movements of the planets, is attested by Aristotle and indirectly by

Plato as being Pythagorean.[64] The Pythagorean 'music of the spheres' is based on Pythagorean musical theory, with its coherence of number and sound. Harmonic intervals correspond to harmonic relationships of distance and velocity, and since a musical tone implies a uniform motion, a Pythagorean system of astronomy, in which the seven planets circle about the earth in uniform movements, at various distances from one another, can be inferred.[65] In Plato's *Laws*, it is stated that of all types of movement, a uniform, circular movement is most closely related to mind or intellect (*nous*) – this movement is perfect and the heavenly bodies move in conformity with it; therefore, soul, as the principle of self-movement, reveals itself in celestial movements; moreover, since these movements are perfect and regular they show that the soul of the universe is intelligent and good.[66] In his work focusing on the Pythagorean way of life, Iamblichus states that Pythagoras purified his followers and made their sleep prophetic through his musical renditions of the harmony of the spheres.[67] According to Iamblichus, Pythagoras' own abilities included hearing the 'harmony of the spheres' directly because of his status as a 'divine' wise sage who had developed his intellectual powers and was close to the gods:

> But employing some ineffable and abstruse divine power, he extended his hearing and fixed his intellect in the heavenly harmonious sounds of the cosmos. He alone could hear and understand, so he indicated, the universal harmony and concord of the spheres, and the stars moving through them, which sound a tune fuller and more intense than any mortal ones. (This harmony) is caused by a movement and most graceful revolution, very beautiful in its simultaneous variety...[68]

Iamblichus says that Pythagoras held that the cosmic sounds, the 'music of the spheres', were audible and comprehensible, but only to himself because of his divine status, the latter of which seems to have been related to Apollo, as the god of harmony and music, given that Iamblichus interprets the story that Pythagoras was son of Apollo as meaning that Pythagoras' soul was sent down to humans under Apollo's leadership and that Pythagoras was therefore closely connected with Apollo.[69] Returning to Iamblichus' explanation of the use of music in possession rituals, it is precisely the divine harmony which permeates the cosmos and which is exhibited on the cosmic level through the harmony of the spheres which underlies divine possession. Iamblichus denies that the body and soul interact with one another or with the music; rather, divine inspiration is aligned with and connected to the divine harmony that pervades every part of the cosmos. However, Iamblichus maintains that this process is not primarily or ultimately dependent on the soul itself:

> But one should not even claim this: that the soul primarily consists of harmony and rhythm; for in that case divine possession would belong to the soul alone. It is better, then, to bring our discourse back to this assertion: before it gave itself to the body, the soul heard the divine harmony. And accordingly even when it entered the body, such tunes as it hears which especially preserve the divine trace of harmony, to these it clings fondly and is reminded by them of the divine harmony; it is also borne along with and closely allied to this harmony, and shares as much as can be shared of it.[70]

Iamblichus does not wish to deny that the soul is somehow connected with the divine harmony, but it is primarily the soul's *memory* of the divine harmony, which it heard before entering the body, that facilitates the experience of possession in these rituals. This is a clear reference to the soul's recollection of the good and the divine realities which Neoplatonist philosophers derived from Plato's *Phaedo* and the myth in the *Phaedrus*.[71] In the latter work, Socrates, in his Palinode speech, recounts a myth about the procession of Zeus and the other gods around the revolution of the heavens and the 'outer' regions of the heaven where the divine realities exist. He recounts how human souls try to follow the procession of the gods through the heavens: the soul who follows the divine can raise the head of the charioteer into the outer region, while less perfect souls see some of the divine realities but not others (because they are disturbed by the unruliness of the horses, representing the faculties of spirit and desire in the soul) and some cannot reach the upper region at all.[72] The human soul needs to recollect those realities which it once 'saw' when journeying with the gods.[73] Proclus reports that both Plotinus and Iamblichus (the latter in his *Commentary on the Phaedrus*) interpreted the 'heaven' towards which Zeus leads the way and all the gods follow is an intelligible entity; that is to say, the 'heaven' referred to in the myth describes the intelligible, invisible realm.[74]

CONCLUSION

In conclusion, the concept of harmony played a central role in Neoplatonic cosmology and metaphysics, as well as in theurgy and soteriology (the salvation and perfection of the human soul). Following Plato's *Timaeus*, these philosophers saw the cosmos as one single living being and harmonious whole, underpinned by order, harmonic ratios, symmetry and proportion. The cosmos is a manifestation or copy of the intelligible, invisible realities and so there is a harmony between the perceptible, visible universe and the invisible 'intelligible' universe; the visible universe is caused by and derived from the intelligible realities. Harmony was intimately linked with music and dance in antiquity; this connection is reflected in the illustrative examples used by Plotinus to demonstrate the organic unity of the cosmos and the harmonious movements of the heaven and the planets in imitation of Intellect. Following Plato's *Cratylus*, late Platonist philosophers considered Apollo to be the god of harmony since within ancient Greek culture one of his traditional roles was as *Mousêgetês*, the leader of the Muses, the latter of whom were considered to inspire and engender poetry, music and dance. Apollo was also traditionally the god of divination, especially oracles (an inspired form of divination) and, consequently, harmony was also an important factor which was considered by Iamblichus and Proclus to enable and facilitate the efficacy of theurgic ritual, which included divination as one of its central ritual practices: they explained the workings of theurgy partially through the concept of cosmic sympathy and partially through divine love or friendship, from which sympathy derived, in their view. Both cosmic sympathy and divine love are intimately connected with the inherent harmony of the cosmos in this worldview. Crucially, theurgy was held to work *through* human souls and *through* nature: the tunes and sounds of music bear a trace of divine harmony, while the gods could be accessed through symbols found in nature. Therefore, the theurgist had to align her or himself with the harmonic nature of the universe in order to attain contact, assimilation and union with the gods. According to Iamblichus, music made by humans contains traces of the original music caused by the harmonious movements of the planets and fixed stars, and of divine harmony itself. Within Neoplatonism, the concept of harmony relates to the providential, relational and connected nature of the cosmos. According to theurgic iterations of this worldview, traces of the divine – and divine 'illumination' itself – surround us everywhere in nature, in the sensible cosmos, and the task of the human being is to establish right relationship, alignment and assimilation to the harmony of the world and the gods. Late Platonist philosophy – and especially theurgy – is based on the idea that we are all connected – to each other, to every part of the cosmos and to the gods, just as the cosmos itself is connected – so that if we develop our full potential, we can remember the divine harmony and 'hear' the music of the spheres, 'dancing' with the cosmos itself and thereby with the gods.

NOTES

1. See Dominic J. O'Meara, *Platonopolis. Platonic Political Philosophy in Late Antiquity* (Oxford: Clarendon Press, 2003), p.3, n.1; Crystal Addey, *Divination and Theurgy in Neoplatonism. Oracles of the Gods* (Farnham: Ashgate, 2014), p. 1, n.2.

2. See Walter Burkert, *Lore and Science in Ancient Pythagoreanism* (Cambridge, MA: Harvard University Press, 1972) for a useful introduction to Pythagoreanism, and Dominic O'Meara, *Pythagoras Revived: Mathematics and Philosophy in Late Antiquity* (Oxford: Clarendon Press, 1989) on the reception of Pythagoreanism in Neoplatonism.

3. See Plotinus, *Ennead* IV.3.10.10-13. All translations of this work are those of A.H. Armstrong, *Plotinus: Enneads*, Loeb Classical Library Vols. I-VII (Cambridge, MA: Harvard University Press, 1966-1988), with my own modifications. Cf. also Iamblichus, *In Philebum*, Fragment 6 (= Damascius, *In Philebum* 227), ed. and trans. John M. Dillon, *Iamblichi Chalcidensis. In Platonis Dialogos Commentariorum Fragmenta* (1973; repr. Westbury, Wiltshire: Prometheus Trust, 2009).

4. Cf. for example, Plotinus, *Ennead* IV.3.10.13-14; V.1.

5. Dominic O'Meara, 'The Hierarchical Ordering of Reality in Plotinus', in Lloyd Gerson, ed., *The Cambridge Companion to Plotinus* (Cambridge: Cambridge University Press, 1996), pp. 66-81; Pauliina Remes, *Neoplatonism* (Stocksfield: Acumen, 2008), p.42.

6. For useful introductions to Neoplatonic metaphysics and 'first principles', see John Gregory, *The Neoplatonists: A Reader* (1991; repr. London and New York: Routledge 1999), pp. 12-15; Remes, *Neoplatonism*, pp. 42-59.

7. Proclus, *Elements of Theology*, trans. E.R. Dodds (Oxford: Oxford University Press, 1963), Proposition 35.

8. Proclus, *Elements of Theology*, Proposition 103.

9. Addey, *Divination and Theurgy*, p. 25.

10. Addey, *Divination and Theurgy*, p. 25.

11. Iamblichus, *On the Mysteries (De mysteriis)*, ed. and trans. E.C. Clarke, J.M. Dillon and J.P. Hershbell (Atlanta: Society of Biblical Literature, 2003). All translations of this work are taken from this edition.

12. Plotinus, *Ennead* II.3.12.32; IV.4.32.4-7 citing Plato, *Timaeus* 30d3-31a1. Cf. also *Timaeus* 33a-b, trans. R.G. Bury, *Plato. Timaeus. Critias. Cleitophon. Menexenus. Epistles* (1929; repr. Cambridge, MA: Harvard University Press, 2014).

13. Iamblichus, *De mysteriis* 4.9 (192.11-13).

14. Plotinus, *Ennead* IV.4.32.4-7 citing Plato, *Timaeus* 30d3-31a1: Πρῶτον τοίνυν θετέον ζῷον ἓν πάντα τὰ ζῷα τὰ ἐντὸς αὑτοῦ περιέχον τόδε τὸ πᾶν εἶναι, ψυχὴν μίαν ἔχον εἰς πάντα αὑτοῦ μέρη, καθόσον ἐστὶν ἕκαστον αὑτοῦ μέρος· μέρος δὲ ἕκαστόν ἐστι τὸ ἐν τῷ παντὶ αἰσθητῷ, κατὰ μὲν τὸ σῶμα καὶ πάντη ...

15. Cf. also Plotinus, *Ennead* IV.3.7.7-8; IV.3.9.46-48.

16. Plotinus, *Ennead* IV.3.9.36-37. Cf. Plato, *Laws* 896d-897b, for the idea that 'soul' moves and controls heaven.

17. Plotinus, *Ennead* IV.3.3.

18. Plotinus, *Ennead* IV.3.6.7-8, 20-25. Cf. also IV.3.9.16-18.

19. Plotinus, *Ennead* IV.3.8.1-4.

20. Plotinus, *Ennead* IV.3.10.10-13, 38-43.

21. Plotinus, *Ennead*, III.2.16.23-27; III.2.17.8-11; IV.4.33. See Anne Sheppard, 'Drama, Dance and Divine Providence in Plotinus, Ennead 3.2 (47).15-18', in E. Volonaki and V. Konstantinopoulos, eds., *ΠΛΑΤΩΝ, ΤΟΜΟΣ 60, 2015: ΠΕΡΙΟΔΙΚΟ ΤΗΣ ΕΤΑΙΡΕΙΑΣ ΤΩΝ*

ΕΛΛΗΝΩΝ ΦΙΛΟΛΟΓΩΝ (Athens: Papazisis Publishers, 2016), pp.287-95; 'Neoplatonists and Pantomime Dancers', in R.L. Cardullo and F. Coiglione, eds., *Reason and No-Reason from Ancient Philosophy to Neurosciences: Old Parameters, New Perspectives* (Sankt Augustin: Academia Verlag, 2017), pp. 65-78, for detailed examinations of the ways in which Plotinus uses imagery from dancing and drama in order to convey his philosophy.

22.   Sheppard, 'Neoplatonists and Pantomime Dancers', p. 66.

23.   Sheppard, 'Neoplatonists and Pantomime Dancers', p. 68.

24.   Plotinus, *Ennead* IV.3.11.8-12.

25.   Plotinus, *Ennead* II.9.16.39-44: Τίς γὰρ ἂν μουσικὸς ἀνὴρ εἴη, ὃς τὴν ἐν νοητῷ ἁρμονίαν ἰδὼν οὐ κινήσεται τῆς ἐν φθόγγοις αἰσθητοῖς ἀκούων; Ἢ τίς γεωμετρίας καὶ ἀριθμῶν ἔμπειρος, ὃς τὸ σύμμετρον καὶ ἀνάλογον καὶ τεταγμένον ἰδὼν δι᾽ ὀμμάτων οὐχ ἡσθήσεται; trs. Armstrong with my own modifications (substituting 'harmony' for 'melody'). Cf. Stephen R.L Clark, *Plotinus: Myth, Metaphor and Philosophical Practice* (Chicago and London: University of Chicago Press, 2016), p.1 08 n. 17, who notes that Armstrong's translation of *harmonia* as 'melody' here is misleading.

26.   Iamblichus, *In Timaeum* II, Fragment 34.

27.   Plato, *Timaeus* 30c-d; 37d.

28.   Iamblichus, *De mysteriis* 1.9 (30.13-15; 31.3-5): ... πάντα ἔξωθεν ἐπιλάμπει, καθάπερ ὁ ἥλιος ἔξωθεν φωτίζει πάντα ταῖς ἀκτῖσιν. Ὥσπερ οὖν τὸ φῶς περιέχει τὰ φωτιζόμενα, οὕτωσὶ καὶ τῶν θεῶν ἡ δύναμις τὰ μεταλαμβάνοντα αὐτῆς ἔξωθεν περιείληφεν ... οὕτω καὶ τῶν θεῶν τὸ φῶς ἐλλάμπει χωριστῶς ἐν αὐτῷ τε μονίμως ἱδρυμένον προχωρεῖ διὰ τῶν ὄντων ὅλων.

29.   Iamblichus, *De mysteriis* I.9 (31.14-32.6): ὅπερ δὴ καὶ ὁ σύμπας μιμούμενος οὐρανὸς καὶ κόσμος τὴν ἐγκύκλιον περιφορὰν περιπολεῖ, συνήνωταί τε πρὸς ἑαυτόν, καὶ τὰ στοιχεῖα κατὰ κύκλον περιδινούμενα ποδηγεῖ, πάντα τε ἐν ἀλλήλοις ὄντα καὶ πρὸς ἄλληλα φερόμενα συνέχει, μέτροις τε τοῖς ἴσοις ἀφορίζει καὶ τὰ πορρωτάτω διῳκισμένα, καὶ τὰς τελευτὰς ταῖς ἀρχαῖς οἷον γῆν οὐρανῷ συγκεῖσθαι ποιεῖ, μίαν τε συνέχειαν καὶ ὁμολογίαν τῶν ὅλων πρὸς ὅλα ἀπεργάζεται.

30.   See also Iamblichus, *De mysteriis* 4.7 (191.4-5) on the harmonious nature and quality of the gods.

31.   Iamblichus, *De mysteriis* I.9 (32.7).

32.   For a detailed examination of the 'Music of the Spheres' in Neoplatonism, see Christine Harris, 'How did philosophers of the Late Antique period consider the concept of the Music of the Spheres, its roles and effects, in relation to ascension to the divine?' (MA dissertation, University of Wales Trinity St David, forthcoming 2019).

33.   Plato, *Timaeus* 40b-d.

34.   Plotinus, *Ennead* IV.4.8.46-62; Clark, *Plotinus*, pp. 106-107; 117-19.

35.   Cf. Clark, *Plotinus*, p. 106.

36.   Plotinus, *Ennead* IV.4.8.56-58.

37.   Plato, *Republic* , ed. and trans. Paul Shorey, 2 vols. (Cambridge, MA: Harvard University Press, 1963-1982), 4.427c; Clark, *Plotinus*, 119.

38.   Iamblichus, *In Timaeum* III, Fragment 55.

39.   Plato, *Timaeus* 33b; 34a.

40.   Plotinus, *Ennead* II.2.1.1; Clark, *Plotinus*, p. 136.

41.   Plato, *Timaeus* 90d; Clark, *Plotinus*, p. 136.

42.   It should be noted that this is a simplification of a complex set of issues, given that Neoplatonic writings on the nature of the soul and its 'spiritual' or 'ontological' journey are copious and that Neoplatonist philosophers often differ in how they understand and describe the essential nature and 'location' of the human soul.

43.   Plato, *Timaeus* 34c-36d (on the formation of Soul); 41d-44d (the formation of

the human soul).

44. Plato, *Timaeus* 47d: ὅσον τ᾽ αὖ μουσικῆς φωνῇ χρηστικὸν πρὸς ἀκοὴνDἕνεκα ἁρμονίας ἐστὶ δοθέν· ἡ δὲ ἁρμονία, ξυγγενεῖς ἔχουσα φορὰς ταῖς ἐν ἡμῖν τῆς ψυχῆς περιόδοις, τῷ μετὰ νοῦ προσχρωμένῳ Μούσαις οὐκ ἐφ᾽ ἡδονὴν ἄλογον, καθάπερ νῦν, εἶναι δοκεῖ χρήσιμος, ἀλλ᾽ ἐπὶ τὴν γεγονυῖαν ἐν ἡμῖν ἀνάρμοστον ψυχῆς περίοδον εἰς κατακόσμησιν καὶ συμφωνίαν ἑαυτῇ ξύμμαχος ὑπὸ Μουσῶν δέδοται·

45. Plato, *Republic* 401d-402a.

46. Proclus, *Commentary on Plato's Cratylus* (*In Crat.*) 176 (= 102, 4-9) ed. G. Pasquali, *Procli Diadochi in Platonis Cratylum commentaria* (Leipzig: Teubner, 1908): ἀλλὰ καὶ ἡ οὐρανία πᾶσα τάξις καὶ κίνησις τὸ ἐναρμόνιον τοῦ θεοῦ ἔργον ἐνδείκνυται· διὸ καὶ αἱ μερισταὶ ψυχαὶ οὐκ ἄλλως τελειοῦνται ἢ διὰ τῆς πρὸς τὸ πᾶν ἐναρμονίου ὁμοιότητος τὸ ἀπὸ τῆς γενέσεως ἀνάρμοστον ἀποσκευασάμεναι· τότε γὰρ τυγχάνουσιν τοῦ προτεθέντος αὐταῖς ὑπὸ τοῦ θεοῦ ἀρίστου βίου. Trans. Brian Duvick, Proclus, *On Plato Cratylus* (London: Duckworth, 2007).

47. Plato, *Cratylus* 405d, ed. and trans. H.N. Fowler, *Plato. Cratylus, Parmenides. Greater Hippias. Lesser Hippias* (Cambridge, MA: Harvard University Press, 1926).

48. Proclus, *In Crat.* 174 (= 98, 10-15), citing Chaldean Oracles Fragment 71: ἡ δὲ μουσικὴ τὸν ἡγεμονικὸν καὶ ἀρχικὸν μᾶλλον ἔχει διάκοσμον· ἐκεῖνος γάρ ἐστιν ὁ καὶ τὸν κόσμον ὅλον ἁρμόζων κατὰ μίαν ἕνωσιν, τὸν τῶν Μουσῶν χορὸν περὶ ἑαυτὸν ὑποστησάμενος, ἁρμονίᾳ φωτὸς γαυρούμενος, ὥς φησί τις τῶν θεουργῶν.

49. See Iamblichus, *De mysteriis* 5.7 (207.8-12).

50. Proclus, *In Crat.* 174 (= 99, 4-7): καθήκουσι γὰρ αἱ τῶν θεῶν δυνάμεις ἄνωθεν ἄχρι τῶν ἐσχάτων, οἰκείως ἐν ἑκάστοις φανταζόμεναι, ἃ καὶ ἡ τελεστικὴ διὰ τῆς συμπαθείας συνάπτειν ἐπιχειρεῖ τοῖς θεοῖς.

51. Iamblichus, *De mysteriis* 5.23 (233.9-13). See Peter Struck, *Birth of the Symbol: Ancient Readers at the Limits of their Texts* (Princeton and Oxford: Princeton University Press, 2004), pp. 204-26.

52. Proclus, *On the Hieratic Art*, 7-13, 32-37, ed. and trans. Brian Copenhaver, 'Hermes Trismegistus, Proclus and the Question of a Philosophy of Magic in the Renaissance', in I. Merkel and A.G. Debus, eds., *Hermeticism and the Renaissance: Intellectual History and the Occult in Modern Europe* (London: Associated University Presses, 1988), pp. 79-112; Struck, *Birth of the Symbol*, pp. 231-32.

53. Iamblichus, *De mysteriis* 10.6 (292.4-14), maintains that theurgy conjoins the human soul individually to all the parts of the cosmos and to all the divine powers pervading them, which leads and entrusts the soul to the universal demiurgic god.

54. Iamblichus, *De mysteriis* 4.2 (184.1-10).

55. On the crucial role of divine love or friendship (φιλία) within theurgy, see Iamblichus, *De mysteriis* 1.12 (42.5-7); 4.3 (184.14-185.2); 5.7; 5.9 (209.9-11); 5.26 (238.6-8). Cosmic sympathy as an auxiliary or subordinate cause of theurgy: *De mysteriis* 5.8. Cf. also 2.11 (97.13-15); 5.5 (206.8-10). On the importance of 'friendship of all with all' (φιλίας δὲ πάντων πρὸς ἅπαντας) (between gods, spirits (*daimones*), humans and animals) in the Pythagorean tradition, the latter of which influenced Iamblichus' conceptions of theurgy, see Iamblichus, *On the Pythagorean Way of Life* 16.69.6-7; 33.229-230. Iamblichus closes the *De mysteriis* by speaking of the 'harmonious friendship' between himself and Porphyry: *De mysteriis* 10.8 (294.4).

56. *Chaldean Oracles* Fragment 39 (= Proclus, *In Tim.* II.54.5-16), ed. and trans. R. Majercik (1989; repr. Westbury, Wiltshire: Prometheus Trust, 2013).

57. Iamblichus, *De mysteriis* 5.20-21; 5.23 (233.1-234.11); 5.25 (257.1-5); 5.26

(240.9-14); 10.5; 10.6 (292.4-14).

58.   Cf. Iamblichus, *De mysteriis* 3.11 (126.10-13); 3.12 (129.1-11).
59.   Iamblichus, *De mysteriis* 3.9.
60.   Iamblichus, *De mysteriis* 3.9 (117.10-118.2).
61.   Iamblichus, *De mysteriis* 3.9 (118.3-12).
62.   Iamblichus, *De mysteriis* 3.9 (118.13-119.11): Μᾶλλον οὖν ἐκεῖνα λέγομεν, ὡς ἠχοί τε καὶ μέλη καθιέρωνται τοῖς θεοῖς οἰκείως ἑκάστοις, συγγένειά τε αὐτοῖς ἀποδέδοται προσφόρως κατὰ τὰς οἰκείας ἑκάστων τάξεις καὶ δυνάμεις καὶ τὰς ἐν αὐτῷ <τῷ> παντὶ κινήσεις καὶ τὰς ἀπὸ τῶν κινήσεων ῥοιζουμένας ἐναρμονίους φωνάς· κατὰ δὴ τὰς τοιαύτας τῶν μελῶν πρὸς τοὺς θεοὺς οἰκειότητας παρουσία τε αὐτῶν γίγνεται (οὐδὲ γάρ ἐστί τι τὸ διεῖργον), ὥστε μετέχειν αὐτῶν εὐθὺς τὸ τὴν τυχοῦσαν ἔχον πρὸς αὐτοὺς ὁμοιότητα, κατοχή τε συνίσταται εὐθὺς τελεία καὶ πλήρωσις τῆς κρείττονος οὐσίας καὶ δυνάμεως. Οὐχ ὅτι τὸ σῶμα καὶ ἡ ψυχὴ ἀλλήλοις ἐστὶ συμπαθῆ καὶ συμπάσχει τοῖς μέλεσιν, ἀλλ᾽ ἐπεὶ τῆς θείας ἁρμονίας ἡ τῶν θεῶν ἐπίπνοια οὐκ ἀφέστηκεν, οἰκειωθεῖσα δὲ πρὸς αὐτὴν κατ᾽ ἀρχὰς μετέχεται ὑπ᾽ αὐτῆς ἐν μέτροις τοῖς προσήκουσιν·
63.   *Chaldean Oracles* Fragments 37, 107, 146. On harmony in the *Chaldean Oracles*, see Fragments 71, 97.
64.   Aristotle, *On the Heavens* II.9, 290b12-291a28; Plato, *Republic* VII, 530d; X, 617b (the myth of Er); *Cratylus* 405c; Burkert, *Lore and Science*, pp. 322 and 350-51, argues that the idea is based on the association of the ancient 'seven-stringed' lyre with the idea that the planets are seven in number.
65.   Burkert, *Lore and Science*, pp. 351-52.
66.   Plato, *Laws* 897c-d, 898a-c. Cf. Burkert, *Lore and Science*, pp. 326, 366.
67.   Iamblichus, *On the Pythagorean Way of Life* (*De Vita Pythagorica*) 15.65.1-5.
68.   Iamblichus, *On the Pythagorean Way of Life* 15.65.11-19: ἀλλὰ ἀρρήτῳ τινὶ καὶ δυσεπινοήτῳ θειότητι χρώμενος ἐνητένιζε τὰς ἀκοὰς καὶ τὸν νοῦν ἐνήρειδε ταῖς μεταρσίαις τοῦ κόσμου συμφωνίαις, ἐνακούων, ὡς ἐνέφαινε, μόνος αὐτὸς καὶ συνιεὶς τῆς καθολικῆς τῶν σφαιρῶν καὶ τῶν κατ᾽ αὐτὰς κινουμένων ἀστέρων ἁρμονίας τε καὶ συνῳδίας, πληρέστερόν τι τῶν θνητῶν καὶ κατακορστερον μέλος φθεγγομένης διὰ τὴν ἐξ ἀνομοίων μὲν καὶ ποικίλως διαφερόντων ῥοιζημάτων ταχῶν τε καὶ μεγεθῶν καὶ ἐποχήσεων, ἐν λόγῳ δέ τινι πρὸς ἄλληλα μουσικωτάτῳ διατεταγμένων, κίνησιν καὶ περιπόλησιν εὐμελεστάτην ἅμα καὶ ποικίλως περικαλλεστάτην ἀποτελουμένην.
69.   Iamblichus, *On the Pythagorean Way of Life* 2.5-8; 15.66.4-6; O'Meara, *Pythagoras Revived*, pp. 36-39; Burkert, *Lore and science*, p. 357, notes that Pythagoras was often portrayed as a 'shaman'; thus, the tradition that he heard the heavenly music is related to the idea that the shaman, in ecstatic trance, can 'travel' to heaven and hear the heavenly music.
70.   Iamblichus, *De mysteriis* 3.9 (120.3-10): Ἀλλ᾽ οὐδὲ τοῦτο δεῖ λέγειν, ὡς ἡ ψυχὴ πρώτως ὑφέστηκεν ἐξ ἁρμονίας καὶ ῥυθμοῦ· ἔστι γὰρ οὕτω ψυχῆς μόνης οἰκεῖος ὁ ἐνθουσιασμός· βέλτιον οὖν καὶ τὴν τοιαύτην ἀπόφασιν ἐκεῖσε μετάγειν, ὅτι δὴ ἡ ψυχή, πρὶν καὶ τῷ σώματι δοῦναι ἑαυτήν, τῆς θείας ἁρμονίας κατήκουεν· οὐκοῦν καὶ ἐπειδὰν εἰς σῶμα ἀφίκηται, ὅσα ἂν μέλη τοιαῦτα ἀκούσῃ οἷα μάλιστα διασώζει τὸ θεῖον ἴχνος τῆς ἁρμονίας, ἀσπάζεται ταῦτα καὶ ἀναμιμνήσκεται ἀπ᾽ αὐτῶν τῆς θείας ἁρμονίας, καὶ πρὸς αὐτὴν φέρεται καὶ οἰκειοῦται, μεταλαμβάνει τε αὐτῆς ὅσον οἷόν τε αὐτῆς μετέχειν.
71.   Plato, *Phaedo* 72e-78b. Cf. also Plato, *Meno* on the argument for learning as recollection.
72.   Plato, *Phaedrus* 246e-248b. Cf. Burkert, *Lore and Science*, p. 365.
73.   Plato, *Phaedrus* 249b-c; 250b-251a. Cf. *Republic*. 402a.
74.   Iamblichus, *In Phaedrum*, Fragment 3 (= Proclus, *Theol. Plat.* IV, 188, 15ff Portus).

THE CONCEPT OF HARMONY IN JUDAISM

David Rubin

THIS CHAPTER WILL ADDRESS THE CONCEPT OF HARMONY IN JUDAISM as it appeared in the Hebrew Bible (*Tanach*), Midrash and Talmud, and later rabbinic writings – the Torah corpus from antiquity to the post-modern period.[1]

For the purpose of this analysis, 'Judaism' is defined as a way of life that incorporates a system of law, ethics and morality based upon the Torah, where *Torah* (lit. 'teaching') denotes the *Tanach*, the rabbinic works of late antiquity and the body of Jewish literature centred upon them, and the precepts and modes of behaviour that flow from them.[2] 'Harmony' (from the Greek term, ἁρμονία – *harmonia*, meaning 'union' or 'fitting together') is used here in its modern sense of a state of agreement between two or more divergent entities, its sense of correspondence and sympathetic resonance between two entities, and its applied concept of agreeable artistic composition, as in the pleasing quality that arises from a combination of contrasting notes, parts, or elements.[3]

INTRODUCTION

Academic research has ascertained the prevalence of cross-cultural engagement and mutual influence between the ancient near east and ancient Greece.[4] Thus, although the term 'harmony' is of Greek provenance and its meanings owe much to ancient Greek mythology and Hellenistic philosophy, a similar idea might conceivably be found in Judaism.[5]

Indeed, many ancient cultures embraced concepts of harmony. Native American peoples emphasised the interconnectedness of all living things, and the harmony of spirit, mind and body.[6] The Japanese term *wa* referred to an individual whose conduct and aspirations were in harmony with society.[7] It might be posited that the type of harmony stressed by a particular culture or society indicated the aspects or elements that society considered important for its survival or success.

Whilst difficult to specify a single (classical) Hebrew word that equated fully with all the concepts of harmony outlined above, three individual terms, שלום *šolōm* (or *šalom*; 'peace'), יחד *yāḥād* ('together'), and תפארת *tif'ereth* ('beauty' or 'splendour'), conveyed separate elements thereof.

The word, *šolōm*, and the noun, *šoleim* (complete), originated from the triconsonantal root, שלם SH-L-M, related to the Akkadian word *šalamu* that had the connotation of 'being complete'.[8] Though usually translated as 'peace', *šolōm* essentially denoted a state of perfection and wholeness, either in wellbeing, as in Gen. 37.14, or through the coherence and agreement of conflicting forces, as evinced in Job 25.2 and Isaiah 45.7.

The word *yāḥād* related to אחד *eḥod* ('one') and was primarily used to denote unity. Hence, in Psalms 133.1, *yāḥād* described brothers dwelling together in social harmony.[9] In Job 38.7, it described the harmonious singing of the morning stars.[10] The term *tif'ereth* (from the root פאר *pe'air*, 'glory') was used, in the early and late modern periods, to depict the beauty of harmonious balance.[11] If *yāḥād* denoted the ontological state of togetherness, and *šolōm* described its wholeness and perfection, *tif'ereth* referred to the appealing quality of that state. However, as the term generally used to express concordance between individual interests was *šolōm*, most of this research will centre on that term.

How Two Become One

In Judaic thought, *šolōm* did not merely indicate agreement between conflicting parties. It frequently entailed the introduction of a third element that caused the opposing sides to blend and meld and cooperate with each other.[12] The resultant state was a conglomeration of ideas, each individual position recognising the validity of the other.[13]

The unifying element was a common focus or purpose, often portrayed as the desire to serve G-d, that incorporated a degree of self-nullification that catalysed cohesivity between opposites. Thus, in the Midrashic reading of Job 25.2, fire and water make peace in heaven.[14] The Biblical phrase 'fire flashing in the midst of the hail' (Ex. 9.24) was understood by the Midrash as fire and water making peace to do G-d's will.[15] The buffalo and lion are natural enemies, but, in the chariot of G-d (described in Ezekiel 1.10), says the Midrash, 'there is peace between them; they love each other'.[16]

In a similar vein, according to the Mišnah (c. 50–200 CE, collection of laws and traditions appertaining to Biblical and rabbinic injunctions), if participants of a learning debate are arguing 'for the sake of Heaven', both sides of the argument will remain valid and endure.[17] Arguing for 'the sake of Heaven', or the absence of self-interest, was the lynchpin that bound the parties, such that 'Torah scholars increase *šolōm* in the world', in spite of – or, indeed, because of – their debates.[18] Thus, R. Samson Rafael Hirsch (1808–1888) taught that true peace between peoples was unachievable unless the parties were in harmony with G-d, their differences dissolving through mutual self-abnegation to the Significant Other.[19]

Self-nullification, to attain spiritual harmony, was especially stressed in the esoteric writings of the modern period (c. sixteenth to twentieth century). Here, self-effacement was a means of attaching to the supernal worlds, to approach a state of harmony with G-d, allowing one's finite being to unite with the infinite Being, causing opposite elements to unite.[20] Thus, R. Elimelekh of Lizhensk (1717–1787) taught, if you are completely self-effacing, you will hear G-d's voice.[21] In the words of R. Dovber, the *Maggid* ('Preacher') of Mezritch (c. 1700–1772), by seeing oneself as nothing, one could 'transcend time, and access the world of thought, where everything – life and death, sea and dry land – is equal'.[22]

Moreover, self-effacement created a receptacle to receive G-d's Presence and blessing: the quintessential *šolōm*.[23] The only antinomian prayer with magical overtones and directed to angels (both aspects frowned upon in normative *halakha*) to survive in the main Jewish liturgy concerned domestic peace and harmony.[24] Clearly, *šolōm*, the harmonic balance of opposing influences, played a key role throughout Judaic culture.

COSMIC HARMONY

Harmony between heaven and earth was an overriding theme in the Midrashic analysis of the Genesis account.[25] Man was created as a product of both the celestial and sublunar spheres in order to ensure *šolōm* (harmony) in the cosmos.[26] His bridging the animal and the divine, through the union of a physical body with a G-dly soul, enabled conflicting elements within the macrocosm to coalesce.[27]

Likewise, when the Midrash ascribed the variation in divine names in the opening two chapters of Genesis (the first chapter used *Elōqim* ('powers' or all-powerful); the second chapter, the tetragrammaton (understood to denote love and mercy) in conjunction with the name *Elōqim*) to G-d's combining loving-mercy with harsh judgement, so as to create the world through *šolōm*, it was asserting *šolōm*-harmony as the underpinning of creation, and the heart of G-d's relationship with the world.[28]

Isaiah associated *šolōm* with harmony in creation when he described G-d as 'forming light and creating darkness, making *šolōm* and creating evil' (Is. 45.7).[29] In the Canaanite pantheon, *Šālim*, represented not only the completion of day, but, as numen of dusk, the blend of light and darkness at sunset.[30] Similarly, *šolōm* denoted the manifestation of a new state, the union of light and darkness: G-d created harmony that married the conflicting elements.[31]

According to the *parallelismus membrorum* in Isaiah 45.7, *šolōm* ('harmony') paralleled light and was the antithesis of evil.[32] The relationship between *šolōm* and goodness was also suggested in Psalms 34.15, where 'do[ing] good' was conflated with 'seek out *šolōm* and pursue it'.[33] The connection between goodness and light, first portrayed in Genesis 1.4 ('The L-rd saw that the light was good'), was underlined by the rabbis' exegesis on the word 'good', in Exodus 2.2, to denote a luminescence surrounding Moses' birth.[34] This is especially significant in view of the connection of light with G-d evinced in Isaiah 60.1-3.

Hence, as well as being essentially good, *šolōm* ('harmony') was associated with the Divine.[35] In Numbers 6.26, the ultimate deific blessing was *šolōm*.[36] In Isaiah 45.7, only G-d had the power to create *šolōm*-harmony.[37] In Job, only G-d could impose harmony in the celestial realms: 'He makes peace (*šolōm*) in His heights' (Job 25.2).[38] Thus, the Midrash insisted that true *šolōm* has to be divinely gifted to the world.[39]

This idea is underscored in the frequent mention of Job 25.2 in the daily liturgy (in the *Qaddish* prayer, in Grace After Meals, and in the thrice-daily silent devotion) where 'He Who makes *šolōm* in His heights' (עֹשֶׂה שָׁלוֹם בִּמְרוֹמָיו) is beseeched to cause harmony and peace to reign over humanity, suggesting that, ultimately, only G-d can affect true peace over humankind.[40]

Isaiah 45.7's parallelism was echoed in the Morning Prayers, where G-d is blessed for 'forming light and creating darkness, making *šolōm* and creating all'.[41] This Tannaic (c. first century CE) liturgic text substituted the word 'evil' – as it appeared in Isaiah 45.7 – for 'all', with the implication that *šolōm* effected the subsequent creation of all.

Yet, in this parallelism, 'creating all' becomes the antithesis of *šolōm* and mirrors 'darkness'. This rabbinic amendment possibly reflected a concept expressed by the Roman poet Ovid (43 BCE–17/18 CE): 'though fire and water fight each other, heat and moisture create everything, and [therefore] this discordant union is suitable for growth'.[42] In other words, if *šolōm* is a state of concordance between opposing factors, its shadow is their discordant union, but it is a chaos that results in creation. The liturgical blessing thus ameliorated the 'evil' expressed in Isaiah 45.7 to a chaos that engenders growth.

GREEK PARALLELS

According to the Midrashic gloss on Job 25.2, G-d 'making peace (*šolōm*) in His heights' involved imposing regulation and order on natural and supernatural forces.[43] In this sense, divine harmony was analogous to the Greek deity, *Harmonia*, that presided over cosmic stability and was responsible for connecting dissimilar items into the κόσμος kosmos, the divinely ordained order of the natural world.[44] The antithesis of šolōm, עֹרֶר rā ('evil'), that paralleled darkness in Isaiah 45.7 and signified disharmony and chaos, was similar to the Greek goddess Eris, daughter of night and darkness, and the antithesis and nemesis of Harmonia.[45]

Like Harmonia and Eris, harmony and disorder were described in Isaiah as ever-present. However, rather than the kosmos being permanently created out of a primordial void (khaos χάος), as in Greek cosmogony, according to the Midrash, the dynamic harmonisation of šolōm – the primeval force that predated the creation of the world – needed to be continually crafted.[46] Cosmic harmony, the equilibrium between natural forces, was not a given. The potential for chaos was ever-present. G-d creates the Chaoskampf and quells it by effecting šolōm.[47]

Furthermore, since G-d is its creator, עֹר ra (evil or chaos) was seen as a necessary part of cosmic balance. As noted in the liturgical text above, chaos can instigate growth. Thus, the Midrash saw pain and suffering, the evil inclination, even death, as so integral to the divine plan as to be alluded to in creation's climactic statement 'G-d saw all that He had made and behold, it was very good' (Gen. 1.31).[48] All were part of an over-arching harmony and cosmic order.

Accordingly, Isaiah declared, 'He Who makes harmony and creates evil: I am G-d, Who makes all these' (Isaiah 45.7).[49] Similarly, Jeremiah declaimed, 'Do not [both] evil and good emanate from G-d?' (Lamentations 3.38).[50] In this theology, G-d creates all: the good and the untoward, harmony and discordance. 'Lightning and hail, snow and sleet, whirlwinds; each carry out His word' (Psalms 148.8).[51] Though šolōm is a divine blessing, its antithesis, discordance, is not a curse. It was viewed as a necessary evil to provoke creation and growth.[52]

Ultimately, all that G-d does was seen as part of His harmonious plan for the good of His creations.[53] It was up to man through his faith, theurgy, or attainment of henosis, to reveal the good inherent in even the most challenging of G-d's actions.[54]

HARMONY WITH THE ECO-SYSTEM

In Judaism, the creation of humanity in G-d's image is considered both axiomatic and far-reaching.[55] As G-d created both harmony and chaos, so it is in humanity's power to engender ecological harmony or, alternatively, to cause chaos and destruction. Thus, the Midrash had G-d saying to Adam,

See how beautiful and splendid My works are! All that I have created is for you. Take care not to spoil or destroy My world, for once you spoil it, there is no-one to repair it after you.[56]

The Midrash placed man firmly at the helm. He alone was responsible for the preservation of the planet. Though the ebb and flow of cosmic harmony and chaos are part of the divine order, man, through his choices and actions, had the power to disturb that balance. Irresponsible neglect of the biosphere, or wanton destruction, could lead to his own devastation. *Ergo*, if he spoilt the world, there would be no-one to restore it.

Precedent for man's charge to live in harmony with nature and his responsibility for the earth's upkeep can be found in Genesis 2.15, where G-d was described as placing Adam in the Garden of Eden for the prime purpose of 'tilling and tending it', and in the psalmist's declaration, 'The heavens are G-d's but He gave over the Earth to man'.[57]

Based upon the Genesis account of creation, the Midrash saw maintenance of the natural environment as a principal element in the principle of *imitatio Dei*.

> After the L-rd your G-d shall you walk' (Deut. 13:5). Can one walk after G-d? [...] 'In Him shall you cleave' (ibid.). Can one [...] cleave to the [Divine] Presence? [...] But, just as, from the beginning of creation, the Holy One, blessed be He, was, before all else, occupied with plantation, [...], so you, when entering the land, occupy yourselves first with planting ...[58]

Since G-d engaged Himself in planting, so must man. World conservation was deemed so important that, according to the rabbis, 'if you have a sapling in your hand and someone tells you that the Messiah has arrived, stay and complete the planting, and only then go to greet the Messiah'.[59]

Biblical ordinances likewise demanded ecological responsibility. In Deuteronomy 20.19–20, the Jewish people were instructed not to destroy fruit trees, an injunction understood by the rabbis to include the unwarranted destruction of any item of value.[60]

> [The righteous] will not destroy even a mustard seed. They are distressed at every ruination and spoilage. If they can, they will endeavour, with all their power, to save anything from destruction.[61]

In a similar vein, the Deuteronomic commandment (22.6–7) to refrain from taking the mother bird together with its young was interpreted by various medieval commentaries as an injunction to preserve the species.[62] Ecological responsibility was also demonstrated in Leviticus 25.1–7, where the Torah was mindful of the fact that if nothing was planted in the sabbatical year, wildlife may suffer.

Accordingly, storing of produce for one's own cattle was only permitted so long as the herbivorous animals of the wild also had sustenance.[63] Indeed, causing pain to animals was largely regarded as Biblically prohibited.[64]

As will be further explored, these and similar ordinances were seen as part of an all-embracing principle that *šolōm* (harmony in all its aspects) was equal to G-dliness. Through performing actions that resonated with harmony, societal or cosmic, one approached the Divine, an idea analogous to Plato's statement, that 'through association with *kosmios* (that which is divine and orderly), [one] becomes divine and orderly'.[65]

HARMONY IN THE G-DHEAD

'What's in a name? That which we call a rose, by any other name would smell as sweet'.[66] A name, William Shakespeare (1564–1616) philosophised, has no inherent connection with its bearer.[67] In contradistinction, Torah saw Gd's name as having innate meaning, expressing the nature of His revelation, allowing humankind to perceive His attributes and relate to Him.[68]

Accordingly, when the rabbis commented (on Jud. 6.24: 'G-d is *Šolōm*') that 'G-d's name is *Šolōm*', they were ascribing harmony as a key to understanding G-d.[69] Though G-d's incomprehensibility is an essential tenet of Judaism, harmony was seen as an integral quality of the Divine.[70]

This quality was apparent, too, in G-d's Ineffable Name. According to Maimonides, as well as the kabbalists, the Ineffable Name denoted G-d's essence, and described His 'absolute existence' and His causing existence.[71] Medieval commentaries saw the Tetragrammaton as deriving from the words 'He was, Is, Will be' (היה הוה יהיה).[72] In the liturgical poem, *Ādōn Ōlom* (אדון עולם), 'Master of the World', this description of G-d's transcendence was described with the adjective *tif'oroh* [תפארה] ('beauty' or 'harmony'), implying that G-d's beauty and harmony becomes apparent through the unfolding of world history.[73] Furthermore, as creation was seen as structured and beautiful, G-d's Name, responsible for its constant manifestation, expressed cosmic harmonic unity.[74]

In the Midrash, G-d's attribute of *šolōm* was seen as harmonising two opposing aspects of the divine nature: severity and loving kindness.[75] Here, too, *šolōm* was analogous to the Greek goddess Harmonia, described by Hesiod (c. 700 BCE) as the product of the god of war (Ares) and the goddess of love (Aphrodite).[76]

In kabbalistic thought, this state of harmony became the objective of theistic worship.[77] G-dly efflux, realised through the performance of *mitsvōth* ('good deeds', lit. 'commandments': Divine instructions of sacred behaviour or ritual conduct, or, according to the mystical conception of the term, 'connections' to G-d), caused a balance of the cosmogonic extremities.[78] This celestial harmony then influenced the *Šekhinah* (lit. 'that which dwells'), G-d's perceived presence on earth.[79]

Similarly, when the *Zohar* identified three 'bonds' in the G-dhead that harmonised polarities of deistic consciousness at three levels of its integrational manifestation (G-d's 'knowing', harmonising His wisdom and understanding; His attribute of truth harmonising His loving-kindness and harsh-might; and *šolōm* – harmonising victory, the projection of self, and majesty, a passive element of the G-dhead), it was identifying harmony as a key aspect of the divinity.[80]

Especially in Lurianic Kabbalah (after its teacher, R. Isaac Luria, 1534–1572), G-d's administration of this world was seen to be the product of a harmonious gestalt of six attributes, the so-called 'extremities' (קצוות *qetsovoth*) of divine conduct, outlined in I Chronicles 29.11: 'To you, O G-d, is the greatness and the might and the glory and the victory and the majesty, indeed all that is in the heavens and earth', where 'all' was viewed as the sixth attribute, the harmonious blending of all elements of creation.[81] Through theurgical practice, involving esoteric intentions and unifications (*yichudim*), the kabbalist sought to harmonise the divine emanations and sweeten the attribute of harsh judgement, to bring about the sublime revelation of *tif'ereth* ('glory' or 'beauty'), the deific 'beauty' inherent in the harmony of cosmic balance.[82]

Though *tif'ereth* was seen as the balance of the six extremities, and the beauty inherent in that concordance, it was specifically the sixth attribute, termed *yesōd* ('foundation'), appended to *tif'ereth* in the kabbalistic 'Tree of Life', that was seen as the element that caused the polarised components to blend and coalesce, the divine transmitter of *šolōm* (harmony and peace) to the lower spheres.[83]

Through *yesod*, G-d infused the *Šekhinah* (Divine Presence) with supernal light, and the Divine Presence interacted with the world. A synonym of *šolōm* in kabbalistic literature, *yesōd* facilitated the transmission of divine love, the spiritual vibration that united Heaven and Earth, the divine element that could connect with the numen of the underworld.[84]

TORAH AS ŠOLŌM

'The heavens recount G-d's Glory; the firmaments tell of the works of His Hands. [...] The L-rd's teaching (Torah) is perfect; it revives the soul' (Psalms, 19.1, 8).[85]

The psalmist contrasted the heavens, which speak of G-d's Glory, with G-d's teaching, which bespeaks a perfection that revives the soul.[86] According to the Sifre's (early Midrash , c. 100 CE) exegesis of this verse, as G-d was perfect, so the Torah was perfect, such that it engendered perfection in its adherents.[87]

Similarly, as G-d was called *Šolōm*, so Torah was called *šolōm*.[88] While Maimonides interpreted this as meaning Torah's engenders peace and harmony in the world, the *Zohar* saw Torah as a metaphysical manifestation of G-d's attribute of *šolōm*.[89]

In this respect, Torah was an abstract *logos*, the inner soul of the written and oral law.[90] This view was based upon an early Midrashic statement that the world was created according to an order and system whose root is the Torah.[91] In the words of R. Menachem Azariah da Fano (1548–1620), 'the Torah is the imprint of G-d, and the world is the imprint of the Torah'.[92] As G-d was *Šolōm*, so Torah was *šolōm*, and thus the cosmos bespoke *šolōm*.[93] *Šolōm*-harmony was the mainstay of the world.

Indeed, harmony was a predominant theme throughout the Torah. The Sinaitic revelation of G-d speaking to the entire population (Ex. 19–20), was only possible through the populace uniting, 'as one man, with one heart'.[94] Laws and ordinances are overridden in order to achieve social harmony and peace.[95] Though truth is the seal of G-d, uttering falsehood an abomination, one is allowed to utter an untruth to preserve matrimonial harmony.[96] G-d even allows His name to be erased for the sake of peace.[97]

Since G-d is called *Šolōm*, and Torah is *šolōm*, through living the Torah's precepts, one achieves a state of harmonic resonance with G-d.[98] Such was the power of harmony, and the resultant resonance with G-d, that even were a society to worship idols, the accuser ('satan') could not affect them.[99]

According to the kabbalists, this harmonic resonance had cosmic consequences, such that sustainment of cosmic harmony was man's responsibility. As R. Meir Leibush Weiser (1809–1879) wrote:

> G-d arranged the world that it be affected through man's actions, for good or otherwise. This has been likened to two harps, a larger and smaller, placed in an auditorium in a specific position. If the smaller is played upon, the larger will vibrate. Similarly, G-d arranged the world, the larger instrument, to play according to the melody played by the smaller instrument, man.[100]

Beyond harmonic resonance with the macrocosm, kabbalists saw the Torah as a manual of how to be in resonance with G-d. The Torah's divine commandments and precepts were seen as examples of *imitatio Dei*, in line with the Talmudic statement that G-d performs the Torah's *mitsvōth*.[101] Their pure performance caused harmonic resonance with aspects of G-d, enabling G-d to dwell within the person, the person to unite in *unio mystica* with G-d.[102]

Aesthetic Harmony in the Torah

Pythagoras (fl. sixth century BCE) was credited with the notion that mathematical harmony underlies the structure of the world and determines beauty.[103] Pythagorean cosmogony was based on geometric harmonic relationships: the universe's various elements related to each other in numeric proportion (*harmonia*), and those proportions and numbers were the bedrock of reality.[104]

A similar view, stressing numbers as the fabric of cosmic structure, appeared in the *Sefer Yetsirah* ('Book of Formation', c. second to fifth century CE) and in later kabbalistic writings.[105] For R. Joseph Gikatilia (1248–after 1305),

> The world exists through letters, the letters exist through *ḥeshbōn* (numbers and mathematics), and *ḥeshbōn* exists through G-d. [...] The existence of all *ḥeshbōn* is G-d's name. [With it] He created the world, and arranged it [the world] according to *ḥeshbōn*. [...] Everything He created is through the exactitude of *ḥeshbōn* and only operates according to *ḥeshbōn*.[106]

Numbers comprise the cosmic metastructure. They exist through the name of G-d. Since G-d's name is *Šolōm*, it follows that the numeric structure underlying the world is one of deep harmony.[107]

For the Pythagoreans, beauty becomes apparent through harmonious proportions.[108] The golden ratio (its recognition attributed to Pythagoras but formally defined by Euclid, third century BCE) was extolled by the Pythagoreans as the most perfect proportion, and therefore, one that conferred an aesthetic pleasing quality wheresoever it appeared.[109]

In the words of Euclid, the golden ratio is realised when, in dividing a line into two segments, 'as the whole line is to the greater segment, so is the greater to the lesser'.[110] This ratio also gives the proportions for the golden rectangle. An approximation of these proportions is known as the 'rule of the thirds' (a ratio widely used in photography).[111]

Gary Meisner noted that the dimensions of both the Ark of the Covenant (Ex. 25.10) and the Copper Altar (Ex. 27.1–2) were of the golden rectangle.[112] The notion of beauty through harmonious proportions, and particularly, the golden ratio, can also be found underlying the rabbinical ideal dimensions of various *mitsvōth*.[113] This is especially noteworthy in view of the rabbis' exegesis, of (Gen. 9:27), 'May G-d expand Yép̱eṭ and dwell in the tents of Šem', that the beauty of Yép̱eṭ (understood by the rabbis to refer to Greece) should dwell in the tents of Šem, by incorporating Greek beauty in Judaism's practical observance.[114]

Indeed, based upon the verse 'this is my G-d and I will beautify him' (Ex. 15.2), R. Ishma'el saw it as a divine injunctive that *mitsvōth* be aesthetically pleasing.[115]

> 'This is my G-d and I will beautify him' (Ex. 15.2). Is it possible for a person to beautify his Creator? But [the interpretation is] I will beautify Him with *mitsvōth*: I will make before Him a beautiful *sukkah* (temporary shaded dwelling constructed for the festival of Sukkoth), a beautiful *lulov* (palm-frond used in ritual service on the festival of Sukkoth), a beautiful *šofar* (typically a ram's-horn), beautiful *tsitsith* (ritual fringes), a beautiful *sefer Torah* (Torah-scroll), beautiful *tefilla* ('phylactery').[116]

As the Jew approaches G-d through *mitsvōth*, beautifying the *mitsvōth* effectively beautifies his experience of G-d.

Each of these *mitsvōth* listed by R. Ishma'el had specific dimensions that could be seen to approximate to the golden ratio or derivatives thereof, or were guided by simple proportions. The minimum size of a *sukkah*, as outlined by the Talmud, was 7 x 7 x 11 hand breadths, the approximate dimensions of a golden rectangle.[117] The minimum size of a *šofar*, 'so that it can be held [in a fist] and seen on both sides', again gave the approximate dimensions of a golden rectangle.[118]

Tsitsith (ritual fringes) were made up of a braided segment followed by free-hanging tassel strings.[119] According to the third-century CE Sasanian sage, Rav, the ideal ratio of the braided segment to the tassels was one third braided versus two-thirds tassel, a rough approximation of the golden ratio, as in the 'rule of the thirds'.[120]

Similarly, a *mezuzah* (lit. 'doorpost') scroll (parchment inscribed with certain Biblical texts affixed to the doorpost of a house) was to be affixed to the base of the top third of the doorpost.[121] This, too, conformed with an approximation of the golden section.

The placement of the *tefillin* ('phylacteries') can also be seen to be in accordance with the golden section and simple proportions. The boxes housing the ritual texts are to be perfectly square.[122] The left strap from the head-phylactery was to reach the navel, whilst the right strap reached the covenant of circumcision, accentuating the golden section proportions of the upper torso.[123] (Others opined that both straps were to reach the navel; this would accentuate the golden section vis-a-vis the entire person).[124] Likewise, the location of the arm phylactery vis-à-vis its strap that reached until the middle finger, accentuated the golden ratio proportions inherent in the arm.[125]

A similar ratio could be seen in the proportions of the minimum sizes of the willow and myrtle branches vis-à-vis the palm-frond (*lulov*): the palm-frond was to be three handbreadths, plus one handbreadth with which it was held, whilst the accompanying foliage was two handbreadths, plus the one handbreadth holding it.[126]

The ideal dimensions of a *sefer Torah* outlined in the Talmud, its height equalling its circumference, adhered to a simple ratio of 1:3, another type of golden or Euclidean rectangle.[127] The horizontal black lines of the Hebrew letters written in ritual texts, in relation to the white parchment, displayed a ratio of 1:3, close to the dimensions of the golden ratio.[128]

Besides the golden ratio, dimensions of other *mitsvôth* conformed with simple proportions, consistent with Marcus Vitruvius Polio's philosophy (c. 80–70 BCE; after c. 15 BCE) that simple mathematical proportions that mirrored the proportions and symmetry in nature were definitively beautiful.[129]

The Midrash, above, presented alternative interpretations of *ve'anvaihu* (Ex. 15.2).[130] Abba Saul saw ואנוהו *ve'anvaihu* as consisting of the words אני והוא *ani vo'hu* ('I and He'); he thus interpreted the verse as an injunction of *imitatio Dei*: 'as He [G-d] is merciful and gracious, so you be merciful and gracious'.[131] R. Jose, son of Durmasqeith, understood *ve'anvaihu* to mean 'I will make Him an abode', from the root נוה *nevai* 'an abode'.[132]

The three opinions might be viewed as variations of the same theme.[133] How does one improve on the appearance of G-d? How can one beautify the Infinite? Through introducing an element of the Infinite into the finite. According to the Midrash, G-d desired an abode in the physical world.[134] Building a sanctuary allowed man to approach G-d and G-d to be immanent, the macrocosm in the microcosm. Likewise, through imitating His attributes, one resonated with the perfection and harmony inherent in G-d, allowing the Whole to be expressed in the part. By beautifying *mitsvôth*, the harmony and beauty expressed in G-d's creation are articulated in the spatial and temporal spheres.

CONCLUSION

This exploration saw harmony as central to the Judaic ethos, manifest in its lore and culture, its Mosaic and rabbinic literature, its esoteric philosophies, and both its nomian and antinomian traditions.

Though related to the terms *yāḥād* ('together') and *tif'ereth* ('glory'), the modern sense of harmony was found to be closest to both the Bible's and the rabbinic notion of *šolōm* ('peace'). Fundamentally good, *šolōm* was seen by the rabbis as the blueprint of creation, the backbone of society, the essence of the Torah, and the name of G-d. Uniting polarities, it was synonymous with spiritual light and G-dly revelation.

Rabbinic works of late antiquity were replete with concepts distinctly similar to the ancient Greek ideas of harmony.[135] The Midrashic description of *šolōm* as representing the union of G-d's loving kindness and strict judgement, and denoting cosmic stability, was observed to be analogous to the Greek deity Harmonia – portrayed in Greek mythology as the product of Ares, god of war, and Aphrodite, goddess of love – who connected dissimilar items into the divinely ordained order of the natural world. In a similar manner, the opposite of *šolōm*, *rā* – 'evil' – was comparable to the Greek goddess Eris, the antithesis of Harmonia.

However, in contradistinction to Greek mythology, *ra* ('evil' or 'chaos') was seen as a part of the divine plan, necessary to engender growth. Though the revealed state of *šolōm* implied cosmic stability, there was a hidden, overarching state of *šolōm* – peace, harmony and goodness – that included evil and chaos. Reflecting that philosophy, the Torah instructed one to choose life (Deut. 30.19), to attach to goodness, whilst channelling any destructive leanings to good.[136]

In its interaction with nature, humankind must be wary of impairing the cosmic balance, a theme underscored in both Biblical and rabbinic ordinances. Accordingly, the Torah was called *Šolōm*, as it engendered harmony in the world and stressed *šolōm* above all else.

In kabbalistic thought, Torah was the manifestation of G-d's attribute of *šolōm* ('harmony'). Termed *yesod* (lit. 'foundation') in kabbalistic nomenclature, *šolōm* – or harmony – was an integral facet of G-d, the divine element of love through which He connected with the lower spheres. By personifying *šolōm*, through the performance of *mitsvōth* and theurgical practice and through its expression in one's social, domestic and marital relations, one achieved harmony with that aspect of G-d.

As both G-d and the Torah were called *Šolōm*, the Torah was seen as an instruction manual for *imitatio Dei* and the attainment, thereby, of *unio mystica*. Though absolute *šolōm* could only be achieved through G-d's blessing, observing the *mitsvōth* caused one to be in harmony with G-d, such that through self-abnegation to G-d and His precepts, and emulation of His attributes, man can resonate with and thus manifest *šolōm*.

Finally, harmony, in its sense of beauty of form and structure, was found to be present in the *mitsvōth*. Based on Biblical exegeses, the rabbis saw beautifying *mitsvōth* as part of the Biblical injunction of *imitatio Dei*. Furthermore, their beautification appeared to conform with Pythagorean concepts, as the dimensions of several *mitsvōth*, both Biblical and rabbinic, were found to be in accord with the golden ratio and rectangle, and simple proportions.

To summarise: in Judaism, man is to attach to *šolōm*, harmony and peace, in all manner of human expression: socially, morally, aesthetically, and through theistic worship. The notion of being in harmony with G-d was central to the rabbis' understanding of the *mitsvōth*. Though, ultimately dependent upon Gd's blessing, through acting in harmony with the Deistic expression, man effects its revelation.[137]

If the Pythagorean philosopher or Native American seek to be in harmony with nature because they value nature above all else, the Jew must seek *šolōm* and harmony in all its forms, because G-d is in *šolōm* and the root of the *mitsvōth* is to realise harmony with G-d.[138]

NOTES

1.   For the purpose of this study, the Bible is defined here as the *Tanach* – the classic Jewish Writings from late second millennium BCE (earliest estimate of the Mosaic Pentateuch) until late first millennium BCE – according to the Masoretic Text (MT). The MT can be found in *Biblia Hebraica Stuttgartensia Liber Gen*, (H. Bardtke: Hendrickson Publishers Marketing, LLC; 2017), available online at https://www.academic-bible. com/en/online-bibles/biblia-hebraica-stuttgartensia-bhs/read-the-bible-text/ [accessed 6 December 2018]. 'Culture', in the context of this paper, is understood in its widest sense as a society's values, ideas, beliefs, customs and social behaviour whose meaning lies in the way they are interpreted by that society; see Geert Hofstede, Gert Jan Hofstede, Michael Minkov, *Cultures and Organizations: Software of the Mind* (USA: McGraw Hill, 2010), Chapter 1; Helen Spencer-Oatey, P Franklin, 'What is culture? A compilation of quotations', *GlobalPAD Core Concepts* (Warwick University, 2012); V. Žegarac, 'Culture and communication', *Culturally Speaking: Culture, Communication and Politeness Theory*, edited by Helen Spencer Oatey (London/New York: Bloomsbury Publishing, 2008), pp. 48-70. Talmudic references are to the Babylonian Talmud (Bavli) unless otherwise specified.

2.   Jacob Neusner, 'Rabbinic Judaism in Late Antiquity', *Encyclopedia of Religion*, edited by Thomson Gale (Detroit: Macmillan Reference, 2005), p. 7583; Asher Maoz, subsection 'Judaism', in 'The Impact of Jewish Law on Contemporary Systems with Special Reference to Human Rights', *Olir* (2004), pp. 1–3. Cf. Daniel Boyarin, *Judaism: The Genealogy of a Modern Notion* (New Brunswick, New Jersey: Rutgers University Press: 2018); Shaul Maggid, 'Is "Judaism" Necessary?: A Response to Boyarin's Judaism', in *The Marginalia Review of books* (24 May 2019). *Tanach* stands for Torah (the Mosaic Pentateuch), *Nevi'im* (Prophets), *Khethubim* (Writings). In Deuteronomy and Joshua, 'Torah' refers specifically to all or parts of Deuteronomy: see Deut. 4.44, 27.3, Joshua 1.18, *Bereishith Rabboh*, 6.9 (Theodor-Albeck edition, pp. 49–50.). In Mishnaic literature (c. 10

CE–200 CE), 'Torah' referred specifically to the Pentateuch (see George F. Moore, *Judaism in the First Centuries of the Christian Era: The Age of Tannaim* (Peabody, Massachusetts: Hendrickson Publishers, 1997). Later, however, especially in the medieval period, 'Torah' may have referred to the entire corpus of traditional Jewish thought, depending on the context. 'Talmud' in this paper refers either to the Jerusalem Talmud, complied in Roman Israel (Palestine) c. 300–400 CE or to the Babylonian Talmud, compiled by the rabbis (*Amoro'im*) of Sasanian Babylon, c. 230–750 CE.

3.   Henry George Liddell, Robert Scott, *A Greek-English Lexicon: Based on the German Work of Francis Passow* (New York: Harper & Brothers, 1846), s.v. ἁρμονία; Edward A. Lippman, 'Hellenic Conceptions of Harmony', *Journal of the American Musicological Society* (University of California Press: Vol. 16, No. 1: Spring, 1963), pp. 3–35, p. 3; G.W. Leibniz, *Philosophical Papers and Letters: A Selection* (Springer Science & Business Media, 6 December 2012), p. 57; Alan Rich, 'Harmony', *The New Encyclopaedia Britannica* (New York: Encyclopaedia Britannica, 1983), available online at https://www. britannica.com/art/harmony-music [accessed 12 May, 2019]; Carl Dahlhaus, 'Harmony', *Grove Music Online*, edited by Deane L. Root (Oxford: Oxford University Press, 2019), available online at https://www.oxfordmusiconline.com/grovemusic [accessed 12 May 2019]. See also, 'symmetry' in *Dictionary.com*, online at https://www.dictionary.com/ browse/symmetry [accessed 20 May 2019].

4.   See e.g., Scott B. Noegel, 'Greek Religion and the Ancient Near East', in *A Companion to Greek Religion*, edited by Daniel Ogden (Oxford: John Wiley & Sons, 1 Feb 2010), pp. 21–37; Jan N. Bremmer, *Greek Religion and Culture, the Bible, and the Ancient Near East* (Leiden: Brill, 2008), p. 102; also, Thomas Francis Glasson, *Greek Influence in Jewish Eschatology: With Special Reference to the Apocalypses and Pseudepigraphs* (London: S. P. C. K., 1961).

5.   See, for example, Edward A. Lippman, 'Hellenic Conceptions of Harmony', pp. 3–35.

6.   Jerry V. Diller, *Cultural Diversity: A Primer for the Human Services* (Boston, MA: Cengage Learning, 2013), p. 271; Derald Wing Sue, David Sue, *Counseling the Culturally Diverse: Theory and Practice* (Hoboken, NJ: John Wiley & Sons, 2015), p. 484; Marcus Colchester, *Salvaging Nature: Indigenous Peoples, Protected Areas and Biodiversity Conservation* (Darby, PA: Diane Publishing, 1994), p. 25.

7.   Robert Whiting. *You Gotta Have Wa* (New York: Random House: Vintage, (1989) 2009), pp. 27–52.

8.   Maximillien de Lafayette, *Vol. 14. Comparative Encyclopedic Dictionary of Mesopotamian Vocabulary Dead & Ancient Languages* (New York and Berlin: Times Square Press, 2014), p. 64.

9.   See R. David Qimḥi, *Peruše Rabbi David Qimḥi* (ReDaQ) *'al hat-tōrā* (Jerusalem: Mossad Horav Kook 1975), Psalm 133.1.

10.   R. Meir Leibush Weiser (Malbim), *Miqra'ei Qōdesh* (Rome, 1891), Job, 38.7.

11.   See: R. Shneur Zalman Baruchovitch of Liadi, 'Iggereth HaQodesh', in *Liqutei Amorim Tānya* (Brooklyn, New York: Kehot Publication Society, (5714) 1954), Epistle 19; R. Eliyohu Munk, *Ascent to Harmony* (New York: Feldheim, 1987).

12.   R. Nachman of Breslav, *Liqutei Moharan*, I, (Jerusalem, 5735 [1974/5]), p. 93b; R. Isaac ben Moše Arama, *Aqeidath Yitschoq* (Salonika, 1522), p. 74. Cf., 'Richard Dagley' in *Instructive gleanings, moral and scientific, from the best writers on painting and drawing, arranged by R. Mainwaring*, edited by Rowland Mainwaring (London, 1832), p. 113.

13.   R. Abraham Isaac Kook, *Siddur Tefilloh im Peirush Ōlath Re'iyoh* (Mossad Horav Kook, 1983), Vol. 1, p. 320.

14. עֹשֶׂה שָׁלוֹם בִּמְרוֹמָיו – 'He makes peace (šolōm) in His heights': see *T. Yerushalmi, Roš Hašanah*, 2:5, *Numbers Rabbah*, 12.8.

15. *Exodus Rabbah*, edited by E. Halevy (Tel Aviv: 1956-1963), *parashah* 12, subsection 6 (Vilna edition). All subsequent references to *Midrāš Rabbah* are to this edition). (Note: the *parashiyyoth* [sections] were based on a triennial cycle of reading the Pentateuch). Cf. Ḥagigoh 12a.

16 . *Midrash Shir HaShirim Zuta*, 1, edited by S. Buber, (Berlin: 1894). Regarding the lion and buffalo's natural rivalry, see *Midrasch Tanchuma*, edited by Buber, (4 vols. Wilna: Wittwe and Gebruder, 1885), *Vayigash*, 43.3.

17. 'Pirqei Ovoth' [Ethics of the Fathers], *Mishnayoth*, edited by P. Blackman, (New York: Judaica, 1964), 5.17. (Henceforth, all mishnaic references are to this edition.)

18. B'rachoth 64a: תלמידי חכמים מרבים שלום בעולם; R. Yonathan Eybeschütz, *Yā'arōth D'vāsh* (Lublin, 1875), Vol. 2, p. 8a; R. Abraham Isaac Kook, *Olath Re'iyoh*, Vol. 1, pp. 320, 435. See also, Yevamoth 14b (also, Mišnah Yevamoth, 1.4; Tosefta Yevamoth, 1.10); Soteh 47b.

19. R. Samson Rafael Hirsch, *The Hirsch Chumash: The Five Books of Torah* (Jerusalem: Feldheim Publications, 2008), Numbers, 25.12.

20. See e.g. R. Dov Baer of Mezritch, *Maggid Devorov L'Ya'aqov, Liqutim Yeqorim* (Jerusalem: Yeshivath Toldoth Aharon, 5731), §8, §122, §213; R. Yehuda Aryeh Leib Alter, *Sefāth Emeth, Qōrāch*, Section 1 (Pe'er, Israel: 1993).

21. R. Elimelech of Lizhensk, *Nō'ām Elimelekh* (Lemberg 5619 (1859)) *Yithro*, s.v. אם תהיו כדבר הנסתר והנעלם [...], רק תהיו נכנעים כאילו אינכם כלל, אז תשמעו בקולי *VeAtoh*.

22 R. Dov Baer of Mezritch, *Māggid Devorov L'Ya'aqōv*, §159. צריך האדם לחשוב את עצמו כאין, וישכח א"ע מכל וכל [...] ואזי יכול לבא למעלה מזמן, דהיינו לעולם המחשבה ששם הכל שוה, חיים ומות, ים ויבשה. – See also, §142,

23. *Sefath Emeth*, ibid.

24. The prayer, *Šolōm Āleikhem*, said before the Friday evening (Šabbath) meal. 'Halakha' here denotes the Jewish rabbinical legal system.

25. *Genesis Rabbah*, 12.8.

26. *Genesis Rabbah*, 12.8.

27. *Zohar* (Zhitomir, 1863), vol. II, 24b; R. Moše Cordovero, *Pardes Rimōnim* (Cracow: 1591), Gate 31, Ch. 1.

28. Genesis Rabbah, 12.15; R. Shlomoh Efraim Lipshits, *'Ir Gibōrim* (Froben: 1580), p. 81a; Sifre, Deut., 26.10.

29. יוֹצֵר אוֹר וּבוֹרֵא חֹשֶׁךְ עֹשֶׂה שָׁלוֹם וּבוֹרֵא רָע [Translations are the author's, unless stated otherwise.]

30. *Dictionary of Deities and Demons in the Bible*, edited by Karel van der Toorn, Bob Becking and Pieter W. van der Horst (Leiden: Brill; 1998), pp. 755-6; *Theological Dictionary of the Old Testament, Volume 15*, eds. G. Johannes Botterweck, Helmer Ringgren, Heinz-Josef Fabry (Grand Rapids: Eerdmans Publishing, 2006), pp. 24ff. Regarding the classic ANE understanding of transcendent deities influencing the lower realms through manifestation of agency in the celestial bodies, see Manfred Hutter, 'Astral Religion', *Religion Past and Present*, edited by Hans Dieter Betz (Brill, 2009); Francesca Rochberg, '"The Stars and Their Likenesses": Perspectives on the Relation between Celestial Bodies and Gods in Ancient Mesopotamia', *What is a god? Anthropomorphic and Nonanthropomorphic Aspects of Deity in Ancient Mesopotamia*, edited by Barbara N. Porter (Chebeague Island, Maine: Transactions of the Casco Bay Assyriological Institute I, 2000), pp. 41-91, 65, 79, 83, 89, 90; E. Frahm, 'Reading the Tablet, the Exta, and the Body: The Hermeneutics of Cuneiform Signs in Babylonian and Assyrian Text Commentaries and

Divinatory Texts', *Divination and the Interpretation of Signs*, edited by A. Annus (2010), pp. 93–141.

31. See Šabbath 34b; R. Dov Baer of Mezritch, *Māggid Devorov L'Yā'aqōv,* § 213; see also, John Curtis Franklin, 'Harmony in Greek and Indo-Iranian Cosmology', *The Journal of Indo-European Studies* (American Academy in Rome: Volume 30, Number 1 & 2, Spring/Summer 2002), p. 4.

32. See, R. Abraham Ibn Ezra, *Peirush Rābeinu Ibn Ezra*, Isaiah 45.7, s.v. ובורא את הרע; R. David Qimḥi, *Redaq*, ad loc; R. Isaac ben Moše Arama, *Aqeidath Yitschoq*, p. 74. Regarding the parallelistic structure of Biblical verse, see Tawny L. Holm, 'Ancient Near Eastern Literature: Genres and Forms', *A Companion to the Ancient Near East* (Malden, Mass: Wiley-Blackwell: second edition, 2007), pp. 269–288, p. 271; Robert Alter, *The Art of Biblical Poetry* (New York: Basic Books, 2011); James Kugel, *The Idea of Biblical Poetry: Parallelism and Its History* (Baltimore: The Johns Hopkins University Press, 1998), Chaps. 1, 2. See also, Robert Lowth, *De sacra poesi Hebraeorum: praelectiones academiae Oxonii habitae* (Oxonii, 1753).

33. סוּר מֵרָע וַעֲשֵׂה טוֹב, בַּקֵּשׁ שָׁלוֹם וְרָדְפֵהוּ. See also, ibid., Psalm 121, 6-7; Psalm 128, 5–6.

34. Gen. 1.4: וַיַּרְא א- לֹהִים אֶת הָאוֹר כִּי טוֹב; Ex. 2.2: וַתֵּרֶא אוֹתוֹ כִּי טוֹב הוּא – 'She saw him that he was good'; Soṭeh 12a. See also Ḥagigoh 12a.

35. See ibid., Psalm 121, 6-7; J. H. Weiss, ed. *Sifra on Leviticus* (Vienna: Jacob Schlossberg, 1862), Levit. 26.5.

36. וְיָשֵׂם לְךָ שָׁלוֹם. See Megillah 18b.

37. יוֹצֵר אוֹר וּבוֹרֵא חֹשֶׁךְ עֹשֶׂה שָׁלוֹם וּבוֹרֵא רָע אֲנִי ד' עֹשֶׂה כָל-אֵלֶּה: – He Who forms light and creates darkness, makes šolōm and creates evil, I am G-d, doer of all these things.

38. Job, 25.2 – הַמְשֵׁל וָפַחַד עִמּוֹ עֹשֶׂה שָׁלוֹם בִּמְרוֹמָיו.

39. *Midrāš Šoḥer Tov* (*Midrash Tehillim*, edited by S. Buber, Vilna, 1891), Psalm 18; *Genesis Rabbah*, 6.7 – והמאורות, רבי . שנאמר, ונתתי שלום בארץ...שלושה דברים ניתנו מתנה לעולם, ואלו הן,; 'Pereq HaŠolōm' in *Mascheth Derech Erets Zuta:* התורה והגשמים זעירא בשם ריש לקיש אמר, אף השלום, אלמלא שנתן הקב"ה שלום בארץ היתה החרב משכל את האדם.

40. עֹשֶׂה שָׁלוֹם בִּמְרוֹמָיו, הוּא יַעֲשֶׂה שָׁלוֹם עָלֵינוּ וְעַל כָּל יִשְׂרָאֵל וְאִמְרוּ אָמֵן -*Siddur Tefillath Yisro'el* (Brody, 1872), pp. 80, 96.

41. יוֹצֵר אוֹר וּבוֹרֵא חֹשֶׁךְ עֹשֶׂה שָׁלוֹם וּבוֹרֵא אֶת הַכֹּל

42. Ovid, *Metamorphosis, Book I*, lines 416-437, trans. by A. S. Kline (Borders Classics, 2004).

43. *Numbers Rabbah*, 12.8; *Jerusalem Talmud*, Roš Hašonoh (Bomberg, Venice, 1523), 2:5; *Zōhar*, Vol. III, 225a; *Midrash Zuta, Song of Songs* 1.

44. 'Philolaus of Croton', in *Introductory Readings in Ancient Greek and Roman Philosophy*, eds. Patrick Lee Miller, C. D. C. Reeve (Indianapolis, Indiana: Hackett Publishing, 2015), p. 7; Sheramy Bundrick, *Music and Image in Classical Athens* (Cambridge: Cambridge University Press, 2005), p. 141.

45. George S. Phylactopoulos, *History of the Hellenic World: The Archaic period* (Athens: Ekdotikē Athēnōn, 1975), p. 185. 'Chaos' here denotes disorder, unlike the Greek χάος, khaos, which denoted a chasm or abyss.

46. 'Philolaus of Croton', in *Introductory Readings in Ancient Greek and Roman Philosophy*, p. 7; Sheramy Bundrick, *Music and Image in Classical Athens*, p. 141; *Genesis Rabbah*, 8.5.

47. See Jon D. Levenson, *Creation and the Persistence of Evil: The Jewish Drama of Divine Omnipotence*, (Princeton University Press, 1994), Chapter 2.

48. וַיַּרְא אֱלֹקִים אֶת כל אשר עשה וְהִנֵּה טוֹב מְאֹד. See *Genesis Rabbah*, 9.7–10.

49. Isaiah 45.7- יוֹצֵר אוֹר וּבוֹרֵא חֹשֶׁךְ עֹשֶׂה שָׁלוֹם וּבוֹרֵא רָע אֲנִי ד' עֹשֶׂה כָל אֵלֶּה

50. Lamentations 3.38 – מִפִּי עֶלְיוֹן לֹא תֵצֵא הָרָעוֹת וְהַטּוֹב.

51. אֵשׁ וּבָרָד שֶׁלֶג וְקִיטוֹר רוּחַ סְעָרָה עֹשָׂה דְבָרוֹ.

52. *Genesis Rabbah*, 9.7.

53. B'rachoth 60b; *Leviticus Rabbah*, 24.2.

54. B'rachoth 60b; Ta'anith 21a; *Zohar*, vol. I, 15a.

55. See Gen. 1:26-7; *Avot d'Rabbi Natan* (Ethrog: 2000), **Chap. 31**, Mišnah 3; *Zōhār*, I, 20b; 71b; R. Ēliyyāhû Ben-Moše De Wîdaš, 'Gate of Awe', *Reishith Ḥokhmoh* (Cracow, 1593), 7:10.

56. *Ecclesiastes Rabbah* 7.13: בְּשָׁעָה שֶׁבָּרָא הַקָּדוֹשׁ בָּרוּךְ הוּא אֶת אָדָם הָרִאשׁוֹן, נְטָלוֹ וְהֶחֱזִירוֹ עַל כָּל אִילָנֵי גַּן עֵדֶן, וְאָמַר לוֹ, רְאֵה מַעֲשַׂי כַּמָּה נָאִים וּמְשֻׁבָּחִין הֵן, וְכָל מַה שֶׁבָּרָאתִי בִּשְׁבִילְךָ בָּרָאתִי, תֵּן דַּעְתְּךָ שֶׁלֹּא תְקַלְקֵל וְתַחֲרִיב אֶת עוֹלָמִי שֶׁאִם קִלְקַלְתָּ אֵין מִי שֶׁיְּתַקֵּן אַחֲרֶיךָ. See also, *Ecclesiastes Rabbah* 6.6.

57. Gen. 2.15: לְעָבְדָהּ וּלְשָׁמְרָהּ; *cf.* Qimḥi, *Redāq*, Gen. 3.17. Psalms 115.16: השמים שמים לה' והארץ נתן לבני אדם.

58. Leviticus Rabbah (Vilna edition) 25:3.

ר' יוחנן ב"ר סימון פתח: (דברים יג) אחרי ה' א-להיכם תלכו. וכי אפשר לבשר ודם להלוך אחר הקדוש ברוך הוא, אותו שכתוב בו (תהלים עז): בים דרכך ושבילך במים רבים, ואתה אומר: אחרי ה' תלכו!? ובו תדבקון (דברים יג) וכי אפשר לבשר ודם לעלות לשמים ולהדבק בשכינה, אותו שכתוב בו (דברים ד): כי ה' א-להיך אש אוכלה, וכתיב (דניאל ז): כורסיה שביבין דינור, וכתיב (שם): נהר דינור נגד ונפק מן קדמוהי ואתה אומר: ובו תדבקון!? אלא, מתחלת ברייתו של עולם, לא נתעסק הקב"ה אלא במטע תחלה, הדא הוא דכתיב (בראשית ב): ויטע ה' אלהים גן בעדן, אף אתם, כשנכנסין לארץ לא תתעסקו אלא במטע תחלה, הדא הוא דכתיב: כי תבאו אל הארץ.

59. Avot d'Rabbi Natan, 31b.

60. Bava Qama 91b; Ḥullin 7b:9; Kiddushin 32a; R. Moše ben Maimon (Maimonides), *Mišneh Tōrāh*, Section: 'Laws of Kings and Battle', 6:10. See also, *Ex. Rabbah* 38.2.

61. R. Aharon HaLevi, *Sefer HaChinukh* (Mishor, 2001), 529:2.

ולא יאבדו אפילו גרגיר של חרדל בעולם, ויצר עליהם בכל אבדון והשחתה שיראו, ואם יוכלו להציל יצילו כל דבר מהשחתה בכל כחם

62. See, for example, Nachmanides (1194–1270), *Ramban Ul HaTorah* (Jerusalem, A. Blum, 5753), Deut. 22.6, s.v. כי יקרא קן צפור לפניך; R. Bahya ben Asher ibn Halawa (1255–1340), *Rabbeinu Bachya Ul HaTorah* (Jerusalem, Ma'ayan HeChochmoh, 5718), Deut. 22.7, s.v. שלח תשלח את האם ואת הבנים תקח לך.

63. See Leviticus 25.7; Niddah 51b.

64. See Deut. 22.4; 22.10; 25.4; *Da'ath Zeqeinim* on Deut. 22.10; R. David ben Zimri (Radvaz), *Mogen Dovid, Mitsvah* 54.

65. See Plato, *Republic*, trans. by Joe Sachs (Indianapolis: Hackett Publishing, 2011), p. 197; *Plato's Republic*, trans. by Benjamin Jowett (Jazzybee Verlag, 1894), p. 279; also, quoted in Porphyry, *Neoplatonic Saints: The Lives of Plotinus and Proclus by Their Students*, trans. by Mark Edwards (Liverpool: Liverpool University Press, 2000), p. 64, n. 54.

66. William Shakespeare, 'Romeo and Juliet', *William Shakespeare: The Complete Works* (*Oxford Shakespeare*), eds. Stanley Wells, Gary Taylor, John Jowett, and William Montgomery (Oxford: Oxford University Press; second edition, 2005), Act II, Scene II.

67. Shakespeare, 'Romeo and Juliet', ibid.

68. See Ex. 3.13-15; R. Joseph Gikatilia, 'Introduction', [שערי אורה] *Sha'arei Orah* (*Gates of Light*) (Cracow, 1883). See also, Máire Byrne, *The Names of God in Judaism, Christianity, and Islam: A Basis for Interfaith Dialogue* (London: Continuum, 2011), p. 12; F. V. Reiterer, s.v. 'Šem', *Theological Dictionary*, p. 134. Zohar, vol. II, 22a; *Pirqei DeRabi Eliezer*, edited by Eliezer Treitel (Hebrew University, 2012), Ch. 3; R. Menahem Azariah da Fano, *Yonath Elem* (Yerushalayim: Mekhon Yiśmaḥ Lev, Torat Mosheh, 5767 [2006/7] Amsterdam, 5491), p. 1.

69. אמר רבי יודן בן רבי יוסי: גדול שלום ששמו של הקב"ה נקרא שלום, ההוא דכתיב (שופטים ו, כד)ויקרא לו ה' שלום.

THE CONCEPT OF HARMONY IN JUDAISM 171

Paul. P. Levertoff, *Midrash Sifre on Numbers* (London: S.P.C.K., 1926), Ch. 42; *Leviticus Rabbah*, 9.9. See also, *Song of Songs Rabbah*, Chap. 1, V 1; Chap. 3, verse 10; Šabbath, 10b.

70. Isaiah 40.25; R. Moše ben Maimon (Maimonides), *Mišneh Tōrāh*, Section: 'Fundamentals of the Torah', 1:8; R. Judah Halevi, *Kuzari* (Prague: 1838), 5:21. See also, Maimonides, *Guide to the Perplexed* (*Moreh Nevuchim*), (London: Pardes Publishing House, 1904), 1:58–59.

71. R. Moše ben Maimon (Maimonides), *Guide to the Perplexed*, I: 61; R. Joseph Gikatilia, *Ginath Egōz* (Yeshi-vath HaCḤayyim VeHaŠolōm, 1989), p. 289; R. Isaiah Horowitz, *Shenei Luḥoth HaBerith* (Amsterdam, 1698), p. 418a; R. Shneur Zalman of Liadi, 'Iggereth HaQodesh', in *Liqutei Amorim Tanya*, Ch. 4.

72. R. Jonah of Gerondi, *Sefer HāYir'oh* (Dubno: 1804); R. Menachem Tsiyoni, *Sefer Hātsiyoni* (County: 1560), p. 105b; R. Bachya ben Asher, *Khād Hāqemāḥ* (Warsaw: 1872), Ch. 1; R. Gershon Shaul Yom-Tov Lipmann, *Tosefeth Yom Tov* (Amsterdam: 1685), Sukkah 4.5.

73. Attr. to R. Shlomo Ibn Gabirol (1021-1058), beginning of Morning Prayers; see *Siddur Tefillath Yisro'el* (Brody: 1872), p. 15.

74. R. Joseph Gikatilia, *Ša'arei Ōrah*, Ch. 6; Gikatilia, *Ginath Egōz*, p. 289.

75. *Genesis Rabbah*, 12.15.

76. Hesiod, *Theogony*, trans. by Hugh G. Evelyn White (1920), lines 933–978; Hesiod, *Hesiod: Theogony, Works and days, Testimonia*, trans. by Glenn W. Most, (Loeb Classical Library: Harvard University Press, 2006), lines 933-978. See also, Lippman, 'Hellenic Conceptions of Harmony', p. 5.

77. R. Ḥayyim of Volozhin, *Nefesh HāḤāyyim* (Vilna: 1874), 1:4, n 4; 1:6; Avinoam Fraenkel, '*Nefesh HaTzimtzum*', translation & commentary of *Nefesh HaḤayyim*, (Israel: Urim Pub., 2015) Vol. 1, p. 126.

78. R. Ḥayyim of Volozhin, *Nefesh HaḤayyim*, ibid. Regarding the term *mitsvōth* (sing. *mitsvōth* מצוה), see R. Isaiah Horowitz, *Shenei Luḥoth HāBerith, Asereth HaDibroth, Yoma, Derekh Ḥāyyim*; R. DovBaer of Mezritch, *Ōhr Tōrāh*, (Koretz: 1804) *T'rumoh*, p. 103b; Marcus Jastrow, *Dictionary of the Targumim, the Talmud Babli and Yerushalmi, and the Midrashic Literature*, (Philadelphia: 1903), p 823, s.v. מצוה; also, B'rachoth 6a; Bava Bathro 21a; *Zohar* III, p. 284a.

79. R. Ḥayyim of Volozhin, *Nefesh HaḤayyim*, ibid.

80. *Zōhār* III, 10b; 142b.

81. I Chronicles 29.11: לך ה' הגדלה והגבורה והתפארת והנצח וההוד כי כל בשמים ובארץ. See R. Moše Ḥayyim Luzzatto, *Kelalej maamar ha-chochma*, 1.7, available online at http://www. daat.ac.il/daat/mahshevt/mahadurot/klaley-2.htm [accessed 19 May 2019].

82. See, for example, R. Ḥayyim Vital, *Pri Eits Ḥayyim* (Korzec: 1785), *Sha'ar Ho'Amidoh*, Ch. 1; R. Elimelekh of Lizhensk, *Nō'ām Elimelekh*, s.v. *Im Beḥoqothai* (p. 66b).

83. *Zohar* III, 31a, 115b; R. Moše Cordovero, *Tomer Devorah* (London: 2003 [Vienna, 1589]), Ch. 5, p. 121;

84. R. Ḥayyim Vital, *Pri Eits Ḥāyyim, Sha'ar Ho'Amidoh*, Ch. 20; Cordovero, *Pardes Rimōnim*, Ch. 23.10, 21.

85. *Psalms*, - הַשָּׁמַיִם מְסַפְּרִים כְּבוֹד אֵ-ל וּמַעֲשֵׂה יָדָיו מַגִּיד הָרָקִיעַ. [...] תּוֹרַת ה' תְּמִימָה מְשִׁיבַת נָפֶשׁ 19.1,8.

86. Qimhi, ad loc.

87. *Midrāš Šoḥer Tov*, 1; *Pesiqta Derav Kahana*, 12:19; *Midrasch Tanchuma*, Ex. 12.1; R. Yehudoh Loew of Prague, *Tif'ereth Yisroel* (Jerusalem, Sifrei Maharal, 5744),

Chapter 2.

88. *Numbers Rabbah*, 11.10; *Pesiqta Zutratha*, Leviticus 52b.

89. Maimonides, *Mišneh Torah,* Section: 'Laws of Chanukah and Purim', 4:14; *Zōhār*, vol. III, p. 176b.

90. R. Isaiah Horowitz, *Shenei Luḥōth HāBerith*, vol. I, p. 343a, vol. II, p. 198a; *Midrasch Tanchuma*, 1.1; *Gen. Rabbah*, 8.2; *Lev. Rabbah*, 19.1.

91. R. Judah Loew ben Bezalel (Maharal) (c. 1512–1609), *Nethivoth Olom* (Warsaw: 1873), p. 3b; R. Moše Ḥayyim Luzzatto (1707-1746), *Adir BāMorōm* (Machon Ramchal: 2018), p. 110.

92. התורה רושם האלקים והעולם רושם התורה - R. M. A. da Fano, *Yonath Eilem*, Ch. 1; see also, *Ex. Rabbah* 32:4; R. Menachem Nochum Twersky, *Meor Einayim*, (Square, N.Y.) pp. 16, 94, 292, 294, 427. Cf. R. J. Gikatilia, *Ginath Egoz*, p. 289: 'the world's continued existence is through letters, existence of the letters is through numerics, and the numerics exist through G-d' (קיום העולם באותיות, וקיום האותיות בחשבון, וקיום החשבון בו יתברך).

93. *Pirqei Ovoth* ('Ethics of the Fathers'), 1.18; *Zōhār*, vol. III, p. 176b; Šabbath 88a.

94. כאיש אחד בלב אחד - Mekhilta, Ex. 19.2.

95. Avōdah Zarah, 26.

96. Ex. 23.7; Yōma 69b; Šabbath 55a; Yevamōth 65b; Bava Metsi'ah 23b.

97. *Leviticus Rabbah*, 9.9.

98. R. Don Isaac Abravanel, *Abravenel* on I Samuel, 3:3; Maimonides, *Guide to the Perplexed*, 1:10, 61; R. Judah Loew ben Bezalel, *Derech Chāïm* (Machon Yerushalayim, 2005), Ch. 4, Mišnah 8.

99. *Midrāš Sifre* on Numbers, Ch. 42.

100. כבר המשילו זאת חקרי לב למה שהמצישו חכמי המוזיקא בשני כלי זמר אחד גדול ואחד קטן העומדים בהיכל. מוכן לכך בסדר ובערך ובגבול ידוע, שאם יפרטו על פי הנבל הקטן, יתן הנבל הגדול זמירות לעומתו, וכן ערך ה' שהנבל הגדול אשר עשה עשה שהוא העולם בכללו, יתן זמירות לנגן נגוני שמחה או עצב, טוב או רע, לפי מה שיפרוט האדם על פי הנבל הקטן שהוא גופו, שאם ישמיע נגונים נגוני שירים טובים ותשבחות כפי מצות ה' וחוקיו ומשפטיו, כן ינגן הנבל הגדול שהוא העולם שירי שמחה וחדוה וחדוה שפע וברכה וכל טוב, ואם ישמיע קול קינים והגה והי בעברו מצות ה' ותורתו, כן יגדל המספד על הנבל הגדול על החרבן וההשחתה שיהיה בכל העולמות *Malbim*, Shemoth 25, *Remozai HaMišhkon*.

101. B'rachoth 6a; Todros Ben Joseph Ha-Levi Abulafia, *Ozār hā-Kavōd (Warsaw, 1879), ad loc.*; R. David ibn Zimra, *Magen David* (Amsterdam, 1723).

102. See R. Moše Cordovero, *Tōmer Devōrah.*

103. See, for example, Klaus Mainzer, *Symmetry and Complexity: The Spirit and Beauty of Nonlinear Science* (New Jersey: World Scientific, 2005), p. 35.

104. Lippman, 'Hellenic Conceptions', p. 8; Günter Berghaus, 'Neoplatonic and Pythagorean Notions of World Harmony and Unity and Their Influence on Renaissance Dance Theory', *Dance Research: The Journal of the Society for Dance Research*, Vol. 10, No. 2 (Edinburgh Univ. Press, Autumn, 1992, pp. 43-70), p. 44, quoting Jula Kerschensteiner, *Kosmos: Quellenkritische Untersuchungen zu den Vorsokratikern* (Munich, 1962). Plato, too, emphasised the harmonic nature of the cosmos: see Plato, *Timaeus*, trans. by R. G. Bury (Loeb Classical Library, 1929), pp. 56-59; also, Plato, *Theaetetus*, trans. by H. N. Fowler (Loeb Classical Library, 1921), pp. 234–6.

105. *Sefer Yetsirah* 1:1 – *Sefer Yetzirah: The Book of Creation: In Theory and Practice*, edited by Aryeh Kaplan (Red Wheel/Weiser, second revised edition, 1997), p. 5.

106. R. Joseph Gikatilia, *Gināth Egōz*, p. 289: 'the existence of the world is through letters, the existence of the letters through numerics, and the numerics exist through G-d' קיום העולם באותיות, וקיום האותיות בחשבון, וקיום החשבון בו יתברך. [...] באמתת שם שהוא יסוד כל חשבון, ובראא (את העולם ותלאו בדרך חשבון [...] והוא לא המציא דבר כי אם מאמתת החשבון ואינו מתנהג כי אם על דרך החשבון).

107. See: R. Judah Loew, *Tif'ereth Yisro'el* (Warsaw: 1835), Ch. 25.

108. *Rhys Carpenter, The Esthetic Basis of Greek Art: Of the Fifth and Fourth Centuries B.C, (Pennsylvania: Bryn Mawr College: 1921),* pp. 107, 122, 128.

109. Alexey Stakhov, *The Mathematics of Harmony: From Euclid to Contemporary Mathematics and Computer Science* (New Jersey: World Scientific, 2009), p. 40; Euclid, *Elements* (Santa Fe, New Mexico: Green Lion Press, 2002), Book VI, Definition III; Mario Livio, *The Golden Ratio: The Story of Phi: The World's Most Astonishing Number* (New York: Broadway Books, 2002), p. 10. See also, Johannes Kepler (1571–1630), *Mysterium Cosmographicum* (Tübingen, 1596) trans. by A. M. Duncan (Abaris Books, 1981): 'Geometry has two great treasures: one is the Theorem of Pythagoras: the other the division of a line into extreme and mean ratio. The first we may compare to a measure of gold; the second we may name a precious jewel'.

110. Euclid, *Elements*, Book VI, Definition III.

111. Ritendra Datta, Dhiraj Joshi, Jia Li, James Z. Wang, (2006) 'Studying Aesthetics in Photographic Images Using a Computational Approach', *Computer Vision – ECCV 2006: Lecture Notes in Computer Science*, eds. A. Leonardis, H. Bischof, A. Pinz (Berlin, Heidelberg: Springer, vol. 3953, pp. 288–301), p. 294; I. Christopher McManus, Fanzhi Anita Zhou, Sophie l'Anson, Lucy Waterfield, Katharina Stöver, Richard Cook, 'The Psychometrics of Photographic Cropping: The Influence of Colour, Meaning, and Expertise', *Perception*, (US: London, Pion: vol. 40, issue 3: 2011), pp. 332–57; Kamila Svobodova, Petr Sklenicka, Kristina Molnarova, Jiri Vojar, 'Does the composition of landscape photographs affect visual preferences? The rule of the Golden Section and the position of the horizon', *Journal of Environmental Psychology* (Elsevier: 2014, pp. 143–152), p. 144. See also, Livio, *The Golden Ratio*, p. 182; Sarah Kent, *Composition* (London: DK Adult, 1995); Alan Pipes, *Foundation of Art and Design* (London: Laurence King Publishing, 2003), p. 222.

112. Gary Meisner, *The Golden Ratio – The Divine Beauty of Mathematics* (New York: Race Point Publishing. 2018), p. 55.

113. Regarding the golden ratio in *tsitsith*, see Mois Navon, 'Rav's Beautiful Ratio: An Excursion into Aesthetics', *Threads of Reason* (Ptil Tekhelet, 2013), pp. 104–114.

114. שֵׁם בְּאָהֳלֵי וְיִשְׁכֹּן לְיֶפֶת אֱ-לֹהִים יַפְתְּ; Megillah 9b; see R. S. R. Hirsch, *The Hirsch Chumash*, Gen. 9.27.

115. "זה א-לי ואנוהו"; Šabbath 133b; Mekhilta de-Rabbi Ishmael, Ex. 15.2.

116. The translation combines the Mekhilta (לבשר אפשר וכי אומר ישמעאל ר' ,ואנוהו א-לי זה) with the Talmudic נאה תפלה נאה ציצית נאה סוכה נאה לולב לפניו אעשה במצות לו אנוה אלא לקונו להנוות ודם) text (ציצית ,נאה ושופר נאה ולולב נאה סוכה לפניו עשה ,במצות לפניו התנאה ,(ב ,טו שמות) "ואנוהו א-לי זה" נאה). תורה ספר ,נאה).

117. Sukkah 2a/b, 7a. The eleven handbreadths comprise ten for the edifice, plus one for the *s'chach* (natural, temporary roof) as implied in Sukkah 2b.

118. Mišnah Roš Hašanah, 3.6. Interestingly, the hand is included in the Lurianic meditations of the ritual of blowing the *šofar* on Roš Hašanah.

119. See Menaḥōth 39a, based upon Numbers 15.38.

120. Menaḥōth 39a; Maimonides, *Mišneh Torah*, Section: 'Laws of *Tsitsith*' 1.8; R. Joseph Karo, *Kesef Mišneh* ad loc; see Mois Navon, 'Rav's Beautiful Ratio: An Excursion into Aesthetics', *Threads of Reason* (Ptil Tekhelet, 2013), pp. 104–114.

121. Deut. 11.20; Menaḥōth 33a.

122. Menaḥōth 35a.

123. Menaḥōth 35b; R. Asher ben Yechiel, (end of) *Laws of Tefillin* (end of Tractate Menaḥōth).

124. Maimonides, *Mišneh Tōrāh*, Section: 'Laws of *Tefillin*', Ch. 3; R. Asher, (end of) *Laws of Tefillin*.

125. Menaḥōth 36b/37a.

126. Mišnah Sukkah, 3.1–3.

127. Bava Bathra 14a. Regarding the Euclidean Rectangle, see Stakhov, *Mathematics of Harmony*, pp. 21–22.

128. Menaḥōth 30a.

129. Vitruvius, *The Ten Books on Architecture*, trans. by Morris Hicky Morgan (1914), edited by Tom Turner (2000), Book 1, Chap. 2.4; Book 3, Chap. 1.1; Livio, *Golden Ratio*, p. 161. (Thomas Aquinas (1225–1274) similarly wrote, 'beauty consists in due proportion, the senses delight in things duly proportioned': Thomas Aquinas, *Summa Theologica*, I.5.4 (Thomas Aquinas, *Basic Writings of St. Thomas Aquinas*: Volume 1 (Indianapolis, Indiana: Hackett Publishing, 1997), p. 47.)

130. Mekhilta de-Rabbi Ishmael, Ex. 15.2.

131. אבא שאול אומר ואנוהו הוי דומה לו מה הוא חנון ורחום אף אתה היה חנון ורחום. Mekhilta de-Rabbi Ishmael, Ex. 15.2; Šabbath 133b; Raši, ad loc.

132. Mekhilta de-Rabbi Ishmael, Ex. 15.2.

133. See R. Ḥayyim Vital, *Ma'amar Pesi'othov Shel Avrohom Ovniu* (Ahavath Šalom, 1998) that opinions in Midrash were not at variance but merely expressed different aspects of an idea.

134. *Midrasch Tanchuma*, 7.16.

135. See, for example, Morton Smith, 'Palestinian Judaism in the First Century', *Israel: Its Role in Civilization*, edited by M. Davis (New York, 1956), p. 71; Henry A. Fischel, 'Story and History: Observations in Greco-Roman Rhetoric and Pharisaism', *American Oriental Society, Middle West Branch, Semi-Centennial Volume* (1969), edited by D. Sinor, p. 82; Martin Hengel, *Judaism and Hellenism: Studies in their Encounter in Palestine during the Early Hellenistic Period* (Wipf and Stock Publishers, 2003); Lee I. Levine, Yiśra'el L. Leṿin, *Judaism and Hellenism in Antiquity: Conflict or Confluence?* (University of Washington Press, 1998).

136. Deut. 30.19, 6.18; Mišnah B'rachōth 9.5.

137. Mišnah, Uqtsin, 3.12; TY, Berachōth 17b.

138. Cicero, *De Finibus Bonorum et Malorum*, trans. by H. Rackham (Loeb Classical Library, 1931), Book III.33, pp. 245, 253; Epictetus, *Discourses*, trans. by Robert Dobbin (Penguin UK, 2008), Book I, Chaps. 2, 4, 6.

The Bible: A Guidebook for Harmony?

Rhodri Thomas

GIVEN THE SERIOUSNESS OF THE SITUATION facing Earth and its inhabitants – the ecological crisis represents what is surely *the* most formidable challenge that humankind is likely to be faced with in the twenty-first century – there is a real and pressing need for a drastic re-assessment of humanity's relationship to the natural world; only once this is undertaken will the crisis truly begin to be resolved. What is required, in other words, is a collective *metanoia*: a radical change in worldview from one clouded by a sense of human beings' superiority and special status to one, in the words of eco-theologian Norman Habel, 'where ecology conditions our thinking'.[1] In short, a worldview shaped by ecology will not elevate or extol certain of the planet's living beings (namely, *human* beings) to the detriment or disadvantage of others, but will acknowledge and appreciate Earth/the cosmos for what it is: a wondrous network of complementary familial relationships. This I define as *harmony*. Becoming aligned with such an understanding of existence on Earth demands, of course, that the Bible be read in a fundamentally different way, that is, from the perspective of fellow members of the Earth community. The question that follows is whether or not the biblical texts, when read in this way, can be said to support or affirm an ecology-oriented worldview. Whatever the answer received, this ecological approach to the Bible must be considered a necessary and worthwhile enterprise; contemporary concern over the perceived threats to the natural world, coupled with the massive influence which the Bible continues to exert over hundreds of millions of Christians around the globe, means that what the biblical texts 'say', or are purported to say, in relation to the natural world cannot easily be ignored.

In the same way, then, as the women's rights movement helped prepare the ground for feminist readings of the Bible – or, for that matter, in the same way as debilitating socio-political conditions in Latin America led to the development of liberation theology – so a greater awareness of contemporary environmental concerns has provided the context for a fresh kind of interaction with the biblical tradition. Even during the reasonably short life-span of the modern environmental movement, attempts made by ethicists and theologians at reading the Bible through a green lens have yielded a truly staggering volume of literature. It is, after all, only natural that, having been confronted with the potential for ecocatastrophe on an unprecedented scale, ecologically-minded scholars should seek to engage biblical material in a bid to discern what (if anything at all) the individual texts might be able to offer a discussion of environmental issues.

While the importance of ecological engagement with the biblical tradition is now commonly recognised, it is Lynn White's excoriating critique of the Judeo-Christian 'dogma of creation' in a 1967 paper entitled 'The Historical Roots of Our Ecologic Crisis', that is usually identified as providing the specific catalyst for the debate over the Bible's ecological legacy.[2] To White's mind, the advent of the Christian faith, particularly in its triumph over pagan religion, initiated a dualism of humanity and nature, promulgating the alien notion that the human race existed above, and apart from, the rest of the created order. The biblical account of creation in Genesis 1-2, writes White, in fact legitimises the exploitation of the natural world; its insistence that humans are created in the image of God, and thus share in God's transcendence of nature, actually set humanity in direct opposition to non-human creation. White's opinion of the environmental legacy of the Christian tradition is as uncompromising as it is controversial: 'Especially in its Western form, Christianity is the most anthropocentric religion the world has seen'.[3] As such, it 'bears a huge burden of guilt' for the contemporary ecological crisis.[4]

Though his analysis of the biblical creation myth remains controversial, the publication of White's paper provided a genuine watershed moment in the context of twentieth century eco-theological discourse: the article is almost single-handedly responsible for instigating a long-standing debate about the ecological value of the biblical material. White's charge – that the Judeo-Christian tradition bears an innate anthropocentric bias that has paved the way for the ecological crisis – elicited a fierce response from scholars keen to demonstrate that the creation story in Genesis does not preserve or promote any 'anti-environmental' sentiment. More than this, it stimulated an intense interest in the environmental credentials of the biblical tradition as a whole. Many works of eco-theology composed in the wake of White's paper were written to show that the biblical texts *do* endorse a positive view of creation, and that these ancient writings *can* be viewed as a relevant and useful tool in generating an ethical stance with regard to green issues.

The Green Bible may be seen as the culmination of this kind of positive approach to the biblical tradition, and its publication is arguably the most momentous occurrence in the history of the Christian environmental movement to date. This recent edition of the New Revised Standard Version takes seriously its claim that the Bible functions as a 'powerful ecological handbook on how to live rightly on earth'.[5] Accordingly, those verses considered to disclose some ecological wisdom are demarcated by green ink; over a thousand passages are highlighted in this way. The inference to be drawn, therefore, is that the individual biblical texts do not simply convey the occasional exhortation to environmental care, but that they are veritably chock-full with ecological insight and guidance, able to 'speak directly to how we should think and act as we confront the environmental crisis facing our planet'.[6] Clearly, the committee of scholars responsible for *The Green Bible* have a quite different opinion of the ecological worth of the biblical material from critics such as Lynn White. What is more, White's charge of an all-pervasive anthropocentrism would appear hopelessly wide of the mark if the 'evidence' of *The Green Bible* is to be taken at face value; one of the stated aims of *The Green Bible* is to testify that the Bible *in its entirety* is bursting with ecological overtones. That is, it can be shown to tender a blueprint for human interaction with the Earth and delineate exactly the kind of role and responsibility humans are called to have in caring for God's creation.

One of the texts highlighted in *The Green Bible*, and a useful illustration of the way in which Christians have strived to ground burgeoning ecological sensitivities in an appeal to biblical material, is the so called 'dominion mandate', the command issued to the first humans by God in Genesis 1:26-28 to 'fill the earth and subdue it (שבכ); and have dominion (הדר) over the fish of the sea and over the birds of the air and over every living thing that moves upon the earth'. These verses are frequently understood by both scholars and lay people as advocating a need for humans to act as responsible stewards or managers of creation. Subscribers to the stewardship model duly reject a notion of 'dominion as exploitation'. To interpret the passage as a divine imperative to the exploitation of Earth, they claim, is to misunderstand the imagery and symbolism at play. Quite the opposite: 'dominion means responsible stewardship'.[7] Stewardship is a widely recurring theme across many Christian denominations and, for millions of Christian believers, is the key to theological engagement with ecological concerns. Upheld as a principal tenet (particularly in Protestant evangelical attempts at formulating an ethical stance towards environmental issues), it is a trope that has resurfaced time and again, often in the declarations and statements issued by organisations such as the Evangelical Climate Initiative and the Evangelical Environmental Network. The pre-eminence of the stewardship model among evangelicals may justifiably be attributed to the fact that it allows the Christian to maintain a high view of the authority and sovereignty of scripture; this approach preserves a formative and definitive role for the Bible – acknowledged by evangelicals as the inerrant

word of God – in articulating a thoroughgoing biblical ecological ethic. The 2006
declaration published by the ECI, for example, states:

> Christians, noting the fact that most of the climate change problem is human
> induced, are reminded that when God made humanity he commissioned us
> to exercise stewardship over the earth and its creatures. Climate change is the
> latest evidence of our failure to exercise proper stewardship, and constitutes a
> critical opportunity for us to do better (Gen.1:26-28).[8]

However, it is doubtful whether the Bible can actually be shown to support a
principle of stewardship at all, at least as far as the Earth-human relationship is
concerned. The idea that humanity has a responsibility to exercise care for non-
human creation, that humans must act as divinely-appointed 'managers' of the
Earth, is not borne out by the biblical material. A brief inspection of the use of
the verbs שבכ and הדר in the biblical texts reveals that a reading of Genesis 1
which acknowledges the supremacy of humanity over creation may actually be
the more faithful to the passage's original meaning: other instances of these verbs
indicate their overtly hostile connotations. The word שבכ, translated in the NRSV
as 'subdue', is repeatedly used in the sense of 'subjugate', while הדר denotes 'rule'
and 'domination'.[9] On this evidence, the reader might be justified in maintaining
the view that Genesis 1: 26-28 does supply a mandate for human dominion
over creation and that these verses do encourage the exploitation of the planet's
resources to suit humanity's ends. In other words, Genesis 1:26-28, in its original
context at least, does not look anywhere near as 'environmentally friendly' as
purveyors of the stewardship model might like to think. The passage leaves a
legacy that is infinitely more ambiguous that that.[10]

Of course, Genesis 1:26-28 is not the only biblical passage to have had
aspersions cast on its ecological credentials. A number of texts, in both the Hebrew
Bible and New Testament, have proven particularly problematic to eco-theologians
keen to demonstrate the ecological worth of the biblical tradition. More often
than not, these are the texts which – at least on a superficial reading – appear to
portend great disaster upon the Earth as a part of their eschatological outlook.
Joel, for example, imagines the Sun turned to darkness and the Moon turned to
blood on the 'great and terrible day of the LORD' (2:30-32); the Synoptic Gospels
envisage ecological and social meltdown as a concomitant of the coming of the
Son of Man (Mark 13; Matthew 24:1-44; Luke 21:5-36); 2 Peter 3:7-13 warns
of an impending conflagration that will consume the elements; while images of

divinely-sanctioned ecological catastrophe abound in the book of Revelation (8:6-9:21; 16). What is more, these images of environmental collapse tend to be accompanied by the anticipation of a new creation or new Earth, a dwelling place radically different in character from the old, and significantly better in quality (e.g. 2 Peter 3:13; Revelation 21-22). Given such eschatological expectation, extracting any sizeable nuggets of ecological wisdom from this type of biblical material can be a laborious task. After all, what possible motivation to environmental care can there be for the believer if the current Earth is destined for the proverbial scrapheap, ready to be replaced by a superior model?

That is not to say that there have been no attempts to recover some ecologically positive meaning from these texts. Usually, this approach involves placing some emphasis on the renewal or transformation of the present Earth, together with the continuity between this world and the one to come, as opposed to its complete dissolution and re-creation. Ernest Lucas typifies this thinking when he suggests that the use of the adjective καινός rather than νέος in the phrase 'new heavens and a new earth' in 2 Peter 3:13 points to 'renewal through transformation, not a total destruction of the old and its replacement by something quite different'.[11] It should be said, however, that the difficulties associated with eschatological texts like 2 Peter 3 have proven too great for some. Barbara Rossing, for example, rejects 2 Peter 3 as ultimately unhelpful in generating biblically-informed attitudes towards the environment, calling it 'the most ecologically problematic chapter in the entire New Testament'.[12]

It would appear, then, that there are serious problems in indiscriminately viewing the biblical tradition as uniformly 'green'. Certainly, an unreserved commitment to the idea of an 'eco-friendly' Bible, exhibited in *The Green Bible* for example, appears misguided, if not completely naïve.[13] It is certainly worth remembering that the Bible is not a collection of eco-centric documents; its texts do not bear the weight of twenty-first century concern over the state of the environment or the well-being of Earth. With that in mind, it is probably healthier to exercise a level of scepticism when assessing the ecological value of the biblical material. In other words, ecological engagement with the Bible must begin with the suspicion that the Bible does not necessarily provide a useful foundation for the construction of a responsible environmental ethic. It involves the admission that the ecological orientations of most texts are more ambivalent than many recent studies have been willing to concede, and that the Bible may, for the sake of harmony (i.e. aligning one's actions with an ecology-oriented worldview), require a label other than 'eco-friendly': the one marked 'bio-hazard'.

NOTES

1. N. Habel, *An Inconvenient Text* (Adelaide: ATF Press, 2009), p. 41.
2. L. White, Jr., 'The Historical Roots of our Ecologic Crisis', *Science,* 155 (1967): 1203-7.
3. White, 'The Historical Roots', p. 1205.
4. White, 'The Historical Roots', p. 1206.
5. DeWitt, 'Reading the Bible through a Green Lens', *The Green Bible* (London: HarperCollins, 2008), pp. I-25.
6. Preface to *The Green Bible*, I-15.
7. C. B. DeWitt, 'Reading the Bible through a Green Lens', *The Green Bible*, I-26. Former US Secretary of the Interior, James Watt, encapsulates this standpoint when he writes: 'The Bible commands conservation – that we as Christians be careful stewards of the land and resources entrusted to us by the Creator'. See J. Watt, 'The Religious Left's Lies', *The Washington Post* (Saturday, May 21 2005).
8. 'Statement of the Evangelical Climate Initiative: CLAIM 3: Christian Moral Convictions Demand Our Response to the Climate Change Problem', http://www.christiansandclimate.org/statement/ [accessed 1 July 2018].
9. For שבכ as 'subjugate' see 14 occurrences in all (e.g. Jos. 18:1; 2 Sam. 8:11; 1 Chron. 22:18; Neh. 5:5; Jer. 34:11; Zech. 9:15) and for הדר as denotes 'rule' and 'domination' see 24 occurrences (e.g. Lev. 26:17; Num. 24:19; Judg. 5:13; 1 Kgs. 4:24; Neh. 9:28; Isa. 14:2). See J.W. Rogerson, 'The Creation Stories: Their Ecological Potential and Problems', *Ecological Hermeneutics: Biblical, Historical and Theological Perspectives*, eds. D. G. Horrell, C. Hunt, C. Southgate and F. Stavrakopoulou (London: T & T Clark, 2010), 21-31, at 25.
10. For a more sustained critique of the stewardship model, see C. Palmer, 'Stewardship: A Case Study in Environmental Ethics', *Environmental Stewardship: Critical Perspectives, Past and Present*, ed. R. J. Berry (London: T & T Clark, 2006), pp. 63-75. See also within the same collection of essays: R. Attfield, 'Environmental Sensitivity and Critiques of Stewardship', pp. 76-91; J. Lovelock, 'The Fallible Concept of Stewardship of the Earth', pp. 106-111.
11. E. Lucas, 'The New Testament Teaching on the Environment', *Transformation*, 16:3, p. 97.
12. B. R. Rossing, 'Hastening the Day When the Earth Will Burn? Global Warming, Revelation and 2 Peter 3', *Compassionate Eschatology: The Future as Friend*, eds. T. Grimsrud and M. Hardin (Eugene, Oregon: Cascade Books, 2011), p. 89.
13. For criticisms specific to *The Green Bible* itself, see D. G. Horrell, '*The Green Bible*: A Timely Idea Deeply Flawed', *Expository Times*, 121 (2010), pp. 180-86.

Harmony in Islamic Cosmology
Subjugation, *Sujūd* & Oneness in Islamic Philosophical Thought

M.A. Rashed

Introduction

THIS CHAPTER INTRODUCES THE DEEPER COSMOLOGICAL AND ONTOLOGICAL SYMBOLISM and significance of three concepts in Islamic philosophy and mysticism that affirm the significance of harmony as both the origin and outcome of universal equilibrium: first, the concept of subjugation, second, the concept of *sujūd* or prostration in prayer and third, the concept of oneness. I shall draw on the work of three noted medieval Islamic thinkers and philosophers: Abū Ḥāmid al-Ghazālī (1058–1111 CE), Muḥyī-d-Dīn ibn 'Arabī (1164–1240 CE) and ibn Qayyim al-Jawziyya (1292–1350 CE).[1] After examining the exoteric and esoteric connotations of *sujūd* and its intimate relationship with the Islamic view of subjugation, I will explore how these two interrelated concepts can elevate the spiritual seeker to a state of 'oneness', a mystical awareness of the harmonious unity of the cosmos.

Although the exact Arabic equivalent of the term 'harmony' is nonexistent in the major lexical works of prominent medieval Arab linguists and philologists, the concept of 'harmony', on the other hand, is ubiquitously present in Islamic theology, philosophy, cosmology and mysticism. The modern Arabic synonym for 'harmony' or 'to be in harmony' is *tanāghum* (تَنَاغُم), a term derived from the root *na-gha-ma* which according to medieval lexicons may either denote musical melody and beautiful tone in singing or enunciation or concealed speech.[2] The definition of na-gha-ma corresponds with the notion of universal harmony depicted in the Holy Qur'ān expressed as the entire universe being subject to (يَعْبُدُ) Allah, in verbal or physical expression, whether perceptible or imperceptible, audible or inaudible:

> The seven heavens and the earth, and all beings therein, declare His glory: there not a thing but celebrates His praise; and yet ye understand not how they declare His glory![3]

Believers are urged by the Qur'ān to be in musical concordance with this cosmic melody sung by all creation, and to be synchronised with the universal cycles in order to attain spiritual bliss:

> Celebrate (constantly) the praises of thy Lord before the rising of the sun, and before its setting; yea, celebrate them for part of the hours of the night, and at the sides of the day: that thou mayest have (spiritual) joy.[4]

A verse reported to have been composed by Alī ibn 'Abī Ṭālib (601-661 CE), the Prophet Muḥammad's cousin and son-in-law, and the fourth caliph after the Prophet, which reads 'A little body thyself thou deem, while the great universe in thee dwells' equally reflects a cosmology that sees the universe as a unified whole, with harmonious affinities, despite apparent randomness, disconnection and dispersion.[5] Similarly, two of the most fundamental principles of Islam, namely 'ubūdiyya (عُبُودِيَّة) or being subject to Allah (which I shall refer to as subjugation), and tawḥīd (تَوْحِيد) or monotheism, reveal to ibn al-Qayyim, ibn 'Arabī and al-Ghazālī a cosmological perception that confirms the harmonious interconnectedness of the inner self of the individual and the outer manifest cosmos.[6]

SUJŪD AND SUBJUGATION:
THE INTERCONNECTEDNESS OF THE CELESTIAL AND TERRESTRIAL

Sajada, the perfect verb from which the infinitive sujūd is derived, is generally defined in medieval Arabic lexicons as the action of placing one's forehead on the ground.[7] In the liturgical sense, sujūd involves the deliberate placement of seven body parts on the ground following the Prophetic tradition: 'I have been ordered to prostrate (asjidu) on seven bones, on the forehead (and then the Prophet pointed towards his nose) both hands, both knees and the toes of both feet'.[8] Sajada and its derivatives are used in the Qur'ān to describe the act of lowering oneself in prostration before Allah as a physical expression of utmost glorification, 'of those whom We guided and chose; whenever the Signs of (Allah) Most Gracious were rehearsed to them, they would fall down in prostrate adoration and in tears'.[9] Originally written as a commentary on the spiritual manual Manāzil al-Sā'irīn or Stages of the Wayfarers by Sufi scholar Imām al-Harawī (1006-1089 CE), Ibn al-Qayyim's Madārij al-Sālikīn or Ranks of the Divine Seekers provides the reader with profound spiritual insights pertaining to the concepts of sujūd and 'ubūdiyya, or subjugation, the latter being the origin from which the former emanates.[10]

According to ibn al-Qayyim, subjugation is a four-tiered concept that reflects absolute, exclusive and unconditional love for Allah through the expression and action of the heart, the former two being the first and second tiers respectively, while the expression of the tongue and the actions of the body parts respectively constitute the third and fourth tiers.[11] The expression of sincere love for Allah via the heart involves an intrinsic and sound belief in the His Divine Names, Attributes, Actions, angels and prophets as described in His revealed words.[12] The expression of the heart must then be solidified through metaphorical actions of the heart, such as entrusting one's soul and destiny to Allah, committing to His commands and abstaining from His prohibitions.[13] Internal sentiments and thoughts are then to be explicitly disclosed through the verbal articulation of the tongue and the physical actions of the limbs.[14] *Sujūd* may hence be understood as the physical, external and final demonstration of subjugation.

ibn al-Qayyim also noticed that subjugation and *sujūd*, in their spiritual connotations, transcend the tangible and individual to the abstract and universal.[15] The latter, according to ibn al-Qayyim, is 'universal subjugation,' a variety of subjugation which comprises 'all creatures of the heavens and earth to Allah, the pious and the impious, the believer and the disbeliever, for this is the subjugation of [Allah's] Coercion and Dominance'.[16] Since subjugation and *sujūd* are nearly synonymous concepts, ibn al-Qayyim added that universal subjugation naturally finds expression through a parallel mode of *sujūd*, similarly universal and ubiquitous as mentioned in the Qur'ān:

Seest thou not that to Allah bow down in worship [*yasjidu*] all things that are in the heavens and on earth – the sun, the moon, the stars; the hills, the trees the animals; and a great number among mankind? But a great number are (also) such as are fit for Punishment: and such as Allah shall disgrace none can rise to honour: for Allah carries out all that He wills.[17]

Clearly, the universal expression of *sujūd* to which ibn al-Qayyim was referring is a holistic form of submission to the Lordship and Authority of Allah that encompasses in its entirety both the perceptible *sujūd* of humans and the metaphorical *sujūd* of everything else. Within this category falls a sub-variety of *sujūd* defined by ibn al-Qayyim as *sujūd al-korh* or coercive *sujūd* illustrated in the Qur'ānic verse 'Whatever beings there are in the heavens and the earth do prostrate themselves to Allah (acknowledging subjection) — with good — will or in spite of themselves: so do their shadows in the mornings and evenings'.[18]

Although one does not find sufficient explanation by ibn al-Qayyim on the unique characteristic of coercive *sujūd*, it seems that he was indirectly equating it to the theological concept of *'ubūdiyyat al-rubūbiyya*, or deistic subjugation, which asserts the compulsive and inescapable subjection of all creation to the Lordship of Allah, His Will and His ordained natural laws. Coercive *sujūd*, as a reflection of deistic subjugation, greatly resonates with the Ghazālian perception of universal subjugation:

> Nature is in subjection to God the most exalted, not acting of itself, but serving as an instrument in the hands of its Creator. The sun, moon, stars and elements are also in subjection to His command, for none of them act in accordance to their own essence.[19]

In contrast, ibn 'Arabī's *al-Futūḥat al-Makkiyya* or *Meccan Revelations* introduces to the reader deeper esoteric and cosmological interpretations of the Qur'ānic verses on the universal and coercive varieties of *sujūd*. In ibn 'Arabī's view, the universal *sujūd* mentioned in the verse above is in truth symbolic of prostration of animate and inanimate beings to the *mashī'a* or Divine Decree of Allah, the Omnipotent.[20] Nevertheless, ibn 'Arabī also noticed that in the same verse, the partitive 'a great number' is used in reference to humans only and no other beings, which means that universal *sujūd* includes only a limited portion of fortunate humans enabled by Divine Decree to participate in the universal harmonious compliance with the universal Laws of the Creator:

> Joining [through willed prostration] thus those [the creatures] that prostate in heavens and [the creatures that prostrate] on earth, and the [prostration of] the sun as it sets, and [of] the moon as it wanes, and [of] the stars in their stations, and [of] the mountains in their stillness, and [of] the trees in their erection, and [of] the quadrupeds in their submission [to humans].[21]

Evidently, ibn 'Arabī saw that what differentiates the *sujūd* of humans from the *sujūd* of the rest of creation is that the former represents the willed surrender to Divine Laws as opposed to the spontaneous and mechanical submission of other entities and bodies. The free will of humans, however, remains determined by, and subordinate to, the superior and ultimate *mashī'a* or Divine Will as affirmed in the Qur'ān: 'Ye shall not will except as Allah wills the Cherisher of the Worlds'.[22] Alternatively, the second Qur'ānic verse on coercive *sujūd*, 'Whatever beings there are in the heavens and the earth do prostrate themselves to Allah (acknowledging subjection) — with good — will or in spite of themselves: so do their shadows in the mornings and evenings,' provides ibn 'Arabī with an alternative, and rather mystical, interpretation.[23] The verse, as ibn 'Arabī considered, delineates the structure of a cosmos inhabited by beings that belong to two distinct groups: mental beings, intelligences or angels, existing in the upper realms, and the terrestrials, which possess material, corporeal, forms.[24] Humans possess the qualities of both mental and terrestrial beings and therefore belong to the celestial and terrestrial realms simultaneously.[25] Since the corporeal form is in actuality the 'shadow' referred to in the same verse, it becomes merely a projection of the true essence of the spirit and the mind.[26] In other words, a physical body in prostration is purely a mirror of the mental prostration of the soul within, the former prostration controlled and coerced by the selective and voluntary prostration of the latter. And so, when the believing servant of Allah develops both facets of their existence equally, they become what ibn 'Arabī described as the 'Human-Angel and Angel-Human prostrating via both will and coercion'.[27]

In comparison, ibn al-Qayyim perceived the *sujūd* of humans as an individualistic form of *sujūd*, which in its turn reflects an individualistic form of subjugation that is based on love and is expressed through devotional compliance with the commands of Allah.[28] Summarising his thoughts, ibn al-Qayyim wrote, 'subjugation and devotion originates from the love of Allah [...] And if loving Him [Allah] was in truth the essence of, and the secret behind, being in subjugation [to Him], [it thus becomes apparent] that it [subjugation] is realised only through complying with His Laws, and abstaining from what He forbade'.[29] In this instance, the individualistic *sujūd* of believers, as specified by ibn al-Qayyim, appears to be reflective of *'ubūdiyyat al-uluhiyya* or theistic subjugation which results from the fulfillment of the four-tiered prerequisites for a perfected personal subjugation.

ONENESS: THE HOLISTIC PERCEPTION

It may therefore be deduced that the 'Angel-Human' places their free-will in subjection to the commands of the Creator, aligning themself to the harmonious cycles of the cosmos, participating actively in the universal subjugation to *al-Fard al-Aḥad*, the One Eternal God. *Tafakkur*, or thoughtful meditation, upon the *malakūt* of Allah, or all that exists in His created universe, is another form of worship which has been described in the Qur'ān as the deed of the wise believers:

Behold! in the creation of the heavens and the earth, and the alternation of Night and Day, – there are indeed Signs for men of understanding.

Men who celebrate the praises of Allah standing, sitting, and lying down on their sides, and contemplate [yatafakkarūn] the (wonders of) creation in the heavens and the earth, (with the thought): 'Our Lord! not for naught hast Thou created (all) this! Glory to Thee! Give us salvation from the penalty of the Fire.[30]

Tafakkur has been equally praised by the Prophet, for 'An hour of thoughtful meditation [*fikra*] is better than the worship of sixty years'.[31] In *'Iḥyā' 'Ulūm al-Dīn* or *Revival of Religious Sciences*, al-Ghazālī discussed how meditative worship through *tafakkur* can lead one to a higher level of ontological awareness of the unparalleled Unique Divinity of Allah, yet unrestrained *tafakkur*, al-Ghazālī warned, could be detrimental to the mental and psychological wellbeing of the contemplator:

The vision [gained through *tafakkur*] of the rest of mankind and [their capacity to perceive] the Majesty and Glory of Allah is comparable to the vision of a bat and [its capacity to perceive] the light of the sun, for it cannot ever endure it, and so instead, it [the bat] hides during daytime and emerges during the night [so that it can tolerate] whatever has remained of the sunrays [as reflected] on the earth [...] And similarly so, looking at His Being [through *tafakkur*] causes bewilderment, perplexity and confusion to the [human] mind.[32]

The only alternative, and tolerable, path towards the profound mystical awareness of His Divinity, as al-Ghazālī's proposed, is *tafakkur* upon the wondrous creations of Allah:

For they all demonstrate His Glory, His Mightiness, His Holiness, His Transcendence, and indicate the perfection of His Knowledge, His Wisdom, and the penetrative force of His Will and His Omnipotence. And so, His Attributes are understood through looking [by means of *tafakkur*] at the effects of His Attributes, for we are not capable of looking [directly and through *tafakkur*] at His Divine Attributes [...] Everything which exists in this [mundane] world is [merely] a [manifest] effect [reflective] of the effects of the Omnipotence of Allah, Exalted He is, and is [merely] a single light [emanating] from the lights of His Being. Indeed, there is no darkness more abysmal than nonexistence and no light brighter than existence [*wujūd*].[33]

Al-Ghazālī was alluding to the transcendental states of awareness known to Sufi mystics as *wuḥdat al-shuhūd* or oneness of witnesses, and *wuḥdat al-wujūd* or oneness of existence, both states corresponding to, respectively, the advanced third and fourth stages of *tawḥīd*, the uncompromising monotheistic belief in the indivisible oneness of Allah as declared by the descendants of Abraham: 'We shall worship thy God and the God of thy fathers — of Abraham Isma'il and Isaac — the one (true) God to Him we bow (in Islam)'.[34] According to al-Ghazālī, *tawḥīd* as an absolute unitarian belief may be classified into four distinct categories, or accumulative experiences, that may be compared to the layers of a walnut.[35] The first stage is superficial and apparent, just like the external shell of a walnut, and the professed *tawḥīd* of a person at this stage is similarly superficial; it is the meaningless utterance of the *tawḥīd* phrase, '*Lā illāha illā Allāh*', or 'There is no God but Allah', by the hypocrite and the imposter; it lacks depth and value.[36] The second stage resembles the inner lining of the shell, and being a secondary shell, it remains somewhat depthless, yet being close to the seed, it reveals that the declared *tawḥīd* reflects monotheistic faith in the heart, however shallow such a faith may be.[37] The third and fourth stages, which al-Ghazālī correlated to the walnut kernel and the walnut oil respectively, reflect a profound ontological realisation of the absolute truth of monotheism seldom experienced by ordinary people.[38] Although the latter two stages are equally complex states of intrinsic awareness, the third stage involves the state known as *kashf* – enlightened perception or revelation – and the fourth stage involves the state of *fanā'* in *tawḥīd*, the complete annihilation of the self and subsequent absorption into the truth of *tawḥīd*.[39]

Before proceeding to further examine *wuḥdat al-shuhūd* and *wuḥdat al-wujūd* as transcendental states of awareness, it is necessary to first understand the Sufi phase or station known as *fanā'*. According to Imām al-Harawī, the author of *Manāzil al-Sā'irīn* or *Stages of the Wayfarers*, when the seeker of the Eternal One embarks upon the path of profound monotheism, they transit through one hundred stations, or states of awareness, that reflect the level of their spiritual progression.[40] Positioning *fanā'* in the ninety-second stage, al-Harawī perceived it as an advanced state of elevated awareness that precedes attaining the highest level of *tawḥīd* possible.[41] Succinctly describing *fanā'*, al-Harawī wrote, '[it is] the fading away of that which is inferior to the Absolute Truth [one of the Names of Allah] through intelligent perception, then [through] negation, then [through] certainty'.[42] In his commentary on al-Harawī's manual, ibn al-Qayyim pointed out that al-Harawī was referring to how *fanā'*, or the annihilation of the perceived world in the mind of the perceiver, develops:

> [All that is created and seen] fades from the heart and awareness through intelligent perception, even if its own being [its physical body] was not at that time decayed or faded, so that the images of the existing [creatures] vanish from the awareness of the servant, as if it has entered into [the state of] non-existence, just as it has been before it was made existing, and what remains is the Absolute Truth [Allah], the Lord of Bounty and Honour, alone in the heart of the witness, just as He has been before the creation of the worlds.[43]

After the process of annihilation of the apparent world from the perception of the viewer, they then go through a total intrinsic negation of all that is not the Creator, and a subsequent refusal to attribute any effects to anything other than Allah, the Omnipotent Cause of all effects.[44] Once that is attained, the seeker witnesses how the celestial and terrestrial worlds are sustained and maintained in a state of harmonious equilibrium by the Powerful and All-Capable Allah as He said in the Qur'ān: 'It is Allah Who sustains the heavens and the earth, lest they cease (to function): and if they should fail, there is none not one can sustain them thereafter'.[45] Ibn al-Qayyim then concluded that the last manifestation of *fanā'* belongs to the people of gnosis who are totally immersed in witnessing the things that testify to al-Haqq, the True and Only Creator.[46]

The first two sub-stages of *fanā'* correspond to the third stage of *tawhīd* described by al-Ghazālī as the point when the seeker sees through *kashf*, or enlightened Divine revelation, that the apparent multiplicity of all that is created testify [*yashhadū*] to the unrivaled mightiness of the One.[47] Such is the state of *wuhdat al-shuhūd*, which Imām al-Qushayrī (c. 986-1073 CE) attempted to depict in his *Epistle on Sufism* via the poetic verses:

> My existence [wujūdī] is to be absent from all that is existing [al-wujūd]
> Through what appears to me through the witnesses [al-shuhūd][48]

Thus, *wuhdat al-shuhūd* is the state of aware intelligent realisation of the holistic and interconnected nature of all creation as witnesses, or *shuhūd*, that testify through their impeccable harmonious synchronicity to the existence of a Masterful and Knowledgeable Creator. The final stage of *fanā'* is equivalent to al-Ghazāli's fourth stage of *fanā'* in *tawḥīd*, the phase when the existence of all things fades away from the consciousness of the seeker, including awareness of his own existence, to be replaced by an overwhelming and all-encompassing monotheistic devotional awareness of the Creator.[49] It is a mystical regression into an era that preceded time, space and creation, when according to the Prophetic saying, 'Allah was, and nothing was except him'.[50]

CONCLUDING REMARKS

I have explored the significance of three Islamic concepts, namely subjugation, *sujūd* and oneness, which assert the importance of harmony as one of the Divine Laws that sustain the equilibrium of the cosmos. It is clear that, to the Islamic mystic and thinker, *subjugation*, and its external manifestation *sujūd*, go beyond the individual and humanistic to the universal and cosmic since all creation is in perpetual metaphorical universal subjugation and *sujūd* before the universal Laws of Allah. The 'Angel-Human' who through his or her free-will performs *sujūd*, and who has been fortunate enough as to be enabled to do so through *mashī'a* or Divine Decree, transcends spatial and temporal limits, and actively participates in the harmony of universal subjugation of all that is created, both celestial and terrestrial. While subjugation and *sujūd* can aid the 'Angel-Human', or the seeker of Divine Truth, in synchronising him or herself with the harmonious order of the cosmos, the act of contemplative meditation, or *tafakkur*, can transport the seeker to a state of *fanā'*, and subsequently to a state of oneness, or ultimate harmony, where all apparent shapes, physical boundaries, and restrictions of the intellect are annihilated to reveal the True One Omnipotent Creator.

ACKNOWLEDGEMENTS

Special thanks are due to Azharite scholars, Dr. Hesham Baharia and Sheikh Rabea al-Qadi, for clarifying and revising some of the rather complex and profound esoteric concepts and Sufi terms referred to in the works of ibn 'Arabī and al-Ghazāli.

NOTES

١. **الفتوحات**، محي الدين محمد بن عربي؛ (٢٠٠٥ ،أبو حامد الغزالي، **إحياء علوم الدين** (بيروت: دار ابن حزم للطباعة والنشر والتوزيع)
المكية، ضبط وتصحيح أحمد شمس الدين (بيروت: دار الكتب العلمية، ب.ت.)؛ محمد بن أبي بكر بن قيم الجوزية، **مدارج السالكين بين إياك**
نعبد وإياك نستعين، تحقيق محمد المعتصم بالله البغدادي (بيروت: دار الكتاب العربي، ٢٠٠٣)

Abū Ḥāmid al-Ghazālī, 'Iḥyā' 'Ulūm al-Dīn or Revival of the Religious Sciences (Beirut:
Dār ibn Ḥazm lil-Ṭibā'a wal-Nashr wal-Tawzī', 2005); Muḥyī-d-Dīn Muḥammad ibn
'Arabī, al-Futūḥāt al-Makkiyya or The Meccan Revelations, ed. by Aḥmad Shamsu-d-Dīn
(Beirut: Dār al-Kotob al-'Ilmiyya, n.d.); Muḥammad b. abī Bakr ibn Qayyim al-Jawzīyya,
Madārij al-Sālikīn or Ranks of the Divine Seekers, ed. by Mohammad al-Mu'tasim bi'llah
al-Baghdadi (Beirut: Dār al-Kitāb al-'Arabī, 2003).

٢. أبو الفضل جمال الدين محمد بن مكرم بن منظور، **لسان العرب** (بيروت: دار صادر، ١٩٩٤)، ١٢/٥٩٠؛ محمد بن أحمد الأزهري،
تهذيب اللغة (بيروت: دار إحياء التراث العربي، ٢٠٠١)، ١٤٢/٨؛ أبو الحسن علي بن إسماعيل ابن سيده، **المحكم والمحيط الأعظم**
(بيروت: دار الكتب العلمية، ٢٠٠٠)، ٥/٥٤٥-٤٦

Abū al-Fadl Jamāl al-Dīn Muhammad b. Mukarram ibn Manẓūr, Lisān Al-'Arab or
The Arabic Tongue (Beirut: Dār Sāder, 1994), XII:590; Muḥammad b. Aḥmad al-Azharī,
Tahthīb al-Loghah or The Refinement of the Language (Beirut: Dār Ihyā' al-Turāth al-
'Arabī, 2001), VIII:142; abū al-Hasan 'Alī b. 'Ismā'īl ibn Sīdah, al-Muḥkam wal-Muḥīṭ al-
'A'tham or The Precise Book on Arabic Philology (Beirut: Dār al-Kotob al-'Ilmīya, 2000),
V:545-46.

3. The Meaning of the Holy Qur'ān, trans. by Abdullah Yusuf Ali and Muhammad
Marmaduke Pickthall (Beltsville: Amana Publications, 1997), 17:44.

4. The Meaning of the Holy Qur'ān, 20:130.

5. علي بن أبي طالب، **ديوان علي بن أبي طالب**، تحقيق عبدالعزيز الكرم (ب.م: ب.ن.، ١٩٨٨)

'Alī b. 'Abī Ṭālib, The Dīwān of 'Alī b. 'Abī Ṭālib, ed. by 'Abdul-Azīz al-Karam (s.l.:
s.n., 1988), p. 45:

«وتَحْسَبُ أَنَّكَ جُرمٌ صَغِيرٌ   وفيكَ انطَوى العالَمُ الأكْبَرُ»

6. ibn al-Qayyim, Ranks of the Divine Seekers, I:118-128; ibn 'Arabī, The Meccan
Revelations, II:195-98; al-Ghazālī, Revival of the Religious Sciences, pp. 1603-7, p. 1810.

7. ibn Manẓūr, The Arabic Tongue, III:204-206; al-Azharī, The Refinement of the
Language, X:300-2; ibn Sīdah, The Precise Book on Arabic Philology VII:261-262.

8. abū 'Abd Allāh Muḥammad b. Ismā'īl al-Bukhārī, Sahih al-Bukhari, trans. by
Muhammad Muhsin Khan, book 10 on Call to Prayers (Virginia: al-Saadawi Publications,
1996), hadīth no. 207:

«أُمِرْتُ أَنْ أَسْجُدَ عَلَى سَبْعَةِ أَعْظُمٍ عَلَى الجَبْهَةِ. وَأَشَارَ بِيَدِهِ عَلَى أَنْفِهِ. وَالْيَدَيْنِ، وَالرُّكْبَتَيْنِ وَأَطْرَافِ الْقَدَمَيْنِ، وَلَا نَكُفِتَ الثِّيَابَ وَالشَّعَرَ.»

9. The Meaning of the Holy Qur'ān, 19:58.

10. ibn al-Qayyim, Ranks of the Divine Seekers, I:118-128.

11. ibn al-Qayyim, Ranks of the Divine Seekers, I:120-121.

12. ibn al-Qayyim, Ranks of the Divine Seekers, I:120.

13. ibn al-Qayyim, Ranks of the Divine Seekers, I:121.

14. ibn al-Qayyim, Ranks of the Divine Seekers, I:121.

15. ibn al-Qayyim, Ranks of the Divine Seekers, I:125-128.

16. The Meaning of the Holy Qur'ān, 19:93; ibn al-Qayyim, Ranks of the Divine
Seekers, I:125-128:

«فَالْعُبُودِيَّةُ الْعَامَّةُ عُبُودِيَّةُ أَهْلِ السَّمَاوَاتِ وَالْأَرْضِ كُلِّهِمْ لِلَّهِ، بَرِّهِمْ وَفَاجِرِهِمْ، مُؤْمِنِهِمْ وَكَافِرِهِمْ، فَهَذِهِ عُبُودِيَّةُ الْقَهْرِ وَالْمُلْكِ، قَالَ تَعَالَى {وَقَالُوا
اتَّخَذَ الرَّحْمَنُ وَلَدًا – لَقَدْ جِئْتُمْ شَيْئًا إِدًّا – تَكَادُ السَّمَاوَاتُ يَتَفَطَّرْنَ مِنْهُ وَتَنْشَقُّ الْأَرْضُ وَتَخِرُّ الْجِبَالُ هَدًّا – أَنْ دَعَوْا لِلرَّحْمَنِ وَلَدًا – وَمَا يَنْبَغِي
لِلرَّحْمَنِ أَنْ يَتَّخِذَ وَلَدًا – إِنْ كُلُّ مَنْ فِي السَّمَاوَاتِ وَالْأَرْضِ إِلَّا آتِي الرَّحْمَنِ عَبْدًا}.»

17. The Meaning of the Holy Qur'ān, 22:18.

18. ibn al-Qayyim, *Ranks of the Divine Seekers*, I:127-128; *The Meaning of the Holy Qur'ān*, 13:15.

19. أبو حامد الغزالي، **المنقذ من الضلال**، تحقيق جميل صليبا كامل عياد (بيروت: دار الأندلس، ٧٦٩١)

Abū Ḥāmid al-Ghazālī, *Al-Munqith Min Al-Ḍalāl Or Deliverance from Error*, ed. by Jamil Saliba and Kamil Ayyad (Beirut: Dār al-Andalus, 1967), p. 83:

«الطبيعة مسخرة لله تعالى، لا تعمل بنفسها، بل هي مستعملة من جهة فاطرها. والشمس والقمر والنجوم والطبائع مسخرات بأمره لا فعل لشيء بذاته عن ذاته.»

20. ibn 'Arabī, *The Meccan Revelations*, II:197-198.

21. ibn 'Arabī, *The Meccan Revelations*, II:197-198:

«الذين التحقوا بمن يبعض سجودهم ممن في السموات ومن في الأرض، والشمس في غروبها، والقمر في محاقه، والنجوم في مواقعها، والجبال في إسكانها، والشجر في إقامها على سوقها، والدواب في تسخيرها...»

22. *The Meaning of the Holy Qur'ān*, 81:29.

23. *The Meaning of the Holy Qur'ān*, 13:15; ibn 'Arabī, *The Meccan Revelations*, II:195.

24. ibn 'Arabī, *The Meccan Revelations*, II:195.

25. ibn 'Arabī, *The Meccan Revelations*, II:195.

26. ibn 'Arabī, *The Meccan Revelations*, II:195.

27. ibn 'Arabī, *The Meccan Revelations*, II:195.

28. ibn al-Qayyim, *Ranks of the Divine Seekers*, I:126.

29. ibn al-Qayyim, *Ranks of the Divine Seekers*, I:119.

30. *The Meaning of the Holy Qur'ān*, 3:190-191.

31. عبدالله بن محمد بن حيان الأصبهاني، **كتاب العظمة**، تحقيق رضاء الله بن محمد بن إدريس المباركفوري (الرياض: دار العاصمة، ٨٠٤١ هـ)، ٩٩٢/١، حديث ٣٤

'Abdulla b. Muḥammad b. Ḥayyān al-Aṣbahānī, *Kitāb-ul-'Athama Or Book of Glory*, ed. by Riḍā' Allāh Muḥammad b. Idrīs al-Mubārakfūrī (Riyadh: Dār al-'Āṣimah, 1408 AH), I:299, ḥadīth no. 43:

«فِكْرَةُ سَاعَةٍ خَيرٌ مِنْ عِبَادَةِ سِتِّينَ سَنَةً»

32. al-Ghazālī, *Revival of the Religious Sciences*, p. 1810:

«أبصارهم بالإضافة إلى جلال الله تعالى كحال بصر الخفاش بالإضافة إلى نور الشمس فإنه لا يطيقه البتة بل يختفي نهارا وإنما يتردد ليلا ينظر في بقية نور الشمس إذا وقع على الأرض [...] وكذلك النظر إلى ذات الله تعالى يورث الحيرة والدهش واضطراب العقل.»

33. al-Ghazālī, *Revival of the Religious Sciences*, p. 1810:

«فإنها تدل على جلاله وكبريائه وتقدسه وتعاليه وتدل على كمال علمه وحكمته وعلى نفاذ مشيئته وقدرته فينظر إلى آثار صفاته فإنا لا نطيق النظر إلى صفاته [...] وجميع موجودات الدنيا أثر من آثار قدرة الله تعالى ونور من أنوار ذاته بل لا ظلمة أشد من العدم ولا نور أظهر من الوجود.»

34. *The Meaning of the Holy Qur'ān*, 2:133.

35. al-Ghazālī, *Revival of the Religious Sciences*, pp. 1603-1607.

36. al-Ghazālī, *Revival of the Religious Sciences*, p. 1603.

37. al-Ghazālī, *Revival of the Religious Sciences*, p. 1603.

38. al-Ghazālī, *Revival of the Religious Sciences*, p. 1603.

39. al-Ghazālī, *Revival of the Religious Sciences*, p. 1603.

40. عبدالله الأنصاري الهروي، **منازل السائرين**، تحقيق علي فاعور (بيروت: دار الكتب العلمية، ٨٨٩١)

'Abdulla al-Anṣārī al-Harawī, *Manāzl al-Sā'irīn Or Stages of the Wayfarers*, ed. by Ali Faour (Beirut: Dār al-Kotob al-'Ilmiyya, 1988)

41. al-Harawī, *Stages of the Wayfarers*, pp. 127-129.

42. al-Harawī, *Stages of the Wayfarers*, p. 128:

«اضمحلال ما دون الحق علماً، ثم جحداً، ثم حقاً.»

43. ibn al-Qayyim, *Ranks of the Divine Seekers*, III:345:

«وَقَوْلُهُ 'الْفَنَاءُ اسْمٌ لِاضْمِحْلَالِ مَا دُونَ الْحَقِّ عِلْمًا' يَعْنِي: يَضْمَحِلُّ عَنِ الْقَلْبِ وَالشُّهُودِ عِلْمًا، وَإِنْ لَمْ تَكُنْ ذَاتُهُ فَانِيَةً فِي الْحَالِ مُضْمَحِلَّةً،

فَتَغِيبُ صُوَرُ الْمَوْجُودَاتِ فِي شُهُودِ الْعَبْدِ، بِحَيْثُ تَكُونُ كَأَنَّهَا دَخَلَتْ فِي الْعَدَمِ، كَمَا كَانَتْ قَبْلَ أَنْ تُوجَدَ، وَيَبْقَى الْحَقُّ تَعَالَى ذُو الْجَلَالِ
وَالْإِكْرَامِ وَحْدَهُ فِي قَلْبِ الشَّاهِدِ، كَمَا كَانَ وَحْدَهُ قَبْلَ إِيجَادِ الْعَوَالِمِ'»

44. ibn al-Qayyim, *Ranks of the Divine Seekers*, III:346.

45. ibn al-Qayyim, *Ranks of the Divine Seekers*, III:346; *The Meaning of the Holy Qur'ān*, 35:41.

46. ibn al-Qayyim, *Ranks of the Divine Seekers*, III:346.

47. al-Ghazālī, *Revival of the Religious Sciences*, p. 1604.

48. عبدالكريم بن هوازن القشيري، **الرسالة القشيرية** (بيروت: دار صادر، ب.ت)، ص.٨٢.

'Abd al-Karīm b. Hawāzin al-Qushairī, *al-Risāla al-Qushairiyya Or Epistle on Sufism* (Beirut: Dār Ṣāder, n.d.), p. 28:

«وجودي أن أغيب عن الوجود   بما يبدو عليَّ من الشهود»

49. al-Ghazālī, *Revival of the Religious Sciences*, p. 1604.

50. al-Bukhārī, *Ṣaḥīḥ al-Bukhārī*, Book 59 of *The Beginning of Creation*, ḥadīth no. 3191:

«كَانَ اللَّهُ وَلَمْ يَكُنْ شَيْءٌ غَيْرُهُ»

The Perennial Philosophy and the Recovery of a Theophanic View of Nature

Jeremy Naydler

The Forgotten Tradition

We suffer from a peculiar kind of cultural amnesia today. Since the time of the Reformation and the Scientific Revolution, we have increasingly lost awareness of the rich wisdom tradition that for hundreds of years nourished the inner life of contemplatives and seekers of truth. This wisdom tradition is often referred to as the *philosophia perennis* or 'perennial philosophy'. In both the West and East it is articulated in manifold works of spiritual philosophy, visionary poetry and mystical literature, harboured within pagan, Judaeo-Christian, Islamic, Hindu, Buddhist and Taoist worldviews, and in the oral traditions of many indigenous peoples. While it is expressed in distinctive and different ways, the perennial philosophy articulates truths that are essentially universal and timeless and which help us to understand our place in the cosmos and the deeper purpose of human life.

Central to the perennial philosophy is the recognition that there is a spiritual dimension of existence that is the primary reality from which all creation derives. All creatures seek to express in their own way this reality, and all creation seeks ultimately to unite with it. The perennial philosophy reminds us that our fundamental orientation as human beings should be towards spirit, that we should revere the natural world as the manifestation of the divine, and that we should affirm the possibility of an ever more conscious union between ourselves and the spiritual source of existence.

It is important to understand that the perennial philosophy is not a 'philosophical system' produced by abstract reasoning. It is primarily an *orientation of the human soul* towards a spiritual dimension that essentially transcends the particular cultures, religious outlooks and historical contexts within which it finds expression. At the kernel of the perennial philosophy is less a set of arguments, concepts or doctrines than a human encounter with the sacred, both in nature and within the human heart. This is why the perennial philosophy is articulated in countless different ways, according to the languages of different religious, philosophical and imaginative milieux. But we nevertheless recognise, shining through these different forms of expression, a deeper level of truth which derives from an authentic spiritual intuition that has touched, and been touched by, a transcendent source of meaning.

The aim of this essay is to consider how certain insights of the perennial philosophy may contribute to the healing of our current disharmonious relationship to nature and to the remembrance of our human purpose within the natural and spiritual orders. In what follows I shall draw mainly on the Western tradition of the perennial philosophy, found in the works of such thinkers as Plato, Aristotle, Plotinus and Thomas Aquinas, and upheld through the ages by Christian contemplatives and by mystics such as Meister Eckhart. The reason for drawing on this Western tradition is that, for those of us living in the West, it is, after all, our rightful inheritance. And it lies so close to the surface of our forgetfulness that it is, perhaps, still within reach of recall.

THE LEGACY OF THE REFORMATION AND THE SCIENTIFIC REVOLUTION

First of all it is necessary to understand how and why the eclipse of the perennial philosophy took place, leading to the collective amnesia that has descended upon us today. Many reasons could be given, but there are two historical occurrences which seem to be of greatest relevance. The first is the enormous upheaval that affected every corner of Western Europe during the Reformation, and the devastating assault on monasticism that was carried out by Reformers in the sixteenth and subsequent centuries. This had the long-term effect of undermining the ideal of the life of prayer and spiritual contemplation, pursued over many generations in the shelter of the monasteries. By providing the protective space in which the inner life could be nurtured, the monasteries had for more than a thousand years fostered a conscious relationship to both the psychic and spiritual dimensions of existence. Their emphasis on moral development, and on the interior life of prayer and contemplation, practised in conjunction with meditation on sacred texts and the discipline of 'holy imagination', had an effect on the whole tenor of medieval society.[1] The monasteries and religious houses were a constant reminder to people to attend to the inner life, to make the inner turn towards soul and spirit. With their destruction, not only did the medieval era effectively come to an end, but the value placed on inwardness also began to be seriously eroded.

The following example might help us to grasp how this erosion of the value of inwardness occurred, with the resultant coarsening of the way in which people approached the understanding of the realities of the spirit. Central to the sacred learning practised in the monasteries was the recognition that there are different levels of meaning and symbolism in Biblical texts. As early as the third century, Origen had argued that just as the human being is composed of body, soul and spirit, so too does all of Scripture have a threefold meaning, deepening as we move from the physical to the psychic, and from the psychic to the spiritual levels of interpretation.[2] During the Middle Ages, a fourfold interpretation of sacred Scripture was widely adopted, according to which no sacred text could be properly understood unless the reader travelled from its literal to its more subtle allegorical, moral and mystical meanings.[3] Since 'the book of nature' was also regarded as a sacred text, the same nuanced sensibility applied to the understanding of nature. When Luther and Calvin asserted that only the literal sense of Scripture is valid, it meant that not only were other levels of meaning in sacred Scripture subverted but so also was the idea that nature, too, could be approached with different levels of understanding that went beyond the merely literal. Thus the Reformation prepared the way for a desacralised knowledge of nature, no longer capable of recognising nature as a manifestation of spirit.[4]

The second historical occurrence which caused the wisdom tradition of the perennial philosophy to be so neglected in modern times was the Scientific Revolution of the seventeenth century. Prior to the Scientific Revolution, there was great reverence for the sages and seers of the past, not only the great prophets of the Old Testament but also pagan philosophers such as Pythagoras and Plato. This is well illustrated in the saying of Bernard of Chartres (who taught in Chartres during the early twelfth century) that we are like 'dwarfs perched on the shoulders of giants ... we see more and farther than our predecessors, not because we have keener vision or greater height, but because we are lifted up and borne aloft on their gigantic stature'.[5] This attitude of respect and humility towards the past was typical of the Middle Ages, as indeed it was typical of much earlier historical periods too. In cultures as diverse as ancient Egypt, Greece, and India, we meet a similar belief that the further back in time one reached, one would find that human beings were nearer to the sources of spiritual wisdom, for in the distant past conditions on earth were more closely aligned to conditions in heaven.[6]

Along with this high estimation of the past there was also, across the same diverse range of cultures and historical epochs, a view of nature as a manifestation of the divine. Whether the divine was conceived in terms of a multiplicity of gods and spirits (as in the great polytheistic cultures both East and West) or as a single divine source (as in the monotheistic religions of Judaism, Christianity and Islam) human relationship to the natural world was essentially 'theophanic'.[7] Nature was never seen as merely physical – it always mediated a sacred presence, whether of gods, spirits or God.

Both the respectful admiration of the past and the theophanic view of nature were anathema to the founding fathers of the Scientific Revolution. They wanted to make a clean sweep of the past and were virulently hostile towards the notion that wisdom could be sought and found in the spiritual philosophy of antiquity, transmitted to us in ancient teachings and texts. The attitude of Francis Bacon is typical. In his *Novum Organum*, he pointedly wrote that we should best regard the ancients as like children compared to us adult moderns, so it was a great mistake to give their ideas any credence whatsoever. Descartes held a similarly disparaging view of the ancients.[8] And so the idea of 'progress' – unheard of before the seventeenth century – was conceived as an alternative to the previous reverence for the past. The assertion of this idea involved the denigration of the wisdom tradition that had been handed down through the centuries. And it eventually led to the attitude of most people today that the spiritual teachings of the past are at best of only marginal relevance to contemporary life. Far from it being a mark of culture to know and revere the 'wisdom of the ancients', it is a sure sign of swimming against the science-driven, future-oriented current of inevitable progress that is sweeping us all forwards towards ever greater material prosperity and technological sophistication.

The underlying reason for Bacon and Descartes' campaign against the wisdom tradition handed down to them was that they wanted to establish a new *kind* of knowledge, that not only cut out all reference to ancient authorities but above all re-established knowledge on the basis of what would be useful to human beings. They did not want a knowledge founded on contemplation and religious piety, that invested the world with religious meaning. They wanted a knowledge that would give human beings power to take control of the physical world and bend nature to the service of human ends. Such a knowledge had to be freed of all symbolic and metaphysical content, and be based on experimental observation, systematic research and analysis, for only then might we become – as Descartes put it in his memorable phrase – 'masters and possessors of nature'.[9] The new knowledge would prove its value not by bringing us to a deeper spiritual understanding but by increasing our ability to manipulate and control nature in ways both practical and useful, to the greater material advantage of human beings.

Over the next four hundred years, the collective energies of the West were directed towards the achievement of the aims of the new knowledge, with the result that today we reap the benefits of hot baths and flushing toilets, washing machines, motorcars, aeroplanes, electric lights, smartphones and all the other paraphernalia that characterise economically 'developed' societies. As more and more countries across the globe seek to claim their share of these benefits, we see ever more clearly the heavy price that the rest of nature pays for them: polluted rivers and oceans, the degradation of the soil to critical levels, forests systematically destroyed, and numerous species of animal and plant in catastrophic decline, with many facing extinction.[10] In this afflicted world, more and more people are crowded into ugly, sprawling cities, with so many areas of modern life infected by a creeping tawdriness. It has been a heavy price, too, for the inner environment of soul and spirit, which in our extroverted mainstream culture is to a large extent starved and neglected, and increasingly denied. Contemporary champions of the new knowledge, such as Richard Dawkins, Stephen Hawking, and Yuval Noah Harari have sought to convince us that there is no reality other than that which has material existence, that human beings are just biological machines, that there is no such thing as the soul, and that there is no transcendent meaning or purpose to human life.[11]

Here, then, is the legacy of the Reformation and the Scientific Revolution. There can be no doubting the impressive material and technological achievements of the last four hundred years, but neither can we ignore the wrecking of the natural environment and the impoverishment of the inner life of human beings that have been the price paid for these achievements. If we have fallen out of harmony with nature and lost our deeper sense of purpose – lost even our sense of the reality of the spiritual order of existence – these are two aspects of a single phenomenon. The one mirrors the other. The ecological crisis is a symptom of sickness and disharmony in the human soul, which nature mirrors back to us.[12] There can be no technological solution to this inner/outer malaise: it lies beyond the purview of the new knowledge because this knowledge is based on ignorance, forgetfulness, or outright rejection of fundamental spiritual truths. Note that these truths, though time-honoured through having been reiterated over millennia, do not belong to the past. They are perennial not because they have been around for a very long time, but because they have a validity that *endures through time*. If we have lost sight of them, it is because we have lost our connection with an order of existence that transcends historical and cultural conditions, and which constitutes a universal and eternal ground of meaning and value.

THE REALITY OF WHOLENESS

We have seen that the theophanic view of nature, as a manifestation of the divine, was rejected during the Scientific Revolution. It was regarded as having nothing to contribute towards reliable knowledge. From the perspective of the perennial philosophy, the rejection of the theophanic view of nature is the root cause of the ecological crisis that we now face, for it led to the treatment of nature as a mere resource to be exploited without restraint. At this time of crisis, it is imperative that the theophanic view of nature is recovered, but this requires a radical shift in the way we perceive the natural world.

The first step towards this change in perception is that we turn our attention towards the intrinsic wholeness of things. Today we have largely lost sight of wholeness. This is because, central to the project of the Scientific Revolution and the mechanistic philosophy on which it was based, there was an attempt to explain wholes in terms of their parts. Living organisms were treated as conglomerations of parts (conceived in due course in terms of their genetic make-up and biochemistry) put together without any inherent unifying principle, and hence to be understood as if they were machines. Taking this view, if anything unifies an organism it would not be a non-material principle of wholeness but a purely physical mechanism. The same approach applies to mountains, rivers, seas and forests. They should be regarded as no more than the sum of the different elements of which they are composed. They do not have an intrinsic identity, character or soul-quality that vouchsafes to them their integrity of being, and for this reason it is all the easier to use, abuse and plunder them.

The wholeness of things was referred to in the older philosophical tradition as their 'form'. The form is to be understood as the guardian of a creature's wholeness, and it is contrasted with the matter out of which a creature is made in the following way: the form gives it *actuality*, whereas the matter exists only as *potentiality* in relation to any given form.[13] A creature is what it is not because of its material components but because of its inherent form. Its form is a non-material organising principle that organises the parts into a coherent unity. As such, it is not reducible to any material determinant (like DNA). Only the matter, not the form, can be subjected to chemical analysis, but the form is none the less real. Indeed, it is the underlying reality which invests all the biochemical and physical aspects of an organism with coherence and meaning, just as it invests the matter out of which other entities in nature are composed with their specific qualities. The entity that we call water, for instance, has characteristics that are quite different from the oxygen and hydrogen atoms out of which water is composed.

The form is the non-material foundation of a being that gives integrity, meaning and identity to all its material parts, and to all its characteristic habits, behaviours and gestures. In the Western philosophical tradition, the form is also referred to as the inherent idea or 'the idea within the thing' (*universale in re*), by which is meant the inner organising principle that constitutes its intrinsic nature, rather than a concept or theoretical explanation we project onto it. Through grasping this formative organising principle or 'idea within the thing' in our thought, and through responding to it in feeling and imagination, we are able to know and relate to the being that stands before us as more than just a conglomeration of material attributes, but as an entity with its own essential integrity.

The recognition of the form is something that comes naturally to poets, and it belongs to the spontaneous, untutored and wonder-filled awareness of nature which many people still have, even though they may feel obliged to consider such awareness unscientific and as making no contribution to real knowledge. From the seventeenth century onwards, the scientific endeavour regarded experimentation, data-collection and analysis, along with complete reliance on the faculty of analytical reasoning (traditionally referred to as the *ratio*) that functions through calculation and logical deduction as the way to acquiring real knowledge. But this kind of reasoning, augmented to such great effect in modern times by the computing power of electronic technologies, is unable to conceive of wholes as anything more than the sum of their parts. It is not able to penetrate beyond the material surface of things. It cannot countenance the notion that there is a non-material aspect to reality, and so it cannot see the wholeness of things as a living energy. For that, a different kind of cognitive faculty – the faculty of intuitive insight – must be brought to bear. In the Western philosophical tradition, this faculty of intuitive insight is referred to as the *intellectus*, but it is far from being 'intellectual' in the modern sense. Thomas Aquinas described the *intellectus* as a faculty of inner perception, for the word *intellegere* means 'to read inwardly'. And so, he explains, 'perception by the *intellectus* penetrates to the very essence of things ...'[14] We experience it every time we gain imaginative insight into another human, or non-human being, and are able to open ourselves to their inner nature.

Unlike the analytical reasoning of the *ratio*, so readily augmented by the binary intelligence of computer technology, the *intellectus* draws on imagination and inspiration to enhance its ability to enter into the inner being of another creature. To approach nature in this way does not mean that we have to reject the results of scientific enquiry, but it does mean that we must approach these findings from a quite different standpoint. To prioritise the whole over the parts is to affirm the existence of a dimension within nature that is essentially inward. It is the first step towards *restoring inwardness* to nature. In so doing, we give cognitive value to our perception of an essentially non-material aspect of reality that cannot be reduced to physically detectable or measurable components. We acknowledge a different *kind* of reality in our midst, in our everyday experience. The wholeness of things calls to us. It cries out to be attended to and to be known and celebrated for what it is in itself, in its inherent integrity.

RECOVERING THE THEOPHANIC VIEW OF NATURE

All of nature cries out for this turn of our attention towards the kind of knowing that is not based on manipulation and control in order to fulfil our utilitarian needs and desires, but issues from a genuine desire to relate to the creatures with which we share the world as they are in themselves and for their own sake. The affirmation of the wholeness and inherent integrity of creatures is, then, a first step towards restoring a more harmonious relationship to nature. We do not have to be scientists to know the inner nature of other creatures: it is a question rather of how open or closed we are as human beings to the forms and qualitative attributes of the multiple natural phenomena that surround us.

But there is a further step that can be taken. By intensifying our focus on something's innate qualities, we may deepen the experience of 'the idea within the thing' to the point at which the numinous ground of its existence becomes present to our consciousness.[15] Then the spiritual source of the idea, or organising principle, begins to speak to us. This source is not in the sense-perceptible world but in the creative energies out of which the sense-perceptible world unfolds into manifestation. What is sense-perceptible is thus revealed to be the exteriorisation of a deeper, spiritual level of existence. For the creative energies that pour into the world belong to a sphere of reality that is intrinsically numinous, and it is within this numinous sphere of reality that the organising principles are rooted. Encountered as creative powers, they are traditionally referred to as spiritual archetypes, or 'ideas *prior* to things' (*universalia ante res*).[16]

This is, of course, a quite different kind of knowledge to that pursued by contemporary science. It is sacred knowledge. In the Western philosophical tradition, the spiritual archetypes are conceived as being 'in the mind of God', a phrase which signifies that their provenance is beyond space and time and that they have a purely spiritual mode of existence before they manifest in any material form. This spiritual mode of existence is as thoughts or ideas within the greater cosmic intelligence, or cosmic *Logos*, which endows them with generative power. As thoughts in the greater cosmic intelligence, they possess a creative potency that human thoughts do not have.

The recognition of the reality of the wholeness of things – as more than just the sum of their parts – thus leads to the recognition of an altogether more interior level of reality, in which mind is understood as the 'container' of matter, rather than the other way round. The notion that there is a universal intelligence at work within creation, that bestows upon it being, order and meaning, does not in itself contradict the findings of science. It only contradicts the philosophical stance of reductionist 'scientism'. To those with eyes to see, our world is not meaningless chaos but it everywhere displays order and harmony. In such a path of knowledge, so different in its intent from the utilitarian and technological objectives of the 'new knowledge' inaugurated during the Scientific Revolution, the contemplation of nature leads to the opening of the doors of perception to the divine ground of being. It was from such a contemplative indwelling of nature, in which all creatures are apprehended as rooted in God, that the great mystical thinker, Hugh of Saint Victor, was able to declare: 'all of nature speaks of God'.[17]

Hugh's was by no means a lone voice. The theophanic view of nature as a manifestation of spirit, and therefore as sacred, was reiterated over and over again throughout the period before the Scientific Revolution.[18] His saying gives expression to a degree of relatedness and attunement to nature that strongly resonates with the mystical praise poetry of the Psalms and other Biblical texts, and also with manifold non-Christian religious views of nature worldwide, which see the order of nature as a direct manifestation of the divine.[19] It also resonates with modern holistic approaches to nature, such as that of Goethe, who pioneered a path of knowledge that leads from immersion in the observable characteristics of natural phenomena to a beholding of the spiritual archetypes present within them. For Goethe, the culmination of the act of knowing is an intuition of the spiritual archetype. It enabled him to affirm in the same terms as Hugh: 'The works of nature are like a freshly spoken word of God'.[20]

If the natural world fails to speak to us of God (as Hugh puts it) or fails to speak to us in God's voice (as Goethe would say), this is because we have submitted to a kind of knowledge that, whilst giving us excessive power, has blunted our ability to enter into a selfless relationship with nature. Accustomed to a diminished view of the world from which the divine has been excised, our eyes no longer see, our ears no longer hear, the reality in which we actually live. And herein lies the root cause of the ecological crisis, which mirrors the obtuseness, the alienation, the self-obsession, of the modern/post-modern soul. That is to say, it is precisely through denying the sacred level of knowledge that the scientific-technological mentality has permitted humanity to turn upon nature with such destructive fury. By conceiving the aim of knowledge primarily as being to equip us with greater power over nature, and to enable us to utilise natural resources more effectively for our own benefit, we commit an offence not only against the theophanic reality of nature but also against ourselves too as bearers of knowledge.

THE CALL TO KNOWLEDGE

The full comprehension of our responsibility as bearers of knowledge constitutes a third step necessary for the recovery of the theophanic view of nature. The traditional understanding of knowledge is that it is essentially a communion of knower and known. The act of knowing, through which we grasp the inner truth of things, cannot occur in isolation from those things but is an actualisation of two potentialities: on the one hand the potentiality of a thing to be known, and on the other the potentiality of the knower actually to know.[21] The first potentiality *of a thing to be known* implies that all things not only have an openness to being known but also that the act of being known affects them, for it raises them in a certain way from a state of potentiality to actuality. When a human being observes them and is able to selflessly contemplate their inner nature, this contributes something to them that no other creature or environmental factor can contribute – the possibility of being perceived as they truly are. The poet Rainer Maria Rilke was intensely aware of the significance of this possibility. In one of his poems, he wrote:

> Nothing was finished before I perceived it;
> What was becoming stood still.[22]

What Rilke here expresses implies that human beings have an obligation towards the world, to bring the thoughts in our minds into conformity with the inner nature of the things we are seeking to know, for then the things known by us achieve a kind of completion that they would not otherwise achieve.[23] There is a certain dependence of the natural world upon being cognised by us. As the Islamic philosopher Averroës declared, the things of this world 'are oriented in their inner nature toward being known by us: for this knowability is an essential determination and belongs to their real nature'.[24]

But the obligation also extends to us as knowers. The second potentiality *of the knower actually to know* implies that if we limit the quest for knowledge to what is accessible to the analytical intellect alone and is simply useful to us, to the point even of defining knowledge as that which demonstrably gives us the power to control and manipulate that which we know, then we fail to realise our true potential as knowers. We fail to fulfil our unique position in the natural order as having the ability to bring into our consciousness an awareness of the essentially sacred being of things. Instead, we create for ourselves an inadequate, diminished view of reality, which may be factually correct, and sufficient for us in the short term to obtain the results that we are looking for but, because our knowledge of things falls short both of *their* full truth and of *our* deeper capacity as knowers, it leads inexorably to disharmony both in the human soul and in nature. Disharmony in the human soul is caused by our ignoring the call to fulfil our own spiritual potential and attend to what is essential, with the consequence that we live blighted by an underlying sense that our lives lack meaning. Disharmony in nature is made manifest in the violent disruption to the natural world due to our treating it simply as a resource to be exploited in order to feed our insatiable hunger – a hunger that is truly for the infinite, but which we falsely identify as lying within the finite domain.[25]

In his book *Harmony*, H. R. H. The Prince of Wales observes that the greatest problem that faces us today is a 'crisis of perception'. He writes:

It is the way we see the world that is ultimately at fault. If we simply concentrate on fixing the outward problems without paying attention to this central, inner problem, then the deeper problem remains, and we will carry on casting around in the wilderness for the right path without a proper sense of where we took the wrong turning.[26]

All things have an inner disposition to be known and stand, as it were, ready and yearning to be known in their truth. There is, then, a specifically human task in the greater ecology of the cosmos to know things in the right way, to know them in the truth of their being, to know them in their divine aspect, '*in divinis*'.

The act of knowing is a transformative act that raises the thing known to a higher level of existence. Just as rain and sunshine make plants grow, so the human act of contemplative knowing, in which we perceive things in their divine depth, promotes them towards their own inner reality. This is a deed of illumination on our part, by which we bring to nature the light of conscious recognition of its sacred ground. The act of human knowing is, in other words, part of nature's ecology, uniquely contributed by us.[27] Through it, we prepare creatures for their return to the divine source of their being. As Meister Eckhart said:

> All creatures enter my understanding that they may be illumined in me. I alone prepare all creatures for their return to God.[28]

Such a statement points to the profound responsibility that human beings have, as cognising beings, to hold each creature in special regard, to perceive all things 'in God'. In this act, the numinous dimension that is at their source becomes present to human consciousness, as in a mirror. Put in theistic language, human consciousness becomes the vehicle or mediator of God's self-knowledge. Here then is a sacred task that the desperate state of nature today calls on us to undertake.

We cannot fruitfully undertake this task if we conceive the scope of human cognition as limited to the merely problem-solving intellect (the *ratio*), that relies upon evidence-based reasoning, data-analysis and logical argument. This level of cognition may endow us with immense technological power, but it does not lead us into the divine presence. Nor does it fulfil our deeper human potential. This can only occur when through the awakened intuitive insight of the *intellectus*, supported by imagination and inspiration, we indwell the sacred realm of spiritual archetypes, powers and presences. For this level of cognition to arise, the discursive, analytical mind must become still; and then, from this point of stillness, the possibility of achieving conscious awareness of nature as theophany can be realised. The circle of God's self-knowledge is then completed through us, and we may hope to begin to restore harmony to the world.

NOTES

1.   For the interior life of prayer, contemplation and the discipline of 'holy imagination', see Jean Leclerq, *The Love of Learning and the Desire for God* (London: SPCK, 1978), Chapter Five. See also Bernard McGinn and John Meyendorff, eds., *Christian Spirituality: Origins to the Twelfth Century* (New York: Crossroad Publishing, 1985), pp. 221-23.

2.   Origen, *De Principiis (On First Principles)*, 4.1.11. in Alexander Roberts, James

Donaldson, and A. Cleveland Coxe, eds., *The Ante-Nicene Fathers*, Vol. 4. (Buffalo, NY: Christian Literature Publishing, 1885).

3.   As, for example, in Thomas Aquinas, *Summa Theologica,* translated by the Fathers of the English Dominican Province (New York: Benziger, 1947) I, Prologue, Q1, A.10.

4.   Peter Harrison, 'The Bible and the Emergence of Modern Science' in *Science and Christian Belief*, vol. 18, no. 2 (October, 2006), p. 118.

5.   The source of this saying is Bernard's pupil, John of Salisbury, *Metalogicon*, 3.4 (Philadelphia: Paul Dry Books, 2009), p. 167.

6.   Mircea Eliade, *The Myth of the Eternal Return* (Princeton: Princeton University Press, 1971), pp. 112-130.

7.   The word 'theophanic' derives from the two words *theos* meaning 'divine' and *phanos* meaning 'appearance or manifestation'.

8.   Francis Bacon, *Novum Organum*, 1.84, in Sidney Warhaft, ed., *Francis Bacon: A Selection of His Works* (London: MacMillan, 1965), p. 356. René Descartes, *Principles of Philosophy* (Radford VA: Wilder, 2008), p. 11, considered that the more people studied the ancients, 'the less fit they are for rightly apprehending the truth'.

9.   René Descartes, *Discourse on Method*, Part 6, in *Philosophical Writings*, edited by Elizabeth Anscombe and Peter Thomas Geach (Sunbury-on-Thames: Thomas Nelson, 1970), p. 46.

10.   See the recent series of authoritative reports from the Intergovernmental Science-Policy Platform on Biodiversity and Ecosystem Services (IPBES) *Regional Assessment Reports*, 26 March 2018.

11.   See, for example, Richard Dawkins, *The God Delusion* (London: Random House, 2007), p. 411, where he characterises human beings as no more than 'chunks of complex matter' capable of thinking, feeling and falling in love with other 'chunks of complex matter'. Stephen Hawking, in an interview with Ian Sample, published in *The Guardian,* 15 May, 2011, said: 'I regard the brain as a computer which will stop working when its components fail. There is no heaven or afterlife for broken down computers; that is a fairy story for people afraid of the dark'. Yuval Noah Harari, *Homo Deus* (London: Vintage, 2015), Chapter Three, sees the concepts of God and soul as redundant, and argues that human consciousness is nothing more than electrochemical reactions in the brain.

12.   As Seyyed Hossein Nasr, *Man and Nature: The Spiritual Crisis of Modern Man* (London: George Allen and Unwin, 1976), p.9, pointed out many years ago: 'The ecological crisis is only an externalization of an inner *malaise* and cannot be solved without a spiritual rebirth of Western man'.

13.   Aristotle, *De Anima (On the Soul)*, II.1.412a, 6-11. See also Thomas Aquinas *Commentary on Aristotle's De Anima*, translated by Kenelm Foster O.P. and Sylvester Humphries O.P. (New Haven: Yale University Press, 1951), II.1, 415, who elucidates: 'Matter is that which is not as such a "particular thing", but is in mere potency to become a "particular thing". Form is that by which a "particular thing" actually exists'.

14.   Thomas Aquinas, *Summa Theologica* II, 2, Q.8, A.1. The distinction between the *ratio* and the *intellectus* was transmitted from ancient Greek philosophy (c.f. Plato's distinction between *dianoia* and *nous*) to the Middle Ages by such writers as Augustine and Boethius. From a superficial point of view, it identified the two stages of acquiring knowledge – by first reasoning things out and then gaining understanding or insight. But from a deeper perspective it indicated two different approaches to the pursuit of knowledge – one through argument and disputation, the other through the insights born of contemplation.

15. The progressive deepening of experience of the form in matter to its numinous ground and spiritual source is described in Plotinus, *Enneads*, V.9 in his treatise *On Intellect, the Forms and Being*. See especially *Enneads*, V.9.3-5. Aristotle outlines the approach in *Physics*, 1.1, where he states that 'the path of investigation must lie from what is more immediately knowable and clear to us to what is clearer and more intimately knowable in itself', for thereby we advance 'toward that which is intrinsically more luminous and accessible to deeper knowledge'.

16. Traditionally, three different modes of existence of ideas (or 'universals') are recognised. The *universalia ante res* are the exemplars or spiritual archetypes that pre-exist the things that exemplify them; the *universalia in rebus* exist within things as the principle of their wholeness; and the *universalia post res* exist as concepts abstracted from things by the human mind. The threefold distinction goes back to Porphyry's *Isagoge*, and the alignment of the *universalia post res* first with the *universalia in rebus* and then with the *universalia ante res* was the basis of what was essentially a theophanic way of knowing. See, for example, Plotinus, n.15 above, Boethius, *Consolation of Philosophy*, V.4: 84-91, Albertus Magnus, *De Praedicabilibus*, 2.17-25 and Thomas Aquinas, *Questiones Disputate de Veritate (Disputed Questions on Truth)*, Q. 3, A. 1.

17. Hugh of St Victor, *Didascalicon*, 6.5: 'Omnis natura Deo loquitur'.

18. For example, his contemporary, William of Conches, who taught at Chartres, described the experience beautifully, when he wrote in his *Glossae Super Platonem*,: 'As a stream is to a spring, all things are from the divine mind'. Quoted in Peter Ellard, *The Sacred Cosmos: Theological, Philosophical, and Scientific Conversations in the Twelfth-Century School of Chartres* (Scranton and London: University of Scranton Press, 2007), p. 94 and p. 99, n.45. In the following century, St. Bonaventure, *The Mind's Road to God*, 2.11-13, described the Christian mystical path as leading from the contemplation of creatures to the contemplation of the spiritual principle that is their divine source. Knowledge of nature could not be separated from knowledge of God. See Saint Bonaventure, *The Mind's Road to God*, translated by George Boas (Indianapolis: Bobbs-Merrill, 1953), pp. 20-21. The same theophanic view of nature is also expressed by Meister Eckhart, for example, when he says 'All things speak God. What my mouth does in speaking and declaring God is likewise done by the essence of a stone'. See Meister Eckhart, *Sermons and Treatises*, vol.1, translated and edited by M. O'C. Walshe (Shaftesbury: Element Books, 1987), Sermon 22, p. 178. For Eckhart's contemplative approach to nature, see Joseph Milne, 'Meister Eckhart and the Purpose of Creation' in *Temenos Academy Review*, 20 (2017), pp. 78-90.

19. See Seyyed Hossein Nasr, *Religion and the Order of Nature* (Oxford: Oxford University Press, 1996), Chapter 2 for a helpful review, from the perspective of the perennial philosophy, of the theophanic understanding of nature in all the major religious traditions.

20. Goethe, Letter to the Duchess Louise von Saschsen, 28 December 1789, quoted in Rudolf Steiner, *Goethe the Scientist*, translated by Olin D. Wannamaker (New York: Anthroposophic Press, 1950), p. 194. For Goethe's scientific method, see Henri Bortoft, *The Wholeness of Nature: Goethe's Way of Science* (Edinburgh: Floris Books, 1996).

21. This is how both Aristotle and Thomas Aquinas understood it. See Aristotle, *On the Soul (De Anima)*, translated by W. S. Hett (London: Harvard University Press, 1975), 3.2 and 3.8. See also Thomas Aquinas, *Commentary on Aristotle's De Anima*, translated by Kenelm Foster, O. P. and Sylvester Humphries, O. P. (New Haven: Yale University Press, 1951), Lectio 2, §591-§596 and Lectio 13, §787-§788. A similar view has been put forward in modern times by Owen Barfield, *Saving the Appearances* (London: Faber and Faber, 1957), Chapters 20 and 21.

22. Rainer Maria Rilke, *The Book of Hours,* translated by Christine McNeill and Patricia McCarthy (Mayfield: Agenda Editions, 2007), p. 27.

23. Knowledge is traditionally understood as an assimilation – literally a 'making equal' or *adaequatio* – of the mind with the thing: *adaequatio rei et intellectus.* Only when this conforming of our minds to the inner nature of things occurs can they be said to be known in the truth of their own being. Thus Aquinas says, 'For every true act of understanding is referred to a being, and every being corresponds to a true act of understanding'. See Thomas Aquinas, *Questiones Disputatae de Veritate (Disputed Questions on Truth)* translated by Robert W. Mulligan, S. J. (Chicago: Henry Regnery, 1952), Q. 1, A. 2. Answers to Difficulties, 1. Aquinas' *De Veritate* is the best introduction to the *adaequatio* teaching, especially Q. 1, A. 1-5. See also Pieper, *Living the Truth*, pp. 29-35.

24. Averroës, *Brief Commentary on Aristotle's Metaphysics,* quoted in Josef Pieper, *Living the Truth* (San Francisco: Ignatius Press, 1989), p. 66. The same understanding lies at the heart of the epistemology of Plato and Aristotle, and also Thomas Aquinas, as Pieper clearly shows (see n. 23). See also Joseph Milne, *The Mystical Cosmos* (London: Temenos Academy, 2013), p. 26, who sums it up succinctly: 'The world is not merely passively 'there' but actively revealing something through being there and manifesting itself, and it is this that man is called to know and to engage with'.

25. Seyyed Hossein Nasr, *The Spiritual and Religious Dimensions of the Environmental Crisis* (London: Temenos Academy, 1999), p. 22.

26. H. R. H. The Prince of Wales, *Harmony* (London: Harper Collins, 2010), p. 6.

27. For a shining example of the work of illumining nature through contemplative observation, see Craig Holdrege, *Thinking Like a Plant* (Great Barrington: Lindisfarne Books, 2013).

28. Meister Eckhart, *Sermons and Treatises,* vol. 2, translated and edited by M. O'C. Walshe (Shaftesbury: Element Books, 1987), Sermon 56, pp. 80-81 (translation adapted).

Harmony and Ecology

Jack Hunter

Notions of Harmony and Balance in Nature

IT IS COMMON IN DISCUSSIONS ABOUT OUR RELATIONSHIP with the living planet – especially with regard to the devastating impact of human activity on the health of our global ecosystem – to hear talk about the need to live in 'harmony' with the natural world. The philosopher Arne Naess (1912-2009), founder of the 'Deep Ecology' movement, for example, defines his notion of 'ecosophy' specifically as a 'philosophy of ecological harmony or equilibrium'.[1] Similarly, David Cadman defines harmony as 'an expression of wholeness', with which comes the understanding that human beings (and our actions) are enmeshed within a wider network of 'connections and relationships'.[2] This chapter will explore what observations of ecological systems can tell us about the nature of harmony, and related ideas such as 'balance' and 'equilibrium'.

As other chapters in this book have explored, the idea that the natural world is in a delicate state of equilibrium, or that there is *harmony in nature*, is a very ancient one, with parallel concepts found in numerous societies right across the world. The Taoist concept of yin and yang, whereby the cosmos is understood as an interconnected whole consisting of balanced binary oppositions, is perhaps the clearest expression of this idea, though there are also parallel concepts in different cultural contexts. As Fritjof Capra explains:

> The philosophical and spiritual framework of deep ecology is not something entirely new but has been set forth many times throughout human history. Among the great spiritual traditions Taoism offers one of the most profound and most beautiful expressions of ecological wisdom, emphasizing both the fundamental oneness and the dynamic nature of all natural and social phenomena.[3]

Indeed, so fundamental have these ideas been to our way of thinking for so long that philosopher of science Gregory Cooper suggests the notion of balance 'usually functions as a background assumption' shaping the way we go about making sense of the natural world. He points out, however, that 'rarely has it been brought forward for explicit study'.[4]

When we observe natural systems we see all manner of simultaneous processes in action – growth, symbiosis, and interconnection (what we might consider 'positive' features of ecology), as well as competition, predation, death and decay (which could equally be considered as 'negative' features of ecology). A call to live in harmony with the living principles of nature could, therefore, be taken in either direction. Ecological principles could just as easily be cited as supportive of a worldview based on 'survival of the fittest' as one based on reciprocal, mutually beneficial, relationships. Furthermore, commentators such as Kristin Shrader Frechette have suggested that those who seek to draw ethical principles from 'natural principles' frequently make 'misguided appeals to ecological laws,' despite a lack of consensus in ecological science.[5] This is very slippery territory, and it is well worth taking a moment to unpack what observations of living systems *can* tell us about harmony. In order to do this we will look at key debates on the dynamics of ecosystem functioning, and ecological theories and models used to explain them, and will conclude with the case of permaculture as a means of living 'in harmony' with natural processes. As we will see, however, the kind of harmony engendered in the natural world is far from clear-cut or straight forward.

ECOLOGY

The field of ecology is a relatively new area of research for science. Its roots go back to the nineteenth century, but it did not reach maturity as a distinct discipline until the middle of the twentieth century.[6] A key concept emerging from the study of ecology is the notion of the 'ecosystem'. Pioneer of scientific ecology Eugene Odum (1913-2002), defines the ecosystem as referring to:

> A unit of biological organization made up of all of the organisms in a given area (that is, 'community') interacting with the physical environment so that a flow of energy leads to characteristic trophic structure and material cycles within the system.[7]

In other words, an ecosystem is a complex system of interactions between living organisms (plants, animals, microbes, fungi) and the non-living environment (water, minerals, gases, sunlight, and so on). Above all, therefore, ecosystems are all about *relationships* – relationships between organisms, as well as relationships between organisms and the non-living environment. From this perspective everything is connected, from the smallest bacterium to the largest trees and mammals, bound together through the reciprocal exchange of vital non-living elements. These interactions include the exchange of energy and nutrients through 'food webs'.[8] Plants (whether we are talking about shrubs, trees or phytoplakton), are referred to as 'primary producers' – they capture energy from the Sun by photosynthesis, which enters into the food chain when consumed by herbivores – 'secondary consumers' – who in turn may be consumed by predators. Thus the Sun's energy is distributed amongst biological organisms in an ecosystem, gradually decreasing as it moves higher up the food chain.[9] Energy and nutrients are also constantly cycling around this system through processes of growth and decay. Energy and nutrients collected and stored by trees, plants and animals are slowly released back into the wider system through the action of decomposers such as bacteria and fungi.

Distinctive ecosystems develop in, and are adapted to, specific geographical and environmental niches, so that we can talk of, for example, saltwater ecosystems, freshwater ecosystems, desert ecosystems, woodland ecosystems, and so on. Groups of ecosystems that share similar environmental characteristics are often referred to as biomes.[10] Another major concept in ecology is the notion of 'succession' in ecosystems. Succession refers to the processes by which living organisms colonise and transform environmental niches to suit their own needs, as well as the needs of successive species. Bare scrub land, for example, is colonised by pioneer species, which transform soil and climate conditions as they develop to allow other plant species to move in. Odum defines succession as referring to three key parameters:

(i) It is an orderly process of community development that is reasonably directional ... (ii) It results from modification of the physical environment by the community; that is, succession is community-controlled even though the physical environment determines the pattern, the rate of change, and often sets limits as to how far development can go. (iii) It culminates in a stabilized ecosystem in which maximum biomass ... and symbiotic function between organisms are maintained per unit of available energy flow.[11]

Each stage of succession is referred to as a *sere*. At each successive sere the plant community tends to become more biodiverse, and so more complex. The process of succession eventually culminates with a relatively stable 'climax community'.[12]

The organisms that make up an ecosystem are, therefore, active in transforming local environmental conditions to suit their own needs. Hardy pioneer species colonise bare land and transform the structure of soils, which in turn creates new conditions for other species to inhabit. *Co-operation* between species in an ecosystem, therefore, seems to be essential (though we cannot ignore the very real role of *competition*). Indeed, organisms often work *mutually* (where one species acts as a host for another, for example, the remora fish, which feeds on the parasites of sharks), and sometimes *symbiotically* with one another (where two organisms live an entirely interconnected life, as in the case of mycorrhizal fungi in the root systems of trees) to create optimum conditions for biodiversity. This observation seems to run counter to the mainstream reductionist Darwinian concept of competition and 'survival of the fittest' as the sole drivers of evolution[13] and is a point of contention amongst ecologists, who often tend towards one or the other interpretation.

Emergentism versus Reductionism

In their paper on the sociology of ecological science, John Bellamy Foster and Brett Clark[14] delineate a tension early in the development of the field between those researchers who assumed an organicist, holistic and teleological interpretation of ecosystem development, and those who assumed a materialist, mechanistic, systems view. This is known as the 'holism-reductionism' debate, or the 'emergentism-reductionism' debate.[15] As an example of an holistic approach, Foster and Clark refer to the work of plant biologist Frederic Clements (1874-1945), who is best known for his research into plant succession. For Clements, the direction of succession towards greater biodiversity and complexity was indicative of a teleological drive, with the climax community essentially understood as a single living organism:

> Clements provided an idealist, teleological ontology of vegetation that viewed a 'biotic community' as a 'complex organism' that developed through a process called 'succession' to a 'climax formation'. He therefore presented it as an organism or 'superorganism' with its own life history, which followed predetermined, teleological paths aimed at the overall harmony and stability of the superorganism.[16]

From this perspective, succession is always directed towards 'harmony' and 'stability' within the ecosystem and is the natural process by which such super-organisms grow to maturity. Understood through the lens of organicism (emergentism), ecosystem development is a harmonic process, with different elements working together for the mutual benefit of the 'superorganism'. James Lovelock's famous 'Gaia hypothesis' is essentially an extension of this general observation about ecosystems to the whole Earth system. The Gaia hypothesis, developed by Lovelock in the 1970s, suggests that the Earth itself is a single living system, composed of multiple inter-related parts (including the chemical and mineral composition of the Earth, as well as all organic life forms), which work together to maintain a stable global system.[17]

This teleological perspective has its critics, however. In his 1982 book *The Extended Phenotype*, outspoken atheist and evolutionary biologist Richard Dawkins argued against the Gaia hypothesis on the grounds that it seems to present a top-down teleological explanation for global homeostasis (i.e. that it is, in some sense, purposeful). He writes:

> A network of relationships there may be, but it is made up of small, self interested components. Entities that pay the costs of furthering the well being of the ecosystem as a whole will tend to reproduce themselves less successfully than rivals that exploit their public-spirited colleagues, and contribute nothing to the general welfare.[18]

Dawkins' view differs from that of Lovelock primarily on the grounds that the former presents a reductionist view based on *competition* of individuals within the system (who have no thought for the 'greater good'), while the latter presents an holistic view based on top-down co-operation between biotic and abiotic components of the Earth system. At its core the holism-reductionism debate represents a clash of paradigms – between blind mechanism and teleological organicism. Such disagreements are characteristic of debates in ecology (as well as most other fields) and are unlikely to ever be fully resolved.

BIODIVERSITY AND ECOSYSTEM STABILITY

Just as there have long been debates between holists and reductionists, so too have there been disagreements between ecologists who suggest that ecosystems become *more resilient* to change the *greater the diversity* of species they contain, and those who suggest that *simpler* ecosystems are more resilient. This is known as the 'complexity-stability debate'.[19] Researchers in the 1950s, such as Eugene Odum, who assumed a broadly organicist view of ecology, argued that greater connections for energy transfer within an ecosystem resulted in that system being less susceptible to the loss of a single species, or to unexpected climate fluctuations. In this scenario, if an element is removed from a food web it can be compensated for by redirecting energy flows, or by drawing energy from other parts of the system, so it makes sense to see an adaptive benefit in having a highly biodiverse ecosystem.[20] With this principle in mind, then, ecosystems were understood to develop towards *increased complexity* and *increased biodiversity*, leading to a greater number of energy pathways within the system. This is known as the 'insurance hypothesis'.[21]

By the 1970s, however, this view was increasingly challenged by a new generation of researchers who held that ecosystems with *fewer* elements were more resilient. Basing their models on Newtonian physics, they argued that the more elements a system contains the more chaotic it becomes, and so the more likely it is to collapse. From this perspective simpler ecosystems were thought to be more resilient to change, while larger more complex systems were thought to be less so. Here, again, we see the re-emergence of the holism-reductionism debate. Those who hold that greater biodiversity in a system leads to greater resilience are adopting an holistic perspective that emphasises complexity and reciprocal interconnections between organisms, while those adopting a reductionist view rather focus in on the micro-level, and emphasise a mechanistic simplicity. In reality, however, the truth is likely somewhere in between these two strong positions.

TROPHIC CASCADES

Other important processes affecting the overall balance and stability of ecosystems, and which offers support to an holistic interpretation of ecosystem functioning, are so-called *trophic cascades*, defined as:

> Reciprocal predator-prey effects that alter the abundance, biomass or productivity of a population community or trophic level across more than one link in a food web ... Trophic cascades often originate from top predators, such as wolves, but are not necessarily restricted to starting only in the upper reaches of the food web.[22]

The classic example of the capacity of trophic cascades to transform ecosystems is the case of the eradication, and eventual re-introduction, of wolves in Yellowstone National Park. In 1995, after seventy years of near-extinction as a result of hunting, wolves were re-introduced to the Yellowstone National Park with remarkable consequences. In the absence of predatory wolves, large populations of red deer had resulted in overgrazing around streams and rivers, which in turn had affected the stability of riverbanks. When the wolves returned, red deer numbers declined and, as a consequence, trees along riverbanks were able to flourish, which in turn re-stabilised the riverbanks. The new larger trees shaded and cooled the river, providing cover for fish and creating new habitats for insects and birds. The effects of the re-introduction of wolves into Yellowstone were seen right the way through the ecosystem, encouraging much higher levels of biodiversity through the creation of new niches for exploitation by other species.[23]

What balance there is in ecosystems, therefore, comes from *both* the 'bottom up' perspective – plants, as primary producers, are the foundation of ecosystems – *and* the 'top down' perspective – through the activities of higher predators and their cascading influence on species lower down the food chain. This effect, which is well documented, resonates with ideas about ecological harmony through biodiversity and complexity. Pace *et al.* note that trophic cascades occur in all manner of diverse ecosystems; 'from the inside of insects to the open ocean … in streams, lakes and the marine intertidal zone … fields, soils [and] forests'.[24] Trophic cascades, therefore, appear to be universal characteristics of ecosystem dynamics and go a long way towards demonstrating that harmony in nature is a dynamic process that is never fully in balance. It is a constantly fluctuating ebb and flow that arises through complex interactions between organisms and the environment.

PERMACULTURE AND HARMONY

In essence, permaculture is a design process for the regeneration of natural systems based on the observed principles of ecology. It was developed in Australia in the 1970s by ecologist Bill Mollison (1928-2016) and his student David Holmgren[25] and has been steadily growing as a loosely organised global movement ever since.[26] The term itself derives from the conjunction of the words 'permanent' and 'culture' (or 'agriculture'), so *perma*-culture could be understood as a design system for creating ecologically rooted 'permanent cultures' that are 'regenerative,' rather than just 'sustainable'. One of the most popular formulations of permaculture makes use of twelve key design principles, drawing from the work of David Holmgren, and in particular from his book *Permaculture: Principles and Pathways Beyond Sustainability*. The twelve principles are:

1. Observe and interact

2. Catch and store energy

3. Obtain a yield

4. Apply self-regulation and accept feedback

5. Use and value renewable resources and services

6. Produce no waste

7. Design from patterns to details

8. Integrate rather than segregate

9. Use small and slow solutions

10. Use and value diversity

11. Use edges and value the marginal

12. Creatively use and respond to change[27]

This is not quite the place to give a full analysis of the twelve principles, but suffice to say that they are inspired by observations of ecosystem functioning, and especially of processes such as succession and its tendency towards increasing biodiversity within a system. Mollison and Holmgren explain that permaculture 'unlike modern annual crop culture, has the potential for continuous evolution towards a desirable climax state'.[28] In a sense, then, permaculture is about allowing natural processes to do their thing, but in a way that can be channelled towards meeting the needs of human beings (such as food production), while also enhancing biodiversity and building resilience. Holmgren's twelve principles are themselves couched within a wider tripartite permaculture ethic of:

13. Earth care

14. People care

15. Fair share

These three simple ethics provide a grounding for work in permaculture: to regenerate and protect the Earth system, upon which all life depends and to care for all people (which might even be expanded along the lines of the 'new animism,' to include non-human persons as well as human persons).[29] The final ethic is grounded in the observation that there is very little waste in natural ecosystems – all resources are constantly cycled and redistributed, and surplus is always invested back into the system. Permaculture is just one example of a practice inspired by observations of natural systems. There are various other forms of agricultural and horticultural practice – such as agroforestry, syntropic agriculture, and other grassroots approaches – that have a demonstrated efficacy in producing abundant yields, enhancing biodiversity and reducing reliance on fossil fuels and pesticides for maintenance. Permaculture and related practices are useful examples of practical methods of harmonising with natural processes, and in so doing having a positive impact on the local environment.

CONCLUSIONS

The kind of harmony we see manifest in the natural world is not directly equivalent to, for example, the harmonious relationships found in geometry, or in music. This form of harmony is largely abstracted from nature (notwithstanding remarkable cross-overs): it is clean, neat and rational. The kind of harmony we see expressed in nature is much messier, and much more chaotic – indeed, environmental researcher Daniel Botkin talks of 'discordant harmonies' in nature.[30] It is the difference between understanding the world in terms of rationality and the order of Newtonian mechanics on the one hand, and as a dynamic organism on the other. This is a point that was made by empiricist philosopher David Hume (1711-1776) in the eighteenth century in his critique of William Paley's (1743-1805) argument for intelligent design, which holds that the world resembles the intricate mechanism of a watch – evidence of a divine designer. Hume writes:

> The world plainly resembles more an animal or a vegetable than it does a watch or a knitting-loom. Its cause, therefore, it is more probable, resembles the cause of the former. The cause of the former is generation or vegetation. The cause, therefore, of the world, we may infer to be something similar or analogous to generation or vegetation.[31]

Biologist Rupert Sheldrake makes a similar point in his suggestion that what physicists refer to as the 'laws of nature' (the speed of light, the universal gravitational constant, and so on), are perhaps best thought of as 'habits' rather than eternal unchanging laws. If, as Hume suggests, the world really does resemble more 'an animal or vegetable' than a mechanism, then we might also expect natural laws to be dynamic, to evolve and change over time. Drawing on evidence such as apparent fluctuations in measurements of the speed of light since the 1920s, Sheldrake summarises:

> The idea that 'laws of nature' are fixed while the universe evolves is an assumption left over from pre-evolutionary cosmology. The laws may themselves evolve or, rather, be more like habits...the 'fundamental constants' may be variable, and their values may not have been fixed at the instant of the Big Bang. They still seem to be varying today.[32]

Much as Sheldrake suggests, harmony, balance and equilibrium in nature are perhaps best not thought of as 'laws of nature,' but rather as something more like habits, or tendencies. Harmony, therefore, can shift and change – it is dynamic, not static – and as Hume reminds us, seems to embody organic rather than mechanistic qualities. Observations of ecological systems, just as any other object of scientific inquiry, require an interpretive framework in order to be understood. Reductionist eco-science alone will do little to reverse the damage we have done to the planetary system. Naess suggests that 'Eco-science (ecology) is not enough. Eco-wisdom (ecosophy) is needed: How to live on Earth enjoying and respecting the full richness and diversity of life-forms of the ecosphere'.[33] Perhaps we need the myth of harmony (whether it is an oversimplification or not) to truly alter our collective behaviour. Trying to construct a new worldview based upon objective scientific observations of the natural world will not work – we need to go a step further.

NOTES

1.   A. Naess, 'The Shallow and the Deep, Long-Range Ecology Movement. A Summary', *Inquiry*, No. 16 (1973): p. 99.
2.   David   Cadman,   'Harmony',   https://www.uwtsd.ac.uk/harmony-institute/ [accessed 26 March 2018].
3.   F. Capra, *The Turning Point: Science, Society and the Rising Culture* (London: Gradton, 1985), p. 458.
4.   G. Cooper, 'Must There Be a Balance of Nature?' *Biology and Philosophy*, Vol. 16

(2001): p. 481.

5.   K. Schrader-Frechette, 'Ecology and Environmental Ethics,' in *Values and Ethics in the 21st Century* (Madrid: BBVA, 2012), pp. 309-11.

6.   G. Dickinson and K. Murphy, *Ecosystems* (London: Routledge, 2007), pp. 8-15.

7.   E.P. Odum, 'The Strategy of Ecosystem Development', *Science*, Vol. 164, (1966): p. 262.

8.   Dickinson & Murphy, *Ecosystems*, p. 11.

9.   Dickinson & Murphy, *Ecosystems*, p. 13.

10.   Dickinson & Murphy, *Ecosystems*, p. 4.

11.   Odum, 'The Strategy of Ecosystem Development', p. 263.

12.   Dickinson & Murphy, *Ecosystems*, pp. 93-95.

13.   Dickinson & Murphy, *Ecosystems*, p. 24.

14.   J.B. Foster and B. Clark, 'The Sociology of Ecology: Ecological Organicism Versus Ecosystem Ecology in the Social Construction of Ecological Science, 1926-1935', *Organization and Environment*, Vol. 21, No. 3 (2008): pp. 311-52.

15.   D Bergandi, 'Multifaceted Ecology Between Organicism, Emergentism and Reductionism', in A. Schwarz and K. Jax (eds.), *Ecology Revisited: Reflecting on Concepts, Advancing Science* (Dordrech: Springer, 2011), pp. 31-43.

16.   Cited in Foster & Clark, 'The Sociology of Ecology', p. 326.

17.   J. Lovelock, *Gaia: A New Look at Life on Earth* (Oxford: Oxford University Press, 2000).

18.   R. Dawkins, *The Extended Phenotype: The Gene as the Unit of Selection* (Oxford: Oxford University Press, 1982), p. 237.

19.   M. Loreau, A. Downing, M. Emmerson, A. Gonzalez, J. Hughes, P. Inchausti, J. Joshi, J. Norberg and O. Sala, 'A New Look at the Relationship Between Diversity and Stability', in M. Loreau, S. Naeem and P. Inchausti (eds.), *Biodiversity and Ecosystem Functioning* (Oxford: Oxford University Press, 2002), p. 79.

20.   Loreau et al., 'A New Look at the Relationship Between Diversity and Stability', p. 80.

21.   S. Yachi and M. Loreau, 'Biodiversity and ecosystem productivity in a fluctuating environment: The insurance hypothesis', PNAS, Vol. 96, No. 4, (1999): pp. 1463-68.

22.   M.L Pace, J.J. Cole, S.R. Carpenter and J.F. Kitcehll, 'Trophic Cascades Revealed in Diverse Ecosystems', TREE, Vol. 14, No. 12 (1999): p. 483.

23.   G. Monbiot, *Feral: Rewilding the Land, Sea and Human Life* (London: Penguin Books, 2014), pp. 84-85.

24.   Pace et al., 'Trophic Cascades Revealed in Diverse Ecosystems', p. 483.

25.   B. Mollison and D. Holmgren, *Permaculture One: A Perennial Agriculture for Human Settlements* (Tyalgum: Tagari, 1990).

26.   B. Taylor, *Dark Green Religion: Nature Spirituality and the Planetary Future* (Berkeley: University of California Press, 2010), p. 157, and R.S. Ferguson and S.T. Lovell, 'Grassroots Engagement with Transition to Sustainability: Diversity and Modes of Participation in the International Permaculture Movement', *Ecology and Society*, Vol. 20, No. 4 (2015): p. 39.

27.   D. Holmgren, *Permaculture: Principles & Pathways Beyond Sustainability* (Hepburn: Holmgren Design Services, 2006).

28.   Mollison & Holmgren, *Permaculture One*, p. 7.

29.   G. Harvey, *Animism: Respecting the Living World* (London: Hurst & Company, 2005).

30. D.B. Botkin, *Discordant Harmonies: A New Ecology for the Twenty-first Century* (Oxford: Oxford University Press, 1992).

31. D. Hume, *Dialogues and Natural History of Religion* (Oxford: Oxford University Press, 1998), p. 78.

32. R. Sheldrake, *The Science Delusion: Freeing the Spirit of Enquiry* (London: Coronet, 2012), p. 108.

33. A. Naess, 'From Ecology to Ecosophy, from Science to Wisdom', *World Futures: The Journal of New Paradigm Research*, Vol. 27, Issue 2-4 (1989): pp. 185-90.

Nature's Fragile Harmonies

Stephan Harding

First Words

We tend to have a rose-tinted idea about harmony in nature as if somehow it involves a state of static perfection devoid of conflict, friction and strife. But this view of harmony may not hold up under close scrutiny, for harmony in nature seems to appear when opposite tendencies and forces – creation and destruction, positive and negative, predator and prey, implosion and explosion – reach a transient state of equilibrium in which, for a while at least, some kind of fragile dynamic balance is achieved. If this view is correct, we would expect to find harmony emerging from the interaction of opposites at every level of existence, from sub-atomic particles to galaxies to galaxy clusters and beyond. Here I'll explore the notion that harmony, forever fragile, emerges from a delicate reconciliation between opposites in a progression that begins with so-called sub-atomic 'particles' and ends with Gaia, the global ecosystem, leaving you, the reader, to ponder whether the principle might apply at larger, more cosmic scales.

My definition of harmony, therefore, is that it arises from in all physical, chemical, biological and geological processes as an interplay of opposites. The dynamics of harmonious systems manifest as complex behaviours which can be described mathematically as either stable, periodic or chaotic in the technical sense in which many ordered patterns are folded into one another, often giving a superficial impression of randomness.

Fields and Atoms

If I have understood what my physics colleagues have told me, there is, in fact, no such thing as solid matter, and therefore no solid sub-atomic particles.[1] They tell me that what appears to us as solid matter is in fact constituted by strange field-like entities whose influence is felt everywhere in the universe. We call these entities electrons, protons, neutrons and so on when they collapse out of their field-like state (the quantum vacuum, which contains impermanent fields and particles that come in and out of existence and interact with each other) whereupon we make simplistic assumptions about a particle-like nature which they don't in fact have.[2] If we extrapolate from this and consider the field state to be a kind of non-existence and if the 'solid' state to represent existence, then here, even at this deepest level of things, we can see two opposites interacting in fragile harmony to give us the density and substance of the familiar world around us. When I am able to look at the world in this way, I feel a softening, a rounding of edges. I feel it all flowing as a vast field of being with its own strange mystery and purpose. I experience a delicious, living, fragile harmony, a glow, a golden feeling – in those brief moments when I'm able to hold and grow these opposites – existence and non-existence, being and non-being within myself, inwardly.

As we return from the mysterious world of the quantum vacuum and re-enter the domain of atoms and molecules, we need to see how fragile harmonies might arise out of the interaction of opposites in this new, seemingly much more solid realm. The obvious thought is that the opposites at this level of reality must be the positive and negative electrical charges that somehow haunt all atoms, no matter what their size and configuration. The positive charge emanates from protons in the atomic nucleus, whilst the negative charge inheres in the electrons that orbit around the nucleus in well-defined orbits or shells as suggested by the Rutherford-Bohr model of the atom, which although far too simplistic, is good enough for our purposes and for understanding much of chemistry.[3] According to this model, the inner shell of all atoms holds a maximum of two electrons, whilst the next shell out holds a maximum of eight of these negatively charged particles.

We humans tend to search for harmony in endlessly complex ways, but most atoms, if we consider them to be much simpler beings, find it by filling their outer electron shells though sharing electrons with their neighbours, creating chemically bonded communities of atoms known to us as molecules. Forging these chemical bonds involves a two-step process.[4] First, the positive charges in the nucleus of one atom and the negative charges in the electrons around another atom attract the atoms into a close and intimate proximity. The second step takes place as soon as the atoms have been brought close by this archetypal attraction between the opposites of positive and negative. Now their outer electron shells overlap so much that electrons can at last be shared, allowing each atom to complete its outer shell, and hence feel complete in itself.

The atomic nucleus is itself held together in fragile harmony by another deeply significant interaction between opposites. Consider this conundrum. Given the well-known principle that opposite charges attract and like charges repel, how is that that several positively charged protons co-exist harmoniously within the nuclei of atoms heavier than primordial hydrogen? Shouldn't they fly apart, fleeing from each other's protonic influences as fast as possible? Indeed, they should, were it not for another powerfully opposing force, known to scientists as the strong nuclear force, which counteracts these repulsive energies, binding protons (and neutrons) together in atomic nuclei. [5] One can get a sense of the power of these opposing forces by considering what happens when scientists contrive ways of destroying the fragile harmony at the heart of the nucleus by unbinding the strong force, producing atom bombs and nuclear power stations. The strong force, once released, can be vastly destructive, or, if carefully controlled, can engender huge amounts of electric power, depending on our motivation and intent. Either way, destroying the fragile atomic harmony at the very heart of matter has serious consequences for the life of our planet.

Perhaps the chemical element which most embodies harmony in the atomic/molecular world is carbon, which seeks four new electrons to join the four already present in its outer shell, making the eight needed for completion. How interesting that the number four appears over and over again in relation to wholeness and harmony in our human realm: the four directions, the four personality types, the four psychological functions, the four-fold nature of mandala images and so on as argued extensively by Jung.[6] Could it be the same in the chemical world? Is it mere coincidence that carbon's very embodiment of the number four makes it the most cooperative element in the periodic table, the one most able to forge immensely long chains of carbon atoms that create those deeply harmonious biomolecular cradles of life such as DNA, polysaccharides and fats?[7]

CELLS

What about the action of opposites amongst these biomolecules which teem within the cells of living organisms? The cell is inhabited by a hugely complex melee of large carbon-backboned molecules somewhat like carbon animals with fleshy muscles of nitrogen, oxygen, sulphur, hydrogen and other atoms, all attached to the carbon backbones by 'ligaments' – that sharing of electrons which constitutes the chemical bonding we considered earlier.[8] There is a constant interplay of creation and destruction as these biomolecules consume and remake each other in the continual dance of metabolism that is a key characteristic of life.[9]

These two opposites – creation and destruction – play out within a vastly complex network of relationships amongst the cell's huge variety of biomolecules. The network as a whole is deeply harmonious, regulating the levels of each biomolecule, all at once, simultaneously. This constant remaking by life from itself and out of itself requiring only inputs of energy and nutrients is what Chilean biologists Maturana and Varela came to call *autopoiesis* (self-making) proposing that if autopoiesis is the primary quality of life, then reproduction is merely a secondary phenomenon.[10] Reproduction involves making a full or partial copy of yourself which lives on into the next generation, and you need to remember who you are in order to do this. Some of this cellular memory is held in molecules of DNA, that massive, hugely beautiful spiralling biomolecule – that staggering demonstration of emergent creative harmony in which DNA is held together by oppositely shaped molecular bases at the core of its being that fit together like lock and key. So here, at the very core of life itself, we find something unexpected: life's fragile harmony gives rise to memory, To be alive, to be a coherent, self-making whole, the cell needs to enclose itself within an exquisitely selective semi-permeable boundary which it makes for itself out of its own autopoietic processes. This is the cell membrane. Can we discover how tensions between opposites give rise to harmony even here at this most fundamental of biological levels?

The cell membrane is composed of biomolecules known as phospholipids, which look rather like tiny molecule-sized tadpoles whose phosphorus-bearing heads adore water and whose lipid (fatty acid) tails find water utterly repulsive.[11] Here again we find attraction and repulsion, those by now familiar opposites, which in this case cause these tiny tadpoles to fall easily into a bi-layered arrangement. The tadpoles' heads stick out of the inner and outer surfaces of the bilayer, where there is plenty of water, whilst their lipid tails point inward where water is absent. In essence, the phospholipids settle into the harmonious arrangement that is the cell membrane because the tadpole heads feel themselves drawn towards water molecules that circulate both inside and outside the cell, whilst the water-hating lipids tails face inward to get themselves a far away as possible from the water molecules which they find so repulsive.

This fragile harmony between attraction and repulsion creates the flexible, semi-porous cell membrane which provides protection for the complex metabolic interactions that go on inside all cells. But the cell membrane does not simply rest in a mere molecular passivity, for it is studded with many kinds of selective molecular channels, making it the locus of an acute and highly active membranous intelligence: by intelligence here I mean the ability to sense molecules in the cellular surroundings and to then decide whether to absorb them or not.[12] This implies a rudimentary style of choice. These channels use their exquisite molecular sensitivity to decide what to allow into and out of the cell. Food in, wastes out. Calcium in, calcium out – and so on for thousands if not millions of further kinds of molecular, atomic and ionic beings. Here, in the very precincts of the cell membrane, we discover how a further fragile harmony between myriad opposites gives rise a style of intelligence within each individual cell, an intelligence very much concerned with the issue of *identity*, since a living cell needs to know the difference between 'me' and 'not me'. Each and every cell needs to consume what is not itself in order to produce more of itself. Destruction is needed to consume food and energy and creativity is required to use these inputs to make more cellular material. The fragile harmony which appears from the tension between these opposites, between creation and destruction at the level of cellular metabolism, gives rise to all the beautifully harmonious forms of living beings: exquisite flowers, iridescent beetles, luminous fungi, great whales, earthworms and bacteria.

And yet, most cells don't live in isolation – many live in some sort of community, some very loosely connected, some more strongly communal. Here again our question must be: does harmonious communication *between* cells emerge, somehow, from a harmonious reconciliation of opposites, allowing them to live successful communal lives? Whenever the urge for identity (that powerful desire to be 'me') becomes collective, language (communication) is required, since individual cells clearly need to converse with each other in order to be communal.

Communication amongst cells probably began as soon as the very first cells (archaea and bacteria) appeared on our planet some 3,800 million years ago.[13] We humans mostly use speech to communicate by vibrating molecules of air with our vocal chords. Bacteria don't bother with sound, which would probably be impractical at their scale of existence. Instead, a key aspect of their language involves a system of chemical signalling known as *quorum sensing* in which each bacterium decides whether or not to commit energy to a specific response, such as producing a toxin, based on chemical signals it receives from its neighbours.[14] In essence, each bacterium sends chemical signals to other bacteria of its own species whilst also absorbing and interpreting chemical signals from its fellows. At low bacterial densities each bacterium receives a small number of signals and does not waste energy in responding. But once the bacterial density increases to a critical level, each bacterium receives a much higher number of signals. Individual bacteria interpret this situation as meaning that there are now enough of its fellows present nearby to make investing energy in a response worthwhile, giving rise to suitably adaptive action from the population as a whole.

A classic example involves the emission of light by the bacterium *Vibrio fischeri* which lives in the open sea and in the Hawaiian bobtail squid's specialised light organs which provide the bacteria with shelter and nutrition.[15] At dawn the squid eject around 90% of their *Vibrio* bacteria from their light organs back into the sea. The squid then hide in the sand all day, slowly feeding nutrients to the few remaining bacteria in their light organs. These *Vibrio* bacteria in the squid's light organ use quorum sensing to signal to each other all the while, but during the day there is no response since their density, although increasing, is still below a critical threshold. But by nightfall the bacterial population has grown so much that each bacterium now receives the critical number of signals from its fellows, which it interprets as meaning, as we might say (this is a human's way of saying exactly how the bacteria are interpreting the meaning of the chemical signals): 'there are now enough of us to make it worthwhile for *me* to produce light'. And sure enough the bobtail squid's light organs become bright with luminescence as each bacterium activates genes for light production, helping the squid to avoid predators by eliminating its shadow on the shallow sea floor on moonlit nights when it is active. At dawn, the squid ejects the bacteria to save energy, and the cycle begins again. Here is a case where language, in the guise of quorum sensing, is used to coordinate behaviours that benefit the members of the bacterial community in the light organ, since in exchange for light the squid gives the bacteria high-quality food and shelter.

But quorum sensing can also be used to opposite effect – to destroy members of competing bacterial species. Most bacteria can detect chemical signals specific to competitors.[16] When these alien signals reach a critical density, the entire population of the bacteria under threat can take a collective decision to emit toxins – often antibiotics – that harm or kill their competitors. One well-studied example takes place right inside our mouths, within the complex bacterial community we know as dental plaque. Most bacteria live in biofilms – thin films of bacteria sticking together embedded in slimy substances attached to surfaces such as our teeth, fallen leaves, hot springs and even ice cold glaciers.[17] Many hundreds of bacterial species live in our dental plaque, each species with around 10^{10} individual bacteria, a thousand times greater than the entire global human population. Bacteria in biofilms coordinate behaviours through signalling pathways such as quorum sensing, creating communicative networks that far exceed the complexity of all our human networks, making what have justly called been called 'bacterial cities'. There are many cooperative interactions between different bacterial species in a biofilm, but there are also competitive interactions, mediated by quorum sensing, which help to promote the diversity of bacterial species within the biofilm. The result is that the bacterial community as a whole is less likely to be taken over by one species. Thus bacterial chemical language – quorum sensing – promotes both intraspecies cooperation and interspecies competition, an interplay between opposites which gives rise to species diversity and thus to a shifting, highly adaptive dynamic harmony within the bacterial community as a whole.

ENDOSYMBIOSIS

But perhaps the most astonishing domain of life where we see a fragile harmony emerging from the reconciliation of opposites is in the cells that make up the bodies of all multicellular beings: the fungi, plants, animals and also the unicellular bodies of some protoctista such as amoeba and paramecium, and in their multicellular cousins, such as seaweeds. These kinds of cells are known as eukaryotes (literally, good nut or kernel), since they carry a 'good nut' within them: the cell nucleus which contains large amounts of DNA surrounded by its own semi-permeable membrane, very similar to the cell membrane we encountered earlier.[18] By contrast, the bacteria and archaea are known as *prokaryotes* – meaning 'those before the nucleus' or literally 'before the nut' or 'kernel' – since their DNA is not located in a nucleus but is found as a freely moving loop within the main body of the cell.

Eukaryote cells have many organelles within them – fine structures that carry out a large variety of metabolic tasks. Amongst these are the mitochondria, the energy producers of the cell, and in photosynthesising eukaryotes, those green solar energy capturers known as chloroplasts. The mitochondria and chloroplasts swim freely within the cell. The mitochondria look like little sausages, the chloroplasts like small minty green lozenges.

The stunning thing is that there is virtually incontrovertible evidence that the ancestors of both these organelles were once, thousands of millions of years ago, free-living bacteria that were engulfed by a larger cell, perhaps a predator, which prepared to digest them. This clearly would have spelt destruction for the engulfed one, but instead of digestion, a creative act of communication took place in which the ingested one and the one ingesting worked out a way of living together. One partner was predatory, the other resisted predation (opposites again) out of which an exchange of services emerged. In the case of the mitochondrion, the service offered by its ancestor to its ingestor was the use of oxygen to release energy from molecules of food, and in the case of the chloroplast the service offered by its ancestor was sugar-creating, water-splitting, oxygen-producing photosynthesis. In exchange, the ingestor offered protection to the ingested within a safe intracellular environment. In both instances this symbiosis from within, or 'endosymbiosis' resulted in a completely new, more complex type of cell with its own new style of fragile dynamic metabolic harmony never before seen on the planet.

One can imagine oneself as the ancestral mitochondrion, a free-living oxygen-respiring bacterium, engulfed by a larger cell. It is not at all clear which cell was predator and which prey. Perhaps the soon-to-be mitochondrion was the predator, or perhaps it was the larger cell. We don't know for sure. But what is clear is that one or the other was intent on abating its ravenous predatory hunger. Here, at the very birth of the endosymbiotic encounter, we once again sense the presence of interacting opposites: destruction and creation. The predator must have been doing its best to digest its victim, whilst the prey must have attempted to resist, during which communication between the two helped them discover a new, mutually beneficial endosymbiotic possibility for both together as a new unity. One of the opposites was the urge to dissolve and absorb the other, to annihilate and dismember it for one's own benefit, the other an impulse to resist being digested. Together this opened a space for communication and collaboration, creating a new, totally unexpected emergent kind of cell, greater and perhaps more aware and intelligent than either bacterial partner in isolation. The details of how this happened are almost too complex to contemplate, but the advantages were so great that soon unicellular eukaryotes swarmed all over the global ocean around 1,700 hundred million years ago.

LIFE GOES MULTICELLULAR

For the following billion years or so, life was mostly unicellular and was lived almost entirely in the oceans. But the first large, multicellular eukaryotic organisms appear as fossils in rocks from around six hundred million years ago. Known as *Ediacarans*, these were mostly soft, frond-like jellies, rooted to the floor of shallow regions of the world's oceans.[19] With no mouth and no anus, they must somehow have absorbed small particles of food from the surrounding seawater.

But what caused the evolution of these multicellular eukaryotic beings? Could opposing forces have been involved even here? The period just before the appearance of the Ediacarans was punctuated by a series of aptly named 'snowball Earths' – cold periods so extreme that the entire planet was covered in ice many kilometres thick, with the possible exception of the tropics where slush, not ice, might have predominated. These extreme conditions may have forced unicellular eukaryotes to evolve the animal style of multicellularity. But why?

The intense cold of a snowball Earth would have been an immense challenge. Small populations of genetically similar single-celled eukaryotes would have found themselves isolated from the extremes of cold in small refuges such as hot springs on land and small meltwater ponds on the ice. Single-celled eukaryotes in these refuges would have been forced to resist the powerful obliterating force of the cold by becoming multicellular, an altruistic condition which would have helped them to most efficiently share metabolic processes and local resources. Because the populations were small and genetically similar, destructive 'cheat' cells were less likely to prevail. One can barely imagine the complexity of the chemical communication that must have gone on amongst those single-celled beings that made it possible for them to join together in the mutually beneficial, highly cooperative arrangement of multicellularity. Successful populations would have survived to pass on the trait of multicellularity to future generations, whilst those that failed to develop the multicellular mode of life were wiped out. If this scenario is correct we see once again how the interplay between the opposing forces of stress and resistance gave rise to a new emergent fragile harmony in the biological world – to a new adventure in being and awareness which we call multicellularity.

By about 545 million years ago, multicellular animals with hard shells had appeared in the shallow oceans.[20] There were even small animals in the oceans sporting what would much later become the vertebrate backbone. By 480 million years ago the first moss-like plants colonised the edges of the continents, and by 400 million years ago these had developed into plants with vascular systems, roots and woody stems. By 460 million years ago some of the invertebrates in the oceans evolved into the first land animals. These were the first spiders, mites, springtails and millipedes. In all these domains, and in all those that were to follow after, all the way to our modern world, ecological relationships involving tensions between myriad opposites held the web of life together.

FUNGI

For an ecological community to be healthy there has to be efficient energy capture, efficient communication and exchange of information, and efficient recycling. Perhaps the fungal kingdom most easily displays some of the opposing forces required for this healthy functioning. Fungi are decidedly strange – they are multicellular eukaryotes which have made themselves into tubes by dissolving the boundaries between their cells so that their nuclei, mitochondria and other cellular constituents flow freely along the tubes, which can extend over long distances and fuse when they meet, forming a complex tubular network known as a *mycelium*.[21]

Mycorrhizal fungi engage symbiotically with plant roots. These fungi are essential for the health of almost all land-based ecological communities. The fungus feeds its plants with otherwise inaccessible soil nutrients and water in exchange for sugars forged high up in the plant's leaves by photosynthesis. This is the fundamental source of energy for much of the living world. Mycorrhizal fungi transfer sugars from plants in the light to plants that struggle to find enough light in the shade, often to plants of a different species. They also transfer signalling molecules from plant to plant. A plant being eaten by herbivores synthesises warning chemicals which it passes to neighbouring plants via the mycorrhizal fungal tubes linked into its roots, allowing neighbours to mount their own chemical defences. Thus the fungi provide biotic communities with efficient means of communication and information transfer.

Other fungi are predators. Their tubes flow into a plant's tissues, sometimes waiting until the plant weakens before killing and digesting it, sometimes killing it immediately. Yet others decompose dead bodies – these are the recyclers of the ecological community. In these interactions we see how the fungal kingdom contributes to the emergence of health and harmony in ecological communities through the interplay of the opposing forces of life-giving communion and death-dealing predation. By helping plants to enhance their energy capture via photosynthesis, by enhancing communication between species and by promoting decomposition and hence recycling, the entire fungal kingdom hugely contributes to the efficient functioning of terrestrial ecosystems.

THE EDGE OF CHAOS

Perhaps two of the most pervasive opposites which we have not yet visited are order and mathematical chaos (in which many kinds of order are layered into each other, sometimes giving a superficial appearance of randomness), and nowhere are these more dramatically manifested than in the lives of social insects such as ants. Most ants live in complex colonies in which no single individual holds the blueprint for how the colony as a whole should behave, and yet the colony exhibits a range of supremely well-adapted behaviours in relation to prevailing conditions both within and beyond the colony such as tending the brood and foraging for food. These behaviours emerge purely from the interactions between individual ants during which they stimulate each other through touch and via chemical signals known as pheromones. It turns out that the density of ants in the colony is of critical importance in evoking the coherent behaviours which have enabled the ants to thrive for almost 100 million years. When there are too few interacting ants, the colony displays chaotic or unpredictable rhythms of activity which are too incoherent to favour survival. A situation with too many ants, on the other hand, leads to strongly ordered rhythmical activity patterns which are also of little use in surviving since they are too rigid to allow adaptation to prevailing conditions. And yet, coherent, harmonious well-adapted behaviours do emerge, but only within a narrow range of densities where rhythmical patterns of activity appear which are a fruitful mixture of order and chaos.[22] This zone, known to science as the 'edge of chaos' gives the colony maximum flexibility to adapt well to new challenges, whilst maintaining sufficient order for a coherent identity.[23]

It seems likely that the entirety of the living world, from populations of individual species to entire ecosystems and possibly the entire Earth, tune themselves to the edge of chaos through such finely adjusted interactions amongst their component beings. Order and chaos – for science these are perhaps the most fundamental opposites through which nature navigates herself into a state of creative yet fragile, dynamic harmony. If this view is right, the upshot is that to survive, life must tune itself to the edge of chaos – that delicate, sometimes deeply elusive, finely balanced domain between those archetypal opposites: order and chaos.

GAIA

Finally, we come to the planetary level, to Gaia herself. If our hypothesis is valid (that harmony appears through the interaction of opposites), we would expect to find interacting opposites here too that allow Gaia to self-regulate important aspects of her surface such as temperature, acidity and the distribution of key elements by means of harmonious dynamic emergent feedbacks between life, rock, atmosphere and water. It was James Lovelock, the originator of the Gaia hypothesis (which later developed into Gaia theory) who proposed that the first truly Gaian temperature regulation might have appeared some 2,800 million years ago when microbes in the purely bacterial biosphere of those times invented water-splitting photosynthesis which removes the greenhouse carbon dioxide from the very air itself.[24] The world then was mostly ocean, so there was no shortage of water, nor indeed of carbon dioxide which was abundant in the atmosphere. Water-splitting photosynthesis was a hugely successful strategy, and soon most of the planet's ocean surface was covered by this new kind of photosynthesising bacterium.

Eventually, these photosynthetic bacteria had removed so much carbon dioxide from the atmosphere that the planetary temperature began to plummet dangerously towards a snowball Earth.[25] The situation was saved by decomposer bacteria living in the sediments at the bottom of the oceans. These microbes liberated carbon dioxide and methane by digesting the dead bodies of the photosynthesisers which had sunk down to them from the ocean surface. These greenhouse gases then travelled through the water column back into the atmosphere, thereby warming the planet, saving it from a snowball state. Here we see how dynamic harmony emerges from the interaction of two opposing yet complementary tendencies. The photosynthetic bacteria cooled the planet but preferred it warm, whereas the decomposing bacteria warmed the planet but liked it cool. Each bacterial type gave the other what it needed but could not provide for itself, giving rise to self-regulation in a way that seems to give us clues about how all opposites should be reconciled. Basing its analogy on horse riding, science knows this phenomenon as 'integral rein control', now thought to be common throughout nature.

There are countless more processes involving pairs of opposites that give rise to states of relative harmony at the Gaian level. Perhaps the most fundamental of these are plate tectonics, which are essential for regulating Gaia's temperature within limits suitable for life because of the way they cycle carbon dioxide in and out of the atmosphere over million-year time scales. Two of the opposites we discern here are the upwelling and downwelling of huge expanses of rock. Upwellings involve the rising up of great plumes of semi-molten rocks from deep within Earth's mantle to the surface, where they solidify into seafloor basalt at the mid-oceanic ridges before spreading away horizontally on either side of a ridge. This seafloor basalt is organised into seven major plates, each quite distinct, atop of which ride the lighter continents, made of granite. It was a shock for science to discover (it all began around the 1960s) that the plates move, pushed sideways by upwelling magma rocks at the mid-oceanic ridges.[26] Wherever plates meet, immense forces of collision produce earthquakes, volcanoes and vast mountain ranges such as the Andes and the Himalayas. At one kind of collision zone, known as a 'convergent margin', one plate is pushed deep down into the Earth's depths where it melts into the magma in the process of downwelling, or subduction. The downwelling slabs of seafloor basalt carry water which they have absorbed from the ocean into the depths. Here, high temperatures and pressures cook some of the basalt and water into molten granite, which rises to the surface, cooling into the solid granite of the continents.

The story continues on the continents. Here life in the form of bacteria, lichens, mosses and higher plants help to weather and dissolve these granitic continents, removing carbon dioxide from the atmosphere, locking up its carbon atoms in molecules of calcium bicarbonate that are washed by rivers into the ocean. Here, marine organisms convert the calcium bicarbonate into their exquisitely crafted shells and carapaces of calcium carbonate: chalk and limestone. When these beings die, these shells sink to the ocean floor, settling on top of the seafloor basalt. Subducting plates of seafloor basalt then carry these chalky shells into the depths, where they melt under intense temperatures and pressures, releasing carbon dioxide which bubbles out into the atmosphere through volcanoes. Here again are two opposites vital for life. The life-assisted weathering of granite cools the earth by removing carbon dioxide from the atmosphere, whereas the release of carbon dioxide via volcanoes from the melting of chalk deep under the surface warms our surface world.

When these opposites have harmonised and the removal of carbon dioxide has been balanced by its return through volcanoes by plate tectonics, the planetary temperature has remained relatively constant for many millions of years.[27] But occasionally such periods of fragile harmony have vanished when the delicate balance between these opposites has broken down. Perhaps the most important of such episodes towards global warming took place some 252 million years ago at the end of the Permian, when it seems that tectonic activity released huge amounts of carbon dioxide into the atmosphere, triggering severe climate change and a whole series of further catastrophic environmental events that lead to the extinction of 95% of all fossilisable life forms in the oceans. This is the famous end-Permian mass extinction[28] which was the most extreme of the five mass extinctions which have occurred during the last 540 million years, until, of course, the sixth, which we are initiating.

An example of the opposite tendency, towards global cooling, took place around 640 million years ago when it seems likely that the first lichens colonised the continents, weathering huge amounts of phosphorus from the rocks which was washed into the oceans via rivers. The phosphorus stimulated huge blooms of photosynthetic algae which removed immense quantities of carbon dioxide from the atmosphere, triggering the series of three or so snowball earth episodes which we earlier encountered. Each snowball ended when carbon dioxide emitted by volcanoes accumulated in sufficient quantities in the atmosphere to melt the ice.

FINAL THOUGHTS

I've given several examples from amongst a vast number that could have been chosen to illustrate the point that I have been trying to make throughout this chapter – that harmony in nature is always tentative, always deeply fragile due to delicate, finely balanced interactions between a whole series of contrasting opposites. At any moment nature's fragile harmony can break down when one or other of the opposites becomes dominant, leading to catastrophic effects that can ripple through a given domain of the natural world until a new fragile harmony is established, often when new sets of opposites interact. One could therefore say that nature's fragile harmonies are the result of strife or tension between opposing forces that reconcile their conflicting tendencies, creating transient states of what I call 'emergent harmonious coherence' at every level of existence, including, in the last resort, within our very own human consciousness within which these enlivening mysteries are contemplated, pondered and, hopefully, acted upon to create a better world for all, human and non-human in turn.

Indeed, when we ponder them from the standpoint of our ordinary conscious minds, the opposites seem irreconcilable. But if one can discover (or be blessed by) a different, more open, state of mind, then they somehow mingle to produce a new state of consciousness, an awareness of wholeness, born out of the tension between them. This experience gives the feeling of grace, of a healing insight that might not last very long, or that might one day be permanent. It all seems to be a matter of inner experience, not at all of intellect, which is of course needed to read or hear the words and to understand them in the first instance. Wholeness needs the opposites and the opposites need wholeness – they cannot exist without each other.

And with this we reach the realm of psyche, the most difficult of all. Here mythology helps us plumb the depth of our own natures and that of wider nature herself. According to the ancient Greek myth, Harmony was born alongside two boys, fear and terror, the result of a liaison between Aphrodite, goddess of love, and Ares, god of war.[29] Thus we are led to understand that Harmony was born from the most extreme opposites imaginable, a tension which therefore runs deeply through the whole of nature's fabric. It seems that, psychologically, consciousness, the awareness for oneself as oneself of what is actually going on around and within one, can only be born out of the tension of opposites. This notion is supported by an even deeper myth in which Aphrodite is born when the lovers Heaven (Ouranos – the sky and stars) and Gaia (Earth) are separated through a complex series of events involving the castration of Ouranos by Time (Chronos), and the throwing of his testicles into the ocean.[30] It was from this separation of opposites – of heaven and Earth – that fair Aphrodite was born in the foam of the sea. She represents a knowing, a consciousness, that loves and cares for all.

Since Gaia is Aphrodite's grandmother, one could say that Aphrodite represents for us a deep erotic love of the earth and all her beings: her rocks, swirling atmosphere, waters and life. Perhaps, if each of us, in our own individual way, can reconcile these opposites – Earth and Heaven – which for me at this moment, appear as earthiness and a sense of the eternal – then perhaps Harmony and her mother Aphrodite, she who reconciles the deepest opposites and thus reconciles them all, can come to comfort, guide and inform us in this time of deepest planetary ecological crisis.

NOTES

1.   E. Schrodinger, *Science and Humanism* (Cambridge: Cambridge University Press, 1951), pp. 20-21.

2.   P. Milonni, *The Quantum Vacuum: An Introduction to Quantum Electrodynamics* (Los Alamos, NM: Academic Press, 2014).

3.   Niels Bohr, 'On the Constitution of Atoms and Molecules', Part I. *Philosophical Magazine*. 26, 151 (1913): pp. 1–24.

4.   Bohr, 'On the Constitution of Atoms and Molecules'.

5.   P.D.B Collins, A.D. Martin, and E.J. Squires, *Particle Physics and Cosmology* (New York: John Wiley & Sons, 1989).

6.   Jung, C.G.. 'Archetypes of the Collective Unconscious', in *The Archetypes and the Collective Unconscious*, Collected Works, Vol. 9, part 1, translated by R.F. C. Hull, 1959 (London: Routledge and Kegan Paul, 1968), para. 715.

7.   Steven Rose, *The Chemistry of Life* (London: Penguin, 1999), pp. 45-54.

8.   Rose, *The Chemistry of Life*, pp. 146-7.

9.   Steven Rose, *Life Lines Beyond the Gene* (Oxford: Oxford University Press, 1997)

10.   U. Maturana and F. Varela, *The Tree of Knowledge: The Biological Roots of Human Understanding* (Boston, MA: Shambala, 1992).

11.   Rose, *The Chemistry of Life*, pp. 79-80.

12.   Rose, *The Chemistry of Life*.

13.   E. Ben Jacob, Israela Becker, Yoash Shapira and Herbert Levine, 'Bacterial linguistic communication and social intelligence', *Trends in Microbiology*, vol, 12, 4 (2004).

14.   E. Ben Jacob et al., 'Bacterial linguistic communication and social intelligence'.

15.   K.L. Visick J. Foster, J. Doino. M. McFall-Ngai and E.G. Ruby, 'Vibrio fischeri lux Genes Play an Important Role in Colonization and Development of the Host Light Organ', *Journal of Bacteriology* (2000): doi: 10.1128/JB.182.16.4578-4586.2000.

16.   E. Ben Jacob et al., 'Bacterial linguistic communication and social intelligence'.

17.   P. Watnik and R. Kolter, 'Biofilm, City of Microbes', *Journal of Bacteriology* (2000): doi: 10.1128/JB.182.10.2675-2679.2000.

18.   L. Margulis and M. Dolan, *Early Life* (Sudbury, MA: Jones and Bartlett, 2001).

19.   M. McMenamin, *The Garden of Ediacaria: Discovering the First Complex Life* (New York: Colombia University Press, 1998).

20.   D Briggs, 'The Cambrian Explosion', *Current Biology*, Vol. 25, 19 (2015): doi:https://doi.org/10.1016/j.cub.2015.04.047.

21.   A. Rayner, 'The Challenge of the Individualistic Mycelium', *Mycologia* 83 (1) (1991): pp. 48-71.

22.   R. Sole, Octavio Miramontes and Brian C. Goodwin, 'Oscillations and Chaos in Ant Societies', *Jornal of Theoretical Biology* 161, 3 (1993): pp. 343-357.

23.   C. Langton, 'Computation at the Edge of Chaos: Phase Transitions and Emergent Computation', *Physica D:Non-Linear Phenomena* 42, 1-3 (1990): pp. 12-37.

24. J. E. Lovelock, *The Ages of Gaia* (Oxford: Oxford University Press, 2000).

25. Lovelock, *The Ages of Gaia*.

26. F. J. Vine and D.H. Mathews, 'Magnetic anomalies over oceanic ridges', *Nature*, 199, 947 949 (1963).

27. J. Lovelock and M. Whitfield, 'Life Span of the Biosphere', *Nature*: 296, (1982): pp. 561–563.

28. C. Changqun et al., *Earth and Planetary Science Letters* 281, 3-4 (2009): pp. 188-201.

29. Hesiod, 'Theogony', lines 936-9 and 976-7, in *The Homeric Hymns and Homerica, including 'Works and Days' and 'Theogonis'*, trans. Hugh G. Evelyn-White, (Cambridge Mass.: Harvard University Press, 1917).

30. Hesiod, 'Theogony', lines 181-99, in *The Homeric Hymns and Homerica, including 'Works and Days' and 'Theogonis'*, trans. Hugh G. Evelyn-White, (Cambridge Mass.: Harvard University Press, 1917).

HARMONY, SCIENCE AND SPIRITUALITY

Rupert Sheldrake

From a talk at the Harmony, Food and Farming Conference, organised by the Sustainable Food Trust, Llandovery College, 10 July 2017.

SCIENCE AND SPIRITUALITY at first seem completely separate categories. Within the mechanistic world view that's predominated since the 17th century, the entire universe is the realm of science and is made up of inanimate matter moving in machine-like ways, whereas spirituality is seen as something entirely subjective going on inside human heads. That's the standard view that we've grown up with, a kind of Cartesian split, where the mind is just confined to us. We're the only things within nature that have minds or consciousness, the rest of nature is unconscious.

But the traditional view of the cosmos in all religious traditions is that the cosmos is not an unconscious machine, but is more like a living organism. This is something that science is bringing back. The Big Bang theory, for example, is rather like those ancient cosmogonies which describe the origin of the universe through the hatching of a cosmic egg, and it gives us the picture of a universe that's been hatched and then growing and developing and evolving for nearly 14 billion years. The real difference between religious and non-religious points of view is that, for the non-religious, or at least the non-spiritual, the universe is has no consciousness. Minds are just inside brains. It's a cerebro-centric view of consciousness. We've all grown up with that. It's a standard default position in our academic and intellectual world.

But all traditional views see that the universe is fundamentally conscious: consciousness underlies the whole universe and permeates it. One way of putting this view theologically is panentheism: God is everywhere and everything's in God, God is in nature and nature is in God. That's the view I myself think is the most reasonable and makes best sense of the world we live in. From that point of view, spiritual practices are not just about things going on inside our brains, they're about ways of connecting with this much wider universe.

For example, gratitude is a spiritual practice that all religions encourage and, indeed, many secular practices encourage it too, and there's now been a lot of research in the realm of positive psychology that shows that being grateful makes us happier. There are very strong correlations between people who are grateful and people who are happy. To such evidence sceptics respond, 'Well, of course these people are grateful: they are grateful because they're happy'. Yet the evidence suggests that when we do experiments in which we take people who are neither particularly happy, nor particularly grateful, and divide then into groups who either do gratefulness practices, or do not, the groups who practice gratitude are much happier than those that don't.[1] It's good for us. In fact, most spiritual practices that have been investigated scientifically turn out to be very good for us.

Now, meditation is another one. In the Buddhist and Hindu traditions and the Christian and the Jewish and the Sufi traditions, meditation is about connecting our minds with the underlying mind of God or the mind permeating the cosmos. Within modern secular world, meditation's about doing some that helps your brain work in such a way that you get neurotransmitters released that reduce stress and so forth. Anyway, the science shows meditation is good for you and, whatever we believe it's doing, it's actually doing us good.[2] But for those who see the cosmos as permeated by consciousness, it's a way of connecting with the greater consciousness.

What I want to do now is think about how spiritual practices relate to food and farming, the focus of this conference. There are established spiritual practices within the Anglican church that relate to that. One is the traditional service of Rogation Sunday, celebrated once a year in parish churches throughout the land. There's a blessing of the plough and there are prayers for the fields. This still goes on. Just a few months ago I was in Compton Dundon in Somerset, on Rogation Sunday, in fact. I went to the parish church, which has an ancient yew tree outside it. It's 1,700 years old, about a thousand years older than the church itself. We went out of the church, processed around it,, and in each of the four directions the priest led a prayer for the fields and the footpaths through the countryside and the woods, and the whole countryside we were in. This is still happening – a way of relating the spiritual life and the spiritual centre – but a lot more could be made of it. The valid and long-established tradition is still there.

Another tradition that we still have is harvest festival, Harvest Thanksgiving. Giving thanks for the harvest and the way our lives depend on the land. Now, in many parish churches, like our own parish church in London, in Hampstead, this has slightly degenerated into being a collection of cornflakes packets and tins of baked beans which are then given to food banks. And I'm sure food banks are grateful for the cornflakes and the baked beans, but it does somehow lose the connection with the land. People think 'well, we're in a city, we get food from supermarkets'. I myself think a lot more could be made of this festival, particularly bringing food in that people have grown in their garden or allotments, because many people have gardens and allotments in cities, and relating it much more to our lives.

But the point I really wanted to focus on is prayer. Prayer is also good for you. As a spiritual practice it's been investigated, and there is evidence to show that people who pray are happier, have less depression and so on, than those who don't. One study in the US took over a thousand people aged 65 and found a matching group of a thousand people with similar economic, social and professional l status. One group prayed regularly, the other did not. Od the people who prayed regularly, six years later 60% had survived better than those that hadn't, had had fewer medical complications and they were less depressed.

There are many studies of this kind that show that prayer is good for the person doing the praying.[3] Now of course, people who pray think it's not just about them, but about the wider world and their relationship to it. I think that what I wanted to bring up now is the Lord's Prayer, which is the most familiar of all the Christian prayers. 'Our Father who art in Heaven'. Now, many people may think that's about heaven as some kind of mythic state in which God exists, but actually when, when Jesus said it, I'm quite sure he meant the sky. The heavens are the sky, and as Nicholas Campion reminded us, the sky is the arena within which all our life takes place: we're in the sky. if God is everywhere, then the vast majority of God is in the sky because there's 99.999% more sky than there is Earth.

So, when people say, 'God's not out there', contemptuously, I think that's a false dismissal. You see, I think God is out there and up there and I think when people pray and look up to the sky, they're looking up to the place where most of God is. Of course, God's probably equally present everywhere but, nevertheless, the fact is the sky is where God is on all traditional views. In the 20th century, there was a vast expansion of our view of the universe through the Hubble space telescope, and through the discovery of galaxies beyond our own. The sky is much vaster and much older than anyone imagined in the past, so the arena of God's abode and activity within the universe is vastly increased.

'Our Father who art in Heaven, hallowed be Thy name. They kingdom come, Thy will be done, on Earth as it is in Heaven'. Or, in the form of the prayer that we use in choral evensong in the Anglican church, 'In Earth as it is in Heaven'. Now, what does that mean? How does the will of God get done in Earth, or on Earth as it is in Heaven? How is the will of God expressed in Heaven? Well, views have changed with changes in science. At the time of Jesus, people thought that the heavens moved in a series of spheres and that they were pretty well fixed in their movements, that the heavens were not a realm of change. We now have an evolutionary view of the heavens, and as Nicholas Campion has reminded us, Johannes Kepler showed that the planets didn't just move in spheres, they moved in ellipses.

In his book about the heavens, *Harmonices Mundi* (The Harmony of the World), which was one of the foundations of modern science, Kepler points out that the movements of the heavens are polyphonic.[4] The harmony of the spheres is not just a static chord. Because the planets move in ellipses, they change their relative speeds as they move, and we live in a kind of polyphonic heavenly world. The world is also, as we know from modern cosmology, evolutionary. So the will of God on Earth as in Heaven, at least on a modern understanding, is evolutionary.

However, it's also very clear, that this is not about making products and then dumping them. The heavens all work on a recycling economy. When stars break down, their stardust turns into planets and part of that stardust turns into us. There's a recycling throughout the whole of the universe, and it does give us a model of how things can work on Earth, which is different from the older understanding.

Then there is 'Give us this day our daily bread'. Presumably, if this prayer is going to go on working, then agriculture has to be sustainable. I think that it's important to rethink this, which is the most familiar of all prayers throughout the Christian and post-Christian world that we live in, because this is a prayer prayed by millions of people every day, and many people pray it very sincerely. But I think reinterpreting the way we pray and what we're praying for can have quite a lot of relevance to sustainable food and agriculture.

I'd like to end just by mentioning one final spiritual practice in which I'm very interested and in which two of Britain's leading proponents are here today, namely pilgrimage. Because, as Will Parsons likes to point out, one word for pilgrim was peregrine, as someone who goes through the fields. The pilgrim is someone moving through the fields, and on pilgrimage, on foot, we travel through the landscape, through the countryside, and relate to the land, connecting holy places with the fields and the land that we move through. Pilgrimage is probably the best way in which we can actually embody a relationship to the land and appreciation of it, and a connection to our own spiritual journey in accordance with the traditions in which we're all rooted. The traditions of our land and the holy places of our land include sacred wells and springs and river sources and cathedrals and churches and ancient trees. Luckily the British Pilgrimage Trust is doing a lot to reawaken this spirit of pilgrimage which our ancestors took part in and indeed took for granted.

NOTES

1.   Rupert Sheldrake, *Science and Spiritual Practices: Transformative Experiences and their Effects on our Bodies, Brains and Health* (London: Coronet, 2017). Rupert Sheldrake, 'Science and Spiritual Practices;, Temenos Academy, 2018, https://www.sheldrake.org/books-by-rupert-sheldrake/science-and-spiritual-practices.

2.   See for example, Alvin Powell, 'When science meets mindfulness, *The Harvard Gazette*, 9 April 2018, https://news.harvard.edu/gazette/story/2018/04/harvard-researchers-study-how-mindfulness-may-change-the-brain-in-depressed-patients/ (Accessed 30 March 2019).

3.   Talita Prado Simão, Silvia Caldeira and Emilia Campos de Carvalho, 'The Effect of Prayer on Patients' Health: Systematic Literature Review', *Religions*, 2016, Vol. 7 no 11, pp 1-11,

4.   Aiton, E.J., A.M. Duncan and J.V. Field, 'Introduction' in Kepler, Johannes, *The Harmony of the World*, trans. E.J. Aiton, A.M. Duncan, J.V. Field, (American Philosophical Society, Philadelphia, 1997), pp. 129-254.

HARMONY, SCIENCE AND SPIRITUALITY

Marc Andrus

From a talk at the Harmony, Food and Farming Conference, organised by the Sustainable Food Trust, Llandovery College, 10 July 2017.

IT IS A SERIOUS UNDERSTATEMENT TO SAY THAT INTERESTING THINGS HAPPENED in the USA in the early Sixties. One of the things that happened was a recurrence of people saying how important it is that the world is integral and interconnected. This was remarkable at the time. Martin Luther King actually said in 1963 that the main thing he wanted to say is that all of life is interconnected. He took his cue from St. Paul's statement that, 'And whether one member suffer, all the members suffer with it; or one member be honoured, all the members rejoice with it.[1] Inspired by this passage, he wrote in his famous letter from Birmingham Jail' that

> I am cognizant of the interrelatedness of all communities and states. I cannot sit idly by in Atlanta and not be concerned about what happens in Birmingham. Injustice anywhere is a threat to justice everywhere. We are caught in an inescapable network of mutuality, tied in a single garment of destiny. Whatever affects one directly affects all indirectly. Never again can we afford to live with the narrow, provincial 'outside agitator' idea. Anyone who lives inside the United States can never be considered an outsider.[2]

The idea of interrelatedness is becoming more commonplace, but it comes as a surprise in a world where some have lost that sense of integrity and interconnection. James Lovelock explored interrelatedness in his tremendously important book, *Gaia*, which was published in 1979, although originating in an idea that he had in 1964, a year after King wrote his 'Letter from a Birmingham Jail'. In one passage Lovelock writes,

> As society became more urbanized, the proportion of information flow from the biosphere to the pool of knowledge which constitutes the wisdom of the city, decreased. Soon city wisdom became almost entirely centered on the problems of human relationships, in contrast to the wisdom of any natural

tribal group, where relationships with the rest of the animate and inanimate world are still given due place.[3]

This is a really interesting idea in the conversation about science and spirituality. There is an information flow, Lovelock is saying, from the world – the whole world – into the human community that has made itself into cities. This starts in the Bible narrative with the first murderer, Cain, who killed his brother, Abel.[4] Cain is the first person to build a city. The first thing he does after God confronts him with the murder of his brother is to go and build a city, and from there the Bible poses a question about the origin and nature of urban existence. One conclusion is that in an urban environment we become more centred on ourselves rather than each other, and the information flow coming in from the world does not sufficiently inform us.

But, we could change this in very practical, easy ways: for example by connecting to local farming communities and farmers markets, buying food which is lovingly produced and more than just a commodity. We could even have a conversation with the people producing the food. When I moved to Virginia, I started work in a dairy that was owned by the university. I became a farmer for that time. We did not own our own farm, and we lived in the town, but the farm workers all gathered each morning at the dairy farm, and we worked together during the day and then went home in the evening. It was remarkable to me that they had conversations about the weather patterns, talking about them with careful observations, like, 'Oh, I noticed, Frank, the storm started over your place last night, and then it moved'. This kind of rich information is available to urbanites like me if we would invite these conversations, and also simply pay attention. So, that's a simple thing, and there are many other ways we could do this. In recent times we have become accustomed to images of the Earth from space, but it is a mistake to equate the world with the Earth. The world is not the Earth. The Earth is within the world, if we think of the world as the 'Cosmos'. The Earth is not only within the sky, but the sky contains the Earth. In ancient models of the world the stars were living, animate beings who were deeply interconnected with each other.

The world looks very different to us now. We know that, physically, we are located far out on one of the arms of the Milky Way. But, as we learn more and more about the cosmos, about the world, we take on the ideas of scientists that the universe acts kind of like a machine. We have to ask what happens to machines? They break down: they don't maintain themselves. So, with a machine model of the universe, we have the idea that the whole universe is breaking down, moving towards entropy, a kind of dead state. Also, although the universe is vast and beautiful it's not alive, it's not conscious, it's not integral. And we've removed God. It's just too hard to maintain the idea of a God that had been developed in the old model of the universe, with the Earth in the centre and God in Heaven, up there.

Yet, current science is proposing radical new ideas, such as quantum entanglement. If we take on a new, scientific vision of the Cosmos, combined with that which is true in the ancient idea of the world, and start to think about dark matter, entanglement, consciousness and love, we discover a world that is both reanimated and far vaster than the ancient world. It is a place where God can be seen. In such a world we may find God anew. Marcus Borg made this point in his beautiful book, *Meeting Jesus Again for the First Time*.[5] Marcus was raised as a Lutheran, and he recounts how when he got into university, he ceased believing. Even though he was a doctoral student studying the New Testament, he ceased believing in Christ or in God or in the Trinity, because fundamentally, the idea of Jesus that he had been taught as a child was no longer adequate to the problems he faced in his life – it hadn't been updated to accommodate an adult who struggles in his life in the new expanded universe.

In the new universe of such concepts as Gaia and entanglement, the world is re-animated. We see the whole world as living, so the God of that world is also something that you and I might stand in front of and ask how can we encounter and receive information from God, not just speak to God. This is what I understand that Lovelock is talking about: the information coming into the city, coming from farms, from the planets, from the vast reaches of space, from the universe, and from God. The question I leave you with is, how can we have spiritual practices that will allow us to receive the information that this loving world, the whole loving world, animated by God, is always willing to give us?

NOTES

1.   Corinthians 12.26.

2.   Martin Luther King, 'Letter from Birmingham Jail' by Martin Luther King Jr., August 1963, https://web.cn.edu/kwheeler/documents/Letter_Birmingham_Jail.pdf [accessed 30 March 2019].

3.   James Lovelock, *Gaia: A New Look at Life on Earth* (Oxford: Oxford University Press 1979), p. 126.

4.   Genesis 4.1-18.

5.   Marcus Borg, *Meeting Jesus Again for the First Time* (London: Harper Collins, 1994).

THE MUSIC OF THE SPHERES:
MARSILIO FICINO AND RENAISSANCE *HARMONIA*

Angela Voss

MOST PEOPLE ARE FAMILIAR WITH THE EXQUISITE PAINTING by Sandro Botticelli
(1445-1510) known as the Primavera. But perhaps it is not so widely known
that the programme of its enigmatic symbolism was inspired by the Neoplatonic
notion of the harmony of creation, reflected in the correspondences of the
mythological characters to both the eight planetary spheres and the eight tones
of the musical octave.[1] It is probably even less appreciated that Botticelli's visual
metaphor for the harmony of the spheres was inspired by the work of one man,
Marsilio Ficino of Florence (1433-99), whose desire to unite heaven and earth in
the soul of the human being found its precedent in the writings of the Platonic
tradition. In restoring 'the divine Plato' to Renaissance Florence, Ficino set out
to 'redeem holy religion' from the 'abominable ignorance' of secular philosophy.[2]

My intention in this paper is to illustrate how music theory and performance
became part of a programme of spiritual development stemming directly from
a symbolic understanding of the cosmos which transcended, and yet embraced,
all quantitative modes of thinking. Such a mode of 'knowing' was conveyed by
Ficino in the Latin word *notio* (from which our word 'notion' is derived) in the
course of his translation of the Neoplatonist Iamblichus' treatise on divination,
De mysteriis, Iamblichus asserts,

> Contact with divinity is not knowledge. For knowledge is in a certain respect
> separated from its object by otherness. But prior to knowledge – as one things
> knows another – is the uniform connection with divinity, which is suspended
> from the gods, and is spontaneous and inseparable from them.[3]

In Iamblichus' explanation of unitive thought, Ficino recognised the ground of
both philosophical speculation and religious piety without which 'knowledge'
becomes dissociated from the primary reality of the world and thus, can be of
little meaning. I believe Ficino's articulation of this insight to be the creative
impulse behind the immense flowering of intellect in the Italian High Renaissance,
inspiring art forms which arose from an intensely erotic relationship between the
individual soul and the beauty of creation. As Ficino himself exclaimed,

Bust of Marsilio Ficino by Sandrea di Piero Ferrucci (1522),
Santa Maria del Fiore, Florence.

This age, like a golden age, has brought back to light those liberal disciplines that were practically extinguished; grammar, poetry, oratory, painting, sculpture, architecture, music, and the ancient singing of songs to the Orphic lyre. And all this in Florence.[4]

The founder of the Platonic Academy, Ficino both translated into Latin for the first time the complete works of Plato, Plotinus and others, and combined his vocation as a Christian priest with active work as an astrologer, herbalist, magician and musician. Philosophically, his life-long project was to bring together what he called 'faith' and 'reason' by marrying Christianity and Platonism. He also translated Zoroaster, Hermes Trismegistus, Orpheus, the *Chaldaean Oracles*, Iamblichus, Synesius and Proclus and read Al-Kindi, the *Picatrix*, the Arabic astrologers, Albertus Magnus and Roger Bacon. Out of these diverse sources emerged his own system of natural magic, centred on the combination of astrology and music, whose efficacy depended on more than theoretical knowledge and technical expertise. This meant, in practical terms, that astrology, talismanic magic, herbal medicine and music-making found their place as dynamic expressions of both deep philosophical inquiry and intuitive inspiration, combined with extensive theoretical knowledge.

In looking more closely at Ficinian music therapy, I want to emphasise two further vital ingredients, which we find continually emphasised in his practical writings. These are the *desire* and *imagination* of the human being, which, when focused on images such as music, stars, or talismans, somehow facilitate an interplay with the cosmos and allow the qualities of a particular moment in time to be seized and recognised. This very process may effect a change in being. When we fashion images, Ficino says, 'Our spirit, if it has been intent upon the work and upon the stars through imagination and emotion, is joined together with the very spirit of the world and with the rays of the stars through which the world-spirit acts'.[5]

It is precisely this subjective element which distinguishes the Renaissance magus from the medieval theorist, for static hierarchical schemes and correspondences between planets and music are transformed into dynamic energies at work throughout creation – energies which can be harnessed and transfused for the harmonising of individual souls. Following Plotinus, Ficino emphasises the necessity of focusing the emotion in an act which depends on both intuition and expertise in order to expand consciousness, 'Whoever prays to a star in an opportune and skilled way projects his spirit into the manifest and occult rays of the star, everywhere diffused and life-giving; from these he may claim for himself vital stellar gifts'.[6]

In the Platonic/Pythagorean tradition, music and the stars are inextricably linked as audible and visible images of an invisible dimension of existence, whose intellectual perception is made possible through the senses of hearing and sight. The foundations of the musical cosmos are established by Plato in the creation myth of his *Timaeus*, which itself maintains a vital connection to Egyptian, Chaldaean and other ancient traditions. In this dialogue, Plato sets up a model for a three-fold musical cosmos where the movements of the spheres, the passions of the human soul and the audible sounds of music are all expressions of a divine intelligence manifesting through the various dimensions of creation.[7] Such a tripartite division was to be differentiated by the fifth-century CE theorist, Boethius, as *musica mundana, musica humana* and *musica instrumentalis*,[8] and it was commonplace for music theorists to work out elaborate systems of correspondences between astronomical distances and musical intervals, between the nature of musical patterns and emotional states, between planetary characteristics and audible sound. The key, in this tradition, to the ordering of the cosmos, whether astronomically or musically, is, of course, number – a discovery which was transmitted to Western thinkers by Pythagoras. Indeed for the Platonists, number determines all things in nature and their concrete manifestation, together with all rhythms and cycles of life. Number revealed by the heavenly bodies unfolds as Time and, as the human soul was seen to be mirrored in the order of the heavens, divination, or aligning oneself to the gods, required the appropriate ritual at a precise time. Iamblichus tells us that the numbers governing nature *are* the outflowing energies of the gods and, if we wish to assimilate ourselves to them, we must use their language – that is, align ourselves with the harmonies underlying the cosmos.[9] According to this view, merely humanly contrived numerical systems, discursive conceptions of number, or numerological theories, cannot reproduce an *experience* of unity which will give rise to true knowledge of first principles.

In the *Timaeus*, we learn that the Demiurge created a substance called the world-soul and inserted it into the centre of the world-body.[10] He then divided up this soul-stuff according to the ratios of the three consonant musical intervals: that is the octave, which resonates in the proportion of 2:1; the perfect fifth, 3:2; and, the perfect fourth, 4:3, continuing, by further division, to create the intervallic steps of the Pythagorean scale. The soul was cut into two parts which were bent around each other, forming the circles of the Same and the Different: the Same containing the unmoving sphere of the fixed stars and the Different containing the moving instruments of Time, or the planets. The Different was then divided into narrower strips which were arranged according to the geometrical progressions of two and three; 1 2 4 8 and 1 3 9 27. Permeating the whole cosmos, the soul connected the physical world with the eternal, being 'interfused everywhere from the centre to the circumference of heaven' and partaking of 'reason and harmony'.[11]

The human soul, also partaking directly of the *anima mundi*, must therefore be regulated according to the same proportions. But due to the passions of the body, the soul on entering it became distorted and stirred up – only the correct kind of education could restore harmonious equilibrium.[12] This education would induce a recognition of the soul's congruence with the cosmos through the audible harmonic framework of the musical scale for, as we have seen, the proportions in the world-soul could be reproduced in musical sound. The numbers one to four (the *tetraktys*) thus, not only form the framework for all musical scales, but also embody this dynamic process of embodiment in the fourfold movement of geometry from point to line to plane to solid; from the unity comes the duality of opposition, the triad of perfect equilibrium and the quaternity of material existence. Each stage both limits and contains the one following and the initiate is warned, in the *Chaldaean Oracles*, 'do not deepen the plane'.[13] That is, extend towards the material world from the perfect condition of the triad, but do not lose your limiting power by letting go of it and becoming lost in the quaternity or chaos of matter. This can be understood musically as the imperative of maintaining the perfect intervals as defining structures. In listening to geometry in sound, the perfect intervals set a framework or limit on unlimited sound and, since the specific arrangement of sizes of tones and semitones within this framework mirror the exact astronomical relationships of the planets, the very fabric of creation is brought to the ear and, in Platonic terms, evokes a memory of the harmonies once heard with the ears of the mind.

From this essential premise, the schemes attributing planets to actual pitches and astronomical distances to musical intervals abounded. In the 'Myth of Er' in the *Republic*, Plato suggests that sirens positioned on the rims of the planetary orbits each sound a pitch, making up a musical scale, much like a Greek lyre projected into the heavens.[14] In another interpretation, found in Cicero's 'Dream of Scipio', the planets produce different tones according to their various speeds of revolution. We are told that 'the high and low tones blended together produce different harmonies' and that 'gifted men, imitating this harmony on stringed instruments and in singing, have gained for themselves a return to this region, as have those of exceptional abilities who have studied divine matters even in earthly life'.[15]

Exactly how to imitate the music of the spheres thus became the question raised by music theorists and the science of harmonics, or the study of mathematical properties of musical ratios, was considered to be the first step. It is very difficult to know how much this highly speculative procedure, considered by Plato to be the highest form of knowledge, influenced the practical music-making of classical times. We are certainly better informed about the connection between *musica humana* and *musica instrumentalis* for, central to ancient Greek musical writings is the concept of ethos, or subtle ethical effects produced in the human psyche by the use of different modes or 'set' combinations of tone-patterns. For example, the Phrygian mode moved men to anger, the Lydian soothed them, the Dorian induced gravity and temperance – each quality being reflected in the character of particular regions. By medieval times, the ancient Greek modes had been replaced by the eight Church modes, but this did not interrupt the association of subtle ethical effects by theorists. One twelfth-century writer notes that 'the modes have individual qualities of sound, differing from each other, so that they prompt spontaneous recognition by an attentive musician or even by a practised singer'.[16]

But what of the connection between ethics and cosmology? Ethical powers were attributed to systems of pitch, while planets were generally associated with single pitches. Thus in the writings of most classical theorists, it is difficult to see how an effective form of *musica instrumentalis* could influence the human soul through direct imitation of cosmic harmony, despite the model transmitted by Plato. Generally speaking, celestial phenomena were made to fit a preconceived notion of musical order, rather than the phenomena themselves being asked to reveal their order as principles of intelligence. Although the Middle Ages produced some great original thinkers in this field, such as John Scotus Eriugena in the ninth century and indeed the influential Islamic school of musical and astrological therapy,[17] it was only in the fifteenth century that the West began to explore the practical means by which the harmonic relationships in the cosmos could be expressed through music, not by literally reproducing astronomical measurement in sound, but by symbolically evoking a unifying principle at work in the manifest and unmanifest worlds. With the music theorists, Bartolomé Ramos de Pareja (c.1440-1522) and Georgio Anselmi (1723-97) we see the seeds being sown for a revisioning of cosmic music.

Anselmi of Parma, writing in 1434, explicitly rejects the literal Aristotelian notion, prevalent in the preceding centuries, that the heavenly bodies could make no sound in their movements and envisions the planets not tied to individual tones but each singing its own song in counterpoint with the others.[18] Like Ficino, Anselmi was an astrologer, magician and physician and, although his work on music does not give practical advice on human imitation of heavenly harmonies, his cosmos is liberated from fixed schemes as the planetary cycles participate in a great cosmic symphony orchestrated by the Blessed Spirits. He also derived an eight-octave planetary scale from the Moon to the Fixed Stars from the periods of the planets' rotation around the Earth, breaking the bounds of contemporary musical practice (all music lay within a three-octave limit).[19]

It is, however, only with his younger Spanish contemporary, Ramos de Pareja, that we find the beginnings of esoteric philosophy applied to music in the operative sense. In Ramos' *Musica practica* of 1482, this is achieved through the original move of attributing the Church modes to the planets and so connecting ethos to planet through practical music. Thus, the Dorian mode, from D to D, relates to the Sun and has the effect of 'dispelling sleep' or the Lydian mode, from F to F, relates to Jupiter and 'always denotes joy'; the eighth mode, or Hypermixolydian, from A to A, epitomises the starry heavens, 'an innate beauty and loveliness, free from all qualities and suitable for every use'.[20] Ramos' planet-mode pairings appear to be without precedent, although he may have been influenced by Arabic traditions in his native Spain. Most importantly, they reveal a fundamental shift in perspective, from 'rational' astronomical attributions of pitches, to planets based on distances or speeds, to the realm of the symbolic correspondence of active imagination. Such heavenly harmonies must now derive from intelligences informed by an ensouled cosmos and Ramos also matches the nine Muses to each mode and planet. In other words, the means of effecting a connection between heaven and earth is a magical one – in a universe of operative affinities and correspondences, modes can be seen as possessing occult properties which bring man into relationship with the stars through sympathetic resonance. *Musica practica* is also revolutionary in that Ramos revises the standard Pythagorean tuning system which was proving increasingly restrictive for practising musicians combining perfect fourths and fifths with consonant thirds and sixths which, in fact, lays the foundation for a system of equal temperament such as we know today. Again for the first time, he is concerned with the demands of practising musicians, not speculating theorists and, in suggesting a possible evocation of planetary meaning through the right use of mode, he opens the doors wide for music to be used in a magical context. Ramos illustrates his analogies (see figure 1) with a complex image of spirals which recall the planets in their spheres and which is curiously reminiscent of Ficino's description of sound as series of spirals in his Commentary on the *Timaeus*. This oroboros image is of course a traditional alchemical one and suggests a hidden dimension of spiritual unity beyond the apparent intellectual game of analogy.

Inevitably, the next step toward a fully-operative musical magic was there

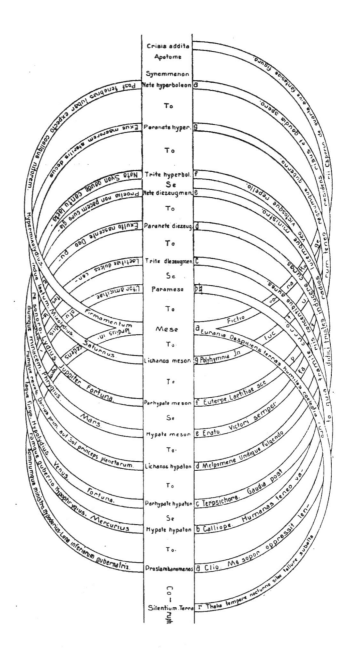

Figure 1. Ramos de Pareja's astronomical-musical analogies from his *Musica practica*,
1482, ed. J. Wolf in *Publikationen der Internationalen Musikgesellschaft*, Beihefte II
(Leipzig: Martin Brelauer, 1901).

to be taken and, seven years after Ramos' treatise, Ficino published his own manifesto for astrological music as a therapy in his *Liber de vita*, the *Book of Life*. Although conceived as a commentary on Plotinus, chiefly to give it philosophical respectability, this work clearly demonstrates Ficino's fascination with the spiritual dimensions of ritual magic and with the potential of astrology as a liberating approach to self-understanding. Having been born with Saturn on the ascendant in Aquarius, Ficino knew only too well the limitations of a melancholy temperament and, the beneficial effects he himself experienced when playing his lyre and singing, led him to formulate a therapeutic system in which music was played in accordance with an individual's horoscope. Ficino gave an account of his horoscope in his letter to Martin Preninger. He was born on 19th October 1433 at 21 hours (that is, 21 hours after sunset the previous day),

> At that time almost half of Aquarius was rising, together with Pisces, I think. Saturn in Aquarius was conjunct the Eastern angle, Mars in the same [sign], in the prison of the twelfth [house], the Sun in Scorpio, together with Mercury in the ninth house, the Moon in Capricorn: Jupiter in Leo in the seventh. Venus in the same [house] in Virgo, Fortune in Aries'.[21]

Mars was in fact in Capricorn and Ficino's error may have been due to inaccurate ephemerides. Ficino identified himself with Orpheus who, not only tamed wild beasts with his song, but also brought back Eurydice from the underworld. At least one of his friends, the poet Poliziano (1454-1494), associated this restoration of the feminine with the 'bringing to light' of esoteric Platonic wisdom.[22]

The third part of the *Book of Life* is entitled, 'How to arrange your life in accordance with the heavens' and concludes Ficino's project to enrich and vitalise the dissociated life of a scholar. In chapter twenty-one entitled, 'The power of words and song for capturing celestial benefits', Ficino suggests that the power of emotionally-charged spoken (or sung) words may intensify the effect of an image, as the Arabs and Egyptians believed, 'they hold that certain words pronounced with a quite strong emotion have great force to aim the effect of images precisely where the emotions and words are directed.'[23] For this process, it is probable that Ficino used as texts the *Hymns of Orpheus* which he himself had translated from the Greek, consisting of epithets to various deities. Indeed, his fellow hermeticist, Pico della Mirandola (1463-1494), affirms that, 'In natural magic, nothing is more efficacious than the Hymns of Orpheus, especially if the correct music, intention of the mind, and all the other circumstances known to the wise are applied'.[24] Combined with these hymns, Ficino composed or improvised a kind of musical accompaniment which appears to draw on the association of modes with stars suggested by Ramos. He writes that,

> tones first chosen by the rule of the stars and then combined according to the congruity of these stars with each other make a sort of common form, and in it a celestial power arises. It is indeed very difficult to judge exactly what combinations of tones especially accord with what sorts of constellations and aspects. But we can attain this, partly through our own efforts, partly by some divine chance.[25]

That is, the confluence of the human and divine, or we might say conscious and unconscious, dimensions may give rise to a creativity which is in essence divinatory as it surrenders to a transpersonal law. This is the 'divine madness' of the poet, lover, priest and prophet and we have one eye-witness account of Ficino himself in the throes of poetic frenzy, 'his eyes burn, he leaps to his feet, and he discovers music which he never learnt by rote'.[26]

Ficino goes further than Ramos in implying the use of particular modes for particular types of people and gives us three 'rules for composition' which require a detailed knowledge of Boethius' three musics, expressed here in terms of astrology, psychology and modal ethos,

> The first rule is to inquire diligently what powers in itself or what effects from itself a given star, constellation or aspect has – what do they remove, what do they bring? – and to insert these into the meaning of our words, so as to detest what they remove and approve what they bring. The second rule is to take note of what special star rules what place or person and then to observe what sorts of tones and songs these regions and persons generally use, so that you may supply similar ones, together with the meaning I have just mentioned, to the words which you are trying to expose to the same stars. Thirdly, observe the daily positions and aspects of the stars and discover to what principal speeches, songs, motions, dances, moral behaviour, and actions most people are usually incited by these, so that you may imitate such things as far as possible in your song, which aims to please the particular part of heaven that resembles them and to catch a similar influx.[27]

The 'power, timeliness and intention'[28] of such a song, says Ficino, will provoke both singer and audience to imitate the qualities it itself is imitating through its action on the airy spirit, which connects man's spirit to the soul of the world. The music-spirit is conceived by Ficino to be like a living animal, composed of warm air, 'still breathing and somehow living'.[29] It carries both emotion and meaning and its influence will depend in part on its congruence with the heavens and in part with the 'disposition of the imagination'[30] of the singer – that is, on a synchronicity between external and internal dimensions of experience. The singer must be a finely-tuned instrument whose spirit has been purified and strengthened through assimilating the properties of the Sun, for such 'vital and animal power'[31] will readily attract the music spirit – particularly if the ritual is also conducted at a suitable astrologically-elected hour. The three essential requirements for the invocation of such a numinous energy are, therefore, the vital solar power of the singer's own spirit, the propitious moment, and the singer's *intention*, which unites the desire of his heart and the focusing of his imagination. Then, Ficino suggests, both mental and physical diseases may be dispelled through sympathetic resonance between music spirit and human spirit, which encompasses, and acts on, both body and soul.

The music spirit, moving through the various planetary spheres, will activate the particular spirit of each, enabling the performer to recreate the music associated with each heavenly body; for example, the songs of Venus are 'voluptuous with wantonness and softness'.[32] Ficino explains that when the petitioned planet is 'dignified' in the heavens, the performer's spirit will naturally attract the response of the planetary spirit, 'like a string in a lute trembling to the vibration of another which has been similarly tuned'.[33]

In a natural magic, based on the neo-Platonic vision of cosmos as harmony, it is not only audible music which may align the soul with the stars. From talismans, medicines, odours through the movements of dancing to immaterial qualities of mind, all may be appropriately used to restore a psychic connection to a particular planet. Thus 'well-accorded concepts and motions of the imagination' lead us to Mars, while 'tranquil contemplations of the mind' are the domain of Saturn.[34] Indeed, Ficino understood true alchemy to be the transformation of the worldly frustrations and hardships traditionally associated with Saturn into the philosopher's gold of intellectual contemplation. Opposing Saturn in Ficino's own chart we find Jupiter and, of course, this aspect symbolises the very polarisation of philosophy and religion that Ficino sought to overcome. His own view of the difficulty represented by the opposition, and the hope that Jupiter might alleviate Saturn, was expressed in his letter to the Archbishop of Amalfi,

You have divined, I think, how much I have long wanted to live my life with someone of a Jovial nature, so that something of a bitter, and as I might say,

Saturnine element, which either my natal star has bestowed on me or which philosophy has added, might eventually be alleviated by the sweet fellowship of someone born under Jove.[35]

Nowhere in Ficino's writings will one find a lack of poetry, metaphor, myth, or imagination, even when he is addressing the most technical subjects. Theory is never divorced from practice, objective 'truth' never distilled from the hermeneutics of spiritual experience – indeed, his criticism falls heavily on those astrologers whose practice merely consists of knowing all the rules and applying them in 'cause and effect' mode. Ficino's very language is a language of sign and symbol, continually pointing towards the inexpressible, yet rich in colour and pithy in content. So it is not surprising that, when considering the subject of *harmonia* or music theory, the concepts of consonance and dissonance are clothed in metaphoric garb. In his letter, the *Principles of Music*, Ficino describes the qualities of intervals in a musical scale as an analogy of the Hermetic procession of the soul from its origins to its final return to God and it is worth quoting this passage in full as an example of Ficino's instinctive ability to bridge the sensible and the metaphysical,

the lowest note, because of the very slowness of the motion in which it is engaged, seems to stand still. The second note, however, quite falls away from the first and is thus dissonant, deep within. But the third, regaining a measure of life, seems to rise and recover consonance. The fourth note falls away from the third, and for that reason is somewhat dissonant; yet it is not so dissonant as the second, for it is tempered by the charming approach of the subsequent fifth, and simultaneously softened by the gentleness of the preceding third. Then, after the fall of the fourth, the fifth now arises; it rises…in greater perfection than the third, for it is the culmination of the rising movement; while the notes that follow the fifth are held by the followers of Pythagoras not so much to rise as to return to the earlier ones. Thus the sixth, being composed of the double third, seems to return to it, and accords very well with its yielding gentleness. Next the seventh note unhappily returns, or rather slips back to the second and follows its dissonance. Finally the eighth is happily restored to the first, and by this restoration, it completes the octave, together with the repetition of the first, and it also completes the chorus of the Nine Muses, pleasingly ordered in four stages, as it were the still state, the fall, the arising and the return.[36]

In his quest for a unifying perspective, Ficino considers the idea of harmony on the three levels of manifestation: the intricacies of specific intervallic relationships in audible music; the relationship of the human senses to specific proportions of fire, earth, air and water; and finally, what he calls the 'astronomical causes of harmony'. Drawing on 'Book Three' of Ptolemy's *Harmonics*,[37] Ficino here seeks to show the congruence between the human experience of musical intervals and the tensions inherent in the angular relationships of the zodiacal signs. But, whereas Ptolemy, as a mathematician, maintains a rigorously objective perspective, aiming to prove that astrology is a true science dependent on number, Ficino is more concerned with the practical experience of musicians and astrologers. Ptolemy focuses on the primacy of the octave, fifth and fourth, but Ficino considers each interval and aspect to be equally as important in contributing to the overall Good, the idea of dissonance merely stemming from the imperfection of the earthly condition. This is justified by music itself – for if the fourth is perfect, how can the square aspect be regarded as discordant? In his commentary on Plotinus, Ficino asserts:

> indeed not only is Cancer not dissonant from Aries, but it is consonant, for those parts are both of one greatly uniform body and of the same nature (which are discordant amongst us by harmonic tempering), and no less in heaven than in musical song are all things consonant among themselves… therefore the union of the planets represents for us the consonance of the octave.[38]

In the *Principles of Music,* Ficino illustrates step-by-step how the intervals and aspects correspond in nature (see figure 2); the second sign 'falling away' from the first as in the dissonant second, the fifth 'looking benignly' on the first in a trine as a model for the perfect fifth.[39] The seventh sign, in opposing the first, is 'very vigorous in its discord' according to Ficino and 'seems in its clear hostility to prefigure the seventh tone in music, which with its vigorous and vehement quality is now most clearly dissonant from the first'.[40] Ficino points out that the eighth sign, traditionally assigned to death, is therefore considered unfavourable. But from a theological perspective, it is quite the opposite, as death frees the soul from the 'dissonance of the elemental world,[41] restoring it to the heavenly harmony'.[42] Thus, its nature is truly represented in the perfect consonance of the eighth tone, or the octave.

Ficino completes his journey around the zodiac by relating the ninth to twelfth zodiacal signs back to the first sign. In this way, the ninth sign relates

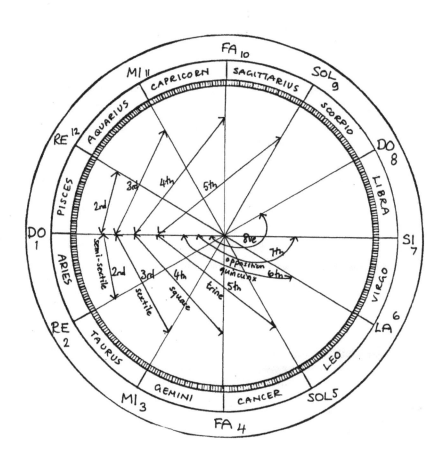

Figure 2. Ficino's musical scale in relation to the zodiac.

to the perfect fifth, the 'tenth sign of human ambition' to the 'human, middling dissonance' of the perfect fourth, the eleventh sign of human friendship to the interval of the third, and the twelfth sign, 'allotted to hidden enemies and prison', recapitulates the extreme dissonance of the interval of a second.[43] It is interesting that whereas Ptolemy, in his concern for perfect mathematical symmetry, compares the opposition aspect to an octave, for Ficino, the dissonance of the seventh, which demands a resolution into the consonant octave, most clearly corresponds to the experienced tension of the opposition aspect in astrological practice. What is more, the resolution of the seventh into the perfect concord of the octave musically embodies a metaphysical potential which, for Ficino, would provide its ultimate justification. From a psychological perspective, it could be seen to represent the resolution of the tension of opposites within the individual into a unified Self; from a spiritual one, it embodies, in sound, the final release of the dissonance and tension of earthly existence into the perfection of heaven through death.

When Ficino played on his Orphic Lyre, which was probably a harp-like instrument, the qualities of the musical intervals he plucked from the strings would thus penetrate the very deepest levels of human experience. Whether he advocates a conscious use of these analogies when he instructs us to find tones or modes which correspond to the pattern of the heavens is difficult to say and, perhaps, not the most important point. Ficino succeeded in bringing the music of the spheres to earth by recognising the uniting and transforming power of symbolic perception – a power whose apprehension depends on a suspension of rational thought and the willingness to be guided by the imagination.

Strongly influenced by Ficino, at the end of the fifteenth century, we find the Milanese music theorist, Franchino Gafori (1451-1522), continuing the theme of harmonic correspondence and elaborating on Ramos' analogies to embrace all aspects of modal ethos related to Muses, planets and signs of the zodiac. If we look at the woodcut he included in two of his works to illustrate this cosmic harmony (see figure 3), we find what Ficino might call 'an image of the world', in which each planet and Muse occupies a sphere or medallion and the alchemical serpent, or three-headed Cerberus, connects the feet of Apollo to the silent, unmoving earth.[44] The serpent can be seen as a bow, drawn across the eight strings of the celestial lyre or, alternatively, as a monochord punctuated by the eight intervals of the octave. Out of the bosom of the earth, where the muse Thalia lies silent, the song germinates. To the lowest string is given the Hypodorian mode; Gafori says 'Persephone and Clio breathe and therefore the Hypodorian is born; here arises the origin of song'.[45] Gafori continues his explanation of the diagram up through the planetary spheres, also associating each planet with its zodiacal sign;

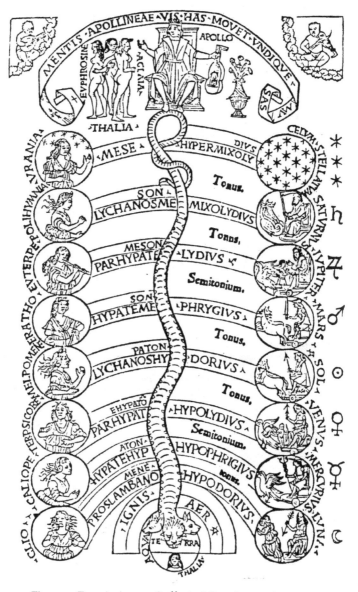

Figure 3. Frontispiece to Gaffurius' *Practica musica*, 1496.

for example, 'the sixth string, parhypate meson, has Jove, home of Pisces and Sagittarius, also Euterpe and the Lydian mode...the Lydian of Euterpe contains also the music of Jove; sounding sweetly, the sixth string rules because a goddess is present.' Above the starry heavens, the whole harmony is governed by Apollo, who directs the dance of the three Graces to his right. Cupids fly overhead, playing a lute and *lira*, while the entire diagram is crowned by an inscription from a poem on the Muses by Ausonius, 'The spirit of Apollo moves these Muses everywhere'. In this animated heaven, the divine spirit manifests through the female principles of poetic inspiration and Edgar Wind suggests that the vase of flowers to Apollo's left probably signifies the celestial *crater* through which the spirit descends to the natural world. [46] Wind also points out that the serpent of Apollo, the Cerberus of Serapis, curls the end of its tail into a loop in an image of eternity or perfection (as in the variant of the serpent biting its own tail), thus evoking the Platonic notion of Time issuing from Eternity as the spheres emanate down from their unchanging source. We also notice that there are six flowers in the vase and three Graces, adding up to the number of Muses, nine. Thus, the number six can be seen as representing the sensory, natural world and the number three, heavenly purity, which may explain why Gafori mysteriously attributes the sixth mode to 'the goddess' or 'feminine' principle of nature. The three 'pure' emanations – the Dorian mode in the centre, issuing forth from Melpomene, Urania, above the planets and Thalia, below the earth – contain and divide the other six 'sensuous' ones into two triads. Wind suggests, 'In the order of the Muses, the triad Urania-Melpomene-Thalia would then emanate from the dance of the Graces, while the six intermediate Muses are "planted" in Apollo's vase'.[47] He adds that the names of the Graces – in contemporary translations – also relate to the celestial sphere (Euphrosyne/*laetitia*), the Sun (Aglaia/*splendor*) and the subterranean seeds of music (Thalia/*viriditas*).

In this way, we can begin to see how this complex illustration is far more than simply a diagrammatic representation of cosmic analogies. Its symbolic content, as an expression of a Neoplatonic vision of an ensouled cosmos, is designed to waken and set in motion the spiritual energy, emanating as musical harmony, from the silent depths of nature. Like the Botticelli's *Primavera*, albeit in a less sophisticated way, it somehow stirs a perception of unity from deep within, in the manner of a talisman. A final quotation from Gafori emphasises the spiritual implications of his *musica mundana*, 'Finally, we did not believe it should be passed over in silence that musical systems contribute much to the perfection of virtue, which some call divination because it is the greatest ornament and salvation for anyone'.[48]

From its humble beginnings in the solitary lyre-playing of Marsilio Ficino, the music of the spheres continued to be heard throughout the following century. For Cornelius Agrippa (1486-1535), whose *Three Books of Occult Philosophy* (1531) provide a compendium of magical theory and techniques, Ficino's sympathetic magic and Gafori's conception of cosmic harmony combine in a system of ritual presided over by the Magus, who shapes and exploits the properties of music in his quest for spiritual truth. In the field of practical music, the great *intermedii* of the Medici court in Florence attempted to recreate the Platonic world of Ideal forms on earth through arousing the wonder of the audience – the enormous scale of the visual spectacle and musical forces were created with the purpose of imitating perfect Beauty. On a smaller scale, the Hermetic revival in Elizabethan England produced such exquisite music as that by John Dowland, whose seven *Lachrimae* Pavans for viols and lute evoke the Neoplatonic descent and ascent of the soul and in whose songs of deepest melancholy lies hidden the gold of the philosophic Saturn.

With the Copernican revolution and the development of physics and astronomy as independent from philosophy, cosmic music became increasingly the domain of the esoteric scientist rather than the practical musician. *Musica mundana* found itself incorporated into ever more complex systems dependent on the unity of all universal elements, by such polymaths as Robert Fludd (1574-1637), Johannes Kepler (1571-1630) and Athanasius Kircher (1602-1680). But despite the attempts of the Florentine Camerata at the beginning of the seventeenth century to revive ancient Greek ideals of musical ethos, *musica instrumentalis* gradually lost its philosophical justification and the Baroque characteristics of formal structure, stylistic nuance and ornamental gesture determined composers' intentions. In esoteric circles, divorced from the enlightened world of rationalism, the invention and elaboration of musical systems explicitly related to occult correspondences continued to be explored and enjoyed a revival in the Romantic period. In the twentieth century, we may think of Rudolf Steiner (1861-1925) or G. I. Gurdjieff (1866-1949) as spokesmen for music as a spiritual discipline, but we do not hear the music of the spheres anymore. The composer, Arnold Schoenberg (1874-1951), at the beginning of this century 'emancipated the dissonance', that is, freed the twelve-note chromatic scale from any notions of internal hierarchy and made all notes equal.[49] Similarly, we no longer have the opportunity to distinguish between musical temperaments or the tuning systems used for specific effects until the eighteenth century. Due to the innovations of J. S. Bach (1685-1750), the music we hear is equally tempered – all perfect fourths or fifths being smoothed away. Astrology too has suffered the fate of becoming divorced from philosophy, forced to attempt to align itself with prevailing paradigms of scientific reality; in a period of confusion over the value of spiritual experience, it is led to deny its roots in *notio*, a contact with the numinous which precedes all processes of differentiation by the mind. However, as Ficino suggests, there may be a very simple way for a human being to re-establish a connection with these roots and once more lend an inner ear to the harmony of the spheres:

> Whenever in your studies you make a serious attempt to postulate that there are many angelic minds beyond heaven, like lights, whose ordering relates them both to each other and to one God, the father of all lights, what will be the point in pursuing your investigations down long winding paths? Just look up at heaven, I pray, Oh citizen of the heavenly realm, at that heaven whose manifestly perfect order so clearly declares God to be its creator.[50]

NOTES

1.   For a full analysis, see E. Wind, *Pagan Mysteries in the Renaissance* (Oxford: Oxford University Press, 1980 [1958]), pp. 113-131.

2.   See Marsilio Ficino, 'De Christiana religione' in *Opera omnia* (Basle, 1576), p. 1, as quoted in J. Hankins, *Plato in the Italian Renaissance* (New York: Brill Academic Publishing, 1991), p. 289.

3.   Iamblichus, *On the Mysteries*, 1.3, trans. T. Taylor, in S. Ronan, ed., *Iamblichus of Chalcis, On the Mysteries = De mysteriis Aegyptiorum* (Hastings: Chthonios Books, 1989), p. 24.

4.   Ficino, *Opera omnia*, p. 944.

5.   Marsilio Ficino, *Liber de vita*, III.20.36, in Marsilio Ficino, K. Caske and J. Clark, *Three Books on Life* (Binghamton: MRTS 1989)., p. 351.

6.   Plotinus, *De rebus philosophicis* libri LIIII f.222r, trans. Marsilio Ficino (Basle 1559), quoted in G. Tomlinson, *Music in Renaissance Magic* (Chicago: University of Chicago Press, 1993), p. 86.

7.   Plato, *Timaeus,* trans. R. G. Bury (Cambridge, Mass. and London: Harvard University Press, 1931), 47b-e.

8.   Boethius, *De Institutione musica* I.2, trans. C. Bower, in J. Godwin, ed., *Music, Mysticism and Magic* (London: Routledge and Kegan Paul, 1986), pp. 46-7.

9.   See G. Shaw, *Theurgy and the Soul* (Pennsylvania: Pennsylvania State University, 1995), pp. 206-7.

10.   Plato, *Timaeus,* 34a-36d.

11.   Plato, *Timaeus,* 37a.

12.   Plato, *Timaeus,* 42e-44c.

13.   Quoted in Shaw, *Theurgy,* p. 214.

14.   Plato, *Republic*, 2 Vols., trans. Paul Shorey, Cambridge Mass., London: Harvard University Press, 1935, *Republic*, X.616-7.

15.   Cicero, *De Republica* VI, quoted in Godwin, *Music*, p. 11.

16.   Johannes Afflighemensis, *De musica* XVI, in C. Palisca, *Hucbald, Guido and John, On Music* (New Haven and London: Yale University Press, 1978), p. 110.

17. See J. C. Burgel, *The Feather of Simurgh: the Licit Magic of the Arts in Medieval Islam* (New York: New York University Press, 1988).

18. G. Anselmi, *De musica*, in G. Massera, ed., *Historiae Musicae Cultores* 14 (Florence: Olschki, 1961), p. 150.

19. See Joscelyn Godwin, *Harmonies of Heaven and Earth* (London: Thames and Hudson, 1987), p. 143.

20. B. Ramos de Pareja, *Musica practica* (Bolgna: Baltasar de Hiriberia, 1482), in J. Wolf , repr. ed., (Leipzig: Brietkopf and Haertel, 1901), p. 59.

21. Ficino, *Opera omnia,* pp. 901-2: my translation.

22. A. Poliziano, *Opere omnia* (Basle: Nikolaus Bischoff Jr, 1553), p. 310.

23. Ficino, *De vita*, p. 355.

24. G. Pico della Mirandola, *Conclusiones nongentae*, A. Biondi, ed., (Florence: Olschki, 1995), p. 121.

25. Ficino, *De vita*, p. 357.

26. Quoted in A. della Torre, *Storia dell'Accademia Platonica* (Florence: Tip. G. Carnesecchi e figli,1902), p. 791.

27. Ficino, *De vita*, pp. 357-9.

28. Ficino, *De vita* 3.XXI, p. 361.

29. Ficino, *De vita* 3.XXI, p. 359.

30. Ficino, *De vita* 3.XXI, p. 359.

31. Ficino, *De vita* 3.XXI, p. 359.

32. Ficino, *.De vita* 3. XXII, p 365.

33. Ficino, *De vita*, p. 361.

34. Ficino, *De vita*, p. 363.

35. Marsilio Ficino, *The Letters of Marsilio Ficino,* trans. members of the School of Economic Science, London, (London: Shepheard-Walwyn, 1975, 1978, 1981, 1988, 1994); Vol IV, pp. 60-1. Also see Ficino, *De vita,* 3 VI. p. 275; 3.XXII. p. 365; 3.XXIV. p. 377; Ficino, *Letters*, vol. 1, p. 161, for Ficino's discussion of planetary meanings.

36. Marsilio Ficino, 'De rationibus musicae', in P. O. Kristeller, ed., *Supplementum Ficinianum,* vol. I (Florence: Olschki, 1937)., p. 55.

37. C. Ptolemy, 'Harmonicorum sive de musica', libri tres III. 16, in A. Barker, ed., *Greek Musical Writings*, vol. 2 (Cambridge: Cambridge University Press, 1989), pp. 275-535.

38. Ficino, *Opera omnia*, p. 1615.

39. Ficino, 'De rationibus', pp. 55-6.

40. Ficino, 'De rationibus', pp. 55-6.

41. Ficino, 'De rationibus', pp. 55-6.

42. Ficino, 'De rationibus', pp. 55-6.

43. Ficino, 'De rationibus', pp. 55-6.

44. As frontispiece to F. Gaffurius, *Practica musicae* (Milan: Gulielmum signer Rothomagensem, 1496), in I. Young, ed., (Madison: University of Wisconsin Press, 1969) and in *De harmonia musicorum instrumentorum opus* (Milan: G. Pontanum, 1518), trans. C. A. Miller (Neuhausen-Stuttgart: American Institute of Musicology/Hänssler-Verlag, 1977)., p. 201.

45. Gaffurius, *De harmonia*, pp. 199-200.

46. For Wind's analysis of this woodcut, see Wind, *Mysteries* , pp. 265-9.

47. Wind, *Mysteries*, p. 268.

48. Gaffurius, *De harmonia*, p. 210.

49. Hans W. Heinsheimer and Paul Stefan, 'Gesinnung oder Erkenntnis', *25 Jahre*

Neue Musik: Jahrbuch 1926 (Vienna: Universal Edition, 1926), trans. Leo Black as 'Opinion or Insight', in *Style and Idea*, Leonard Stein, ed., (London: Faber and Faber, 1975), pp. 258-64 at 260.

50. Marsilio Ficino, *Liber de sole, (Opera Omnia)*, ch. II, trans. G. Cornelius, D. Costello, G. Tobyn, A. Voss & V. Wells, *Sphinx: A Journal for Archetypal Psychology, 6* (London Convivium for Archetypal Studies, 1994), p. 127.

Harmony and a Phenomenology of Liquid Skies

Ilaria Cristofaro

HARMONY IS AN ELUSIVE CONCEPT, one which I have approached by exploring the reflective quality of water, mainly from a visual perspective: water can actually bring us closer to the sky in the sense that we can see the sky reflected in water, even touch it or, more accurately, touch its reflection. I refer to such reflections as 'liquid skies' and in such observations of the sky, intermediated by the action of water, feelings of harmony can be evoked: receptivity, meaningfulness, wholeness, balance and integration. And, as David Cadman said when referring to harmony, 'when we look, we see that Nature expresses qualities such as wholeness, relationship, diversity, living within limits, cooperation, wastelessness, impermanence and thriftiness'.[1] My research provides an example of encountering principles of harmony in nature, and in particular, in liquid skies or reflections of the sky in water. Another time I have used the term 'water-skyscape', assuming that 'the sky is the ultimate focus of the experience and water becomes an affecting medium to access the skyscape vision'.[2]

In 2016-18 I conducted a research project into such liquid skies, keeping a journal of my observations and relating them to poetry, material culture and theories on perception. My observations were made in the United Kingdom, Italy and Greece during 2016 and 2017, and were gathered together into a sky journal in which I recorded my reflexive considerations in the form of notes, poems, drawings and photographs, aiming to bring my perceptions and unconscious feelings into focus. I recorded a description of the ecosystem around me in terms of place, time and significant phenomena, as well as of my sensations, thoughts and emotions in response to the environment. My observations were focused on the qualities and properties of the phenomenon I experienced, by exploring different layers of the relationships between land, sky, and water depending on the time, place and weather.

The Prince of Wales introduced *Harmony: A New Way of Looking at Our World* (his book with Tony Juniper and Ian Skelly) by stating the need to face what he described as the 'crisis of perception' in the current empiricist way of looking at the world.[3] The mechanist approach, he continued, has overwhelmed spiritual and religious dimensions of looking at nature to our detriment. The notion that there are two different ways of knowing is deeply embedded and runs throughout

Western thought. For example, Raymond Corbey looked at the division between the sciences and the humanities and quantitative and qualitative research, after then underlying assumption that the two different types of substance, mind and matter, follow different principles.[4] The Prince of Wales's analysis reflected Max Weber's definition of modernity as characterised by disenchantment, where 'the unity of primitive image of the world, in which everything was concrete magic, has tended to split into rational cognition and mastery of nature, on the one hand, and into "mystic" experiences, on the other'.[5] A series of later thinkers then developed the concept of 'enchantment' as the alternative to disenchantment. Enchantment is another elusive concept, but comes close to Lucien Lévy-Bruhl's 'participation mystique', the feeling we have when we are wholly and completely a part of, and engaged with, our surroundings, rather than separate to them.[6] For Edward Slingerland and Mark Collard, the dichotomy between the two ways of knowing can be overcome by, in other words, trying 'to see the realm of the human as coextensive with the realm of nature', which perhaps be similar to approaching enchantment.[7]

Following similar assumptions, Maurice Merleau-Ponty argued that the phenomenological method can help to perceive the essence of things.[8] I was deeply influenced by Merleau-Ponty's view of phenomenology in which, by immersing ourselves in the natural world we can explore what he called 'beings in depth, inaccessible to a subject that would survey them from above, open to him alone that, if it be possible, would coexist with them in the same world'.[9] I was also drawn to Martin Heidegger's *Dasein* or 'being-in-the-world' in which the subject (i.e., me) is intrinsically and physically involved in the surrounding world when trying to understand it, but without being able to fully transcend it.[10] In other words, we are inside the world we try to study and can never be separate from our study of it. With this in mind, I set out to explore the world of water and liquid skies.

An undervalued feature of phenomenology is the extent to which we assume that elements of the natural world have 'agency', meaning that they possess power, personality and identity. Some hold the opinion that only humans can have agency. Timothy Darvill defined it as 'the proposition that human beings think about the intentional actions they perform and the resources they need to achieve their ends'.[11] Nevertheless, attempts have been made to attribute agency to nonhuman entities. For example, Alfred Gell argued that agency can be present in material culture, but he distinguished between primary agency, which is the preserve of human beings, and secondary agency found in other entities when infused by humans to materials.[12] Similarly, Michael D. Kirchhoff highlighted the qualitative distinction between human and nonhuman agents since, he argued, material agency can only be enacted by the agentive capacity of humans, concluding that 'agency does not reside in matter'.[13] One area where a different perspective has emerged is that of ecology. For example, Bjørnar Olsen, who questioned the distinction between people and environment when advocating what he called a 'return to things', stated that 'landscapes and things possess their own unique qualities and competences that they bring to our cohabitation with them'.[14] In this sense, people, things, spaces and places all cohabit: they live together.

However, a wide variety of opinions have been expressed about the relationship between people and place. Ralph Waldo Emerson, deeply influenced by the Platonic view in which matter emerges from consciousness, wondered whether the material world was no more than a projection of the human mind: 'Have mountains, and waves, and skies no significance but what we consciously give them,' he asked, 'when we employ them as emblems of our thoughts? The world is emblematic. Parts of speech are metaphors, because the whole of nature is a metaphor of the human mind'.[15] Similarly, in her study of contested meaning in landscapes, Barbara Bender looked how one place may mean different things to different people and wondered whether landscapes have boundaries, describing them as 'experimental and porous, nested and open ended'.[16] These perspectives may imply a never conclusive and always contested interpretation. Christopher Tilley and Kate Cameron-Daum, were not so far from Bender's position when they pointed out how the complexity and multivocality of contested landscapes meant that they might not represent any one thing, stating that 'landscapes are … an active presence' but they produce different responses from the people who experience them.[17] Finally, Tim Ingold rejected the division between mind and matter, between inner and outer realities as irrelevant when considering our relationship with place, and highlighted the recognition of the temporality of the environment and its cycles, which is embodied within the individual who dwells within it by an effect of what we can call resonance.[18] For Ingold, 'the landscape … is a story' and meanings are stored within a landscape and revealed through the process of being-in-the-world, to borrow from Heidegger.[19]

Assuming that lands, seas and skies have agency, it follows that they will have different ways to express it. Powerful expressions might include a thunderstorm, an erupting volcano, a starry sky or, in terms of my interest, a glitter path, all of which are natural entities which may prompt strong emotional effects such as awe or fear. My approach to such things embraces Ingold's statement that entities are *possessed by the action* (his italics): having agency, he says, does not mean that they possess intention.[20] In his view, agency is distributed along the field of interweaving relations and represents a dynamic property which is prior to any distinction we can make between object and subject.[21] As Franz Breuer and Wolff-Michael Roth argued, 'the knower and the known form a dialectic unit' so that knowledge forms somewhere in the relationship between the subject and the phenomena observed.[22] As applied to my research, in the tension knower-known, I had to reduce myself as far as possible in a position in which I allowed land, sky, water, and all things in the natural world, to express their own power and agency in terms of the qualities of time and place.

In his final work *The Visible and the Invisible,* Maurice Merleau-Ponty introduced the idea that the 'thickness' of the corporality of the body is the only means of communication by which to reach the essence of things through a dialogue between the researcher and the object of study.[23] Merleau-Ponty stated that it is thanks to a coexistence between the researcher and the object of study that there is a possibility of dialogue.[24] For Merleau-Ponty, perception implies a dualistic movement: as the outer world is perceived by the researcher, so the perceiver sends out intention.[25] But this twofold action is negotiable: the researcher's intentions, thoughts and judgments can be suspended and 'bracketed off' for a short time. This bracketing allows the researcher, hopefully, to experience phenomena on their own terms. As Kathy Charmaz pointed out, preconceptions can arise from the researcher's social and cultural background which influence the research process.[26] Our task as researchers is to free ourselves, as far as possible, from such preconceptions. We can also talk about Edmund Husserl's 'reduction', a process understood as setting aside the intention within the self so the researcher can become a silent testimony of the natural world.[27] In particular, I tried to emancipate myself from distractions such as the few people, cars and planes which passed by. The practitioner may allow the phenomena of the outer world to fully be experienced as an inner reality. This otherness can inhabit the body by moving it, speaking or just being there with its own qualities. Similar to the embodiment of characters in drama, a landscape can be integrated within one's self through the distillation and assimilation of its inherent qualities and temporalities. Using this technique, at times I reached a state of full immersion into the liquid sky observed, which would have lasted for only a few seconds: my body resonated with the observed phenomenon in a status of enchantment. The problem, though, is that control by individuals over their own intention is often very limited by distractions. My own process was aided by my ten-years training in drama, Butoh performance and self-discovery. I have also practised Vipassana meditation, in order to develop self-awareness.

The phenomenological approach also requires reflexive considerations, necessary to contextualise the product of the observations. Personally, I was already biased by having read about the topic on the symbolic role of the water and the sky within ancient Mediterranean cultures: I already had preconceptions relating to the idea of water as a sort of underworld. Once I had started my research, the systematic observation of liquid skies changed me by provoking an intense emotional response. I was deeply affected by the vision of the Sun or Moon reflected in water, as if their intangible presence on land was very near and intimate. Manifestations of sunlight in water became more and more attractive, charged with a sort of charisma and visual magnetism. I became engaged in my own participation mystique. As I wrote up my journal and reflected on my experiences, five themes emerged: receptivity, meaningfulness, wholeness, balance and the integration of opposites.

It will be considered the ways in which individual forms can be seen transformed and dissolved on the dynamic layer of the water surface, echoing the idea of wholeness. Sky and water frame the land holding the tension between the above and below in a state of balance as they mirror each other. Finally, a particularly striking effect is produced when the Sun is low on the horizon and reflections of sunlight in the form of glitter paths seem to unite the above and the below in a bridge of light.

RECEPTIVITY

The surface of the water is a liminal space, a boundary zone which simultaneously separates and connects air and water. When I resonate with water, I am fostered to have a fluid, open and receptive inner condition. Objects floating on a water space can modify the reality of the reflected phenomenon: in Figure 2 the Sun image is distorted by the presence of algae. In a similar way, memories can affect perceptive experiences. Henri Bergson stated that 'perception is never a mere contact of the mind with the object present; it is impregnated with memory-images which complete it as they interpret it'.[28] Bergson implied that perception is always conditioned by memories, but he did not consider the perception of an infant who has no or little memories. For this reason, phenomenology is here understood as a return to a primordial and naïve way of looking at the world, not unlike the perspective of a child. Practicing reduction or bracketing intention is like cleaning water from algae: the researcher aims to achieve a status characterised by resonance, transparency, primordial receptivity, neutrality, participation and empathy, similar to the mirroring of a clean water surface, as evoked in Figure 1. In summary, for me, the peculiar reflective quality of clean water inspired states of transparency and receptivity which in turn foster an empathic condition.

Figure 1. Clear water may be a metaphor for neutral perception: the receptive quality of water fosters a condition of empathy. Heath Farm, Chipping Norton (UK) 30 July 2016. Photo by I. Cristofaro.

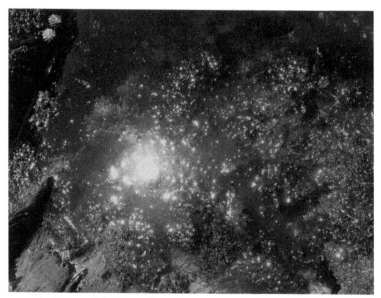

Figure 2. The image of the Sun reflected in a sea pool is distorted by the presence of algae, as memories distort pure perception. Pembrokeshire (UK), 1 July 2016. Photo by I.Cristofaro.

MEANINGFULNESS

Since 'observations are all made from somewhere', as David J. Hufford pointed out in his discussion about reflexivity, I realised the extreme importance of my position in space and time in relation to the reflection observed.[29] For most of the time, I had to move around to find the point in space whose angle allowed me to see the phenomena I wanted to observe: a few times I felt that this was a forcing act in respect to allowing the water and the sky to let me see what they offered to me, at other times I felt as if I was driven by an instinctive force to find the right place from where to look. Therefore I noticed, when looking at small pools of water, that the size of the water space constrained what I could and could not see. In throwing a piece of wood into a small canal covered by pond scum at Oxford Christ Church Meadow, I noticed how the sky was opening up in the water as a reaction to the dispersion of the algae: my notes record that 'it looks like a telescope image, in the sense it reflects only a small portion of the sky' and, within that selected section 'all seems to be embedded of meanings'. The piece of liquid sky is as much limited as the water space which contains it. This reminds me how Etruscan augurs would define a particular part of the sky called *templum* in order to limit their observations and to receive omens from the celestial deities.[30] Blocking off most of the sky enables the selected portion to be received as a meaningful experience: the sky vault is so vast and overwhelming that sometimes I feel lost since I do not know which clouds, stars or phenomena to observe. The fact of being able to see only a little part of the sky imposes a restriction, and yet it seems to have meaningful significance: in that precise moment in time I felt that such sky existed just for me. Figure 3 shows this idea: by reflecting the sky in a bowl of water, the Sun can be found contained within it. Small amounts of water, such as puddles or ponds, have this peculiar characteristic: by reflecting a small portion of the sky vault, they seem to reveal to the observer a meaningful and intimate celestial reality. And I am reminded that, for David Cadman, living in a meaningful and purposeful cosmos can be considered one of the conditions of harmony.[31]

WHOLENESS

Sometimes I felt as if I had turned myself inside-out in such a way to allow the water to be contained in my own self. I noticed that in watery reflections nothing is itself anymore, and liquid skies dissolve the shapes and identities of celestial objects. On 20 July 2016 I wrote 'the reflecting light of the Moon is dispersed

Figure 3. Finding the Sun in a bowl of water. Matera (IT), 28 August 2016.
Photo by Alessandro Scillitani.

and fragmented'. Water spaces appeared as doorways to another realm where the definitions of identities are put under discussion. In my notes from 26 July 2016, I wrote:

> The water, as consciousness reflected
> and blended by the collective circumstances.
> World below the sea
> Made by inconsistent changeability of form limits
> Convex, expressed, merged,
> In the clean slate
> of the wind, the painter
> master of the brush,
> of agile water of beings in motion.

A disintegrating effect on individuality can be visually seen in liquid reflections: the water surface provides a metaphor for the concept of 'wholeness'. Indeed, in the liquid world it is possible to see with eyes how, in David Cadman's words, 'the whole and its parts are always interconnected and related'.[32] The fluid propriety of water dissipates borders, merges shapes and dissolves identities: it can be associated with the fragmentation of the self and subsequent merging in the unity of the world. All shapes lie on the same layer, on the water surface, and they fluently interact with each other, as illustrated in Figure 4. This reminds me of C. G. Jung's statement that 'the meeting between the narrowly delimited, but

Figure 4. The water surface merging shapes. River Thames. Oxford (UK), 25 July 2016.
Photo by I. Cristofaro.

intensely clear, individual consciousness and the vast expanse of the collective unconscious is dangerous, because the unconscious has a decidedly disintegrating effect on consciousness'.[33] In opposite, in Jungian terms, *individuation* is the process by which 'individual beings are formed and differentiated'.[34] The watery surface seems to provide an ambiguous and uncertain space, inspiring a liminal status of passage. Indeed, not having a certain identity means to have freedom and a huge potentiality towards all identities, and therefore the acquisition of new power. My observations from 22 July 2016 reported on the Sun reflected in a natural tidal pool, as if the Sun itself were performing a ritual dance by disentangling itself from its spherical shape (see Figure 5).

The reflection of the Sun
the light is caught in this electric fluid dance:
Frenetic twitches of dazzles
Rapid movements, almost imperceptible
Of wind's whispers
…
Rapid dance, deformation of the circle
Opening of worlds
Rediscovery of hidden treasures.

More insights make me understand that the sky-waterscape polarity can correspond to the binary tension between the human tendency towards individualism and towards being part of a collective whole: the water surface is comparable to the idea of collectivity, and the sky to that of individuality. In particular, Arthur Koestler defined the dual tension within a 'holon' (an element which is paradoxically both a whole and a part within different hierarchical wholes) as both self-assertive, separating itself off, and inclined to integrate.[35] For Koestler this essential, internal opposition of tendencies is present in every type of biological, sociological or ecological complex system.[36] He adds that the integrative tendency can be expressed by individuals as cooperation and altruism, attitudes central to John Fullerton's suggestions for a regenerative economy under the principles of harmony.[37] David Cadman also wrote about how the long history of humans damaging their environments is, in his words, based on 'the proposition that the world is made up of separate parts, ever in competition one with another'.[38] Observing reflections of light in water reminded me of the interconnections between all things and how the liquid world seems to be a metaphor for the unification of single identities in the layer of wholeness.

Figure 5. The deformed shape of the Sun seen from a boat in the Mediterranean Sea
in four subsequent moments. 30 August 2017.
Photos by I. Cristofaro.

BALANCE

Water spaces create a continuation of the sky on the land: waterscapes and skyscapes together constitute a dualistic reality of specular realms. In a scenery where the heaven above and the water below mirror each other, the land is framed in the line of symmetry as evoked in Figure 6. Similarly, Henry David Thoreau, who spent many months in a wooden cabin in the forest near a pond, described the watery space as 'intermediate in its nature between land and sky'.[39] Visions of water and sky can inspire feelings of balance, another element of harmony in David Cadman's analysis.[40] This was also the principle of the Emerald Tablet of Hermes, 'that which is above is like to that which is below, and that which is below is like to that which is above, to accomplish the miracles of one thing'.[41] As inspired by Figures 7 and 8, the world can be seen in balance when the sky is mirrored in a water space.

INTEGRATION OF OPPOSITES

During the phenomenological fieldwork, some qualities of the glitter path, the long reflection of the Sun in water when the Sun is low on the horizon, were observed. This phenomenon extends upon the water a luminous cloak of sunlight. At its full elongation, the glitter path gives the impression of indicating a particular direction, a path to and from the Sun, resembling a luminous bridge connecting the celestial body with the viewer on the seashore. It reaches its maximum elongation around the times of sunrise and sunset when the string of light, as reflected in the water, united the littoral with the horizon, creating a channel of integration between the above vault of the sky and the below expanse of water, as if the two worlds were connected. For me, the glitter path can resemble a road towards the Sun. On 9 July 2017, I wrote:

> Dawn is the epiphany for excellence
> Light coming out from darkness
> Especially when the water is calm
> The glitter path seems like a road
> Connecting myself with the sun
> But it is already very bright
> And I cannot look at it for a long time
> … and it is already day
> The water transports the light, the Sun towards me
> It is connecting.

Figure 6. Above and below mirroring each other, in respect to the line of symmetry
of the land. River Thames. South Stokes (UK), 10 August 2016.
Photo by I. Cristofaro.

Figure 7. A sea pool as intermediate between land and sky.
Pembrokeshire (UK), 1 July 2016. Photo by I.Cristofaro.

Figure 8. In the stillness of the Venice lagoon, a view of above and below as one thing. Venice (IT), 22 December 2016. Photo by I.Cristofaro.

Similarly, on 13 August 2017, I recorded the following observations:

Like a fiery dance
You unwind on the waters
like a golden lace, you hook me
like a bridge of light
you surpass the realms
like an exuberant creature you manifest yourself
body of light
from the Sun you are emanated
from the sea you have emerged
unique and unrepeatable …

Even if intangible and fluid, the impression is of a deep-rooted and consistent connection between the viewer and the Sun, manifesting itself in a dramatic way, shining as a bronze way crossing the sea. This reminds me of James B. Kaler's statement that such reflections can make us feel 'as if you could walk into space'.[42] In my notes, I described the glitter path as 'a bridge of light', a 'road', a 'channel'

Figure 9. The glitter path at dawn seems to connect the sky and the sea with a bridge of light. Milatos Paralia, Crete (GR), 26 July 2017. Photo by I. Cristofaro.

Figure 10. The glitter path as a road of light to reach the Sun. Milatos Paralia, Crete (GR), 9 July 2017. Photo by I. Cristofaro.

toward the Sun, as evoked in Figures 9 and 10. It suggests a sense of connection and integration: by creating an intangible luminescent roadway, the glitter path can unite the seashore with the horizon, bridging the spectacular world of sky and water in a harmonious relationship of opposites.

CONCLUSION

At the end of the process, I realised to have found some principles of harmony by observing Nature. I felt that the clear water surface is transparent, empathic, and receptive to external influences, fostering the viewer (me in this case) to achieve a similar inner condition. Thus, clean water offered me a metaphor for describing neutral perception against distorting memories. In a reflection, the sky is visible in well-defined portions which can become meaningful experiences against the overwhelming celestial valut. Reflected on the liquid world, shapes and identities seem to be transformed, disintegrated and merged together on a layer of dynamic connections, evoking, in my experience, a dimension of wholeness and integration, versus individualism. Another principle of harmony holds that everything in the universe operates in a state of balance: sky and water frame the land in a symmetrical opposition of reversed worlds. Finally, when the Sun is low on the horizon, reflections of sunlight in water provoke a sense of merging opposites, bringing together the two worlds above and below by a road of light. By looking at the sky through reflections on water it is possible to perceive a reality where the principles of harmony manifest themselves: empathy, meaningfulness, wholeness, balance and the unity of opposites are all conditions necessary for cohabiting in harmony with the natural world. Thus, my body, as if a receptacle of water, resonated with the sky in a status of enchantment.

NOTES

1. David Cadman, 'Harmony & Integration', (paper presented to the UNESCO conference organised by the University of Wales TSD and the World Fellowship of Buddhists and the Kingdom of Thailand, Paris 27th-29th September 2017), p. 4.

2. Ilaria Cristofaro, 'Reflecting the Sky in Water: A Phenomenological Exploration of Water-skyscapes', *Journal of Skyscape Archaeology*, 3, 1 (2017), 112-126, pp. 113-114.

3. HRH The Prince of Wales, Tony Juniper and Ian Skelly, *Harmony: A New Way of Looking at Our World* (New York: Harper Collins, 2010), p. 6; see also pp. 7-11.

4. Raymond Corbey, *The metaphysics of apes: Negotiating the animal-human boundary* (New York: Cambridge University Press, 2005).

5. Max Weber, *From Max Weber: Essays in Sociology*, translated and edited by H. H. Gerth and C. Wright Mills (London: Routledge, 2009), p. 282.

6. Lucien Lévy-Bruhl, *How Natives Think* (Princeton: Princeton University Press,

1985), pp. 69-104.

7. Edward Slingerland and Mark Collard, 'Introduction. Creating Consilience: Toward a second Wave', in *Creating Consilience: Integrating the Sciences and the Humanities,* edited by Edward Slingerland and Mark Collard (Oxford: Oxford University Press, 2011), p. 11.

8. Maurice Merleau-Ponty, *The Visible and the Invisible,* edited by Claude Lefort, translated by Alphonso Lengis (Evanston: Northwestern University Press, 1968).

9. Maurice Merleau-Ponty, *The Visible and the Invisible,* p. 136.

10. Martin Heidegger, *Basic Writing from Being and Time (1927) to The Task of Thinking (1964),* edited by David Farrell Krell (New York: HarperCollins, 1993), pp. 48-49.

11. Timothy Darvill, 'Agency', in *The Concise Oxford Dictionary of Archaeology* (Oxford: Oxford University Press, 2009).

12. Alfred Gell, *Art and Agency: An Anthropological Theory* (Oxford: Oxford University Press, 1998), pp. 17-21.

13. Michael David Kirchhoff, 'Material Agency: A Theoretical Framework for Ascribing Agency to Material Culture', *Techné,* Vol. 13, 3 (2009), pp. 206-220.

14. Olsen Bjørnar *In Defense of Things: Archaeology and the Ontology of Objects* (Lanham: AltaMira Press, 2010), p. 10.

15. Ralph Waldo Emerson, *Nature* (Boston: James Munroe and Company, 2009), p. 54.

16. Barbara Bender, 'Time and Landscape', *Current Anthropology,* Vol. 43, Supplement (2002), S103-S112, p. S137.

17. Christopher Tilley and Kate Cameron-Daum, *An Anthropology of Landscape: The Extraordinary in the Ordinary* (London: University College London Press, 2017), p. 288.

18. Tim Ingold, 'The Temporality of the Landscape', *World Archaeology,* Vol. 25, 2 (1993), 152-174, p. 163.

19. Ingold, 'The Temporality of the Landscape', pp. 152, 172.

20. Tim Ingold, *Being Alive: Essays on movement, knowledge and description* (Abingdon: Routledge, 2011), p. 214.

21. Ingold, *Being Alive,* p. 68.

22. Franz Breuer & Wolff-Michael Roth, 'Subjectivity and Reflexivity in the Social Sciences: Epistemic Windows and Methodical Consequences', *Forum: Qualitative Social Research,* Vol. 4, 2 (2003), Art. 25.

23. Merleau-Ponty, *The Visible and the Invisible,* pp. 135-136.

24. Merleau-Ponty, *The Visible and the Invisible,* p. 136.

25. Maurice Merleau-Ponty, *The Phenomenology of Perception,* translated by Colin Smith (London: Routledge 2003 [1945]), pp. 3-14.

26. Kathy Charmaz, *Constructing Grounded Theory: A Practical Guide Through Qualitative Analysis* (London: SAGE Publications, 2006), p. 67.

27. Edmund Husserl, *Ideas: General Introduction to Pure Phenomenology,* translated by W. R. Boyce Gibson (London: George Allen & Unwin, 1962 [1913]), pp. 172-173.

28. Herni Bergson, *Matter and Memory,* translated by N. M. Paul and W. S. Palmer (New York: Zone Books, 1991), p. 133

29. David J. Hufford, 'The Scholarly Voice and the Personal Voice: Reflexivity in Belief Studies', *Western Folklore,* Vol. 54, 1 (1995), pp. 57-76, p. 58.

30. Marcus Terentius Varro, *Opere,* edited by Antonio Traglia (Torino: UTET, 1974), VII, 6-13; Titus Livius, *Books I and II,* translated by B.O. Foster, Vol. I. (Cambridge, MA: Harvard University Press, 1967), I, xviii, 7-10.

31. David Cadman, 'Principles of Harmony' (2017), p.1. http://www. harmonyinitiative.net/images/David_Cadman_Principles_of_Harmony_Jan_2017.pdf. [accessed 8th April 2019].

32. Cadman, 'Harmony & Integration', p. 2.

33. Carl Gustav Jung, *Psychology and the East* (London: Ark Paperbacks, 1978), p. 31.

34. C.G. Jung, *Psychological Types, The Collected Works*, Vol. 6, translated by R.F.C Hull (London: Routledge and Kegan Paul, 1971), para 757.

35. Arthur Koestler, *Il principio di Giano*, translated by Libero Sosio (Milano: Edizioni di Comunità, 1980), p. 76.

36. Koestler, *Il principio di Giano*, p. 77.

37. Koestler, *Il principio di Giano*, p. 81; John Fullerton, *Towards a Regenerative Civilization: Reconnecting our Economics with Harmony Principles* (Greenwich, CT: Capital Institute, 2018) http://capitalinstitute.org/wp-content/uploads/2018/05/Fullerton_ Harmony-In-Regenerative-Economy_FINAL_4.11.pdf. [accessed 8th April 2019].

38. Cadman, 'Harmony & Integration', p. 2.

39. Henry David Thoreau, *Walden*, edited by Jeffrey S. Cramer (New Haven: Yale University Press, 2006), p. 183.

40. David Cadman, 'Harmony' at https://www.uwtsd.ac.uk/harmony-institute/ [Accessed 19 March 2019].

41. Robert Steele and Dorothea W. Singer, 'The Emerald Tablet', *Proceedings of the Royal Society of Medicine*, 21, 3 (1928), 485-501.

42. James B. Kaler, *From the Sun to the Stars* (New Jersey: World Scientific, 2017), p. 11.

Peace from the Perspectives of Harmony

Scherto Gill

A GOOD UNDERSTANDING OF HARMONY CAN HAVE PROFOUND IMPLICATION FOR PEACE. What I want to explore is the understanding of peace from the perspective of harmony, and the opportunities that such a conception might bring to the process of creating peace and peacefulness globally. More specifically, from reading Confucian and Greek classics it interprets harmony as containing the idea of right relationships amongst all, and argues that it is dynamic and proactive, rather than static and merely complying to a predefined universal order. In this reading, I find harmony to be harmonising, a co-creative and generative process that can help transform tension and conflict towards an infinite possibility of relation and interconnection.

These insights lend a holistic understanding of peace which includes a generative and proactive process that is beyond the mere presence of inner peace and the absence of war and violence. Like harmony, peace itself becomes an action embodying principles of harmony in myriad relationships, for example, between humans and spirits; amongst people, communities and nations; between people and the planet, as well as our relationships with ourselves. This holistic understanding of peace does not separate humans, spirits or the world (Confucius' humanity, heaven and earth), nor does it see our being as fragmented. Through the lens of harmony, we can bring together the multiple dimensions of peace as a whole. Thus, in peacebuilding practices, the focus would be on nurturing dialogue, listening and collaborating, encouraging the co-being and co-acting of all agents, and embracing conflict and tension as creative forces for innovation and transformation.

Harmony

In Greek thinking, harmony is situated within (cosmic) order, forms and (mundial) structure, and it is also a balancing force that helps reconcile, for instance, virtues and vices, opposites (eg. in music), and ensures justice.[1] The metaphysical base of harmony is established through a contemplative endeavour and epistemic process which taps into greater wisdom or principles inspired by the Ultimate/ Transcendent.[2] Harmony is intimately connected to beauty and goodness (eg. the

Form of the Good which explains what is just and beautiful), and brings integrity, beauty and goodness within different realms, in one's body, within the body-mind-soul connection, in rational thought and moral reasoning, between the sacred and the profane, and in the cosmic.[3] To the ancient Greeks, harmony is the foundation to life, and hence the principle of creation.[4] In this way, harmony opens up in love, acting to connect the separate elements in the world towards an integrated whole through bonds and relationships.[5]

In the Confucian tradition, the concept of harmony is fluid: as John Berthrong pointed out, Confucian philosophy has been evolving over the course of last two thousand years, through four different chapters, integrating Buddhist thinking and teaching, as well as being placed alongside Western philosophy along the way. [6] What I discuss here is mainly based on the lenses of the four classic books written by Confucius, and scholars' interpretations of these classic texts. Starting with a contrast between harmony on the one hand, and sameness/uniformity on the other, the Confucians argue that at the core of harmony there is a dynamic process of harmonising which includes sustaining differences without rendering them to sameness/uniformity. As we read in the *Analects*, 'The gentleman harmonizes [*he*] without being an echo. The petty man echoes [*tong*] and does not harmonize'.[7] This core notion suggests two combined aspects of harmony: the first concerns one's self and other people, that is, one can be open to harmonise or coordinate with others, but without relinquishing those of one's own (ideas, thoughts, habits, preferences, and so on); and secondly, it recognises constituting parts of wholeness/totality as diverse, at times opposite, each being attributed its worthiness and its place within the whole through interacting with others.[8] These two aspects combined suggest that, to have harmony within wholeness, the different constituents, including those opposites, do not have to lose their distinctness in order to harmonise with others. In fact, 'inclusive opposites', such as *yin* and *yang,* are fundamental to the harmonising process, and through harmonising, co-constituted elements can be brought into a dialectic relationship in which they mutually constrict, contradict but also mutually expand and reconcile towards renewed relationships.[9] In this dynamic process, there is mutual transformation, renewal and advancement. Thus, harmony is perceived as containing within itself a perpetual creative force. Similar to the Greek conception, harmony in Confucian thoughts is comprehensive and multi-faceted, applicable in all dimensions of human life and the natural world. In fact, the perpetual creative tension places Nature and humanity as one in spirit.

The major contention between the classic Greeks and Chinese conceptions of harmony appears to lie in whether there is a pre-existing order imposed upon the world, either through normative description or through divine prescription. In other words, the major contention is the question of whether there is an ultimate ideal of harmony.[10] One argument is that whilst the ancient Greeks recognise a deep transcendent order and structure that can inspire harmony in the world, the ancient Confucian scholars perceive the force to have come from the harmony or harmonising process itself that encompasses spatial, temporal, metaphysical, moral and aesthetical dimensions of dynamic interplay.[11]

There is also a common understanding of harmony between Greek and Confucian schools of thought. For instance, both see harmony as 'the coming-together of different things', and believe that tensions, opposites and conflicts are not only part of harmony, but also the creative driver of harmony.[12] This has led some scholars to reconcile the differences between the Greek and Chinese conceptions of harmony and arrive at some common understanding in the following: harmony is an on-going dynamic and fluid process of integrating different forces towards coordination, collaboration and congruent relationships amongst all that is, including things in nature, amongst humans, within a person him/herself (by harmonising the various parts of his/her body, and body-mind-heart-soul relationship), in society and within social institutions, in relations amongst communities and nations, globally and cosmically.[13] This conception of harmony includes Love which connects and binds all the parts in an integrated whole.[14] Implied in this understanding of harmony is that it is not only desirable in itself but is intrinsically valuable, and also contains in it the values of aesthetics, beauty and goodness.[15]

This reading of harmony reveals at least five important insights that may shed light on our understanding of peace:

First, harmony is both a process and action of harmonising and a quality of what is harmonious. The former is strongly supported by the Confucians; that harmony means harmonising, a process of responding to, coordinating and co-creating with others (plant, animal, human, structure, system). The latter is found in the Greek explanation that harmony is a metaphysical and a moral concept that both describes and prescribes: it describes how the world is and operates (especially in the non-human world) and prescribes how human beings ought to be and act in the world. As I shall discuss, harmony is neither an abstract notion nor a rigid postulation, but rather it expresses the qualities of being harmonious in actions and processes, in relationships, in the traits and characteristics of being harmonious, and in the state of our being within the cosmos.

Second, as already indicated, harmony is multidimensional and the processes and actions of harmonising take place at different levels and in all dimensions at the same time. Thus, harmony assumes far-reaching symbiosis and interdependence within a totality. This symbiosis and interdependence are found in the elements and energies constituting our world, and in movements and changes in nature and human activities, such as music making, making a livelihood, developing social institutions and governing.

A closely connected point here is the relational nature of harmony, applicable to all the afore-mentioned relationships, for example, between the human world and natural universe, and amongst individuals, groups, societies, structures and systems. Harmony likewise involves intrapersonal and interpersonal relationships, the relationships amongst sciences, religions/spirituality, cosmos, culture, economics, technology, politics and civilisation, as well as general relatedness and interdependence of all things within a greater whole.

Third, within the harmonious whole, differences are not suppressed. Instead, harmony contains and holds tension, differences and contradiction in these diverse aspects, elements, and dimensions, forming an interactive 'sphere' of co-constituted components. As expounded by the classical Confucians, within this 'sphere', all relationships are mutually constricting and mutually emancipating at the same time, whilst supporting an optimal space for all to flourish. In this way, the world can accommodate and nurture 'myriad things'.[16] For instance, such an interpretation can be applied when we are engaging with cultural and religious differences, or clashes between value systems. In this case, a value of pluralism is often favoured. Likewise, within international relations, this interpretation of harmony can support mutual accommodation, resulting in 'relativity without relativism'.[17]

Fourth, harmony evokes aesthetic expressions and embodies inherent goodness in these symbiotic relationships. Plato's form is an illustration: where there is harmony, there is beauty, goodness and truth, which in turn epitomise harmony. Similarly, Confucians place great significance on the practice of ritual in everyday life. Practising rituals involves living out the values as the core of an ethical life.[18] The emphasis is not on the repetition of certain practices, such as prayers in religions, etiquette in the family, or ceremony in the community, but instead, the rituals reinforce moral propriety and goodness in a certain way of relating and acting, which further leads to cultivating good human beings and a harmonious world. Thus, harmony entails acting mutually responsively but in appropriate ways. What is appropriate is also aesthetically pleasing, such as music as a ritualistic action.

Finally, harmony is reflective. It provides philosophical guidance about how to be and act in relationship with another, including human, non-human and divine others. Equally, it articulates a spiritual worldview where all is located within a wider framework of meaning and interconnection, and is part of a larger whole. Through harmony and in harmony, there exists the possibility and desirability of flourishing within the totality. From this philosophical and metaphysical base, we can see that harmony within one's self is inseparable from harmony amongst things beyond one's self, such as our relationship with other people and other beings in the world.

Therefore, harmony can be embodied pragmatically in our everyday being and actions, for instance, in our being (or co-being/co-humanity as per the Confucians), language (which reflects our forms of life within communities), attitudes, beliefs and values, our personal, cultural and institutional practices, socio-economic and political structures, in the ways that we appreciate our experiences in the world, and, above all, in the way we educate and nurture people who are capable of living and acting in harmony with each other and with the world, and who are motivated to create harmony in the world.

So far, in interpreting the notion of harmony in classical Greek and Confucian philosophies, I have outlined five important perspectives through which we may further develop a framework to understand and appreciate peace.

UNDERSTANDING PEACE FROM THE PERSPECTIVES OF HARMONY

From the above classical vantage points, there is clearly a connection between peace and harmony. This connection is already obvious from the etymology of the term 'peace' in both ancient Greek and classical Chinese. 'Eirene' is the Greek root word for peace. It describes the following qualities: harmonious relationships between humans, nations and between God and humanity; freedom, welfare, order, friendliness, and a sense of rest and contentment. The Greek conception (and in its use in the New Testament) stresses the quality of peace in relational harmony, highlights the importance of loving connection between people, communities and states and underlines both the inner and outer qualities of being human. In the Confucian conception, especially in the Doctrine of Mean, peace is referred to as *zhong* (中), a state of being centred, and *he* (和), harmony where everything is in its proper place in unity with others and through dynamic and creative tension.[19] The centred harmony is peace itself: equilibrium in accordance with the way of the universe, towards which all strive, and through which all thrive, the divine destiny.

In both traditions, peace is intimately connected with harmony in three aspects. The first is harmony's relationship with peace in a threefold distinction. In other words, harmony is seen as not just the state of peacefulness itself, it is also an expression or outcome of peacefulness, and equally a source or condition of peacefulness. This threefold distinction is helpful in understanding peace holistically. The second is the relational nature of harmony that can enable us to appreciate peace as right relationships. The third is the part that human proactivity can play in promoting harmony, and hence peace can embody proactivity.

A. *Understanding peace holistically*

The threefold distinction reminds us that when we understand peace, we must take into account all three aspects and consider such questions as: (1) what counts as peacefulness in oneself, in the community, within society and globally; (2) what contributes towards peace; (3) what are the expressions of peace. [20] All three aspects are constituted in peace, even though (1) is clearly the primary meaning of the word and (2) and (3) are derivatives of peace, and they are helpful in directing our attention to the holistic nature of peace. Indeed, the perspectives of harmony suggest that peace has rich meanings and proactive qualities, ranging from one's state of mind to interpersonal dynamics, and from community relations and international relations to current global situations and climate change. In all these cases, we can refer to peace, and despite the differences in its meaning in each context, and that what counts as peaceful in one context would not necessarily be perceived in the same way in another, the concept of peace is neither ambiguous nor fragmented. The threefold distinction allows a semantic unity within these contexts. [21]

As the etymology of peace shows, like harmony, peace is not a single value and is instead a much *thicker* notion than any single paradigm can capture. Any understanding of peace must unify its different forms and qualities without denying their variety. It requires a similar exercise to that which we undertook with harmony, which is to consider: is peace primarily a thing, a value, a relation, a process, an action or all of these? Like with harmony, we could be thinking of peace primarily not as an abstract object, but as a quality. In other words, 'peaceful' qualifies something else, such as a state of being (spiritual, psychological, worldly), relationship, social conditions, processes and actions. In this way, we can understand peace not as an absolute ideal or a single idea but from its holistic qualities.

A holistic understanding of peace is important because it provides an opportunity for us to integrate the diverse layers and dimensions of peacefulness from a unified vantage point. It also unites the intrinsic and instrumental values of peacefulness. The intrinsic value of peace lies in the meaningfulness in and of itself; and the instrumental value of peace refers to the goodness that peace can bring to the world, such as flourishing life and thriving communities.[22] Above all, it can further help us overcome limitations in the two most common conceptions of peace: the inner/outer conception of peace and the positive/negative conception.

The inner/outer axis postulates that peace is an inner state, and there can only be peace in the world when we are at peace within ourselves, or when we embody peacefulness.[23] In this view, peace is defined as inner tranquillity or harmony that is found in one's feelings, mind and body, arising from a spiritual state. In the positive-negative axis, peace is regarded as having positive qualities, or prudent features (for example, values) rather than just the absence of negative features (for example, violence).

The inner and positive conception would seem to tend more towards an absolute pacifist position. This is because insofar as peace is an inner state (primary definition), peacebuilding actions are oriented towards the practices of mindful living and meditation. This involves the promotion of positive states of mind, which can help reduce and resolve anger, fear and other negative emotions, thus bringing peacefulness into one's body, mind and feelings, and from there, extending peace towards others as a disinterested love, for example for animals, plants and the natural environment. The inner and positive conception is not focused on the socio-political economic conditions necessary for the emergence of peace. Instead, under this positive conception, the focus lies with cultivating individuals who can serve as peace agents and who will bring peace to the world because of their inner peacefulness.

In the outer and negative conception of peace the default *de facto* natural state is egoistic and one of aggression, war or violence and peace is opposite to this default perpetual state of violence. In the classic view as portrayed by Hobbes the state of nature as one in which all self-interested individuals are in perpetual potential violent conflict with each other in competing for limited natural resources as well as limited positional goods, such as honour or glory.[24] According to this conception, the pacifist state is not a natural state of being human, and as long as there is no violence between peoples, groups and nations, there is peace. According to this view, peacebuilding will stress combating, removing or suppressing violence and its root causes. Hence, peace-oriented processes including democratic governance, the rule of just laws, a fair system of economic distribution, institutions that ensure equal respect for all, and free exchange of thought, among others, are only directed symptoms of violence.

There are a number of limitations in the inner/outer and positive/negative conceptions of peace: one sees peace as an absolute entity or single ideal, the same in all circumstances and contexts; the other that the peaceful state remains at the levels of the psychological and the individualistic; a third is the separation of spiritual peace from worldly peacefulness. This means that institutional and systemic peace must be conceived of in purely negative terms, leaving no space for peace to be conceived in a non-individualistic or integral sense.

Through the lens of the threefold distinction of peace drawn from the perspectives of harmony, these limitations can be overcome. It is helpful to develop a holistic conception of peace, an understanding that connects the idea of peacefulness as a (spiritual) way of being, a psychological state, a relationship between persons and groups, a condition of community, a feature of political economic structures and international relations, and a collective process and co-action for the greater good.

We can see two examples of the holistic conception of peace in contemporary history: one is Gandhi's theory of peace that presupposes peace as both a spiritual value and at the core of worldly peace processes. Non-violence as a spiritual practice of peacefulness is therefore an active struggle against injustice and violation of human rights. It denounces any form of violence and enemy-making. Gandhian peace contains compassion and truth as self-transformation and has important social implications because the idea of active struggle sees peace as a continuous integral process in which the ends cannot be separated from their means.

Another example is that of is Lederach's proposal of building peace through strengthening community relationships and reconciliation in post-conflict societies. Likewise, being peaceful is both a spiritual attribute and socio-economic and political endeavour.[25] Such a holistic conception of peace will make sense of the differences between peace in multiple diverse contexts and conditions without denying such variations.

B. Peace including right relationships

Compelling and convergent ideas have emerged in the perspectives of harmony, especially in the light of harmony's embodying symbiosis amongst all things within a totality. It suggests that our being is fundamentally relational, that peace reflects an interdependence of our being. We are all part of the whole and instead of having 'us' and 'them', there will always only be 'we'. In this sense, peace contains the idea of we-ness and that our being is always already being-with. The word 'with' here is the most important and it really captures the essence of our being – a genuine *coesse*, or co-being, a mutual presence, found in love, trust and respect for each other, and for all there is. At the heart of co-being lies the right relationships.[26]

Through the perspectives of harmony, we are aware of the beauty and goodness and truth in these myriad relationships in the world, and these three core qualities of harmony can in turn help us reflect on the qualities of our relatedness and relationships. For instance, the propriety, appropriateness and different kinds of good must be contained in the right relationships.

For our discussion, I distinguish four kinds of co-being or being-with which are all characterised by peacefulness and harmony: (1) being-with one's self; (2) being-with others; (3) being-with the Divine other; (4) being-with the world, including the social and natural world. I will briefly discuss each below:

Being-with one's self. As human beings, we also have intrapersonal relationship. Some call it inner peace, which is beyond a mere feeling of calm, but instead it is a state of being that is the fruit of harmonising all aspects of ourselves, including the ways we identify ourselves (eg. gender, race, sexuality, family, clan, nationality), the narratives of our past, present and future, our emotions, dispositions, wants and desires, and so forth. This includes our relationships with our past traumas and our aggressive reactions towards trauma and victimhood. Peace and peacefulness is when our being-with our self is a harmonious relationship. and this peacefulness can engender joyful tranquility that silences the potentially aggressive nature of all intra-personal conflict. As we shall see later, this inner/spiritual state is when one is appropriately connected to the transcendent, or divine or sacred, reality in a way that constitutes part of one's development and has moral fruits.

Being-with others. Being human is being aware that we are finite, and our ways of being, our practices, values and worldviews are always situated in our histories, memories, collective wounds, religious teachings, cultural traditions and communal journeys. So it is imperative for us to engage with others including other people and other beings in the world, and to be in a relationship with others in ways that are mutually transformative and transcending.[27] As illustrated in the harmony discussion, this means that our growth is not only enriched by those others we encounter, but also co-dependent on the growth of others and the development of humanity as a whole.[28] Thus our being as co-being has others already constituted in it instead of outside of it, and it challenges the predominant Western individualism, a mentality which tends to accentuate the self, giving priority to the individual's self-interest and self-actualisation.

This interpersonal relational vision suggests that the self should never be understood as a singular bounded individual. Instead, each person experiences their self as a relational being, the meaningfulness of whose existence is intimately connected to that of others.[29] This way of being-with is implicit in the notion of harmony, and hence peace is the *fellowship* of men and women which is depicted in the essence of love.

Being-with Divine. Our being consists not only in the fellowship of men and of women, but also in the communion with the Divine, the sacred, or the transcendent. Divine qualities are only realised in their revelation of presence to humans and all that is, such as in the saying 'Heaven is the author of virtues in me'.[30] To be in peace is to seek, to remain and to sustain this communion with the Divine which affirms our being and becoming. Different traditions have given this state of being-with different names, such as the soul, the true self, atman.

Many non-confessional wisdom traditions point out that divine quality is not limited to God's being, but is also found in all animated beings in the world. So, there is divine nature in animals, in trees, and in other beings. Indeed, most indigenous traditions have always maintained that nature itself is divine. Recognising the divine nature of all beings is to affirm the oneness of all life. That is also to say that by communing with the divine, peace, justice and flourishing can spring from this collective core – our spirit or spirituality.

Being-with the world. The life in peace and harmony is furthermore to be in fellowship with the world, including the social, spiritual (inner) and natural world. This view rejects the concept that 'man is abstract, isolated, independent, and unattached to the world' or that 'the world exists as a reality apart from people'.[31] As we have seen, being in peace is the living out of our human qualities in the world with a view to transforming the world and ourselves in it. Thus, peace is not only constituted in human fellowship and solidarity but also provides a context within which each individual proactively pursues co-being together. This way of understanding peace offers an understanding of the nature of work – work is no longer just a means to gain a livelihood, it is also an expression of who we are and enriches the community's life, strengthens fellowship amongst all and serves the goodness in the world.

As defined by the Earth Charter, peace is intimately connected to the notion of right relationships with 'oneself, other persons, other cultures, other life, Earth, and the larger whole of which all are a part'.[32] The two key concepts here are wholeness and right relationships. Peace and harmony, when so defined, is found in the relationship with all that is.

C. PEACE EMBODYING PROACTIVITY AND CARE

When understanding our way of being from such relational perspectives, peace, and harmony, is communing, being in interdependent and mutually constitutive relationships. Peace entails that we are beings *with* other beings, and we are beings *for* other beings, and these relationships are realised through a form of care, respect and deep concern for each other and for the world. So the caring is reflected in our collective inquiry around the question: 'What kind of being do we want to become?' rather than merely 'What kind of being are we?' This is about what it means to be and become human and to live well with others and with the world. Indeed, peace is also the fruit of our collective endeavour, life's gift we offer to each other through our being and availing ourselves to others, through our growth and development, our synergetic relationship with each other in the service of goodness in the world. As Parker J. Palmer wrote, we, as ourselves, are the only gift we can offer to others.[33] This reflective proactivity can also be directed at one's self, and whenever we attend to and listen to our self and give it the care it needs, we are at the same time attending to, listening to and caring for others.

Like harmony, peace also contains and holds tensions, differences and contradictions, and peace does not rule out conflict. In fact, tension, differences and conflict are essential to human life because without encountering differences, there will be little dialogue, innovation or transformation. Peace must consider the potentially enriching effect that conflict can have on creating constructive and desirable change within wider social processes.

Whilst conflict is a normal part of human relationships, when unaddressed it can become a driver for hostility and can catalyse violence. Thus, as part of building peace and developing harmony, it also necessitates our continued encounter, dialogue and negotiation of differences, tension and contradiction, and invites the proactivity of all people to take the initiative to transform conflicts and bring peace and harmony to the community and to the world. Such proactivity and care are examples of integrating human rights with our responsibilities.[34]

So peace embodies the conscious and enduring striving of humanity. Apart from the already mentioned individual and collective (co)-action, dialogue, coordination and co-creation, peace is also reflected in our endeavour to integrate ethical values in the practices of institutions as a key to governance; and likewise, peace is found in the ways that we structure our economy, develop national political systems, international relations and even global governance. Above all, peace includes cultivating our habits and capabilities of listening, dialogue and collaboration which is one of the primary tasks of education. Education can further help us develop our capacity to navigate through complex socio-economic and political landscapes and all forms of relationship, without losing sight of the goodness, beauty and truth.

CONCLUSION

I have attempted to understand peace from the perspectives of harmony through a reading and interpretation of classical Greek and Chinese ideas. This leads to a more holistic understanding of peace as relationally harmonious, ethically just, structurally dignified, aesthetically pleasing, and that peace can hold tensions and conflicts creatively, and integrate the wholeness of all beings in a mutually transformative way. This understanding of peace highlights that peace is not just a state of being, an intrinsic value or a condition for flourishing life, but it is also a process, a structure, and an action towards a greater good. At the core of this understanding is that peacefulness qualifies our mutual flourishing experience in the world, including how we live our lives, pursue activities, engage in relationships and make a contribution to the well-being of others and the goodness in the world. This also means that any socio-economic structures, political systems and institutional practices that count as peaceful must be directed at enabling mutual flourishing.

In particular, I have highlighted two ideas of peace – one is that peace contains right relationships amongst all and the other is that peace embodies human proactivity and the value of care and caring. Right relationships suggest that peace is constituted in the peaceful or harmonious relationships between people and groups, cultures and communities, amongst nations, and between the human, spiritual and natural worlds. Human proactivity evokes that peace compels us and inspires us to live out our humanity and pursue a flourishing life together, through our being, relating and acting in a caring way, as well as through the design and development of humanising and peaceful pillars in societal, economic, political, ecological systems and structures, and value-based governance. Only when we care can we actively live out right relationships and act with integrity in all aspects and dimensions of life.

Globally, for the first time, humanity has come close to living together in ONE peaceful and harmonious community on the planet earth. Such a global community of peace and harmony will be possible so long as we see love as the promise of belonging and bonding beyond the boundaries of cultures, religions and ideologies; community as the commitment to co-being and co-action; human spirit as a mutual experience of transcendence.

NOTES

1. See, for example, *Plato: Republic* I, Books 1-5 (Cambridge, MA: Harvard University Press, 2013).

2. See *Plotinus: Enneads*, Loeb Classical Library Vols. I-VII (Cambridge, MA: Harvard University Press, 1988).

3. See Pythagoras, trans. André Laks and Glenn W. Most, *Early Greek Philosophy* IV (Cambridge, MA: Harvard University Press, 2016); Plato, *Republic* 401d-402a, 411a; H. Northwood, *Harmony and Stability: Number and Proportion in Early Greek Conceptions of Nature* (University of Alberta, UMI. 1997).

4. See Plato, *Timaeus* 32b-33a,

5. See, for example, *Chaldean Oracles*, ed. and trans. Ruth Majercik (1989; repr. Westbury, Wiltshire: Prometheus Trust, 2013); Iamblichus, *On the Mysteries* (*De mysteriis*), ed. and trans. E.C. Clarke, J.M. Dillon and J.P. Hershbell (Atlanta: Society of Biblical Literature, 2003); Kuznetsova, A. (2006). The concept of harmony in ancient philosophy, www.nsu.ru/classics/eng/Anna/dissertation.htm

6. As noted by J. Berthrong, 'Confucian Formulas for Peace: Harmony 和, *Soc*, 2014. 51:645–655 (https://slideheaven.com/confucian-formulas-for-peace-harmony-.html),

7. Confucius, *The Analects* (New York: Penguin, 2014)13:23

8. Xinzhong Yao, 'The Way of Harmony in the Four Books', *Journal of Chinese Philosophy*, 40 (2), 2013, pp. 252-268.

9. Chenyang Li, 'The Confucian Ideal of Harmony', *Philosophy East and West*, vol. 56, no. 4, 2006, pp. 583–603, p. 587.

10. Chenyang Li, 'The Confucian Ideal of Harmony', p. 594.

11. Chenyang Li. 'The Ideal of Harmony in Ancient Chinese and Greek Philosophy', *Dao*, 7:1, 2008, pp 81–98

12. See Li, 2008, 90; Guthrie, W. (1962). *A History of Greek Philosophy*, vol. I. Cambridge: Cambridge University Press.

13. Li, 'The Ideal of Harmony in Ancient Chinese and Greek Philosophy'.

14. See *Chaldean Oracles* Fragment 39; Iamblichus, *De mysteriis* 1.12 (42.5-7). The capital L in Love is to stress that this is not the love in a sentimental and romantic sense.

15. S. Chen, 'Harmony' in S. Lopez (ed.) *The Encyclopaedia of Positive Psychology* (Oxford: Blackwell, 2009).

16. Li, 'The Ideal of Harmony in Ancient Chinese and Greek Philosophy', p. 86

17. Li, 'The Confucian Ideal of Harmony', p. 599.

18. H. Wettstein, 'Ritual', *The Routledge Encyclopaedia of Philosophy* (New York: Taylor and Francis, 1998).

19. Zhongyong (中庸), in Roger T. Ames and David L. Hall, Dao De Jing (New York: Ballantine Books, 2003).

20. See Scherto Gill and Garrett Thomson. *Understanding Peace Holistically* (New York: Peter Lang, 2019), where we used Aristotle's model health as an example to illustrate this. We argue that, according to Aristotle, medicines, athletes, complexions and diets can all be called 'healthy' but in different ways. The qualities a thing needs to be healthy varied in each of these cases; the term is applied to each of these things in virtue of those different qualities. However, this doesn't mean that the word 'healthy' is ambiguous; it has semantic unity. This yields an account that provides unity to diversity by separating derivative and primary uses of the term.

21. Gill and Thomson, *Understanding Peace Holistically*.

22. Gill and Thomson, *Understanding Peace Holistically*.

23. See Thich Nhat Hanh, *Being Peace: Classic teachings from the world's most revered meditation master,* (London: Rider Publishing, 1987),

24. Thomas Hobbes, *Leviathan, or the Matter, Forme, and Power of a Common-Wealth Ecclesiastical and Civil* (Harmonsdworth, Middlesex: Penguin, 1968).

25. J. P. Lederach, *Building Peace: Sustainable Reconciliation in Divided Societies*(Washington, D.C.: United States Institute of Peace Press, 1997).

26. Earth Charter Initiative, 'The Earth Charter', http://earthcharter.org/discover/the-earth-charter/ [Accessed 9 June 2019].

27. Hans-Georg Gadamer, *Truth and Method* (London: Bloomsbury, 1969).

28. Also see Hans-Georg Gadamer, *Philosophical Hermeneutics,* trans. D. Linge (Berkeley: University of California Press, 1976); Paulo Freire, *Pedagogy of the Oppressed* (New York: Herder and Herder, 1970)

29. Kenneth J. Gergen, *The Saturated Self: Dilemmas of identity in contemporary life* (New York: Basic Books, 1991).

30. Confucius, *Analects,* 7.23.

31. Also see Paulo Freire, *Pedagogy of the Oppressed* (London: Blomsbury, 2000)

32. Earth Charter IV:16:f on http://earthcharter.org/discover/the-earth-charter/ [Accessed 9 June 2019].

33. Parker J. Palmer, *Let Your Life Speak: Listening for the Voice of Vocation* (New York: Jossey-Bass, 1993).

34. Also see Berthrong, 'Confucian Formulas for Peace'.

Rethinking Women and Leadership in Myanmar: A prerequisite of a harmonious society

Sneha Roy

In this chapter I will examine notions of harmony and leadership in relation to the role of women. I propose that, firstly, the discourse and outcome of harmony will always be fragmentary without the active inclusion of women and, secondly, that women who commit to leadership roles, especially in a post-war context, can create pathways for attaining harmony. My definition of harmony is taken from David Cadman, and, as he puts it, 'asks questions about relationship, justice, fairness and respect in economic, social and political relationships'.[1] I will address the question of how the gendered functioning of societies hinders women's participation in peace processes and ask why and how women's voices matter by using Myanmar as a case study. Women in Myanmar face discrimination in terms of education, economic liberties, family codes (including the preference for sons), property rights, health, decision-making and political engagement. I shall focus on women in peace-making processes in Myanmar, how they take on leadership roles despite all odds, and of their vision for a better future. There is no systematic solution to the deep-rooted problem of gender inequality in Myanmar and the trajectory of women's leadership will be defined by several socio-political indices and by the variety of roles that women play in this culture. With a feminist orientation, I identify women as potential agents of change, capable of building a society that is more inclusive, sustainable, and harmonious. I also address factors that may pose the greatest challenges, including social systems, structures, and gender inequalities, and practices for the advancement of a harmonious society involving gender equality and the realisation of human rights.

Socio-Political Background: A gendered paradigm

There are indications from Myanmar that the country's semi-civilian government is pursuing democracy with restrictions, after five decades of military rule. Since 2011, the spotlight has been on nationalist movements that blatantly eschew certain sections of society – including the half of the population made up of women. Ironically, in the mainstream global media, the face of Myanmar politics (indeed, the face of Myanmar itself) is often a woman – Aung San Suu Kyi; but this, unfortunately, changes nothing about the gendered functioning of the country and the lack of women's representation in political and social leadership. The sovereign state in Myanmar is a post-colonial construction and the issue of national identity continues to undergo major upheavals from its multi-ethnic and religious make-up, which has played a significant role in nation-building since Myanmar gained independence from Britain in 1948.[2] Fashioning a modern democratic country from multiple indigenous communities has been a daunting task for the ruling elite and has been fraught with tension originating from the relationship between the majority Burmese population and other minority ethnic groups, especially regarding the issue of autonomy for the latter and their grievances against the government.

The creation of a national identity has been made more difficult by unstable governance and an economy in recession that was catalysed by half a century of strife brought about by ethnic rebellion. Attempts to prevent, manage and resolve ethnic conflict by successive governments of Myanmar, through constitutional arrangements, military suppression and peace overtures, have thus far failed to produce satisfactory results. The present military regime's offer of a radical formula for harmony is too recent to yield measurable results. The first truce in 1990 failed to adequately support the harmony process due to factors such as a lack of active participation from both social stakeholders and international donors, and from militant uprisings.[3] Peace processes have been underway since 2011, however, as Stephen Gray emphasises, the models for harmony are continually undermined by the tensions between religious and/or ethnic communities which result from a lack of trust and positive interaction.[4] Gray advances his proposition by explaining that popular social rhetoric against Muslims, for example, states that Islam is inherently an aggressive religion which creates fear and suspicion among non-Muslims. A radical formula for harmony, in my opinion, should begin by challenging such stereotypes and educating people, using counter-narratives that challenge misplaced perceptions of religious and ethnic communities. Typical issues include human rights violations, forced labour, internal displacement, rape, political repression, and cultural misappropriation.[5] One of the most worrying outcomes of Myanmar's socio-political disputes resulted in the Rohingya crisis of 2016.[6]

It is not widely known that the longest civil war in the history of human civilisation is that of the ongoing 'Karen' conflict in Myanmar. The term, Karen, subsumes approximately a score of ethnic groups that belong to the same language family and represent the second largest minority group in the country. The Karens are well-distributed across the length and breadth of the country and the majority espouse Buddhism, although some follow Christianity, Islam or animism.[7] Though the Karens are diverse in their traditions, dialects, religions, and regions of origin, they are united in a pan-Karen identity and a collective experience of oppression under the reign of the Burmese kings, and subsequently, under British rule: since 1948, the Karens have demanded an independent state. Gravers reports that there was a mass Christianisation of the Karens which helped them etch out an identity of their own, not only different from other Burmese, but also from the remaining Karens.[8] The colonial powers further inflamed conflict, laying the foundation for resistance and hostility, resulting in ethnic disputes.

Section 347 of the Constitution of Myanmar guarantees all persons equal rights before the law and equal legal protection, while Section 348 states that the government shall not discriminate against any citizen on the basis of sex. However, in May 2008, an examining committee expressed its concern on issues that are three-fold: how women within the constitutional framework are frequently referred to as 'mothers'; how local and national governing bodies rarely have any agenda, measures or policies aiming for improved participation of women in public life; and. finally, how the ethnic norms of gender biases continue to be dominant for the majority of the people over the constitutional guidelines.[9] Women in Myanmar have not only been the greater victims of the seventy years of conflict, but have also been left bereft of the most fundamental human rights.[10] Being excluded from participation in political decision-making processes and being subjected to gender-based bigotry impacts, of course, on women's ability to have both a voice and agency, which increases women's insecurity and reduces the possibility of instituting a sustainable society. The most ironic part is that the sectors in which women are better represented have opened doors for women, not because they deserve it, but because a woman with an inactive public life is rendered 'a wasted national resource'.[11]

GENDER INEQUALITY AS AN INGRAINED FACET OF SOCIAL UNFOLDING

In Myanmar, the family is the functional unit of society and governs cultural values and rights. However, this value is patriarchal and asserts that 'men are born with *phon* (power, glory, holiness) but women are not'.[12] It is believed that phon created the family, communal, and social order of Myanmar – in other words, men are superior and women, inferior.[13] Thus, the gender hierarchy of men over women in Myanmar has a long and consistent history. Phon validates the cause of gender grading with the result that women are oppressed, marginalised, excluded, and discriminated against in social, political, economic, and religious spheres. This is a founding principle of social order, religious discourse, and political hierarchy in Myanmar.

The United Nations ranks Myanmar 150th out of 187 nations in its graph of gender disparities.[14] It recommends that state and non-state actors, together with civil society, have to work in the fields of education, health, business, leadership, and policy-making to oppose gender stereotypes. As J. Hedstrom states, the inclusion of women in all fields of social life is imperative, not only for women to avail themselves of fair rights and opportunities, but also to check the number of oppressed women who are now joining extremist or militant groups.[15] Insecurity and failed social systems often leave women in war-torn societies with the perilous option of joining rebellious groups. Hedstrom cites examples of women in IDP (internally displaced person) camps having their needs ignored because the men make decisions on behalf of all camp dwellers.[16] However, men may not understand that women may have different personal hygiene requirements, may need more water for their daily chores, or may require certain health aids and better protection. This aggravates women's feelings of insecurity, forcing them to leave the camps and fall prey to landmines and/or being shot and/or raped.

From religious ideologies to religious practices, phon governs the identities of men and women, in which the former is perceived as 'holy' and the latter has to follow strict restrictive codes. Not only are women excluded from positions of authority in matters pertaining to sacredness, their participation is sometimes restricted in rituals in certain holy spaces. A. Nwe, for example, reported a case in which a devoted and responsible minister, who assisted the pastor in the church, did not qualify to become a pastor herself because she was a woman.[17] In many communities, women are not allowed to partake in religious ceremonies, including ordination. In an authoritarian system, women are denied a voice.[18] Their qualities are not considered in politics, nor is their way of approaching democratic vision through harmonious means and non-violence.[19] A non-violent, particularly female, approach inherently calls for peaceful change and criticises the 'misuse of power that cause violation of human rights and repression'.[20]

WOMEN TAKING UP LEADERSHIP ROLES IN MYANMAR: A PREREQUISITE FOR HARMONY?

One important consideration, as G. Hoogensen and B. O. Solheim put it, is if 'the number of women leaders compared to men across the world is so low, it begs the question why'.[21] Patriarchal social structures all around the world have largely precluded the participation of women at advanced levels and social change has not yet been sufficient to allow a significant number of women politicians into prominent leadership roles; therefore, measures, such as reservation and quota systems have been implemented, both to increase the numbers more rapidly and to make women's leadership more normal.

It has been estimated that, globally since the late 1970s, women have been increasingly taking on leadership roles.[22] However, research by P. Minoletti suggests that women account for only 4.42% of MPs in Myanmar's national parliament.[23] This figure is extremely low in comparison with other ASEAN (Association of Southeast Asian Nations) countries and globally. At the subnational level, women's representation is even lower than at the national level, with women accounting for only 2.83% of MPs at state and regional levels, 0% of administrators at township levels, and 0.11% of village heads. Women generally have a secondary role within Myanmar's various armed groups and their associated political parties. The state and regional parliaments that have women ministers are Kachin State (the Minister for Social Affairs and the Minister for Shan (National Race Affairs)); Yangon Region (the Minister for Finance); and, Ayeyarwady Region (the Minister for Social Affairs). Women's participation is also typically narrow in other political parties, as well as in religious organisations. However, women's participation is at its peak within civil society in which women are engaged, not only in large numbers, but also often occupy senior positions with real managerial command over matters. For example, nearly 40% of judges in regional courts and nearly 50% of judges and judicial officers at the district and township levels are women, indicating proactive participation in the judicial spheres.

In non-governmental institutions, women head several organisations with the aim of combating conflict and working towards social welfare. Many village tract/ward administrators (VT/WAS) are women and the United Nations Development Programme's report on Myanmar in 2015 rightly remarks that having both male and female representation at union and rural levels helps a society progress steadily.[24] The VT/WAS work as an interface between the people and the government and have helped women emerge both as great leaders and flag-bearers for a better tomorrow. Initially, there were hindrances to women taking leadership roles, including cultural stereotypes, lack of experience and confidence, time constraints, religious interpretations, and lack of skills. Currently, there are forty-two female VT/WAS who have been elected at regional levels. One of the most significant strategies of their leadership has been to introduce a bottom-up model for policy implementation instead of the usual top-down model. Though negative perceptions surrounding women in leadership continue to exist, these women are breaking down barriers and challenging stereotypes in their own ways.

This chapter would be incomplete without mentioning the popular public figure and Nobel Peace recipient, Aung San Suu Kyi. Hailing from a politically well-known family, her life has been one of upheaval and struggle. Despite experiencing a series of home-arrests and rebellious campaigns against her, as well as being an easy target for the opposition, she consistently advocated peaceful means of protest and a high percentage of Burmese follow her and her ideology. In 2015, her party, the National League for Democracy (NLD), a strong advocate of democracy and equal rights, emerged victorious in the elections.[25] Under her government, hopes were high that women would play a greater role in ending the conflict that has plagued Myanmar for more than half a century but, Suu Kyi, who ardently speaks and writes on gender equality, did little to ensure women's participation in the current parliament. Of eighteen ministers, there is merely one woman in the cabinet: Aung San Suu Kyi, herself.

Myanmar's *de facto* leader, Ms Suu Kyi once had her name listed alongside those of Mahatma Gandhi, Nelson Mandela, Mother Teresa, and Saint Joan, for her unprecedented struggles to attain justice and harmony for the people of her country.[26] Since 2015, even though her name is still prominent, the reasons have changed drastically. The electoral victory of the NLD in November 2015 offered renewed hope that her authority would improve matters; however, the hopes have been shaken partially in Myanmar and greatly abroad.[27] She has faced condemnation across the globe for her inability to defend the freedom of the press as well as for one of the most serious crises in recent times: the treatment of the Rohingyas. Some scholars write that, as an insider, Suu Kyi is limited in her authority to take action.[28] However, many others feel that the lack of visible action taken against what the United Nations and the United States have termed an 'ethnic cleansing' deeply questions her worth as a leader.[29] Amnesty International accused her of commissioning or perpetuating several human rights violations and revoked its highest honour, the Ambassador of Conscience Award, which she was awarded in 2009.[30] It seems that she is unable to resist the prevailing ideology of the military government which Charles Petrie describes as,

[the] age-old vision of domination by a nationalist Buddhist elite, who have difficulty accommodating any form of dissent, and demonstrate very little

regard for the aspirations of the other ethnic groups, with whom they should be negotiating an end to decades of armed conflict.[31]

Women Spearheading Harmony in Post-War Societies

At the beginning of the twenty-first century, we have, perhaps for the first time, realised the full meaning of war and the possibilities of peace. The current world is a mosaic of regions, some involved in protracted wars, and the outcomes are heart-wrenching, to say the least. Indeed, high on the agenda for a better tomorrow is harmonious coexistence which may be influenced and encouraged by women. As Betty Reardon wrote,

> It was women who formulated and propelled this agenda. It is women who continue to envision humane alternatives for world society. It is women whose resistance to war and struggle for social justice and human rights have in fact provided many of our concepts of positive peace, of the conditions of human society that permit all to live authentically human lives.[32]

In ancient Greek literature, the story of Lysistrata provides a classic example of how women have always been the harbingers of harmony, despite being exploited and mistreated in social spheres.[33]

The popular newspaper, *Myanmar Now*, interviewed female peace activists and victims of ethnic conflicts and reported that it is particularly egregious that women suffer disproportionately in wars, but are excluded from peace negotiations. The lack of women's voices in conflict resolution is now a topical issue: there are tales of peace-building in Myanmar that are not only conceived by women, but are also spearheaded by them and, in my view, such initiatives are important points of departure for embracing harmony. There are various networks working towards this goal. One is the Women's Education for Advancement and Empowerment (WEAVE), an initiative which works with children and women, especially the marginalised and those in the IDP camps, to help them with memory-healing counselling and education, and seeks to empower them economically and socially: it is a network of women for women, run by women. Another is the Sunflowers Group Social Enterprise which believes in 'doing good on the ground' and has dedicated platforms, both online and offline, to bring together the artists of Myanmar and train them, provide them with viable resources, and exhibit and sell their products (ranging from textile goods to paintings).[34] Sunflowers empowers

people economically and also conserves culture and heritage. Their efforts have
been successful in bridging divides among several ethnic groups who perceive
art and craftsmanship to be beyond ethnic identity and they provide a common
platform from which people can engage with each other culturally, irrespective
of their background. Phan Tee Eain (PTE) is another organisation that creatively
participates in conflict management by supporting and facilitating women. Their
primary mandate is to empower women by enhancing their knowledge and
capacities; to provide consultancy and social business services, conduct research,
strengthen social networks, and advocate decision making. The UN Women report
of 2015-2016 states that, since the implementation of UN Resolution 1325, there
has been more awareness among people about the need to be inclusive of women
to ensure that measures of conflict resolution, peace, and security are sustainable
long-term.[35] The Indigenous Women's Development Centre (IWDC) was set up by
a Karen woman to cater to the needs of indigenous women who were victims of
war crimes. Over the years, IWDC has trained, educated and supported women to
be 'strong and self-reliant' and to help their families and the community. The Mon
Women's Organization came into existence in 1988 when military attacks were
at a historic high in their region.[36] This group exhorted women to be politically
aware and active and catered specially to those in refugee camps or those who were
victims of war-rape and abuse. Even today, they conduct capacity development
workshops to resolve conflict through peaceful means.

The tales of success are many, but the stories of struggle are unending and
depressing. The women of Myanmar have faced so much at the hands of violence
and bigotry, and yet, instead of sitting down and giving up the divide, they are
determined to face challenges and build bridges.

CONCLUSIONS

The narrative is never black and white. Ethnic minority armies continue to
perpetuate division and vulnerable groups continue to be exploited by the military.
The fight is not yet over between those who have signed the ceasefire and those
who have not. Aung San Suu Kyi has visibly compromised on ethics and human
rights issues. She is, perhaps, going to remain the *de facto* president even though
every item of legislation she tries to pass can be, potentially, vetoed by the military.
Besides the grave political climate, the nation has also been affected by a series of
natural disasters. Hence, there is a long way to go before Myanmar can experience
peace and prosperity. However, it can start by securing the rights and protection

of its civilians and take one step at a time to build a nation in which diversity is respected and celebrated.

The World Bank's 2012 *World Development Report* claims, compellingly, that women's ability to influence society and policy is a decisive aspect of their agency.[37] Following M. Nussbaum and A. Sen's pioneering work on 'capabilities', agency is held to have core bearing for women's individual well-being and quality of life, with a person's ability to make effective choices and exercise control over their own life being a key dimension of well-being.[38] In times during which the world is facing the consequences of grave mistrust, prejudice, and fear, we often commit the mistake of not including people from all communities. I am not claiming that women are better skilled than men at making decisions and implementing policies, but that they must have equal participation. After centuries of human civilisation, we are still stuck at 'why women?' It is time we reflect on 'why not women?'.

Given the unnerving situation of the world today, there is a need to redress the way we define harmony. The essence of harmony is not to encourage homogenous thoughts and experiences, but rather to recognise differences and attempt to create something beautiful out of them. We live in a world in which attributes like freedom are treated not as a right, but as a privilege. Universal human rights and ethical absolutes continue to be at best *ad hoc*. From the Far East to the extreme West, harmony is construed and exhibited in different manners, yet what binds it is the intent of what David Cadman calls 'justice and fairness in economic, social and political relationships' and what Nicholas Campion calls 'ideas of universal balance and the integration of all things [that] occur in many cultures'.[39] Harmony, in my opinion, can serve, not just as a tool, but also as a way to transform ourselves and the lives around us for the better. Understanding harmony may not offer easy answers to the conflicts that have engulfed our global community, but it can certainly provide normative frameworks to help us engage with others and inculcate the values of peaceful coexistence and mutual understanding.

NOTES

1.   David Cadman, 'Living on Earth: Charting a Course for Harmony', (2019), http://www.harmonyinitiative.net/event-harmony-conference-14-march-2018.php [Accessed 10 April 2019].

2.   M. Gravers, *Exploring Ethnic Diversity in Burma* (Copenhagen: NIAS Press, 2007).

3.   C. Petrie and A. South, 'Peace-building in Myanmar', in Mikael Gravers and Flemming Ytzen, eds., *Burma/Myanmar: Where now?* (Copenhagen: NIAS Press, 2014), pp. 223-49.

4. Stephen Gray, 'Unpacking the Complex Causality So-Called Religious Violence in Myanmar', Adapt Peacebuilding, (22 May 2018), https://adaptpeacebuilding.org/blog/unpacking-the-complex-causality-so-called-religious-violence-in-myanmar [Accessed 19 March, 2019].

5. K. Snitwongse and W. S. Thompson, *Ethnic Conflicts in Southeast Asia* (Singapore: Institute of Southeast Asian Studies, 2005).

6. Anon., 'Myanmar military leaders must face genocide charges – UN report', UN News, (27 August 2018), at https://news.un.org/en/story/2018/08/1017802 [Accessed 19 March 2019].

7. A.M. Thawnghmung, *The 'Other' Karen in Myanmar: Ethnic Minorities and the Struggle without Arms* (New York: Lexington Books, 2012).

8. Gravers, *Exploring Ethnic Diversity*.

9. Human Rights Watch, 'Vote to Nowhere: The May 2008 Constitutional Referendum in Burma April 30th 2008', https://www.hrw.org/report/2008/04/30/vote-nowhere/may-2008-constitutional-referendum-burma [Accessed 28 March 2019].

10. Human Rights Watch, 'Burma: Events of 2015', Human Rights Watch World Report 2016, https://www.hrw.org/world-report/2016/country-chapters/burma [Accessed 19 March 2019].

11. D. Kandiyoti, 'Identity and its contents: Women and the nation', *Millennium*, 20(3) (1991): p. 10.

12. A. Nwe, 'Gender hierarchy in Myanmar', *Rays*, no. 10 (2009): p. 135.

13. A. Nwe, 'Gender', p. 131.

14. UNFPA, 'Gender equality', UNFPA report 2017, UNFPA Myanmar at https://myanmar.unfpa.org/en/node/15284 [Accessed 28 March 2019].

15. J. Hedström, *A feminist political economy analysis of insecurity and violence in Kachin state* (Singapore: ISEAS-Yusof Ishak Institute, 2016).

16. Hedström, A feminist political economy.

17. Nwe, 'Gender', p. 131.

18. Nwe, 'Gender', p. 134.

19. S. Richter-Devroe, 'Gender, culture, and conflict Resolution in Palestine', *Journal of Middle East Women's Studies*, no. 4(2), (2008): pp. 30-59.

20. A. San and S. Kyi, *Freedom from Fear* (London, New York: Penguin, 1991), p. 270.

21. G. Hoogensen and B. O. Solheim, *Women in Power: World leaders since 1960* (Westport, CT: Praeger Publishers, 2006), p. 10.

22. S. Campbell, D. Chandler, and M. Shabaratnam, *A Liberal Peace?: The problems and practices of peacebuilding* (London and New York: Zed Books, 2011).

23. P. Minoletti, *Women's Participation in the Subnational Governance of Myanmar* (Myanmar: Asia Foundation, 2014).

24. UNDP, 'UNDP Myanmar Annual Report 2015', ReliefWeb, (27 September 2016), at https://reliefweb.int/report/myanmar/undp-myanmar-annual-report-2015 [Accessed 19 March 2019].

25. Charles Petrie, 'The Verdict is In: Aung San Suu Kyi is an authoritarian', *Washington Post*, (11 January 2019), www.washingtonpost.com/opinions/2019/01/11/verdict-is-aung-san-suu-kyi-is-an-authoritarian/?utm_term=.fd6e317498a3 [Accessed 19 March 2019].

26. T.M. Kwang, *The Paradox of Identities and Politics: Aung San Suu Kyi in Myanmar/Burma and the Globalised World* (Doctoral dissertation submitted to NUS

Repository, 2018).

27. Z. Barany, 'Burma: Suu Kyi's Missteps', *Journal of Democracy*, 29(1) (2018): pp. 5-19.

28. E. Hoffman, & K. Christie, 'Managing the crises at home: the role of Suu Kyi's mediation efforts in Myanmar', in Jonathan Wilkenfeld, Kyle Beardsley, and David Quinn, eds., *Research Handbook on Mediating International Crises* (UK: Edward Elgar, 2019), pp. 252-64.

29. S.B. King, 'The Ethics of Engaged Buddhism in Asia', in Daniel Cozort and James Mark Shields, eds., *The Oxford Handbook of Buddhist Ethics* (Oxford: Oxford University Press, 2018), p. 479.

30. Amnesty International, 'Amnesty International withdraws human rights award from Aung San Suu Kyi' at https://www.amnesty.org/en/latest/news/2018/11/amnesty-withdraws-award-from-aung-san-suu-kyi/ [Accessed 10 April 2019]

31. Petrie, 'The Verdict'.

32. B. Reardon, *Women and Peace: Feminist visions of global security* (New York: SUNY Press, 1993), p. 2.

33. Jay M. Semel, 'Sexual Humor and Harmony in "Lysistrata"', CLA *Journal* Vol. 25, No. 1 (September 1981): pp. 28-36.

34. Anon, 'About Sunflowers' (2019), https://www.sunflowersgroup.org/about-sunflowers [Accessed 19 March 2019].

35. UN Women, 'UN Women report of 2015-2016' at http://www.unwomen.org/-/media/annual%20report/attachments/sections/library/un-women-annual-report-2015-2016-en.pdf?la=en&vs=3016 [Accessed 28 March 2019].

36. Edith T. Mirante, 'Update: empowering indigenous women of Burma', *Cultural Survival Quarterly Magazine*, March 1995, at https://www.culturalsurvival.org/publications/cultural-survival-quarterly/update-empowering-indigenous-women-burma [Accessed 28 March 2019].

37. The World Bank, 'Gender Equality and Development', *World Development Report 2012* PDF, (Washington, DC: The International Bank for Reconstruction and Development/The World Bank, 2011), p. 150, https://siteresources.worldbank.org/INTWDR2012/Resources/7778105-1299699968583/7786210-1315936222006/Complete-Report.pdf [Accessed 28 March 2019].

38. M. Nussbaum and A. Sen, eds., *The Quality of Life* (Oxford: Oxford University Press, 1993).

39. Nicholas Campion, 'Harmony', The Third Harmony Conference, University of Wales Trinity Saint David Lampeter Campus Arts Hall, 14 March 2018, http://www.harmonyinitiative.net/images/Harmony-programme-14-march.pdf [Accessed 22 March 2019].

HARMONY, NOSTALGIA AND A SENSE OF PLACE

Louise Emanuel

THE PHYSICAL, INTELLECTUAL AND EMOTIONAL INTERACTION between mankind and nature has been the focus of research and debate across a wide range of disciplines and has often revolved around whether we can consider ourselves to be set apart from nature, embedded in nature, as Thomas argues, or as Gold suggested, linked to it in some way so that its meaning keeps changing in relation to us.[1] This relationship has been subject to constant evolution as changes in culture, science and economics have led to a re-evaluation of whether we are subjugated to nature, in nature, or over nature.[2] Many authors, such as Baker, point to the emergence of industrial capitalism and its misinterpretation of the Genesis idea of dominion as being the single most important factor in stimulating a re-evaluation of nature, as Smith suggested, and a shift towards a more anthropocentric world view in which humanity started conquering time and space through industrial growth and innovation, and in conquering time and space, believed we were conquering nature.[3] This disconnection between humanity and nature, characteristic of the Anthropocene epoch, has often been blamed for contemporary global environmental and humanitarian crises; in misinterpreting our place in the world we have upset the balance and have taken on nature as a challenge, instead of realising our wholeness within it.

We are now reaching a point when it is imperative to rediscover the connection, to find harmony in being within nature, not apart from it. The question is – are we ready to be re-connected? Our knowledge of the disconnection is not new – early authors often cited include Thomas Malthus in 1798, George Perkins Marsh in 1864, and Rachel Carson in 1962, all of whom warned us of the impact that mankind was having on the earth's resources – but perhaps most of us who had access to this knowledge were not ready to listen.[4] Today, there is undoubtedly a growing concern with our negative impacts on the earth as both through policy and personal action we try to respond to the UN's Sustainable Development Goals and the media has fuelled our understanding of climate change, plastics, population, poverty and humanitarian crises.[5] However, as we come to an understanding of some of the negative impacts we have had on the natural world, are we looking to find a solution to the problems in hand, or do we realise that we need to re-evaluate our relationship with the natural world for ourselves and start thinking of the earth in its wholeness, as the subject, not the object of being? Within this chapter, I would like to start exploring some of what I believe to be the outward manifestations of our disconnection and consider how these can be interpreted as a call for harmony – a return to a more connected way of being.

The first of these is the idea of 'nostalgia'. Described by the Cambridge Dictionary as 'a feeling of pleasure and sometimes slight sadness at the same time as you think about things that happened in the past', it is often considered to be a mixed emotion similar to homesickness.[6] Nostalgia can be personal or collective – almost a need to retrace an individual path or a return to a perceived better time. The growth of a collective nostalgia is evident, as Lowenthal expressed in 1985 – 'if the past is a foreign country, nostalgia has made it the foreign country with the healthiest tourist trade of all'.[7] The past fifty years have seen a significant growth in the demand for heritage; the same fifty years have seen an increasingly globalised world, and the emergence of real global environmental, socio-cultural and economic concerns. In an era when we have become increasingly detached or disconnected from nature, the past can often offer a way of offering us meaning, purpose and value – it can be a means of reasserting our identity as Lowenthal expresses: 'the past is integral to our sense of identity – the sureness of 'I was' is a necessary component of the sureness of "'I am'"'.[8] The past can offer us escapism, an alternative to an unacceptable present, a response to a growing dissatisfaction with today.[9] In being nostalgic are we searching through personal or collective time to find something to connect to that is missing from our contemporary lives? Is it an acknowledgement of a loss of connection or a lack of wholeness?

Nostalgia is often thought of as a negative emotion, perhaps because of its obsession with the 'other' that we no longer we have, a world that is no longer in our grasp. A preoccupation for the past has also often been blamed for a lack of attention to the present –the postmodern plundering of time and space in an eclectic mix frees us from immediate concerns and offers us the world on our doorstep. However, more recently authors have pointed to the idea that 'far from being a feeble escape from the present, nostalgia is a source of strength, enabling the individual to face the future'.[10] Instead of thinking of nostalgia as a malaise which stops us moving forward, can we start to think of it as a call to action – to find what is lost, to reconnect and become whole? An increasing number of studies point to the positive associations of nostalgia in terms of its ability to help people find meaning by increasing social connectedness and self-continuity.[11]

A second, related manifestation of disconnectedness is a sense of place. In recent years concepts such as sense of place and place making have increasingly come to the fore as the realisation that places are socially constructed by the people who live, work and visit there has transitioned from academic to more practical domains such as planning, economic development and health. Places embody the wholeness of humans, habitat, environment, community through their intertwined relationships and as such offer us opportunities for meaningful connections with each other and with our environment. The complexities of place are interwoven into our lives without us even realising – what is often 'out there' is also 'in here' as the boundary between the individual and place is indefinite as that between mankind and nature – place is born out of our relationships with the environment and with each other.

In the study of place we again come across issues of identity and belonging as Malpas writes:

> 'Sense of place' refers us, on the face of it, both to a sense of the character or identity that belongs to certain places or locales, as well as to a sense of our own identity as shaped in relation to those places—to a sense of 'belonging to' those places.[12]

Fundamentally, 'sense of place' is about people, the way in which throughout time they have stamped their mark on the landscape, the way they have interpreted their personal and social history and the way in which they have interacted, and continue to interact with each other, and with their locality, 'a way of seeing, knowing and understanding the world'.[13] A growing body of work has also recognised the relationship between sense of place and community wellbeing: DeMiglio and Williams point to the relationship between sense of place and sense of well-being as being mediated by variables which affect the relationship between people and places.[14] As such Eyles and Williams contend that 'sense of place is an important link in the pathway that translates population health determinants to health outcomes'.[15]

So why this contemporary concern for place? Again, we return to the issue of industrial growth and technological development. A resurgence in concern for identity, belonging and community has emanated in part from increasing globalisation – as places experienced an apparent de-differentiation, technology and communications annihilated space through decreasing the time it takes to transport ideas, information and people around the globe.[16] With this homogenisation of place, comes a loss of identity that in turn leads to a search for one's sense of place, something we have lost in our disconnection from physical places and communities. If we take Creswell's definition, sense of place is the way in which we know and understand the world, so to truly understand the world we need to reconnect with place in its entirety – as the web of interrelationships that embody our physical, social, cultural and emotional being.

In this short chapter I have briefly discussed nostalgia and sense of place, which I suggest are two manifestations of the same phenomena – the human impulse to search for wholeness, the need to connect with another place or another time that offers meaning. Our search through time and space is a call to reconnect, to find the wholeness that we have lost through our separation from the natural world. Rather than understanding the underlying impetus to reconnect, our response has often been to create marketing opportunities to enable individuals and groups to temporarily satisfy their search for meaning through the creation of tourism opportunities and experiences. The focus of research on both sense of place and nostalgia is now shifting to illustrate how these concepts relate to wellbeing, community cohesion and environmental relationships. Instead of trying to find economic value in all opportunities, we need to find real value and realise that sense of place and nostalgia provide opportunities to reconnect – we need to see the opportunities laid before us. In the words of George Perkins Marsh:

> To the natural philosopher, the descriptive poet, the painter and the sculptor, as well as to the common observer, the power most important to cultivate, and, at the same time, hardest to acquire, is that of seeing what is before him. Sight is a faculty; seeing, an art.[17]

NOTES

1.   Lewis Thomas, *The Lives of a Cell* (New York: Bantam, 1975); Mick Gold, 'A history of nature' in Doreen Massey and John Allen (eds), *Geography Matters!* Milton Keynes: Open University, 1984).

2.   For culture see Yi-Fu Tuan, 'Man and nature', *Landscape,* vol 15, part 3 (1966): pp. 30-36 and Yi-Fu Tuan *Topophilia: A Study of Environmental Perception, Attitudes and Values* (Englewood Cliffs, New Jersey: Prentice Hall, 1974); for science see Clarence Glacken, *Traces on the Rhodian Shore* (Berkeley and Los Angeles: University of California Press, 1961); for economics see Neil Smith, *Uneven Development: Nature, Capital and the Production of Space* (Oxford: Blackwell, 1990), and Andrew Sayer , 'Epistemology and concepts of people and nature in geography', *Geoforum*, vol 10, part 1 (1979): pp. 19-44 and for the question of whether we are dominant over nature see Florence Kluckhorn, 'Dominant and variant value orientations' in Clyde Kluckhorn, Henry Murray and David Schneider (eds), *Personality in Nature, Society and Culture* (New York: Knot,1953).

3.   John Baker, 'Biblical attitudes to nature' in Hugh Montefiore (ed.), *Man and Nature* (Glasgow: William Collins and Sons, 1975) and Smith, *Uneven Development*.

4.   Thomas Malthus, *An Essay on the Principle of Population* (London: Johnson, 1798); George Perkins Marsh, *Man and Nature* (New York: Scribner, 1864), and Rachel Carson, *Silent Spring* (Boston: Houghton Mifflin, 1962).

5.   United Nations, *Transforming our world: the 2030 Agenda for Sustainable*

Development, A/RES/70/1 (2015), https://sustainabledevelopment.un.org/post2015/transformingourworld/publication

6. 'Nostalgia', Cambridge English Dictionary, https://dictionary.cambridge.org/dictionary/english/nostalgia (Accessed 1 November 2019).

7. David Lowenthal, *The Past is a Foreign Country* (Cambridge: Cambridge University Press,1985), p. 4.

8. Lowenthal, p. 41.

9. Robert Hewison, *The Heritage Industry* (London: Methuen, 1987).

10. C. Sedikides and T. Wildschut, 'Past forward: Nostalgia as a motivational force', *Trends in Cognitive Sciences*, 20:5 (2016): p. 321.

11. C. Sedikides and T. Wildschut, 'Finding meaning in nostalgia', *Review of General Psychology*, 22:1 (2018): pp. 48-61.

12. Jeff Malpas, 'New Media, Cultural Heritage and the Sense of Place: Mapping the Conceptual Ground' *International Journal of Heritage Studies*, vol. 14, no. 3 (2008): p. 199.

13. Tim Creswell, *Place: A Short Introduction* (Oxford: Blackwell, 2008), p.11.

14. Lily DeMiglio and Allison Williams, 'A Sense of Place, A Sense of Wellbeing' in John Eyles and Allison Williams (eds), *Sense of Place, Health and Quality of Life* (Aldershot: Ashgate Publishing, Ltd, 2008).

15. John Eyles and A. Williams, 'Introduction' in Eyles and Williams (eds), *Sense of Place*, p.1.

16. David Harvey, *The Condition of Postmodernity*, (Oxford: Blackwell,1989).

17. Marsh, *Man and Nature*, p. 15.

Harmony of Place: An Ecological Community in Post-Apartheid South Africa

Eve Annecke

As a Bertha Fellow, Founding Director of the Sustainability Institute for sixteen years and co-founder of the Lynedoch EcoVillage, Stellenbosch, South Africa, this chapter is a personal reflection on a two-decade journey rooted in place, transformative learning and radical change during the early years of a democratic South Africa. It is just over 21 years since my family and I were drawn to a small valley called Lynedoch, nestled at the entrance to the Cape Winelands, Stellenbosch. The Eerste River runs through this valley and, in the distance, the Hottentots Holland Mountain range creates a vast portal to the African continent stretching northwards.

Both these natural features are named by our Dutch colonisers. The names provide reminders of complex and painful histories that include British settlement, land appropriation, killing of the Khoi Khoi and San first peoples, and slavery. Colonisation has included change of languages, cultures and ways of being. The notorious system of *apartheid,* the iniquitous institutionalisation of racial divide privileging those born with white skin, had fertile ground on which to build its ultimate in human modernist divides.

Lynedoch, so-called by Scottish missionaries who arrived in the early 1900s to tame the wildness of the supposedly 'primitive', is now a messy mixture of high quality wine farms, massive wine-storing buildings, storage units, a few dilapidated structures, and five-star tourist destinations. It is a place where the word 'harmony' may not easily sit, at least not at first sight. David Cadman writes,

> Harmony is an expression of wholeness, a way of looking at ourselves and the world of which we are part. It's about connections and relationships. The emotional, intellectual and physical are all connected. We are connected to our environments, both built and natural; and all the parts of our communities and their environments are connected, too. Harmony asks questions about relationship, justice, fairness and respect in economic, social and political relationships. As an integrative discipline it can be expressed in ideas and practice.[1]

Lynedoch Village. Photo by Sustainability Institute

Woodlands sunrise. Photo by Sustainability Institute

The connections and relationships to which David Cadman refers have a tendency to sound neat, apparently with little room for the discord so firmly entrenched in the structural inequities rife in this area. It is true too though that some landowners in this valley are involved in radical worker-equity schemes, biodynamic farming, playing classical music to grape vines and competing for a selective tourist market. From histories of deep exploitation of land and farmworkers alike, this is change indeed; a place with stories of deep divide and disconnect, along with the dynamic tensions of creating new beginnings.

I had imagined in 1997 what seemed a fairly straightforward project of arriving to set up an African leadership institute on the Spier Wine Farm in the Lynedoch Valley. While the Spier Leadership Institute was formalised as a non-profit trust in 1999, the question that arose rapidly was 'leadership for what?'. Here began the real complexities of context. Questions of social justice, ecological integrity, poverty and landless peoples rapidly raised their heads and our focus for the institute fairly obviously became 'leading for sustainable futures'. In 2002, we chose to align the Institute fully as a 'living and learning centre for studies and experience in ecology, community and spirit' and changed the name to the Sustainability Institute.[2] The focus on building leadership capabilities became implicit in all ensuing activities from that day to this. During the infancy of South Africa's new democracy, there seemed little more important and, in the glow of a Mandela-led government, I felt progressive, radical, and aching to contribute to the building of a different nation.

With the strange agency that place seems to have, it became abundantly and immediately clear that before any 'thought-leadership institute' could emerge, the most pressing calling in this particular context was, in fact, a school to serve the needs of extremely poor farmworker children. In addition to being an area where foetal alcohol syndrome is at one of the highest levels in the world due to the past 'tot' system of paying farmworkers a portion of their wages in alcohol, the previous government had never made education for black children compulsory. Not only had many children dropped out of school, but those who did attend were subjected to appalling conditions, corporal punishment and little in the way of learning. The existing ramshackle prefabricated school was on land owned by a white farmer and had been for 30 years. He was ready to bulldoze the building to reclaim this land.

In a way, this place in a post-apartheid South Africa seemed to resemble a war zone. Not the global apocalyptic images with which we are familiar from CNN, but rather an insidious erosion of childhood. Poverty, families for generations without land tenure and exploited as farmworkers, alcohol abuse rife, and sexual exploitation and violence as common occurrences. The next generation was so clearly bearing the costs.

We rapidly discovered through social and community conversations that a new, ecologically-designed school for farmworkers' children would not solve all the other issues that simultaneously arose through our focus on children. Poverty, malnourishment, lack of housing, poor community structures, and little in the way of state support were all connected to their everyday lives and, on reflection, I only now see the harmony arising from paying attention to the rhythms and shifting patterns in the daily detail of building the fabric of social and ecological justice. Suddenly, the dream of a leadership institute was firmly embedded in our local context, dealing with local issues; as grounded leadership needs to be.

In the year 2000, we opened a Montessori nursery for 75 enthusiastic pre-schoolers and, in 2002, we completed the ecological renovation of one of the dilapidated buildings, to be used as a primary school for 500 farmworkers' children. Building homes together with people who were being evicted from farms was the next challenge. The small Lynedoch EcoVillage, home to 32 families, was born as the first mixed-income ecological development in the country. Threaded into the countless tiny decisions affecting what was to become everyday life, were questions of appropriate design, building materials, renewable energy, re-use of water in a water-scarce country, sanitation, road surfaces, and pedestrianisation where possible. In integrating social and ecological justice, it became evident that very simple choices such as solar water geysers and careful insulation would make both qualitative differences to daily living as well as financial sense. Poorer people were able to afford their homes because of the home loans that we could make available when banks would not loan to the poor, a tiered levy system and ecological building technologies which simply save money each month.

'How would nature do it?' was the question we attempted to answer in creating the infrastructure for fairer living. Now, the Lynedoch EcoVillage incorporates 100% water recycling through a horizontal wetland; solar water geysers; adobe, sandbag or recycled brick homes; large indigenous woodlands in green spaces; and food gardens. All of these may not have cost less in design nor building, but they do cost less in maintenance, upkeep and monthly payments.

Social dynamics have always been exactly that: dynamic. They remain so. Historical inequity, blind privilege and entrenched racism have demanded commitment, humour and constant conversation. There is no handbook for living in an emerging community, where the tensions between people who live, work and bring up children together play themselves out in sometimes rather unpredictable ways. Participating together in daily lives, traversing a multiplicity of roles, requires levels of skill that we do not always have. Trust built over time, organically and against significant odds, seems to be forming ways of engaging that are sometimes softer and more gentle; not always, but sometimes. We cannot ignore the fact that the Lynedoch EcoVillage also reflects the wider system of which we are part in a country with serious challenges in unemployment, equity, violent crime and state capture. What does seem to constantly connect us in the entire valley is putting children, safety and nature first.

My experience with Fritjof Capra and Satish Kumar at Schumacher College, in 1993, and the Gorée Institute, Senegal, was profound in terms of transformative and transgressive learning. It added a beautiful mix of science and spirituality to my background in the Montessori approach and I became convinced then that the work that may have the greatest opportunity to de-colonise learning in a post-apartheid South Africa would be the space for learning that touches the yearning in peoples' hearts and souls. A yearning that is for connectedness, through studies and experience in ecology, community and spirit. It was our version of Satish Kumar's beautiful trinity of 'soil, soul and society'.[3]

Seven years later, I co-created our transformative approach to learning at the Sustainability Institute.[4] I have taught in our master's level programme accredited by Stellenbosch University and held at the Sustainability Institute since its inception in 2003.[5] One of beauties is the intricate connectedness between what we learn from our practice and the classroom: organic farming, land reform, alternative education for children, appropriate building technologies, renewable energy, water recycling, and on-site waste disposal. Connecting the children and students to ways of living that create direct experience for an intricate weaving between head, heart and hands or theory, practice and critical reflection rooted in nature, has become a golden thread of difference in multiple realities that ground the institute as unique.

This means beginning each day with a morning circle outside under giant Ficus trees with the mountains in the distance. A sense of wholeness emerges through gentle body work, meeting as a learning community, dividing into work groups to assist with gardening, waste collection, cleaning the Institute, cooking for the Lynedoch Children's House, and generally caring for and nurturing ourselves, each other and the place that holds this small, vibrant band of explorers. The children, from the time they can toddle up until adolescence, are frequent visitors to the woodlands and food gardens; simply as a beautiful place to play, or as avid diggers, weeders, planters and harvesters. The Montessori Lynedoch Children's House holds nature as the first teacher and all learning is rooted in reverence for the natural world.

Meditation walk youth. Photo by Sustainability Institute

Morning stretches. Photo by Sustainability Institute

Weaver nest. Photo by Sustainability Institute

Food garden. Photo by Sustainability Institute

Growing garden. Photo by Sustainability Institute

Learning in the garden. Photo by Sustainability Institute

Children in food garden. Photo by Sustainability Institute

Art in nature made by students. Photo by Sustainability Institute

Students making art. Photo by Sustainability Institute

Getting ready to plant. Photo by Sustainability Institute

Children dancing. Photo by Sustainability Institute

Starting with a focus on transformative learning for children has thus led, in the small seven-hectare property, to the development of a pre- and primary school serving 600 children and adolescents; to Stellenbosch University undergraduate and master's level programmes in Sustainable Development, held at the Sustainability Institute for 75 students annually; and, to 32 newly-built homes. Lynedoch EcoVillage is thus a live case study for students in their daily experience, participative learning and research.

The local issues such as poverty, food insecurity, violent crime, land reform, racial divides and massive inequity that are dictated by this place, time and historical context involve messiness and the gritty building of community. It is intriguing now to note that the experience and practice of our early work has become bound into the modules in the undergraduate and postgraduate programmes. The taught postgraduate modules include, Sustainable Development; Complexity Theory; Leading Transitions and Environmental Ethics; Sustainable Cities; Ecological Design, Food Security and Globalised Agriculture; Food System Transitions; and, Biodiversity and Ecosystem Services. In co-creating the structure for this set of learning pathways, it was our own hands-on experience that led to the conceptual development of the ideas to be researched and courses to be taught. This year is the seventeenth year of the postgraduate diploma and, so far, we have approximately 600 graduates.

In founding what became the Sustainability Institute in the Lynedoch EcoVillage, we have become activist-academics in ways that have created connections between seemingly entrenched fragments and divides'. Over the two

Graduation celebration in labyrinth. Photo by Sustainability Institute

decades, a harmony of place seems to be emerging. This is not a sanitised claim to some form of development nirvana. It is, instead, a recognition of the poetics of place as teacher and of seemingly infinite connections and rhythm within transformation, individually and in learning communities. While the questions David Cadman asks in his definition of harmony hold a sense of depth and elegance, these are perhaps added to by the grittiness and pain within the practice of transformative community-building. Focusing on learning connected to living issues has made the possibility of an institute for the studies and experience of ecology, community and spirit something that is sensed, felt, seen, touched, heard and smelt; and, not always easily described.

Acknowledgment: all photographs are credited to the Sustainability Institute

NOTES

1. David Cadman, 'What is "Harmony"', (23 May 2017), https://www.uwtsd. ac.uk/harmony-institute/ [Accessed 9 May 2019].
2. Sustainability Institute, https://www.sustainabilityinstitute.net (Accessed 27 September 2019).
3. Satish Kumar, *Soil, Soul, Society: A New Trinity for Our Time* (Lewes: Leaping Hare Press, 2013).
4. 'The Sustainability Institute' at https://www.sustainabilityinstitute.net/ (n.d.) [Accessed 9 May 2019].
5 . 'Master's Program in Sustainable Development' at https://www. sustainabilityinstitute.net/learn/us-pgd-mphil (n.d.) [Accessed May 2019].

Learning by Heart:
Harmony and my Journey as a Pythagorean

Kayleen Asbo

I OFFER THE FOLLOWING CHAPTER ABOUT HARMONY AND PYTHAGORAS from a deeply personal perspective of how those teachings and philosophies have been both a mirror of my life and a source of continual inspiration as a professional musician and independent scholar.

I begin with a dream I had when I was three years old:

There was a shiny black grand piano on a stage, nestled in the alcove of a beautiful and gracious building with beveled glass windows that looked out upon a grove of rustling birch trees. I began to run with excitement to the piano, but I fell down, wounded. Each step was excruciating, but I got up and continued on. Weaker with each movement, I fell down again and again. Exhausted, I finally crawled the last few feet to the piano until collapsing, I touched the keys. I had to. It was very clear in the dream: playing the piano would save my life.

The dream spoke truth to me: music did indeed save my life, over and over again. It gave purpose, inspiration and meaning to a painful childhood and adolescence. In times of grief, music has often been the only consolation I have found. It became the foundation of a multifaceted career that has encompassed composing, historical research, performing as a pianist, teaching pedagogy, and lecturing for symphonies orchestras, opera companies and music festivals. Most importantly, music has been my unfailing guide to discovering the Good, the True and the Beautiful throughout the centuries and for awakening my own spiritual and humanitarian impulses.

As a twelve-year old pianist, I made my debut performing a Mozart Concerto with a local community orchestra. I tried everything I could to have more practice time – including feigning illness. I dreamt of finding a school in nature where musicians would live able to devote themselves completely to their studies without the gross materialism, competition, shallowness and ugliness that defined my adolescent world. I resented every moment spent in mathematics class, where I resisted learning ancient geometric formulas that seemed inert and lifeless.

When I complained that as a concert pianist, I would never use the Pythagorean formula, my hapless teacher regrettably agreed. Indeed, he added, especially since I was female.

By the age of sixteen, I had fallen under the spell of philosophy, mythology and comparative religion. My maths homework took me away not just from practicing Mozart and Bach, but from reading Plato, Dante and Joseph Campbell, and from chanting with nuns in a Vedanta Temple. I could hardly wait to be finished with my required mathematics courses, which consisted of memorizing a seemingly endless string of formulas taken from worn out, ugly and colorless books.

How ironic it is, then, that the man who stands behind the formula of $A^2 + B^2 = C^2$ was not just a seminal figure in the history of numbers, but also the father of musical theory, the founder of Western philosophy, a student of Eastern religions and the hidden root of Western monasticism, who founded an egalitarian utopian community dedicated to the pursuit of wisdom and beauty.[1]

It is Pythagoras who lies beneath the surface of so many things that enchanted me during those dark and troubled days of adolescence: Renaissance art, Gothic architecture, Greek philosophy, comparative religion. It is Pythagorean ideas that were reawakened in Leonardo da Vinci's notebooks; it is his theories of the music of the spheres which rippled forth to influence Hildegard of Bingen, Johannes Kepler and Dante. It is upon Pythagoras's giant shoulders that Plato and Aristotle stand, and it is his likeness that is inscribed in stone above the scholar's doorway of Chartres, the most glorious Gothic Cathedral in Europe, another Utopian school dedicated to the Good, the True and the Beautiful.

If only one of my mathematics teachers had known that Pythagoras and his followers believed fervently in a worldview that does not divide life into art / math or religion/science. Instead, the Pythagorean perspective held that there is a divine pattern that permeates every dimension of our universe and that Harmony is the result of a dynamic, ever fluid weaving of the seeming opposites. The Pythagoreans believed that if we search, we can find expressions of this pattern, this *logos,* in every facet of the world around us. I believe that we can discover it in the unfolding of the smallest chambered nautilus as well as in majesty of the spiral galaxy above us. If we have the eyes to see and ears to hear, we will find manifestations of the divine pattern in sound, number, form, and astronomy. If I had known the story and philosophy of Pythagoras, I would have been eager to embrace and explore the mathematical theories of geometry class.

Pythagoras was born circa 570 BCE in Samos, off the coast of Turkey. His passionate search for knowledge led him to travel to all the cities of high learning

Figure 1. Pythagoras and his monochord, Chartres Cathedral.
Photo by Jean-Louis Lascoux, public domain (retrieved from Wikimedia commons).

in the ancient world. It is said that he imbibed the teachings of the Persian Magi, mastered the interpretation of dreams with the Hebrews, immersed himself into the hieroglyphic symbolism of the priests of Egypt and the absorbed the sagacity of the Chaldeans. When called the most perceptive of all persons, it is said he demurred, modestly calling himself only a lover of wisdom, literally a 'philosopher'.

At the end of all his far-flung travels, Pythagoras returned to the Mediterranean lands of his birth where he established a school in Southern Italy that was devoted to both scientific and ethical study: both aspects were thought to be necessary in the pursuit of wisdom. For the first few centuries after his death, Pythagoras's greatest fame actually rested not upon his mathematical formulas but as the founder of an ascetic way of life that emphasized dietary restrictions, religious ritual and rigorous self-discipline. He revered the principle of unity in all things and taught that the soul was eternal. The *telos*, or object, of life was to purify oneself to become like god: to become a source of harmony in service of the cosmic order. This would be done in a succession of lifetimes. Through spiritual practice and study, Pythagoreans believed that one could both purify oneself, and also come to remember previous incarnations.

Pythagoras believed that we were intimately connected with one another in a matrix of creation. Nothing stood outside of this, and so all aspects of the earth should be treated with reverence. Animals deserved empathy; plants were seen as having their own form of intelligence. Conversation with and reverence of all creation, rather than subjugation and dominion, was the guiding principle. A life of wisdom could not dismember the pursuit of the mind from the care of the body or the cultivation of the spiritual senses. In order for wisdom to flourish in the individual, all aspects must be nourished and integrated in concert with one another and within the bonds of friendship. In order for wisdom to flower in community, the ethical, intellectual, political and spiritual must be all aligned and connected.

The Pythagoreans saw the human being as akin to a musical instrument. Like a great violin in an orchestra, we can become discordant by being out of sync with the vibrations around us. What the Pythagorean community sought was a way of life that would allow them to come into tune with the harmony of all creation.

Remarkable for its egalitarian spirit, members would enter and donate all their possessions to be held in common. Both men and women were free to join and to leave – and if they did depart, they would be given double what they had contributed to the community. A rigorous course of life was prescribed, built on a foundation of intellectual inquiry, music, physical exercise and above all, silence. In the mornings, the community would gather outside in silence to sing hymns to Apollo as the sun rose. After a day's labor of individual work punctuated by walks in nature, they would gather together again in the evening to listen to the lectures of the master before dancing and singing hymns once more, ending in silence. It is said that, before drifting off to sleep at night, each member would conduct a moral inventory of three questions: Where did I go wrong today? What did I accomplish? What obligation did I not perform?

Students in the first stages of instruction were known as *auditores*, or listeners, who were not allowed to ask questions but were to hear and memorize certain sayings, many which have the nature of zen koans. After successful completion of five years of silent study and initiation, they became *esoterics,* initiates into the inner circle of adepts. Aristotle reports that there were two rival schools of Pythagoreans: the *akousmatikoi* focused on ritual life, while the *mathematekoi* concentrated on the study of mathematics. Both circles pursued a lifestyle of simplicity where vegetarian diet, philosophical study and the disciplining of the passions were essential in the pursuit of an ethical life. The community flourished in Italy, where twelve hundred years later Benedictine monasticism took root. The parallels between the two lifestyles are striking, with their twin ascetic paths of study, silence and song. Both were in essence monotheists, though the Pythagoreans called God by the name of Apollo.

The Pythagorean curriculum comprised what later came to be known as the *Quadrivium*: arithmetic, geometry, music and astronomy. In this fourfold path, what was sought was the underlying principle of harmony that governed the cosmos. In each subject, the Pythagoreans sought to find the Divine Pattern. How was the sacred expressed in number? In geometric form? In sound? In astronomy? The Pythagoreans were searching for nothing less than God's DNA – a sacred fingerprint that left its mark in all creation.

A central tenet of Pythagorean life was the importance of balancing the opposites and the avoidance of excess. Vitality, beauty and well-being were to be discovered by harmonizing the ten opposites articulated by the Pythagoreans:[2]

finite, infinite
odd, even
one, many
right, left
rest, motion
straight, crooked

light, darkness
good, evil
square, oblong
male, female

The Pythagorean rhythm of life seemed to also balance the multi-faceted needs of the human being. The community was composed of intellectuals who set aside time for exercise and dancing and abstract mathematicians for whom music and walks in nature were central spiritual practices. It was a life of shared community in which silence was an integral part. All of these point to a delicate and conscious balancing of seeming opposites.

Across the centuries, even before he was associated with the mathematical theorem that bears his name, Pythagoras was connected to music theory and sound healing. Like Orpheus, Pythagoras was identified with the lyre and with his capacity to use melody as a vehicle for transformation. The ancient Greek world did not see music as mere entertainment. Rather, it was a force that shaped the world and crafted character. It was a moral energy that bound together society. The correct application of music could heal – or harm. Most of Western music, whether classical or popular, confines itself to one of two modalities: either a major or a minor key. Most pieces that are meant to express happiness, contentment or joy are written in a major key; those that express anger, melancholy or despair are in a minor key. The Greeks were far more flexible and nuanced. There were seven modes, each the province of a different god, each associated with a different psychological state. Both Plato and Aristotle, influenced by Pythagoras, agreed that the modes of music used in educating the young should be carefully chosen.[3] The application of the wrong mode could deform the character of an individual and even jeopardize the state.

Live music was not only essential in daily life, where it was to be found at the gymnasium as athletes practiced and competed. It was integral to well-being. In the ancient *Asclepeion*, the centers of healing scattered around the ancient world, music played a foundational part of any cure.[4] A patient who was suffering in mind, body or spirit upon arriving would undergo ritual purification. They would participate in the catharsis of a dramatic production and then incubate a dream in the temple. The following day, the physician would attend them. Part of the prescribed cure would be medicine – herbs and tinctures. But an equally important part of it would be the application of music therapy: a homeopathic immersion into the mode that would draw out the ailment. For mania, immersion into one mode would be called for; for depression and listlessness, a sound bath in another would be in order. Legends of Pythagoras emphasize his astonishing capacity as a sound healer. Through his lyre, he was able to create radical psychological shifts. In one famous story, he encountered a young man enraged and intent on murdering his unfaithful lover. Pythagoras, it is said, began to play in a mode that matched the cuckold's rage, but modulated to a more serene and accepting tune. Having received this infusion of musical homeopathy, the would- be assailant put down his tools of destruction and placidly left the scene.

According to another folk tale, Pythagoras was walking past a blacksmith's shop when he heard a ringing of beautiful sound. Perpetually tuned to finding the Divine Pattern in the cosmos, he investigated whereupon he discovered that the smith's hammers were precisely constructed in perfect mathematical proportions of 2:1. Pythagoras replicated this ratio with bells, flutes filled water glasses and on a monochord's strings and discovered that when mathematically pure ratios were present, the sound was harmonious indeed. If I had only learned that this legend was behind the final movement of Handel's the enchanting *Suite in E Major* ('Harmonious Blacksmith') , my two worlds (school and music) would have been harmonized themselves, becoming more integrated and meaningful. And if only I had learned in Geometry class that Pythagoras believed that music could heal the world, cure illness, quell rage and violence and lead to spiritual awakening, I might have developed an interest and even a fascination in searching for the *logos* inside mathematics. Instead, it took a Medieval Cathedral to open my heart to the power and beauty of Pythagoras.

Walking through the Scholar's Door of Chartres Cathedral for the first time at the age of 35, I was unaware of the encyclopedia of history and philosophy inscribed in stone and illuminated in glass. I was bathed in blue light cast from the windows above, I only knew that there was something profoundly beautiful and powerful that was creating a powerful effect in my body. The top of my head was vibrating, my pulse was racing and my breath began to change. Without knowing it, *I myself was being tuned.*

Over the course of the next few days, my mind happily absorbed a flood of information. I was able to give names to the figures depicted in the windows and statues and to analyze features of Gothic architecture. I had my first fascinating introduction to Sacred Geometry and I began to learn about the Cathedral School established by Bishop Fulbert in the 11th century.[5] All of this was interesting to a mind that delighted in history.

But something profoundly deeper happened at the private labyrinth walk. As the crown of our week together, the Veriditas Labyrinth Organization took us after hours into the depths of the darkened crypt. We were instructed to sit in

Figure 2. Chartres Cathedral.
Photo by Jean-Louis Lascoux, public domain (retrieved from Wikimedia commons).

Figure 3. The interior of Chartres Cathedral.
Photo by Emma Asbo

Figure 4. The crypt of Chartres Cathedral.
Photo by Emma Asbo

silence for what seemed like an eternity. As the cathedral bells chimed, a poem was read, calling us to enter into ritual space. One by one, each pilgrim rose and followed a pathway of 3,000 candles that led past Fulbert's altar with its the statue of Notre Dame Sous Terre, past the ancient Druid well, and down the long cool corridor. From far away, as if underwater, the sound of voices singing Gregorian chant beckoned. Eventually, I came to two human sentinels holding lit torches by the baptismal font. Tremulously, I ascended up the staircase into the nave.

The cathedral was bathed in blue light as the last of the sun's rays streamed through the Medieval glass. Ahead lay the ancient labyrinth, surrounded by white votives. I took my place in line, awaiting my turn to step onto the worn path, placed here at the dawn of the 13th century by anonymous masons. As my bare feet kissed the cold contours of the stone, a woman's voice pierced my heart: her song buried its way into the knots of my being, causing shivers to run up my spine and tears to run down my cheeks. Music, architecture and ritual wove together to create an experience unlike anything I had ever known. I sensed in the depths of my being that something had shifted and I would never be the same.

That night at Chartres marked a threshold of transformation. Though I couldn't see it clearly while it was happening, the seeds of everything I have done for the fifteen years since were sown that evening. As a scholar, this experience awakened my interest in mythology and Medieval history. I read hundreds of books (and even collected two more degrees) because of the thirst for knowledge this ritual imparted. I learned in ever more fascinating detail how the labyrinth had been placed as a pathway of peace and reconciliation during a time of war, an alternative pilgrimage of inner transformation during the very years when the Albigensian Crusade and Inquisition was being launched by the Church of Rome. This experience conveyed to me in a way that my classical musical training had never addressed that music as *ritual* – and not passive entertainment- has a power and potency to alter destiny and shift consciousness. It also encodes information like clues, that when followed and decoded, contain *memory* and the *seeds of awakening and healing.*

When I returned to the United States, my old life was suddenly far too small. I had been a well-respected piano teacher with an elite studio and a professor of pedagogy at the San Francisco Conservatory of Music, but my appetite for the world of competition suddenly had vanished. My pleasure in either listening or performing Classical music in traditional ways waned. Now, I wanted to create transformative experiences that would be illuminating and healing, the way the labyrinth walk had been for me. I was drawn to creating rituals and multidimensional experiences where music, story, poetry and embodied experience came together and where there was no longer a division between 'performer' (entering wordlessly onto a stage) and 'audience' (sitting passively in chairs). Over time, a new pattern of life emerged: three months of study and research and pilgrimage around the world; nine months living in California where the focus of my work would be to take the strands of what I had gathered and weave them into classes and events that would offer more than information: they would be communal events of inspiration, hope and integration.

The music I heard at the labyrinth walk sung by the Braslavksy Ensemble immediately had brought a lump to my throat and tears to my cheeks. In Benedictine monasticism, such tears in response to beauty are considered a *chrism*, a divine gift and a path to be followed. Over the next ten years, I searched out the stories behind the songs and found in each one of them a history that, in the true spirit of Pythagoras, reconciled opposites. *Stella Splendens* led me to Montserrat in Catalonia where I beheld the Black Madonna. Here I learned about the centuries of pilgrims who had come to this remarkable mountain for hope and healing singing the *Libre Vermell*, an astonishing manuscript of the 14th century that weaves together the seeming opposites of Gypsy songs and Gregorian chant. *Santa Maria Stella Do Dia* led me to the Camino Santiago Compostella where this song was composed by a team of Jewish, Christian and Islamic musicians as a hymn in praise by the enlightened courts of King Alfonso El Sabio. The tears that bathed my cheeks with *Benedicti e Laudate,* ran all the way to Italy, where I scoured Florence, Cortona and Assisi to discover the hidden story of the Franciscan Confraternities and the Roman Noblewoman, Lady Jacqueline de Settesoli Frangipane whom Francis nicknamed 'Brother Jacoba'. Because of *Caritas Abundant*, I traversed to the Rhineland to walk in the footsteps of the mystical Benedictine Abbess, Hildegard von Bingen. The *Salve Regina* brought me to monasteries throughout France, where I learned about Troubadours who became Cistercians, and discovered how this song sung each night at the end of Compline represents a musical reconciliation between two enemies: Peter Abelard and Bernard of Clairvaux.

All of this knowledge was rich and wonderful in and of itself, but my own calling was to carry these stories and songs back and create ritual experiences for others so that they could experience an echo of these sacred sites themselves. Through concerts, classes and workshops, the songs and stories took root, not just as something performed on a stage, but as something shared in together in community.

My labyrinth walk at Chartres marked my conversion to being a teacher and musician in the Pythagorean tradition. Like the Pythagoreans, I devote my days to a rhythm that begins with music, meditation and yoga. My long days of teaching and researching are punctuated by walks in nature. I am growing into the other practices so important to the Pythagoreans: care for the earth through a sustainable lifestyle, periods of sustained silence, and trying to hold the tension of the opposites. Gradually I am growing to see the world with eyes that try to embrace all creation as 'friend', and the works of all I do with the company I created, Mythica, are dedicated to the healing and service of the world.

In times of crisis – after the deadly fires that devastated Sonoma County, after a wave of suicides, after the shootings at the Synagogue in Pittsburg and the massacre at Christ Church, I've gathered the community together at labyrinths throughout the San Francisco Bay Area. The combination of ritual and song has worked time and time again to transform grief into hope and despair into creative action. I returned to Chartres in June 2019 for the tenth time. As a keynote speaker for a Veriditas pilgrimage, I wove stories of my favorite poets, the history of the Middle Ages, circle dancing and music together in a rhythm of life that is deeply consonant with Pythagorean philosophy. In a program called 'Beyond the Boundaries of Loss and Grief: Love and Friendship as A Transformative Path', I explored the Greek myths of Orpheus and Eurydice, myths that are intimately connected with Ancient Greek rituals and that may have even originated with the Pythagoreans. Amongst the songs that we sang each day was one written down by followers of St Francis in the town of Cortona, a Tuscan village that boasts a landmark known as 'Tonella di Pythagoras' (The Tomb of Pythagoras). In our tours of the Cathedral, we entered the scholars' door under which Pythagoras presides in order to share the story of how the Chartres School thrived as a beacon of learning and shined for centuries as a Christian temple of wisdom because it taught the works of both Plato and Aristotle – books that had been preserved by Islamic scholars. And together, as the heart of our week, we walked the labyrinth in a private ceremony of blessing, honouring a tradition that began as a movement for peace during a time of war, in the very same ritual that marked the beginning of my journey as a Pythagorean.

As an aspiring concert pianist in middle school, how I would have been astonished if you had told me that the re-enchantment of my life, its greatest flowering and fulfillment and its most long lasting inspiration would occur in a Gothic Cathedral based on the very mathematical formulas that seemed so dry and insignificant on page. How I would have marveled to learn that the man I only knew as a mathematician was not only the hidden source of beauty and wonder across the centuries, but a beacon of light that might guide us to a more hopeful future from our tumultuous and troubled age.

Firmly rooted in the stance that there is an organizing pattern that is both intelligible and beautiful, the Pythagorean approach to life is suffused with reverence for all creation. It tells us that we do not need to choose between science or spirituality, art or geometry, ethics or politics. Each one is an essential and radiant expression of the One. Mathematics is not a merely rational dissection and explanation of dry formulas, but a cosmological perspective that affords an every expanding landscape of wonder.

A Pythagorean perspective sees nature as an expression of the divine and shrinks from any exploitation of animals or land. A Pythagorean approach suggests that the greatest wisdom flourishes when we can honor the seeming contradictions within our human existence. From silence and humility, within community and friendship with all beings, we can find a meaning far deeper than our individual pursuits.

Throughout the centuries, the ideals of the Pythagoreans have bubbled up as if from some neglected and forgotten fount, and when they did so, the world was re-enchanted. In the first few centuries in Alexandria, Egypt and in the School of Chartres during the 12th century, a holistic approach led to astonishing advancements in both science and architecture. The Pythagoreans provide a pattern for renewal and inspiration to reclaim a life saturated with beauty where science and the sacred are both in the service of humanity. Finally, what the legends of Pythagoras teach us is an echo of my dream as a three-year old: music can heal us when nothing else can, and if we can reach the place where black and white sound in harmony, we will be saved. I would like to end with my own poem, 'Pythagoras Said':

> *…There is a symphony that never ceases*
> *Stretching out its ecstatic shimmer*
> *Through the waves of the sky*
> *Across the ripples of the late summer's shore*
> *And into the depths of your own being.*
> *If you are still,*
> *You can ride the crest of its vibrations.*
> *If you quiet your mind,*
> *You will know the voices of the stars.*
> *Tune, tune, tune your heart*
> *To hear the eternal celestial song.*

NOTES

1.   For what follows on Pythagoras' life and teaching, see Kenneth Guthrie, *The Pythagorean Sourcebook and Library*, pp 137 – 140, Grand Rapids, Michigan, 1987; G.S. Kirk, G.S., J.E. Raven, M. Schofield, *The Presocratic Philosophers*, 2nd edition, (Cambridge: Cambridge University Press, 1983) pp. 214-239; Charles H. Kahn, *Pythagoras and the Pythagoreans* (Indianapolis: Hackett, 2001). For an introduction to Pythagoreanism, the philosophy emerging from Pythagoras see the excellent article by Carl Huffman, 'Pythagoreanism', *Stanford Encyclopaedia of Philosophy*, revised 31 October 2014 https://plato.stanford.edu/entries/pythagoreanism/ [accessed 31 March 2019].

2.   Aristotle, *Metaphysics*, 2 Vols, Vol. 1 trans. Hugh Tredennick, (Cambridge Mass., London: Harvard University Press, 1933), 986 a.23.

3.   Plato, *Republic*, 2 Vols., trans. Paul Shorey (Cambridge Mass., London: Harvard University Press, 1935), III, 398-403; Aristotle, *Politics*, trans H.Rackham (Cambridge Mass., London: Harvard University Press, 1943) VIII.iv.3.

4.   Emma J. Edelstein and Ludwig Edelstein. *Asclepius: a collection and interpretation of the testimonies*. 2 vols. The Publications of the Institute of the History of Medicine. (Baltimore: Johns Hopkins University Press, 1998):

5.   For an introduction to the cosmology of Chartres see Bernadette Brady, 'Chartres Cathedral and the Role of the Sun in the Cathedral's Platonist Theology', in Nicholas Campion and Patrick Curry (eds.), *Sky and Psyche: the Relationship between Cosmos and Consciousness*, Edinburgh: Floris Books 2006, pp. 59-76.

Harmony, Cosmos, Ecology and Politics

Nicholas Campion

*Life begins with the process of star formation. We are made of stardust.
Every atom of every element in your body except for hydrogen has been
manufactured inside stars, scattered across the universe in great stellar
explosions, and recycled to become part of you.*[1]

Harmony

The concept of Harmony suggests that, for the universe to function, there must
be an essential balance in its energy and structure. The notion that balance is
fundamental to the cosmos is found throughout traditional and indigenous
cultures and continues in the our modern world. What I want to do here is pull
together various statements and claims about cosmology, both traditional and
modern, and discuss what they may mean for ecology and politics.

Modern science has identified balance in the universe in novel ways. For
example, the law of conservation of energy states that energy can neither be
created nor destroyed, only converted to another form: when energy appears
in one form an equal amount of energy must disappear in another. As Richard
Feynman put it, 'there is a certain quantity, which we call energy, that does not
change in the manifold changes which nature undergoes'.[2] We also talk about the
'Goldilocks Zone', named after the story of *Goldilocks and the Three Bears* and
the search for the perfect bowl of porridge. Like Goldilocks's bowl of porridge,
the Goldilocks zone which the Earth occupies is close enough to the Sun to
be warm enough, but not too far away to be too cold, and is therefore perfect
for organic life. If these conditions are altered slightly then most biological life
will come to an end. This, of course, is the threat posed by climate change: that
humans will alter the perfect natural conditions that maintain equilibrium on our
planet. Even though we are now discovering that life can survive in more extreme
circumstances than thought only a few decades ago, the basic principle that we
humans depend on special circumstances still remains.[3]

The physicist John Barrow played a leading part in highlighting the idea
that the close relationship between cosmos and consciousness is an integral part

of this order, as claimed by extreme versions of the Anthropic Principle.[4] When Barrow was awarded the prestigious Templeton prize in 2006, news reports drew public attention to his ideas:

> Life as we know it would be impossible, he and others have pointed out, if certain constants of nature – numbers denoting the relative strengths of fundamental forces and masses of elementary particles – had values much different from the ones they have, leading to the appearance that the universe was 'well tuned for life,' as Dr Barrow put it.
>
> In a news release, the prize organizers said of Dr Barrow's work: 'It has also given theologians and philosophers inescapable questions to consider when examining the very essence of belief, the nature of the universe, and humanity's place in it'.[5]

We may distinguish two notions of harmony. There is one with a capital 'H', which assumes the existence of an ordered structure which is inherent in the fabric of the universe. And there is another with a small 'h' which deals with our ability to peacefully get along with each other and the world. I shall use both Harmony and harmony in this chapter, even though it is not always easy to distinguish one from the other. In all the ancient philosophies which deal with it, capital 'H' Harmony is embedded in cosmic cycles which transcend humanity and so is not concerned with humanity's welfare. The entire universe, this model tells us, is subject to the infinitely recurring cyclical creation and destruction of the universe, an idea central to some classical Greek and Indian cosmologies.[6] Small 'h' harmony, on the other hand, is understood as the maintenance of balanced, equitable and peaceful relationships between individuals, societies and the rest of life on Earth. These ideas usually place human welfare as the central concern, focussing on the maintenance of harmonious relationships between people, society and nature, and do not require any philosophical or physical explanation of the universe (although advocates of harmony may well subscribe to such a notion).

A classic use of small 'h' harmony was that of E. F. Schumacher in 1973, in his classic and highly influential work, *Small is Beautiful*. He wrote that:

> In the excitement over the unfolding of his scientific and technical powers, modern man has built a system of production that ravishes nature and a type of society that mutilates man. If only there were more and more wealth, everything else, it is thought, would fall into place. Money is considered to be all-powerful; if it could not actually buy non-material values, such as justice,

harmony, beauty or even health, it could circumvent the need for them or compensate for their loss.[7]

Harmony, for Schumacher, is clearly used in the sense of peace and balance, alongside justice, beauty and health, as values directly opposed to the unrestrained accumulation of money.

Debates concerning the extent to which the universe operates according to a fundamental order have been of deep significance from ancient times up to the present. The classical view of the relationship between order and disorder was set out by Empedocles in the fifth century BCE. He established the classical foundation for theories of cyclical history and argued that all historical, natural and individual processes in the cosmos oscillate between two opposites: strife and love.[8] Whenever one of these reaches universal dominance then the cosmos unravels, only to be born again, in an inevitable and infinite cycle of creation and decay. There is a clear apocalyptic narrative here which was to later feed into accounts of the life and death of the universe.

The idea of a universe based on inherent harmonious mathematical ratios was developed by the Greek philosophers, Pythagoras and Plato.[9] The theory of the harmony of the spheres, for which Pythagoras is widely held to be responsible, was based in the idea that, as each planet travels through space in its own geometrical relationship to the Earth, it makes a sound. Together, these sounds produce a beautiful celestial harmony which, even if we can't hear it, forms part of the cosmic environment which surrounds us.[10] Influenced by this notion, Plato himself expressed the proportions of the cosmos in terms of the musical scale in *Timaeus*, his great work on the origin of the universe.[11] Each of the seven planets orbited the Earth on its own sphere, he argued, with the stars on another, making a total of eight spheres. As the planets move through their interlocking cycles, he claimed, they weave the patterns of life which constitute our destiny, accompanied by beautiful, if inaudible, sounds:

> And the spindle turned on the knees of Necessity, and up above on each of the rims of the circle a Siren stood, borne around in its revolution and ordering one sound, one note, and from all the eight there was the concord of a single harmony.[12]

The question posed by this model is how we respond to this celestially driven fate. One of Plato's central concerns was to reconcile the cyclical life of the cosmos in terms of capital 'H' Harmony with the welfare of humanity in terms of small 'h' harmony. He accepted that catastrophes could occur and that the cosmos might also collapse to be born again, but believed that this took place as part of a series of cycles, patterns and rhythms.[13] Capital 'H' Harmony might well hit a crisis in a cosmic cycle which we experience as deeply unharmonious, and we then have to respond by maintaining small 'h' harmony, as best as we can. Spurred on by the crises of his own time, Plato took on this challenge. His political theories were intended to minimise the disastrous effects of periods of collapse, and he devised an entire system of education and politics designed to preserve stability in the immediate future, and to minimise disruption when a crisis in the cosmic cycle hit.[14]

Nowadays, the word harmony is generally substituted amongst astronomers and astrophysicists by more familiar terms such as beauty and symmetry, words which are themselves connected with the search for the universal force which, it is hoped, will underpin the four recognised fundamental forces (gravity, electromagnetism and the weak and strong nuclear forces). Yet, even here, as physics looks for the single force which holds the universe together, order in the form of the increased complexity that comes with evolution, competes with disorder in the form of the disintegration that results from entropy.[15]

THE COSMOS

Cosmos, from the Greek *kosmos*, may be understood as fundamental and universal order (universe being the Latin equivalent of cosmos), although it is often translated as 'beautiful order' (and the root of our word cosmetic).[16] A typical modern definition describes cosmology as 'the science, theory or study of the universe as an orderly system, and of the laws that govern it; in particular, a branch of astronomy that deals with the structure and evolution of the universe'.[17] It is this understanding of cosmos as order that allows astronomers to construct models for measuring distances over huge areas, such as the so-called 'Cosmological Distance Ladder', which envisages the distances between near and far objects in the cosmos as being proportional in a significant manner.[18]

Two schools of Greek cosmology have exerted an enduring appeal in addition to the Platonic: the Aristotelian and the Stoic, the latter founded by Zeno of Citium. There are significant differences between the three schools, but they all shared a belief that the cosmos is one unified organism in which all things – material, emotional, psychological, and spiritual – are necessarily interconnected and interdependent. People, planets, stones and stars all exist in one immense, marvellous, symbiotic entity. In Aristotle's opinion,

The whole terrestrial region then is compounded of these four bodies and it is the conditions which affect them which, we have said, are the subject of our inquiry. This region must be continuous with the motions of the heavens, which therefore regulate its whole capacity for movement. For the celestial element as source of all motion, must be regarded as first cause.[19]

The Stoic worldview was described by the great biographer Diogenes Laertius, who we believe lived in the third century CE, and on whom we rely for much of our knowledge about the Greek philosophers. The Stoics, Diogenes tells us, insisted that nothing is outside nature. He adds that they argued that the cosmos '… has no empty space within it but is one united whole', and '… is a living being, rational, animate and intelligent'.[20] That is pretty stunning for the average modern world view (if such a thing exists), for it completely challenges the comfortable assumption that some things are alive and animate, and others are inanimate and dead. For the Stoics a stone may have different qualities to a swan, but it is still alive. They believed that human nature is identical to the nature of the universe and that the purpose of life is therefore to live in agreement with nature. Not only that, but nature is essentially good, and guides those of us who live in agreement with it to a virtuous life: a virtuous life is a natural life and vice versa. This what Diogenes wrote around seventeen hundred years ago:

> our individual natures are parts of the nature of the whole universe. And this is why the end may be defined as life in accordance with nature, or, in other words, in accordance with our own human nature as well as that of the universe, a life in which we refrain from every action forbidden by the law common to all things, that is to say, the right reason which pervades all things, and is identical with this Zeus, lord and ruler of all that is. And this very thing constitutes the virtue of the happy man and the smooth current of life, when all actions promote the harmony of the spirit dwelling in the individual man with the will of him who orders the universe.[21]

We then get a proposition which runs something like this: if nature is ordered, benevolent, rational and virtuous, it therefore makes common sense to live in accord with it. And if we are wondering what virtue is, it is 'the state of mind which tends to make the whole of life harmonious'.[22] So there is a feedback loop in which, if we as individuals are virtuous, the entire cosmos benefits. Another Stoic, the philosopher Cleomedes, who lived sometime between the first and fifth centuries CE, gave us the standard definition of the Cosmos. He wrote that,

"Cosmos" is used in many senses, but our present discussion concerns it with reference to its final arrangement, which is defined as follows: a cosmos is a construct formed from the heavens, the Earth, and the natural substances within them. This [cosmos] encompasses all bodies, since, as is demonstrated elsewhere, there is, without qualification, no body existing outside the cosmos... And that the cosmos has Nature as that which administers it is evident from the following: the ordering of the parts within it; the orderly succession of what comes into existence, the sympathy of the parts in it for one another; the fact that all individual entities are created in relation to something else; and, finally, the fact that everything in the cosmos renders very beneficial services.[23]

We can take two key concepts from Cleomedes' account. First, the cosmos is benign. As developed by the Neoplatonists in the third and fourth centuries CE, evil, far from being an essential quality of the cosmos, is a consequence of human error and ignorance.[24] Evil can thus be banished through education and a correct and virtuous life, a view which was to be central to the development of educational and correctional theory in the nineteenth and twentieth centuries. Second, the cosmos is everything: it is me, it is you, it is the chair I am sitting on, the food I eat, the place in which the food grew, the energy which sustained it, and the star from which that energy came.

Thus, the defining features of the classical Harmonious universe are relationality, connectedness and interdependence. If everything is interdependent we can never absolutely stand outside the cosmos or dissect it in a laboratory. We do not stand outside nature and we do not act on it as independent, separate beings, either as its guardian or its destroyer. The ramifications of this worldview take us into some controversial territory. For example, the concept of the Anthropocene, the geological age which we are now supposedly entering, is dependent on the notion that humanity is now exerting a profoundly negative influence on the planet. But, if we are inside nature, how can we be seen to be outside it and acting on it?[25] This is a problem raised by James Lovelock, when he asked 'Is technological man still a part of Gaia or are we in some or in many ways alienated from her?'[26] In other words, how do we tell whether the things that we do are natural or unnatural? Perhaps the lesson is that we should not labour theory too much, but take a pragmatic approach, looking at practical solutions.

The classical Harmonious cosmos proved remarkably persistent, and was to be a staple of Renaissance thought. It received perhaps its most comprehensive treatment from the great seventeenth-century astronomer Johannes Kepler, who followed Pythagoras and Plato when he argued in his great work, the *Harmonices Mundi – The Harmony of the World* – that the universe could be understood literally in terms of musical scales.[27] His ultimate concern was to understand how we act as autonomous individuals in a universe which operates according to divinely inspired mathematical laws. Like Plato, he believed that, as long as human beings successfully live in tune with the celestial harmonies, then peace and order may be maintained. Harmony was carried into Kepler's era by various texts, one of the most popular being the late Roman classic, Boethius's *The Consolation of Philosophy*. As Lady Philosophy speaks, or perhaps sings, she addresses the creator in the following words:

> The elements by harmony Thou dost constrain,
> That hot to cold and wet to dry are equal made,
> That fire grow not too light, or earth too fraught with weight.
> The bridge of threefold nature madest Thou soul, which spreads
> Through nature's limbs harmonious and all things moves.[28]

William Shakespeare had Ulysses say, in *Troilus and Cressida* (published in 1609, ten years before the *Harmonices Mundi*), 'Take but degree, away, untune that string, and hark, what discord follows'.[29] And in the *Merchant of Venice*, Lorenzo calls for music to be played, initiating both sweet harmonies and musings on their effects, evoking the harmony inherent in the soul, if not the body:

> How sweet the moonlight sleeps upon this bank!
> Here will we sit and let the sounds of music
> Creep in our ears. Soft stillness and the night
> Become the touches of sweet harmony.
> Sit, Jessica. Look how the floor of heaven
> Is thick inlaid with patens of bright gold.
> There's not the smallest orb which thou behold'st
> But in his motion like an angel sings,
> Still choiring to the young-eyed cherubins.
> Such harmony is in immortal souls,
> But whilst this muddy vesture of decay
> Doth grossly close it in, we cannot hear it.[30]

Lorenzo continues with thoughts on how the man who has no music in himself is not to be trusted:

> Therefore the poet
> Did feign that Orpheus drew trees, stones, and floods
> Since naught so stockish, hard, and full of rage,
> But music for the time doth change his nature.
> The man that hath no music in himself,
> Nor is not moved with concord of sweet sounds,
> Is fit for treasons, stratagems, and spoils.
> The motions of his spirit are dull as night,
> And his affections dark as Erebus.
> Let no such man be trusted. Mark the music.[31]

For the classical Platonist, Aristotelian, or Stoic and their Renaissance followers, the material and psychological connections between all things are *real* because there is no more difference between one individual and the surrounding world than there is between parts of the individual body. Therefore, if one part changes, all parts change, and the individual in nature has a real effect on the natural world, just as the natural world has a real effect on the individual. The relationship is real, mutual and continuous.

This much was evident to Isaac Newton, who was deeply immersed in the esoteric currents of the eighteenth century. He was well aware of the Hermetic texts, which had been composed around the second century and had a huge impact on the Renaissance world after they were translated into Latin in the mid-fifteenth century. Newton himself made his own translation of one of the most famous Hermetic texts, the *Tabula Smaragdina*, or Emerald Tablet. We now know that this was written in Arabic around the eighth century, but in Newton's time it was thought to be one of the foundations of ancient wisdom. Newton's translation expresses with great clarity his belief in the unity of heaven and earth, with the sun and moon, metaphorically speaking, in parental roles.

> That wch is below is like that wch is above & that wch
> is above is like yt wch is below to do ye miracles of one
> only thing
> And as all things have been & arose from one by ye
> mediation of one: so all things have their birth from this
> one thing by adaptation.
> The Sun is its father, the moon its mother.[32]

The Emerald Tablet's point is that there is a reciprocal relationship between change 'down here' and change 'up there'. If human beings make adjustments in their material or spiritual lives, even on a tiny level, then the material and spiritual life of the entire cosmos is affected. Newton's extraordinary quest to understand and explain the laws of nature was driven by his belief that the movement of the stars and planets were an image of God's creation.[33] Eighteenth century thinkers were quick to draw conclusions. Prominent amongst these was Colin McLaurin, Professor of Mathematics at the University of Edinburgh:

> Our views of Nature, however imperfect, serve to represent to us, in the most sensible manner, that mighty power which prevails throughout, acting with the force and efficacy that appears to suffer no diminution from the greatest distances of space or intervals of time; and that wisdom which we see equally displayed in the exquisite structure and just motions of the greatest and subtlest parts. These, with perfect goodness, by which they are evidently directed, constitute the supreme object of the speculations of a philosopher; who, while he contemplates and admires so excellent system, cannot but be himself excited and animated to correspond with the general harmony of nature.[34]

Newton's discovery that a single law – gravity – governed the whole universe was to have a supportive impact on radical politics, including on the American revolutionaries of the 1770s.[35] The core principle held that, just as one single law governs the entire universe, so human society must also be governed by the same single law: no king should be above the law any more than any commoner, and the arbitrary exercise of political power was condemned as contrary to natural law. The political implications were articulated in what came to be known as Natural Rights philosophy: life, liberty and the pursuit of happiness, it was claimed, were natural, while tyranny, by complete contrast was unnatural: Nature therefore dictated that unfair laws and oppressive regimes must be opposed. Thomas Paine, one of the leading radicals of the 1770s, who played a prominent role in persuading the American colonists to break with Britain, believed that the perfect order of the planets was a profound demonstration of the truth of God's natural creation, and, quoting the radical French aristocrat, the Marquis de Lafayette, he wrote how the truths which Nature had engraved on the heart of every citizen carried an innate love of liberty.[36] That Natural Rights belong to people through the fact of existence and freedom, he claimed, is the default position of the Newtonian universe.[37] And the force which manages society for the best, Paine said in 1776 – paraphrasing Newton in the very year that American independence was declared – is like a 'gravitating power'.[38] It is irresistible: it must succeed. Newtonian harmony emphasised order, stability, regularity and the rule of a law under which all men were, in theory, equal.

The last major explication of Harmony in the west was set out by Gottfried Leibniz, Newton's contemporary and rival. In Leibniz's version of 'Pre-existent Harmony', things were not maintained in a state of balance by their internal natures but by their relationships.[39] Leibniz was criticised for assuming a complete separation between soul and body, unlike mainstream harmony thinking.[40] Yet, if we extend his views to politics then we may fairly conclude that Harmony depends on mutual respect in relationships. To use another familiar metaphor from traditional cosmologies, in the ordered universe earth and sky mirror each other. We can see the one in the other because they are part of the same system. Native North American cosmology talks of a 'patterned mirroring between sky and earth'.[41] And as Xiaochun Sun said of Chinese cosmology, 'The universe was conceived not as an object independent of man, but as a counterpart of and mirror of human society'.[42] In such a perspective it then becomes the duty of all individuals in a cosmologically-aware society to ensure that their actions are matched with the motions of the heavenly bodies. In traditional Chinese culture this is accomplished by living in balance with the energies that flow all around. Harmony is maintained through a whole series of practices including *feng shui*, the art of harmonising the human environment to *chi* (the energy which pervades space). As Nathan Sivin wrote,

> ... macrocosm and microcosm became a single manifold, a set of mutually resonant systems of which the emperor was indispensable mediator. This was true even of medicine... Cosmology was not a mere reflection of politics. Cosmos, body, and state were shaped in a single process, as a result of changing circumstances that the new ideas in turn shaped.[43]

There is, then, a pattern in events, and some periods of time are qualitatively different to others. Again, speaking of China, this is Richard Wilhelm from his commentary on the *I Ching*:

> Events follow definite trends, each according to its nature. Things are distinguished from one another in definite classes. In this way good fortune and misfortune come about. In the heavens phenomena take form; on earth shapes take form. In this way change and transformation become manifest.[44]

The notion of relationality in a living cosmos occurs in the great Maya creation epic, the *Popul Vuh*. The Maya cosmos itself came into being as a gradual emergence of order, of something from nothing. The opening lines, in English translation, express the still beauty of a calm morning, the dawn of everything:

> Now it ripples, now it still murmurs, ripples, still it sighs, still hums and it is empty under the sky.
> Here follows the first words, the first eloquence:
> There is not one person, one animal, bird, fish, crab, tree, rock, hollow, canyon, meadow, forest. Only the sky alone is there; the face of the earth is not clear. Only the sea alone is pooled under all the sky; there is nothing whatever gathered together. It is at rest; not a single thing stirs. It is held back, kept at rest under the sky.
> Whatever there is that might be is simply not there: only the pooled water, only the same sea.[45]

And then, the text continues, deep within the dark, the Plumed Serpent, the Aztec Quetzalcoatl, the Maker and Modeller of all, stirs and speaks with the Heart of Sky, also known as Hurricane, and the creation begins. The cosmos of the great Mesoamerican and Andean civilisations was alive. Its structure was that of a living body. In this sense it is not necessary to ask what something is made of, or how it moves, but what one's relationship with it is. Another creation story, this time from Australia, is described by the travel writer Bruce Chatwin:

> ... legendary totemic beings... wandered over the continent in the Dreamtime, singing out the name of everything that crossed their path – birds, animals, plants, rocks, waterholes... so singing the world into existence.[46]

When we look at traditional cultures we find a variety of perspectives. In some the universe is infinite, in others it is enclosed within tight limits; for some it is highly ordered, while others allow much more room for spontaneity. However, what seems to be universal is a notion of balance. A concise summary of the practical application of the concept of balance and relationality in traditional cosmologies comes from Clive Ruggles, who argues that among modern indigenous communities there is evidence of a belief in direct interconnectedness:

> Modern examples include the Barasana of the Colombian Amazon, who understand that the celestial caterpillar causes the proliferation of earthly

caterpillars; the Mursi of Ethiopia, for who, the flooding of the river they call *waar* can be determined, without going down to the banks, by the behaviour of the star of the same name; and those modern Hawaiians who still carry on the ancient practice of planting taro and other crops according to the day of the month in the traditional calendar (i.e., the phase of the moon).[47]

MODERN PHYSICS

The lineage from classical Stoicism extends down through the centuries to Albert Einstein, the most iconic of all western scientists. Einstein was profoundly influenced by the seventeenth century Stoic philosopher Baruch Spinoza: 'I believe in Spinoza's God', Einstein wrote, 'who reveals Himself in the lawful harmony of the world, not in a God who concerns Himself with the fate and the doings of mankind'.[48] In simple terms, Einstein's theory of relativity assumes a complete integration between space and time, and if there is an alteration in one, there is an alteration in the other. In addition, as a Stoic, Einstein believed in a universal order which transcended human interests. He made this abundantly clear:

> But the scientist is possessed by the sense of universal causation. The future, to him, is every whit as necessary and determined as the past. There is nothing divine about morality; it is a purely human affair. His religious feeling takes the form of a rapturous amazement at the harmony of natural law, which reveals an intelligence of such superiority that, compared with it, all the systematic thinking and acting of human beings is an utterly insignificant reflection. This feeling is the guiding principle of his life and work, in so far as he succeeds in keeping himself from the shackles of selfish desire. It is beyond question closely akin to that which has possessed the religious geniuses of all ages.[49]

Whereas Einstein placed himself in a continuity with the past, quantum mechanics, the other main strand of revolutionary twentieth century physics, envisages not a continuity with the past but a break. Werner Heisenberg, one of the key pioneers of quantum mechanics, himself had views of harmony. He saw himself as representative of a break in cosmology that occurred after Kepler. In Kepler's *Harmony of the World*, Heisenberg mused, the world could not be understood as independent of God and humanity's personal relationship with him. As Heisenberg tells it, first Galileo, who saw the real universe through his telescope, and then Newton, who created a law of gravity which (Heisenberg thought) had no need of God, allowed humanity to stand outside the natural world and, as he wrote,

'... separate out individual processes of nature from their environment, describe them mathematically, and thus "explain" them'.[50] Heisenberg is not quite right in this, for both Galileo and Newton preserved a place for God, but his is a common and influential view. Nonetheless, he continued with his potted history of science, reaching the point in the nineteenth century when electrical theory concluded that force fields rather than matter are the fundamental building blocks of the universe. It is hard to think of a more revolutionary shift in modern science.

The implications for our understanding of our uncertain and complex relationship with the world around us were explored by C. P. Snow in his 1934 novel, *The Search*. Such fictionalised accounts are important because they tell us how far the new physics had penetrated the general intellectual consciousness. In a scene set around 1917, the fictitious narrator's science teacher sets out the personal consequences of the new atomic theory.

> That's all you'd come to in the end. Positive and negative electricity. How do things differ then? Well, the atoms are all positive and negative electricity and they're all made on the same pattern, but they vary among themselves, do you see? Every atom has a bit of positive electricity in the middle of it – the nucleus, they call it – and every atom has bits of negative electricity going round the nucleus – like planets round the Sun. But the nucleus is bigger in some atoms than others, bigger in lead than it is in carbon, and there are more bits of negative electricity in some atoms than others.[51]

The imaginary teacher then comes close to the idea of the reciprocity of the world above with the world below which we found in the Emerald Tablet:

> It's as though you had different solar systems, made from the same sort of materials, some with bigger suns than others, some with a lot more planets. That's all the difference. That's where a diamond's different from a bit of lead. That's at the bottom of the whole of this world of ours.[52]

We ourselves are composed of electricity, as is the entire universe, as Snow's imaginary science teacher reports to his enraptured students, and are intimately connected to the universe – exactly as the Stoics claimed. Heisenberg himself goes on to consider the even more destabilising effects of quantum mechanics.

> The old compartmentalisation of the world into an objective process in space and time, on the one hand, and the soul in which this process is mirrored, on

the other – that is, the Cartesian differentiation of a *res cogitans* [the world of the mind] and *res extensa* [the world of matter] – is no longer suitable as the starting point for the understanding of modern science. [53]

Having challenged the conventional distinction between mind and matter, Heisenberg proposes instead a world in which they are interrelated, and the scientist is an active participant in this relationship, not a disinterested observer:

> In the field of view of this science there appears above all the network of relations between man and nature, of the connections through which we as physical beings are dependent parts of nature and at the same time, as human beings, make them the object of our thoughts and actions. Science no longer is in the position of observer of nature, but rather recognises itself as part of the interplay between man and nature.[54]

The split between mind and body which followed Descartes' new thinking in the seventeenth century is, according to quantum mechanics, a misapprehension. The same goes for the supposed distinction between humanity and nature. Heisenberg continues:

> The atomic physicist has had to come to terms with the fact that his science is only a link in the endless chain of discussions of man with nature, but that it cannot simply talk of nature "as such." Natural science always presupposes man, and we must become aware of the fact that, as Bohr has expressed it, we are not only spectators but also always participants on the stage of life.[55]

In Heisenberg's quantum world it is impossible to imagine that one can interfere in the natural world without having consequences for the rest of life, a concept which not everyone could accept. It is well known, for example, that Einstein himself initially rejected quantum mechanics, which brings us to perhaps his most famous statement (often misquoted as 'God does not play dice'). In a letter to Max Born in 1926, Einstein rejected the inherent uncertainty of quantum mechanics:

> Quantum mechanics is certainly imposing. But an inner voice tells me that it is not yet the real thing. The theory says a lot, but does not really bring us any closer to the secret of the 'old one'. I, at any rate, am convinced that *He* does not play dice'.[56]

Ever since then, physicists have struggled to find a meeting ground between the certainty of relativity and the uncertainty of the quantum world. One option, according to Partha Ghose, 'provides a clear and beautiful harmony of classical waves and particles'.[57] Referring to Niels Bohr, another pioneer of quantum mechanics, Partha Ghose added,

> We have also seen how the concept of waves and particles with all their subtleties and dichotomies have evolved since the inception of quantum theory, and also the principal attempts to harmonize them. Bohr viewed them as mutually exclusive but complementary aspects of a quantum entity whereas Einstein, [Louis] de Broglie and [David] Bohm preferred a more inclusive harmony.[58]

Theories of the origin of the universe are equally dedicated to a vision of interrelationship. The core conceptual model is Big Bang theory in which everything is necessarily connected to everything else, however remotely, thanks to a shared origin in a compressed ball of energy.[59] Fred Hoyle, a great populariser of astronomy, who supported the 'Steady State' theory, in which the creation of matter is continuous and the universe neither begins nor ends, wrote that:

> our everyday experience even down to the smallest details seems to be so closely integrated to the grand-scale features of the Universe that it is well-nigh impossible to contemplate the two being separated.[60]

Relationality has also found a new place in quantum physics. According to John Wheeler:

> The system of shared experience that we call the world is viewed as building itself out of elementary quantum phenomena, elementary acts of observer-participancy. In other words, the questions that the participants put – and the answers they get – by their observing devices, plus their communication of their findings, take part in creating the impressions which we call the system: that whole great system which to a superficial look is time and space, particles and fields.[61]

We may also turn to the physicist Paul Davies, for inspiration. Seeking support for the philosopher Karl Popper, Davies wrote that,

An increasing number of scientists and writers have come to realise that the ability of the physical world to organise itself constitutes a fundamental, and deeply mysterious, property of the universe. The fact that nature has creative power, and is able to produce a progressively richer variety of complex forms and structures, challenges the very foundation of contemporary science. 'The greatest riddle of cosmology', writes Karl Popper... 'may well be ... that the universe is, in a sense, creative'.[62]

Classical Harmony required a divine creator, even if an impersonal one, who set the cosmos in motion. That creator may be a mind or a consciousness. We might characterise it as a god, or God. Perhaps we have no need of God, but rather of pattern, or order, or laws of physics. Frances Crick's 'Astonishing Hypothesis' is a recent and influential statement of the materialist, atheist position. As defined by Crick, '... your memories and your ambitions, your sense of personal identity and free will, are in fact no more than the behaviour of a vast assembly of nerve cells and associated molecules'.[63] Crick's rhetorical bias, though, is evident in his use of words such as 'no more than...' Surely the arrangement of atoms into minds and eventually thoughts is astonishing. When, we may ask, does a group of atoms arrange itself so as to reflect on its existence? We might question whether, if every particle of matter in our bodies has already passed through three stars, including our Sun, since the Big Bang, and if, as Crick argues, consciousness is a property of matter, at what point in this process does matter develop the ability to inquire into itself? In other words, where is the boundary between matter and consciousness?

Both Einstein's and Heisenberg's theories have an immediate significance for ecological thought. In Einstein's universe the complete and harmonious integration of time and place removes any distinction between humanity and nature, and the same outcome flows from Heisenberg's argument that we are all participants within nature, rather than external observers influencing it. Both Einsteinian and Heisenborgian perspectives lead down the path to 'deep ecology', which is discussed below. As an additional consideration, Heisenberg's universe, being inherently uncertain, is therefore complex. There can be no simple solution to any problem, and therefore no quick fix. The universe is characterised by relationality, which means that if I move here, something else moves there, and I may not know what that something is until I have moved. To move carefully, and with regard to the consequences of one's actions, is the essence of Harmonious cosmology in action.

However, the Harmonious cosmos does not guarantee human beings a comfortable life. After all, nature, as Tennyson wrote in 1850 often seems to be, 'red in tooth and claw', hardly harmonious in a placid, peaceful sense, as man, the transcendent creature, is ravaged by the reality of death:

> She cries, "A thousand types are gone:
> I care for nothing, all shall go.
> 'Thou makest thine appeal to me:
> I bring to life, I bring to death:
> The spirit does but mean the breath:
> I know no more." And he, shall he,
> Man, her last work, who seem'd so fair,
> Such splendid purpose in his eyes,
> Who roll'd the psalm to wintry skies,
> Who built him fanes of fruitless prayer,
> Who trusted God was love indeed
> And love Creation's final law--
> Tho' Nature, red in tooth and claw
> With ravine, shriek'd against his creed--[64]

The Stoic perspective is at home with the rawness of nature, and acknowledgment of it leads to a calm soul and moderate action. But Tennyson's nature, 'red in tooth and claw', sits uncomfortably with romantic views of nature as essentially benign.

HARMONY AND ECOLOGY

This all takes us to the question of shallow ecology and deep ecology, the two versions of the ecological movement as defined by Arne Naess in his seminal lecture in 1972.[65] In Naess's distinction, shallow ecology is functional and pragmatic and focuses on the fight against pollution and resource depletion. It is completely anthropocentric. We might consider it Platonic, in terms of classical psychology, in that it depends on the independence of humanity from nature, and sees nature as a separate entity which humanity can manipulate. By contrast, deep ecology is completely Stoic, in that it assumes that there is no distinction between humanity and nature. The first principle of Deep Ecology, with capital letters, as defined by Arne Naess, is: 'Rejection of the man-in-environment image in favour of the relational, total-field image. Organisms as knots in the biospherical nest or field of intrinsic relations'.[66] We can imagine the man-in-environment as the Romantic artist, a poet, or painter, receiving spiritual sustenance from being immersed in nature, wandering through meadows or wooded valleys, or climbing hills beneath sun, clouds or stars. Naess says, however, that we must go beyond imagining the artist as inspired by nature: he or she is a part of it, and has no existence separate from it.

Naess also had a deep interest interest in Einstein, which became explicit in 1985 when he met the physicist Øyvind Grøn, with whom he collaborated on an attempt to make the mathematics of relativity accessible to a general audience. Naess and Grøn considered the consequences of relativity for vector fields. Using wind as an example, they wrote:

> When we are out doors in the wind, the moving wind fills the region around us. There is a measurable velocity of the air everywhere in the region... The velocity [of the wind] has a magnitude and a direction. It is a vector. Thus a velocity field is linked conceptually with every point of the region. These abstract vectors are everywhere. If one can think of God as omnipresent, then one might also be able to think of the factors as omnipresent. In such a region there is said to be a vector field.[67]

Naess and Grøn then consider the conceptual problem that we normally have when using flat weather maps, which must be inherently misleading because they do not represent multi-dimensionality. Of a two-dimensional map, which they use as an example, they write: '... the vectors representing the wind velocity was (sic) implicitly assumed to exist in a flat three-dimensional region. However, the space-time of general relativity is curved'.[68] The conclusion we may draw is that, as opposed to normal Euclidean space, the space of our daily perception, in which arrows are drawn as straight lines without end, Einsteinian space bends arrows, which eventually, if we follow this logic, return to where they started.

To apply this notion to deep ecology, every natural relationship curves in on itself, creating feedback loops which may exacerbate individual trends, adding a chaotic quality. Chaos theory, as defined by James Gleick, still assumes pattern and order, unlike the random confusion we normally associate with chaos, although of a continuously evolving and open-ended kind, rather than operating according to fixed rules.[69] In relation to harmony. the chaos of chaos theory is patterned, emergent and complex, rather than ordered and repetitive as in the classical theory of harmony. Naess's use of wind as an example of complexity, for example, also has a precedent in Heisenberg's use of wind and ocean currents to describe wider contexts:

With the seemingly unlimited expansion of his material might, man finds himself in the position of a captain whose ship has been so securely built of iron and steel that the needle of his compass no longer points to the north, but only toward the ship's mass of iron. With such a ship no destination can be reached; it will move aimlessly and be subject in addition to winds and ocean currents. But let us remember the state of affairs of modern physics: the danger only exists so long as the captain is unaware that his compass does not respond to the Earth's magnetic forces. The moment the situation is recognised, the danger can be considered as half-removed.[70]

Again, having identified the problem, Heisenberg suggests a solution:

For the captain who does not want to travel in circles but desires to reach a known – or unknown – destination will find ways and means for determining the orientation of his ship. He may start using modern types of compasses that are not affected by the iron or the ship, or he may navigate, as in former times, by the stars. Of course we cannot decrease the visibility or lack of visibility of the stars, and in our time perhaps they are only rarely visible.[71]

Naess was clearly a devotee of logic. More to the point, we may see his entire portrayal of deep ecology as a definition of harmony, consisting of seven organisational, political and philosophical components : (1) an understanding that humanity exists completely within nature, not outside it; (2) biospherical egalitarianism or respect for all life; (3) an anti-class posture, which he sees as essential for conflict resolution; (4) the fight against pollution and resource depletion, which is shared with shallow ecology; (5) recognition that the world is complex and that multiple factors must be taken into account to understand the functioning of any single system, if indeed there is such a thing as a single system; and (7) local economy and decentralisation. Perhaps, to continue the shallow/deep ecology dichotomy, we could imagine two versions of harmony, a shallow version which focuses on harmonious relationships (and is no less important for that) and another, capital 'H' Harmony, which is deep in that it envisages the total immersion of humanity in nature.

Having outlined the seven components of deep ecology, Naess then considers three additional principles and consequences. The first is that the principles of deep ecology are not derived from ecology by logic or induction but more from intuition (although that is not a word he uses). This is a position which would be acceptable to the Platonist for whom rationality suggests contact with the divine, and also the Stoics, for whom knowledge comes from immersion in the natural world.

Secondly, Naess deals with the problem of values. Thirdly, and significantly for this chapter, he develops the notion of eco-philosophy, or ecosophy, which he describes as a philosophy of ecological harmony or equilibrium.[72] By harmony he means what he calls 'a kind of *sofia*, which includes norms, rules and values which are human in origin, along with hypotheses concerning the state of affairs in a universe'.[73] He continues: '... wisdom is policy wisdom, prescription, not only scientific description and prediction'.[74]

The question he raises, then, is how we act politically. There is no standard political programme, and neither can there be, for ecosophical priorities vary according to time, place and culture. Referring to Aristotle and Spinoza for support, he states that 'an ecosophy is expressed verbally as a set of sentences with a variety of functions, descriptive and prescriptive (in which)... The basic relation is that between the subsets of premises and subsets of conclusions'.[75]

Naess later proposed eight policy principles and actions:

1. All life has value and human and non-human life are interdependent.
2. Diversity of life forms is valuable.
3. Humans have no right to reduce diversity except to satisfy *vital* (Naess's emphasis) needs.
4. The human population of the Earth needs to decrease.
5. Current human interference with the non-human world is excessive.
6. Social, technological and economic policies must change, and the result will be a 'more joyful experience of the connectedness of all things'.
7. The key ideological change should be towards an emphasis on the quality of life.
8. Individuals who recognise points 1-7 have an obligation to actively work for them.[76]

At face value, few people would disagree with points 1 and 2, which have become conservative orthodoxy in some quarters. Even that symbol of the British media establishment, *The Sunday Times*, recently carried a supplement headed 'Don't try to "solve" diversity, celebrate it'.[77] Most would agree with 5 and, increasingly, with 7 – witness the increasingly popularity of wellbeing and happiness indices. However, number 4 is undoubtedly problematic. Who is to say which people should cease having children, and how would this be policed (given the problems associated with China's one-child policy)? And is number 6 reminiscent of the long – and failed – tradition of religious utopianism in the Christian world, and number 8 suggestive of the authoritarian communitas of modern revolutionary politics? These are all profoundly difficult questions.

HARMONY AND POLITICS

The question is where all this leaves us in terms of contemporary politics. The risk of deep ecology is that it leads directly to an authoritarian perspective; witness the awful example of green Nazism, in which ecology was an adjunct to racial cleansing.[78] The issue is critical at the present time, largely due to the Chinese government's use of Harmony as a guiding political concept. For example, at the opening ceremony of the Belt and Road Forum for International Cooperation, the Chinese President Xi Jinping gave a speech, recalling that,

> The ancient silk routes brought prosperity to these regions and boosted their development. History is our best teacher. The glory of the ancient silk routes shows that geographical distance is not insurmountable. If we take the first courageous step towards each other, we can embark on a path leading to friendship, shared development, peace, harmony and a better future.[79]

The former president, Hu Jintao, declared that a socialist harmonious society 'will feature democracy, the rule of law, equity, justice, sincerity, amity and vitality', resulting in 'an honest and caring society, and a stable, vigorous and orderly society in which humans live in harmony with nature'.[80] This is all very well, but doesn't really match the government's oppression of the Tibetans and Uighurs, which frankly resembles traditional imperialism rather than any lofty ideal. In addition to which we have the example of China's attempts to create an economic empire.[81] The issue can be reduced to a debate concerning the application of Confucian cosmology to the management of the state. Does the Confucian concept of harmony require the sacrifice of individual rights to an overarching order? Or does order depend on the recognition of individual rights? And does it promote economic justice? Even if the Chinese government employs harmony as a rhetorical device to justify its totalitarian instincts, the consensus amongst scholars is that Confucian political theory respects individual rights and sees harmony as emerging from the balance of relationships between people.[82] In response to claims that harmony and human rights are incompatible, Stephen Angle argues that a commitment to both is, as he says, both 'coherent and desirable'.[83] Angle's conclusion is based on an understanding of the word *he* (和) – which in Mandarin Chinese can mean harmony, together with, peace, or union – as recognizing not uniformity, but the complex relationships between all things. In his view, 'The value of harmony comes precisely from its ability to preserve and respect differences, because we are all better off when these differences are meshed, to whatever appropriate degree, rather than flattened'.[84] This is the polar opposite of the Chinese government's position, and a suitable model for a democratic understanding of harmony.

The political-harmony problem was resolved in European thought by Johannes Kepler, who engaged in a major debate with that other major advocate of harmony, Robert Fludd. Kepler's problem with Fludd was that his cosmology was drawn from the kind of abstract mathematical and metaphysical structures constructed over the previous two thousand years and dating back to Pythagoras. Kepler's own theory was based on nature, as demonstrated by his astronomical observations. Kepler wrote of Fludd that: 'He seeks harmonic proportions in degrees of darkness and light, without respect to any motion: I seek harmonies only in motions'.[85] Fludd's harmony can be portrayed, perhaps unfairly, as static and other-worldly (this was certainly Kepler's view of Fludd's position). Kepler saw the cosmos as a single community consisting of the relationship between what he referred to as the 'family relationships of the wandering stars [i.e., planets]' and the 'family relationship of sounds'; his harmonious universe was both dynamic and existed in a continuous state of development.[86] Every moment brought its potential risks and opportunities which might be reflected in one's engagement with God, but also with politics.[87] The ebb and flow of harmony as represented by astronomical cycles had an immediate this-world impact, which would be encountered in political crises when planetary patterns indicated stress. Kepler's harmonic political theory envisaged that the threat of popular disturbance at such moments should be met partly by repression – as normal in his world as it is modern times – but also, radically for his time, reform.[88] In this sense, Kepler's state therefore resembles a modern social-democracy, in which individual rights are respected but, at the same time, regulated by law. If there is a way, then, to practice harmony within a Harmonious cosmos, it is to make mutual respect the default position. No doubt there are situations in which one individual or group may feel that they are not being given the respect that they deserve, in which case the law will step in – not the law of physics (or nature), but the law of society (as correctly arising from nature).

This brings us to the final issue, the need for an ethical framework, 'an ethics of harmony, a true earth ethic', as Haydn Washington calls it.[89] A fundamental distinction in current 'Green' ethical thought distinguishes ' environmentalism', which depends on a managerial approach to environmental problems, from 'ecologism', which argues for 'radical changes in our relationship with the non-human natural world, and in our mode of social and political life'.[90] According to this view, environmentalism is shallow ecology by another name, an inferior activity concerned with saving the planet on behalf of humanity, while ecocentrism and deep ecology subsume humanity within the ecological system. As Stan Rowe put it:

> Because "environment" means that which encircles something more important, literal "environmentalists" are willy-nilly anthropocentric, placing less value on the surrounding world than on humanity and self. If that causes uneasiness, the central position of the self can be retained painlessly by redefining it as a broad field-of-care embracing Earth. But this is an ineffectual gesture if, when push comes to shove, humanity is always accorded top billing. The question of priorities is critical.[91]

Rowe goes on to discuss the consequences of this argument:

> Should our loyalty embrace the entire "field-of-care" or does sympathy fasten first and always on the starving family metaphorically ploughing the "field" into oblivion? The whole field should command our allegiance, say I. It is time to eschew human self-interest and recognize the inherent worth and surpassing values of Earth's miraculous ecosystems whose workings we do not understand. *Anthropocentrism says we know how to control and manage them; ecocentrism says "not yet; maybe never."* (original emphasis).[92]

Ecocentrism is, then, in classical terms, a restatement of Stoicism, in which humanity is completely embedded in nature, itself conceived as a single entity, to the extent that the possibility of effective human action is limited. Stoic ethics has a number of characteristics. First, it is teleological: it assumes a goal which already has a kind of existence in the future and which lends existence in the present a purpose; second, it equates ethics with wisdom.[93] A significant recent contribution to the debate has been made by Patrick Curry, who has extended classical 'virtue ethics', into what he calls a Green Virtue Ethics (GVE), which escapes the anthropocentrism which prioritises human interests alone, and instead recognises humanity as embedded in nature. He suggests that we should acknowledge that 'nature takes its proper place at the heart of *all* beings, not merely an add-on extra to make us feel better humans'.[94] An ecological virtue ethic, he concludes, should take into account the good of the entire Earth community. As Stan Rowe said, 'We are Earthlings first, humans second'.[95] Wherever such ideas lead, the necessary precondition is, as the United Nations states, in a context of social justice.[96] Or, as the eighteenth-century radicals proposed, the law of the universe impacts on everyone equally and, in the harmonious universe, everyone's rights must therefore be equally protected.

NOTES

1. John Gribbin, *Stardust: the cosmic recycling of stars, planets and people* (London: Penguin, 2001), p. 1.

2. Richard Feynman, *The Feynman Lectures on Physics* (New York: Basic Books, 2010 [1963]), Vol. 1, p. 4.1.

3. NASA Science, 'The Goldilocks Zone', https://science.nasa.gov/science-news/science-at-nasa/2003/02oct_goldilocks [accessed 2 May 2020].

4. John Barrow and Frank Tipler, *The Anthropic Cosmological Principle* (Oxford: Oxford University Press, 1996 [1986]).

5. Dennis Overbye, 'Math Professor Wins a Coveted Religion Award', *New York Times*, 16 March 2006, at https://www.nytimes.com/2006/03/16/science/math-professor-wins-a-coveted-religion-award.html [accessed 27 January 2018]. For reports and comment on the award of the Templeton prize to Barrow, see Tim Radford, 'The gods of cosmology', *The Guardian*, 21 March 2006, p. 33.

6. Nicholas Campion, *The Great Year: Astrology, Millenarianism and History in the Western Tradition* (London: Penguin, 1994).

7. E. F. Schumacher, *Small is Beautiful: A study of economics as if people mattered* (London: Abacus, 1993 [1963]), p. 246.

8. G.S. Kirk, J.E. Raven and M. Schofield, *The Presocratic Philosophers*, 2nd edition (Cambridge: Cambridge University Press, 1983), fr.346–7 and p. 286 fr.346, p. 287. See also Denis O'Brien, *Empedocles' Cosmic Cycle: A Reconstruction From the Fragments and Secondary Sources* (Cambridge: Cambridge University Press, 1969).

9. Kenneth Sylvan Guthrie, *The Pythagorean Sourcebook and Library* (Grand Rapids, MI: Phanes Press, 1987); Flora R. Levin (trans.), *The Manual of Harmonics of Nicomachus the Pythagorean* (Grand Rapids, MI: Phanes Press, 2004).

10. For histories of harmony in music see Jamie James, *The Music of the Spheres: Music, Science and the Natural Order of the Universe* (New York: Copernicus Springer Verlag, 1993); Joscelyn Godwin, *Harmonies of Heaven and Earth from Antiquity to the Avant Garde* (Rochester, VT: Inner Traditions International, 1995 {1997]); and William Allaudin Mathieu, *Harmonic Experience: Tonal Harmony from Its Natural Origins to Its Modern Expression* (Rochester, VT: Inner Traditions, 1997). See also the helpful discussion in John D. Barrow, *The Artful Universe* (Oxford: Clarendon Press, 1995), pp. 199–204.

11. Plato, *Timaeus*, trans. R.G. Bury (Cambridge, MA, London: Harvard University Press, 1931), 35A–36B

12. Plato, *Republic*, 2 Vols, trans. Paul Shorey (Cambridge, MA, and London: Harvard University Press, 1935), X.617B.

13. Plato, *Timaeus*, trans. R.G. Bury (Cambridge, MA, and London: Harvard University Press, 1931), 35A–36B.

14. Campion, *The Great Year*; Plato, *Laws*, 2 Vols, trans. R.G. Bury (Cambridge, MA, and London: Harvard University Press, 1934); Plato, *Republic*.

15. See the discussions in Chris Impey, 'Truth and Beauty in the Cosmos', in Chris Impey and Catherine Petry (eds), *International Symposium on Astrophysics Research and on the Dialogue between Science and Religion* (Notre Dame, IN: University of Notre Dame Press, 2002), pp. 40–53; and Brian Greene, *Until the End of Time: Mind, Matter, and Our Search for Meaning in an Evolving Universe* (London: Penguin, 2020).

16. For background see Nicholas Campion, 'The Importance of Cosmology in Culture: Contexts and Consequences', in Abraao Jesse Capistrano de Souza (ed.), *Cosmology* (Rijeka: InTech Open, 2015), pp. 3–17; and Nicholas Campion, 'Cosmos and Cosmology', in Robert

Segal and Kocku von Stuckrad (eds), *Vocabulary for the Study of Religion* (Leiden: Brill, 2015), pp. 359–64.

17. Norris Hetherington, *The Encyclopaedia of Cosmology: Historical, Philosophical and Scientific Foundations of Modern Cosmology* (Oxford and New York: Routledge, 1993), p. 116.

18. Michael Rowan-Robinson, *The Cosmological Distance Ladder: distance and time in the universe* (New York: W.H. Freeman and Co., 1985).

19. Aristotle, *Meteorologica*, trans. H.D.P. Lee (Cambridge, MA, and London: Harvard University Press, 1937), 339a 19–24.

20. Diogenes Laertius, 'Zeno' in *Lives of Eminent Philosophers*, trans. R.D. Hicks (London: William Heinemann, 1925), Vol. 2, pp. 110–263, L VII.143, 142.

21. Diogenes Laertius, 'Zeno', Vol. 2, pp. 110–263, VII.88.

22. Diogenes Laertius, 'Zeno', Vol. 2, pp. 110–263, VII.89.

23 Cleomedes, 'On the Heavens', 1.3.3, 1.3.11, *Cleomedes' Lectures on Astronomy*, trans. Alan C. Bowen and Robert B. Todd (Berkeley, CA: University of California Press, 2004).

24. See for example, Plotinus, 'On Whether the Stars are Causes', Ennead II, 1–3, Vol. 2, trans. A.H. Armstrong (Cambridge, MA, and London: Harvard University Press 1929), 2.15–20.

25. Clive Hamilton, *Defiant Earth: The Fate of Humans in the Anthropocene* (Oxford: Polity Press, 2017).

26. J. E. Lovelock, *Gaia: A New Look at Life on Earth* (Oxford: Oxford University Press 1979), p. 120.

27. Johannes Kepler, *The Harmony of the World*, trans. E.J. Aiton, A.M. Duncan, J.V. Field (American Philosophical Society, Philadelphia, 1997); Johannes Kepler, 'Harmonies of the World', in Johannes Kepler, *Epitome of Copernican Astronomy and Harmonies of the World*, trans. Charles Glenn Wallis, in *Great Minds Series* (Amherst, NY: Prometheus Books, 1995), pp. 167–245. See also *Aviva* Rothman, *The Pursuit of Harmony: Kepler on Cosmos, Confession, and Community* (Chicago, IL: the University of Chicago Press, 2017), pp. 8–19.

28. Boethius, *The Consolation of Philosophy*, trans. V.E. Watts (Harmondsworth, Middlesex: Penguin, 1969), p. 97.

29. William Shakespeare, *Troilus and Cressida*, Act 1 Scene 3, 109–10.

30. William Shakespeare, *The Merchant of Venice*, Act 5 Scene 1, 52–63.

31. Shakespeare, *The Merchant of Venice,* Act 5 Scene 1, 75–86

32. Isaac Newton, 'Tabula Smaragdina, Hermetis Trismegistri Philosophorum Patris', in Keynes MS 28, in *The Chymistry of Isaac Newton,* King's College Library, Cambridge University, lines 1–4, http://webapp1.dlib.indiana.edu/newton/mss/dipl/ALCH00017 [accessed 4 May 2020].

33. Stephen D. Snobelin, 'The Myth of the Clockwork Universe: Newton, Newtonianism and the Enlightenment', in Chris L Firestone and Nathan A. Jacobs (eds), *The Persistence of the Sacred in Modern Though*t (Notre Dame, IN: University of Notre Dame Press, 2012), pp. 149–84.

34. Colin McLaurin, *An Account of Sir Isaac Newton's Philosophical Discoveries* (London, 1775), cited in Carl L Becker, *The Heavenly City of the Enlightenment Philosophers*, (New Haven, CT, and London: Yale University Press, 1966 [1932]), pp. 62–3.

35. Carl Becker, *The Declaration of Independence: a study in the history of political ideas* (New York: Vintage, 1958), pp. 59–60.

36. Thomas Paine, *The Age of Reason* (Mineola, NY: Dover Publications, 2003) pp. 191–2.; Thomas Paine, 'The Rights of Man', in Michael Foot and Isaac Kramnick (eds),

The Thomas Paine Reader (London: Penguin 2003), p. 207. See also Nicholas Campion, *The New Age in the Modern West: Counter-Culture, Utopia and Prophecy from the late Eighteenth Century to the Present Day* (London: Bloomsbury, 2015), chap. 3.

37. Thomas Paine, 'The Rights of Man', p. 217.

38. Thomas Paine, *Common Sense* (London: Penguin 1986), p. 66.

39. Brandon C. Look, 'Gottfried Wilhelm Leibniz', *Stanford Encyclopaedia of Philosophy*, 22 Dec 2007; *substantive revision Wed 25 Jul 2013*, https://plato.stanford.edu/entries/leibniz/#PreEstHar [accessed 15 Feb 2020].

40. Colin McLaurin, *An Account of Sir Isaac Newton's Philosophical Discoveries* (London, 1775), p. 94, https://archive.org/details/anaccountsirisaoomurdgoog/page/n101/mode/2up [accessed 2 April 2020].

41. Trudy Griffin-Pierce, *Earth Is My Mother, Sky Is My Father: Space, Time and Astronomy in Navajo Sandpainting* (Albuquerque, NM: University of New Mexico Press, 1995), p. 63.

42. Xiaochun Sun, 'Crossing the Boundaries between Heaven and Man: Astronomy in Ancient China', in Helaine Selin (ed.), *Astronomy Across Cultures: the History of Non-Western Astronomy* (Dordrecht: Kluwer Academic Publishers, 2000), p. 425.

43. Nathan Sivin, 'State, Cosmos, and Body in the Last Three Centuries B. C.', *Harvard Journal of Asiatic Studies*, Vol. 55, No. 1 (June 1995): p. 6.

44. Richard Wilhelm, *The I Ching or Book of Changes* (London: Routledge and Kegan Paul, 3rd edition 1968 [1951]), p. 280.

45. Dennis Tedlock (trans.), *Popul Vuh* (London: Simon and Schuster, 1996), p. 64.

46. Bruce Chatwin, *The Songlines* (London: Picador, 1988), p. 2.

47. Clive Ruggles, *Ancient Astronomy: An Encyclopedia of Cosmologies and Myth* (Santa Barbara CA: ABC CLIO, 2005), p. 27.

48. Virgil G. Hinshaw, 'Einstein's Social Philosophy', in Paul Arthur Schilpp (ed.), *Albert Einstein: Philosopher-Scientist* (Cambridge: Cambridge University Press, 1949), pp. 649–61 (pp. 659–60).

49. Albert Einstein, *The World As I See It* (No Place: Snowball Publishing, 2014), p. 35.

50. Werner Heisenberg, 'The Representation of Nature in Contemporary Physics', *Daedalus*, Vol. 87, No. 3, Symbolism in Religion and Literature (Summer 1958): pp. 95–108 (p. 96).

51. C.P. Snow, *The Search* (Harmondsworth: Penguin Books Ltd., 1965 [1934]), p. 18.

52. Snow, *The Search*, p. 18.

53. Werner Heisenberg, 'The Representation of Nature in Contemporary Physics', p. 107.

54. Heisenberg, 'The Representation of Nature in Contemporary Physics', p. 107.

55. Heisenberg, 'The Representation of Nature in Contemporary Physics', p. 100.

56. Albert Einstein to Max Born, 4 December 1926, in Irene Born (trans.), *The Born and Einstein letters: correspondence between Albert Einstein and Max and Hedwig Born from 1916 to 1955 with commentaries by Max Born* (London: Macmillan, 1971), letter 52, pp. 90–91. https://archive.org/details/TheBornEinsteinLetters/page/n55 [accessed 5 May 2020].

57. Partha Ghose, 'The Unfinished Search for Wave-Particle and Classical-Quantum Harmony', https://arxiv.org/abs/1502.03208 abstract; see also pp. 2, 21. Ghose is former Professor at the S.N. Bose National Centre for Basic Sciences in Kolkata. The full passage reads, 'The Koopman-von Neumann Hilbert space theory based on complex wave functions underlying particle trajectories in classical phase space, is an important step forward in that

direction. It provides a clear and beautiful harmony of classical waves and particles'.

58. Partha Ghose, 'The Unfinished Search for Wave-Particle and Classical-Quantum Harmony', p. 22.

59. For an accessible introduction see Martin Rees, *Before the Beginning* (New York: Basic Books, 1998).

60. Fred Hoyle, *Frontiers of Astronomy* (BiblioBazaar, 2011), p. 304.

61. John Wheeler, quoted in Martin Rees, *Before the Beginning* (London: The Free Press, 2002 [1997]), p. 257.

62. Paul Davies, *The Cosmic Blueprint: Order and Complexity and the Edge of Chaos* (London: Penguin, 1995), p. 5, citing Karl Popper and John Eccles, *The Self and its Brain* (Berlin: Springer International, 1977), p. 61.

63. Francis Crick, *The Astonishing Hypothesis: the Scientific Search for the Soul* (London: Simon and Schuster, 1994), p. 3.

64. Alfred Lord Tennyson, 'In Memoriam A.H.H.' (London: Edward Moxon, 1850), Canto 56.

65. Arne Naess, 'The Shallow and the Deep, Long-Range Ecology Movement', *Inquiry*, Vol. 16, No. 1–4 (1973): pp. 95–100.

66. Naess, 'The Shallow and the Deep, Long-Range Ecology Movement', p. 95.

67. Øyvind Grøn and Arne Naess, *Einstein's Theory: A Rigorous Introduction for the Mathematically Untrained* (New York: Springer, 2011), p. 3.

68. Grøn and Naess, *Einstein's Theory: A Rigorous Introduction for the Mathematically Untrained*, p. 47.

69. James Gleick, *Chaos: The Amazing Science of the Unpredictable* (London: Vintage, 1998). See also Paul Davies, *The Cosmic Blueprint: Order and Complexity at the Edge of Chaos* (London: Penguin 1995 [1987]).

70. Heisenberg, 'The Representation of Nature in Contemporary Physics', p. 108.

71. Heisenberg, 'The Representation of Nature in Contemporary Physics', p. 108.

72. Naess, 'The Shallow and the Deep, Long-Range Ecology Movement', p. 99.

73. Naess, 'The Shallow and the Deep, Long-Range Ecology Movement', p. 99.

74. Naess, 'The Shallow and the Deep, Long-Range Ecology Movement', p. 99.

75. Naess, 'The Shallow and the Deep, Long-Range Ecology Movement', p. 99.

76. Arne Naess, *Ecology of Wisdom* (London: Penguin, 2008), pp. 111–2.

77. 'Diversity and Inclusion', *Raconteur*, 8 March 2020. See also raconteur.net.

78. See Franz-Josef Brüggemeier, Mark Cioc and Thomas Zeller (eds), *How Green Were The Nazis? Nature, Environment, and Nation in the Third Reich* (Athens, OH: Ohio University Press, 2005); Janet Biehl and Peter Staudenmaier, *Ecofascism Revisited: Lessons from the German Experience* (Porsgrunn: New Compass Press, 2011).

79. 'Full text of President Xi's speech at opening of Belt and Road forum', *Xinhuanet*, 14 May 2015, http://www.xinhuanet.com/english/2017-05/14/c_136282982.htm [accessed 15 February 2020].

80. Cited in Julia Tao, Anthony B. L. Cheung, Martin Painter and Chenyang Li, 'Why governance for harmony?', in Julia Tao, Anthony B. L. Cheung, Martin Painter and Chenyang Li (eds), *Governance for Harmony in Asia and Beyond* (London: Routledge, 2010), pp. 3–11 (p. 7)

81. Max J. Zenglein and Anna Holzmann, 'EVOLVING MADE IN CHINA 2025: China's industrial policy in the quest for global tech leadership', *Mercator Institute for China Studies*, No 8, July 2019, https://www.merics.org/sites/default/files/2019-07/MPOC_8_MadeinChina_2025_final_3.pdf [Accessed 19 April 2020].

82. See for example, Joachim Gentz, 'Chinese he 和 in many keys, harmonized

in Europe', in Yuri Pines and Wai-yee Li (eds), *Keywords in Chinese Thought and Literature* (Hong Kong: Hong Kong University Press, in print); Chenyang Li, *The Confucian Philosophy of Harmony* (London: Routledge, 2014); Chenyang Li, 'The Confucian Ideal of Harmony', *Philosophy East and West*, Vol. 56, No. 4 (2006): pp. 583–603; *Kam-por* Yu, 'The Confucian conception of harmony', in Julia Tao, Anthony B. L. Cheung, Martin Painter and Chenyang Li (eds), *Governance for Harmony in Asia and Beyond* (London: Routledge, 2010), pp. 15–36; Liwen Zhang, 'Harmony and Justice', *Frontiers of Philosophy in China*, Vol. 10, No. 4 (December 2015): pp. 533–546; Jie Yang, 'Virtuous power: Ethics, Confucianism, and Psychological self-help in China', Critique of Anthropology, June 2019, pp. 1–22. For a rhetorical denunciation of the Chinese government see Kai Strittmatter, *We Have Been Harmonised: Life in China's Surveillance State* (London: Old Street Publishing, 2019).

83. Stephen C. Angle, 'Human Rights and Harmony', *Human Rights Quarterly*, Vol. 30, No 1 (Feb. 2008), pp. 76–94 (p. 77).armonyH

84. Angle, 'Human Rights and Harmony', p. 80.

85. Kepler, *The Harmony of the World*, p. 507.

86. E. J. Aiton, A.M. Duncan and J.V. Field, 'Introduction', in Johannes Kepler, *The Harmony of the World*, trans. E.J. Aiton, A.M. Duncan, J.V. Field (Philadelphia, PA: American Philosophical Society, 1997), p. 503.

87. Nicholas Campion, 'Harmony and the Crisis in Early Modern Cosmology: the Political Astrology of Jean Bodin and Johannes Kepler', in Charles Burnett and Ovanes Akopyan, eds, *Astrology versus Anti-Astrology in Early Modern Europe: Changing Paradigms in the History of Knowledge* (London: Routledge, Society for Renaissance Studies series, 2020), forthcoming; Nicholas Campion, 'Johannes' Kepler's Political Cosmology, Psychological Astrology and the Archaeology of Knowledge in the Seventeenth Century', *Mediterranean Archaeology & Archaeometry (forthcoming); Aviva* Rothman, *The Pursuit of Harmony: Kepler on Cosmos, Confession, and Community* (Chicago, IL: the University of Chicago Press, 2017).

88. 'Johannes Kepler's on the More Certain Fundamentals of Astrology Prague 1601', trans. Mary Ann Rossi with notes by J. Bruce Brackenbridge, *Proceedings of the American Philosophical Society*, Vol. 123, No. 2 (April 1979): pp. 85–163, Thesis 71, p. 104.

89. Haydn Washington, 'Harmony – not "theory"', *The Ecological Citizen*, Vol. 1, No. 2 (2018): pp. 203–10 (p. 208).

90. Andrew Dobson, *Green Political Thought*, 4th edn (London: Routledge, 2007), pp. 2–3.

91. Stan Rowe, 'Ecocentrism: The Chord that Harmonizes Humans and the Earth', *The Trumpeter*, Vol. 11, No. 2 (1994): pp. 106–7, http://www.ecospherics.net/pages/RoweEcocentrism.html [accessed 19 May 2014].

92. Stan Rowe, 'Ecocentrism: The Chord that Harmonizes Humans and the Earth', pp. 106–7.

93. For the relationship between classical ethics and ecology see Patrick Curry, *Ecological Ethics: An Introduction*, revised edition (Cambridge: Polity Press, 2011).

94. Curry, *Ecological Ethics: An Introduction*, p. 45.

95. Stan Rowe, *Earth Alive: Essays on Ecology* (Edmonton: NeWest Press, 2006), p. 21.

96. United Nations, 'Transforming our world: the 2030 Agenda for Sustainable Development', 'Declaration, para. 8.

FINDING SEA LEVEL:
THE CONCEPT OF HARMONY IN THE CULTURE OF THE KÁGGABA (KOGI) PEOPLE OF COLOMBIA

Alan Ereira

THIS CHAPTER EXPLORES THE NOTIONS OF BALANCE AND HARMONY expressed by the Colombian Kággaba people, generally known as Kogi. 'Kággaba', their word for themselves, means 'people'; they are better known in Colombia and throughout the world as 'Kogi', a lineage name that places them as descendants of the jaguar. They have become well known and respected in recent decades over much of the world through their representation of themselves as authorities on living in harmony with nature.[1]

I became part of this process when, in 1990, I helped them make a film for the BBC, *From the Heart of the World, The Elder Brothers' Warning*, in which they appealed for viewers to pay careful attention to their advice. The impact of their appearance and words was powerful; the film was immediately repeated, a version was shown on PBS in the USA as the tenth-anniversary presentation of the series *Nature*, and the book I wrote about the experience of working with them was given the Green Book Award. I remained in contact, with occasional visits, and in 2009 they asked me to help them repeat the appeal, with a film called *Aluna*.[2]

The Kogi people of Colombia are profoundly and permanently alarmed by the capacity of human beings to throw the world dangerously out of balance, threatening its very survival. Today, their energy and efforts are very properly concentrated on the harm caused by the ruthless resource-hungry non-indigenous population that invaded their territory five hundred years ago and is ever more busily plundering. But according to the epic myth of their own history, the story of Kabiúkúkwi, people were wrecking the balance of nature there long before Columbus. It is a form of primaeval flood story in which the earliest people themselves upset the balance of land and sea. They then had to restore the harmony of nature using the same processes which the Kogi use today.

The Kogi sages who work to restore balance and harmony to the world today are called *Mamas*. The word is a form of the word for the sun, so could be translated as 'enlightened'. In the early 1990s they gave me a printed version of Kabiúkúkwi's tale as told by Mama Shibulata.

Image 1: Mama Shibulata.

They had organised its translation into Spanish to provide their children with Spanish-language indigenous teaching material for new government-sponsored classrooms.[3] It has not been written down in their own language because the Kogi do not trust literacy and do not use it. It is the nature of myth to be preserved by speaking rather than writing, and for the audience to be a knowledgeable restraint on novelty and invention. The ancient myths that we know from Sumerian, Hebrew and Greek texts, such as those of Jason, Gilgamesh and Noah, were being told a thousand years or more before a script was invented for them. Unlike those heroes, Kabiúkúkwi is remembered for being a dangerous fool. Like them he belongs to pre-historical time; the final section of his name appears to signify 'ancient stone', indicating that he is one of the early ancestors whose lives ended not in death but in turning to stone.

The Kággaba are one of four indigenous communities of the Sierra Nevada de Santa Marta who self-identify as inheritors of the pre-conquest tradition there, known today as 'Tairona'. The other three groups are the Wiwa (or Sanka), Kankuama (or Kagkui) and the Ika (or Arhuaco), each speaking their own language but with a shared cosmology and sense of identity. The adult men all dress in loose garments of undyed cotton, use similar bags woven by women, and chew roasted coca leaves to which they add powdered lime carried in a small gourd or *sugui*, commonly referred to as a *poporo*. In the literature of the first half of the twentieth century they were all spoken of as sub-groups of Arhuaco.[4] They were described at the time as a diminishing society of perhaps 1,500 people, too simple and deferential to be authentic descendants of the indigenous urban culture encountered by early colonisers and subsequently apparently annihilated.

The indigenous people of the Sierra regard it as the original Eden from which all life emanates, the core and controller of the planet ('the heart of the world'). They understand themselves to be descended from the first humans, and Europeans to be descended from a second creation of humanity, which was expelled from the heart of the world to protect it. They call us their 'Younger Brothers'. So the story of Kabiúkúkwi is a tale of what happened when the world was nothing more than the Sierra and the sea.

The Kogi are the least acculturated of the indigenous Sierra groups. In 1950 Gerardo Reichel-Dolmatoff began producing better-informed descriptions of them and their cosmological understanding.[5] He presented them as a people with a sophisticated ecological philosophy which demands, as a moral and practical imperative, that they take care of nature. He described the training of Mamas as being conducted in darkness for up to 18 years from infancy. They learn to become conscious participants in a life of dynamic exchange between all living beings, which they believe is driven and defined by a transcendental consciousness, *Haba* (Mother) *Aluna*, that underlies material reality.[6] The Kogi have no formal religious organisation and Mamas are not priests; they perform a role more like that which Roman sources ascribed to Druids – though without the human sacrifices. That is to say, they are trained to communicate with the world beyond, largely through divination, and act as authorities on law, as adjudicators, experts on lore, history and nature, medical practitioners, and political advisors whose authority transcends individual communities.[7] In fact the similarities to Druids seem rather remarkable, extending from divining from clouds and birds to Caesar's statement that they are uncomfortable with literacy as damaging to the power of memory (a position explained to me by a Kogi Mama), and even to the Druids being reported early in the 1st century BCE as educating chosen youths for up to twenty years in a cave.[8]

Their understanding of the world is fundamentally holistic, and all elements of it must be understood as not simply connected but as different ways of seeing the same totality. For example, Kogi women are constantly stitching bags for their men to wear, and these are understood to be a constant renewal of the work of *Haba Aluna* in conceiving the world.

> Performed constantly, even while traversing steep and rocky paths with a baby on their backs, stitching is understood to focus the mind in synchrony with the Mother's rhythmic motions when she 'wove' the world, recreating Her 'thought'. In fact, on another level the cosmos is also seen as a bag. It is equally knitted from an origin-centre, the position of the Cosmic Pillar, in a gradually expanding circle which at one point starts rising to form the container, i.e., the cosmic levels. A bag recreates the Mother's life-giving cosmic 'womb' that carries the world in it, a force also contained in women (the word for bag also relates to one for the female reproductive organ).... Following the Kogis' analogical thinking, this weave imbues and connects all things (throughout the bag), and has one common origin in Sé[9] (the bag's centre). ...Kogi bags are the cosmos itself.[10]

Image 2: A family moving with goods and cattle; she is making a bag.

Reichel-Dolmatoff interpreted Kogi culture through the philosophical presentation of its cosmology. This is based on their view that the cosmos consists of inherent dualities, including light/dark, male/female, above/below, interior/

exterior, matter/spirit, sea/land. In fact these are all aspects of the same thing; the role of the Mamas is to understand the world in these terms and to actively work to maintain balance between complementary dualities where necessary. The Kogi word *yuluka* signifies balance and more: it is 'the capacity of an individual to be in harmony with what is around them'.[11] It can signify a person's capacity to live in equilibrium by rising above what appear to be grave obstacles (a natural disaster, sickness…) and transform them. When I first went to the Kogi to discuss making a film for the BBC, and they needed to understand what use I might be, I found my *yuluka* being deliberately tested in various ways, including my capacity to rise above obstacles.

The Mamas place great emphasis on maintaining natural harmony through a dynamic of exchange which they see as essential to all life. This takes place at every level of physical existence. It includes the water cycle, with snow-melt descending through thirty river valleys to tropical lagoons and returning by evaporation and precipitation to the peaks. It also includes the constant movement of animals, birds and insects as they work through their life cycles; I was given an educational lecture by one informant on the roads made and used by leaf-cutter ants, which were compared directly to human movement in the Sierra.

Constant human movement around the Sierra is a necessity, driven by the geographical facts of life on their mountain. The massif is the world's highest coastal mountain. It is an equilateral triangular pyramid, each side measuring 80km. The north face rises steeply from some 600m below the surface of the Caribbean to two permanently snow-capped summits 5,770m high just 42km inland. It is a 'biogeographical island' at a collision point between the Caribbean and South American tectonic plates, sharply bounded on the north by the sea, on the west by the plains of the Magdalena river valley and on the east by the narrow valley of the Cesar river and the Guajira desert. The variety of temperatures (tropical to arctic), precipitation (rainforest to desert) and saturation (mangrove swamp to sparse tundra) creates a dense variety of micro-climates, ecological niches that largely mirror the range of life-supporting land environments over the rest of the planet.

The Kogi live in valleys associated with their lineages, where they grow crops on small family farms. Each household normally maintains a number of farms at different altitudes, growing crops appropriate to the elevation. Before colonisation, the sea provided fish and the shore gave the shells needed for lime. That region now has to be visited rather than inhabited; the indigenous people have been largely forced away. The bottom of the valleys is the banana zone;

further up, cotton and sugar cane grow up to about 1000m, and above that are coca, coffee, fruit and beans. This way of life requires them to move constantly up and down the mountain, carrying goods and materials between their own farms and exchanging produce with their neighbours.

Walking up and down the mountain is itself invested with moral significance. It is obviously central to the act of exchanging objects around the Sierra, and exchange is a key part of maintaining the flow and balance of life. Objects and produce are moved around not simply for consumption, but also to manifest and maintain the fact of connection. Sustaining harmony here is not a passive contemplative duty (though Mamas do meditate deeply for days and nights at a time) but an active, dynamic exercise involving steadily striding up and down the steep valleys, carrying bags of stuff. Meeting a Kogi man on the path and enquiring amiably how he is, one is often answered 'I am well seated'. The reference is to being on a simple wooden bench, considering what needs to be done. In his mind that is what the Kogi is doing, in spirit. But in reality he would be travelling around three times as fast as I can walk these paths. In my experience, the ability to keep moving is a requirement of any proper person; I know of one ex-governor of their political organisation, the son of a great Mama, who was relieved of his post partly because he was felt to have become too stout to do the requisite walking. A sedentary man is a defective man, unable to maintain harmony with the world effectively.

I fall far short on this measure of proper action. The Kogi actually seem to have been quite comfortable with having identified my inadequacy; perhaps, if I am a typical 'younger brother', it helps to explain why we have failed so seriously to maintain harmony in the world. In my early days of involvement with them, as a TV producer 30 years ago, they called me 'the BBC' and would ask 'Don't you know how to walk? In your cars, in your aeroplanes, don't you walk?'. The obvious answer was 'no', and as I stumbled on, the cry would go out 'The BBC is dying!' But at least I was carrying my goods to exchange. These were carefully chosen, with the advice of Gerardo Reichel-Dolmatoff; mostly sea shells for the men and stout needles for the women. The shells, which I brought from the UK but which came from the Pacific, were received with particular enthusiasm. I was told that these were forms of shell which had been brought to the Sierra in the past, but are now only found in burials, and were very welcome.

The constant exchange being carried on is not trade; there is no market and no currency. It may be considered a form of the 'vertical archipelago' posited by John Victor Murra to describe the Ayamaras' pattern of exchange in the Andes in sixteenth-century Peru.[12] It depends on a shared understanding of reciprocity. The Kogi word for this is *zhigoneshi*, which expresses a symbiotic and balanced relationship of giving and receiving.

Zhigoneshi represents exchange in the form of information as well as physical material. In 1987 the Kogi established the Organisación Gonawindua Tayrona (OGT) to represent the Mamas' opinions to the Colombian government[13], and then established a documentation centre to hold all the material they could gather which showed how they had been represented. They named the centre *Zhigoneshi*. It was run by Arhuacos who could read and write in Spanish, and published an occasional journal with this title. The journal has gone, and the Zhigoneshi documentation centre now produces videos and photographic exhibitions.

For the Kogi physical possession of a text or image also opens the possibility of engaging with it and affecting its place in the world. This was explained to me when they first said they wanted copies of the images I made. I had assumed that this was the courtesy deposit that a film-maker leaves, but the Kogi certainly believe that their intellectual property in the product extends to an on-going power over it, exercised through divination by Mamas.

As with *zhigoneshi*, harmony and the balance of *yulúka* is achieved through dynamic effort. Kabiúkúkwi's story, like most Kogi stories, begins at the beginning, when the world emerged into the light as the realisation of a slowly constructed conception in the mind of *Haba* Aluna.[14] Today there are freshwater lakes of snow-melt high in the tundra, and fresh water flows down to the sea and salt-water mangrove lagoons below. But in the Mama's telling, at this time the level of the sea was undefined, and there were lagoons of salt water high in the tundra. 'But we did not want them. We saw that these lakes would be bad.'

The sea, as in so many creation myths, is the amniotic fluid from which the world emerged. There is a dichotomy between the feminine sea and the masculine physical reality of the created world. *Haba* Aluna conceived the world, *Haba* being Mother in the sense not just of a birth-mother, but also as the sustainer of what she has produced. It was given its actual physical form by *Hate* Seizhankua. *Hate* means Father in the same enlarged sense that *Haba* means Mother. Father Seizhankua was one of the first 'sons' of the Mother, and the first problem of harmonisation of the world was finding the correct balance between these newly invented categories, masculine and feminine.

In my experience, the Kogi understanding of the world's complementary cosmic forces is one in which balance is extremely delicate, and imbalance is a constant threat and a real danger to life. The whole world is structured on the

Image 3: The balance of masculine and feminine, Hate and Haba,
and the future of the world.

basis of the complementarities of up and down (the levels or worlds above the
earth are mirrored below it), and the most basic dichotomies are between what
exists in reality and what in spirit, and between masculine and feminine.

A Mama once led me into the issues surrounding this by asking rhetorically
'Where does the idea of earth come from? Where do we get the idea of water?
Why do we have a word for it?' [15]

The words he was using for 'water' and 'earth' also invoke 'spirit' and
'matter'. In a dialogue that seemed to echo Socrates, he said that these were
ideas that preceded the material existence of the world; they were formed in the
cosmic mind, aluna. Other Mamas spoke to me of material forms being traces
in reality of ideas that precede materiality. 'These things were conceived in the
beginning. Before the dawning, these were ideas – water and earth. We can only
have ideas of things that have already been conceived in *aluna*.' The need to
achieve balance between land and sea and between masculine and feminine is
the basis of Kabiúkúkwi's story, which explains what went wrong in the earliest
times and what had to be done to bring them back into harmony.

The complementarity of masculine and feminine extends even to communication. Kogi discursive practice, *gwiabawashi*, sets out the rules governing the correct use of speech (and, by extension, of still and moving images). According to local anthropologist Julio Marino Barragán, this mediates relationships between the community, their authorities and nature to achieve harmony *(yuluka)* and balance *(zhigoneshi)*. It acts as a mechanism of social control and spiritual development. The object is for communication governed by these concepts to seek harmony in all areas of spiritual and material existence, in social relations and in all dealings between people.[16] It also requires a balance between masculine and feminine elements of speech, and I have sat in on long discussions about the gender of specific syllables. The Kogi use complex compound expressions to refer to their world and Mamas need to understand the gender balance of their words. Men have to remain constantly aware of the way they relate to women both specifically and in principle. Kogi men will avoid looking a woman in the face, as she is the Mother manifest. The gourd each man carries, his *sugui*[17] containing lime dust made from sea shells, represents a womb.

Image 4: A Mama with his sugui.

He is given it at puberty as part of his education in manliness, which means preparation for starting a family. He is meant to think about his actions and obligations while dipping a stick into the *sugui* and lifting some of the lime to his mouth, so having his masculinity tempered. Complementarity is everything. Men are the weavers, women the spinners.

The Mother spun the thread of her thought on the spindle which is the Sierra, but she needed sons to make the world real, and men weave the threads spun by their wives. A thread spun with a left-hand twist, with the hand moving downwards over the thigh, is female: this is the warp thread, the thread which spans the four corners of the loom, the link between the cardinal points. This marks out the nature of the cloth, defines it. It is passive, productive, timeless; it

Image 5: A man weaves, lifting the comb and passing the black shuttle through the
opening in the female threads

spells out shape and form. It is an aspect of the Mother.

A thread spun with a right-hand twist, with the hand brushing upwards across the thigh, is male. This is the weft, which travels between the cardinal points as the sun travels across the sky, moving in time, with a past and a future.[18] The threads of the warp open in a shape we call the 'shed', to receive the shuttle, which is made of black wood and carries the meaning of blackness, a state that precedes creation. The shuttle penetrates, the comb of the loom presses the white thread it carries down into its place. And thus a cloth dawns which harmonises male and female, ordering and binding together the world.

When men gather in the *nuhue*, their meeting house, they are inside a structure whose architecture is a trace of the structure of the cosmos. Its pressed earth floor is beneath the four worlds that are shown in circles around the conical thatched roof, and which are mirrored in the four subterranean worlds below. It is an exclusively male space, where men have their benches and hammocks around four fires, quartering the world. But between the circles of men and each fire is a ring of stones, which sit there in the place of women.

The men's house conveys the sense of a council and education centre, devoted to lengthy speeches and occasional ritual drumming and, sometimes the playing out of ritual performances. There are no children there, except on special occasions when an infant may be brought to dance. Dance is thought of as fundamental to life; they say that 'we dance to keep us from dying', and as soon as a tiny boy being raised to be a Mama is able to toddle, he makes his first appearance in the men's communal house, the *nuhue*, brought dancing by the Mama caring for him. Mamas are also trained in the use of masks for dancing, each Mama being associated with a specific character. According to Reichel-Dolmatoff, these figurative masks transform the wearers and enable them to resolve struggles between complementary forces. 'Each mask represents a particular supernatural force which 'sees' the other masks in another dimension during the dance. There is generally an underlying concept of opposition (sëlda, in Kogi), of a struggle between two large categories of opposing but complimentary forces represented by masks....and so there are a great number of dances in which masked figures fight together or appease each other and finally re-establish a balance (yulúka).'

There is an equivalent women's house, with a single fire; it has more of the sense of a domestic space, with babies and small children, and much livelier music and dancing which is regarded as recreational. Music, song and dance brings harmony and health, and is therefore good for children. When I first filmed with the Kogi my British crew entertained themselves by setting up a music system in the large Kogi house where we kept camera equipment. Children naturally gravitated to it and I worried that the Kogi might find it an annoyance or worse. Instead, the women clearly decided that our pleasure in music, dance and playfulness was a positive for the happiness, welfare and good health of children, and watched them taking part. Social harmony, the welfare of nature, the health of individuals and the community are all seen as aspects of the same condition, and damage to any part of this biological totality is expected to produce symptoms including disease, natural disaster and social breakdown.

Which is not to say that everything in the Garden of Eden was understood to be paradise. Every aspect of harmony is constantly endangered. The Kogi see life ever teetering on the edge of chaos, birth always endangered by death, unchecked vitality risking the appearance of parasites, disease, chaos. Female sexual energy can be understood to be threatening as well as life-enhancing, and the salt sea which gives life to the world may also threaten intrusion. Then, it is seen as the source of disease, pollution and of course invasion. The Sierra offers a high degree of stability between night and day (every day is an equinox, at least to within about 30 minutes) and the rise and fall of the sea (the maximum tidal range is less than 60 cm.) But the stabilization of sea level, a fundamental element in the harmony of the world, was quite a saga.

When, at the beginning of time, lagoons of salt sea-water existed in the high tundra of the Sierra, it was not a good sign. The Mother of these lagoons was the first female child of the Great Mother, called Nabobá, and she was considered dangerous because she was in the wrong place. The tundra was meant to be black earth, not salt sea.

And so the Mama's story unfolds. [19]

Image 6: Seeing the sea from the high tundra.

We saw that these lakes would be bad. But who thought that? He who thought that was Kabiúkúkwi.

The wonderfully modern arrogance of Kabiúkúkwi was immediately followed by practical action.

People said 'Take these lakes down, they must not stay in the high tundra because they will begin to eat up the land and cause problems. It would be very bad to keep them here'. And they also said: 'We know everything, with the power we have we can obtain all that the lakes have to offer. It is better to send them down to the bottom'. Then the Mother, said 'All right, since you say they are no use here, take them down below'. And so they began to lower the lakes, and they said, 'Take them down further, because they are no use to us here. They have a very bad smell'.' They insisted that they could find shells to make lime and shrubs to grow cotton. 'The Great Mother said, 'So you think the lakes are of no use to you?' And they replied, 'Yes, that is what we think'. But the Mother thought, 'that is what they say, but they are wrong'. But she allowed them to drain the salt lakes, and the trees and plants that depended on them moved down the mountain.

And still not content, these human engineers of the world lowered the sea itself, drying up a great deal of the mountain.

Everything, everything went down with the sea. They took it yet further and the Mother asked once again, 'Have you done?' 'No, lower still'.

The sea dropped to the lower reaches of the Sierra. They considered leaving the

Image 7: A Mama makes an offering in *aluna*.

sea there, but decided against that and lowered it farther still.

> *In every place the sea rested, it left its traces. Still the Great Mother said, 'That is fine, lower it as far as you wish'.*

> *'The sea was flung as far as Jarxsuzaluaka, which today is on the far side of the sea.* (This was probably one of the islands of the Caribbean, which exchanged materials with the Tairona). *And the wisest of them inquired of those who were lowering the sea, 'Is it a good thing that you have done this, that you have lowered mother sea this far?' 'Yes, it is a good thing!' the ancients replied.*

> *But the Mama says they now saw that the sea was very far away, and understood that the sea was necessary to them.*

So Kabiúkúkwi's name became a byword for foolish arrogance, and they had to find a way to bring back the sea, to re-balance the world.

> *They recovered it by making offerings, because they saw that without the sea they could not obtain seashells or the many other objects only the sea can provide. And if the sea was not brought back closer again, they would soon*

use up all the shells on the mountain. and they would be left without. That is why the sea was brought back to where it is today. The ancients had to struggle hard to bring it nearby once again.

The Mama says, all that I have said about the sea, about the Great Mother, I have heard with my ears, and I have learned.[20]

The point of the story is of course that the world needs to be kept in balance by firstly, understanding what they call 'the original law', behaviour required to sustain the healthy balance of life on earth, and secondly, knowing how to make *pagamentos,* payments at appropriate places to compensate for damage that has been done and to restore the harmony on which all life depends.

These *pagamentos* are made to the immaterial essences of things that exist in the material world. After centuries of exposure to Christian terminology, they have settled on using the term 'spiritual', though these essences are ideas with personalities, and the connotations of 'spiritual' seem inappropriate. They are also very different from Platonic essences, because they have the qualities of personal identity. For the Kogi, the ideas manifested in material forms have individual life. *Haba* Aluna, the original Mother, did not conceive lifeless abstractions. She gave birth to non-material beings, whose traces appeared in the world we inhabit at the dawn of time. These are the recipients of the *pagamentos.*

In the words of the Kogis' unpublished explanatory text, *Shikwakala,*

Pagamentos are based on the fact that everything that exists in nature had its origin in an invisible space that we call Sé. Everything that we see today was a person in essence: the trees were people, the water was woman, the hills men. These were the original people. At the materialization of the world they became the elements of the reality we see today, but on Sé's level, they continue being invisible persons, owners and caretakers of the materialized structure. All the elements of nature are beings, all are Kággaba, they are people like us, with whom we must maintain communication so that there is balance and harmony in this world that we inhabit. Therefore, it is our duty to give back to the essence-beings, for everything we use, for permissions they give us: for example to the trees to cut beams, to make a house, a seat, a door. The trees in essence are people like us, and we must make pagamentos to them.[21]

Those essence-beings are the *Habas* and *Hatas,* Mothers and Fathers, of the material forms of which they are essences. The *pagamentos* consist of precisely the objects which were so threatened by Kabiúkúkwi's Cunning Plan: shells and fragments of cotton thread, leaves and secretions which have to be brought from one part of the Sierra and offered in another, in a set of meaningful acts accompanied by serious thought about what has gone wrong, and why, and what restitution must be made. The work is performed at, or directed towards, what they speak of as 'sacred spaces', the places where they believe they can have direct access to the Fathers and Mothers of what they describe as 'all the elements of nature that we use and of all human actions and thoughts from the beginning of humanity'.

Since the *pagamentos* are necessary to maintain harmony and balance, the Mamas are desperately concerned to protect the 'sacred spaces' from physical damage wreaked by tourists, soldiers and commercial development. They believe that all life depends on what is done in these locations. When I have discussed with them what Sierra lands they want restored to their management, the Mamas speak only and always about 'sacred spaces', not about territory for crops or habitation. Where an offering-place has been damaged, they study how to clean and restore it, but say 'This can only be done when the damage is curable, and not where there have been major interventions, such as megaprojects (dams, ports, mines), which cause irreversible damage. That permanently affects the health of the territory and the Kággaba.'.

The level to which the sea was restored after *Kabiúkúkwi*'s intervention, the correct place for sea level, is now represented by the Black Line with which the indigenous people of the Sierra have circumscribed their territory. In 1973, a group of Mamas walked this invisible path around the whole massif. In 1999, OGT contributed to an official declaration of the need for protection of the territory because it is filled with 'sacred spaces'.[22] At the end of 2018 they achieved government recognition by decree of the territory within the Black Line as filled with interconnected living features which are essential to their lives, and connect 'the spiritual principles of the world and the source of life'.[23] The sea has many meanings, and we are one of those meanings. The Younger Brothers, who began their invasion of the Sierra at the start of the sixteenth century, are themselves the destroyers and polluters of the land, who must be held off – not too far away, but not too close. Above all, we must be kept away from the delicate places that cover the Sierra, and which become more vulnerable the higher you go.

But government declarations and the Mamas' offerings do not seem to be enough to hold the world in harmony, and ensure that sea level is in the right place. In January, 2019 it became clear that the geopolitical context has deteriorated suddenly on the North side of the Sierra. In that month several social leaders and an employee of the Sierra Nevada National Park were assassinated in broad daylight. It seems that José de los Santos Sauna, the Kogi governor of the OGT, has been directly threatened. He had to withdraw to high-altitude villages for protection. The peace agreement with FARC has been repudiated by the new government; armed paramilitary bands have returned and the region is unsafe once again. People who act as intermediaries between NGOs and the Kogi have been directly threatened in the past months, and field missions in Kogi territory (the north of the Sierra) have been suspended.

The Kogi remain committed to the belief that the world can be saved, and that their work to restore harmony is vital and remains purposeful. The Mayor of Santa Marta, deeply concerned by the situation, has appealed directly to the Ministry of Interior to try to cope with this worrying upsurge of violence. And since the Kogi view the world as a single living entity, they are quite clear that we and the Mother are as much as risk at themselves. That is the meaning of their concept of harmony.

NOTES

1. E. Kemf, 'The law of the mother', *People Planet*, 1.3 (1992): 16-17.; G.E. Rodríguez-Navarro, 'Indigenous Knowledge as an Innovative Contribution to the Sustainable Development of the Sierra Nevada of Santa Marta, Colombia', *AMBIO: A Journal of the Human Environment* 29(7), (1 November 2000); J. Reddi, 'What Colombia's Kogi people can teach us about the environment', *The Guardian* 29 Oct. 2013; https://www.theguardian.com/sustainable-business/colombia-kogi-environment-destruction consulted 31/03/2019

2. Letter from the Kogi Mamas, February 2, 2011, http://www.alunathemovie.com/letter-from-the-kogi-mamas/ consulted 31/03/2019

3. Part of the text can be found in M. Rocha Vivas, *Antes el Amanecer: Antología de las literaturas indígenas de los Andes y La Sierra Nevada de Santa Marta*, 2. Nación desde la raíces, Ministerio de Cultura, Col., Biblioteca Básica de los pueblos indígenas de Colombia (2010), 551-55, https://en.calameo.com/read/0057303416ef60470e32f consulted 18/3/2019

4. T.D. Cabot et al., 'The Cabot Expedition to the Sierra Nevada de Santa Marta of Colombia', *Geographical Review*, Vol. 29, No. 4 (Oct., 1939): pp. 587-621.

5. G. Reichel-Dolmatoff, *Los Kogi: Una tribu indígena de la Sierra Nevada de Santa Marta, Colombia*, Book 1, (Bogotá: Revista del Instituto Etnológico Nacional, 1950), Book II, (Bogotá: Editorial Iqueima, 1951).

6. G. Reichel-Dolmatoff, 'Training for the priesthood among the Kogi of Colombia', in J. Willbert (Ed.), *Enculturation in Latin America: An Anthology* (Berkeley: University of

California Press, 1976), 265-88.

7. J. Caesar, *Gallic Wars*, VI, 13-14; Cicero, *De Divinatione* I, xli 90; Strabo, *Geographica*, IV, 4 c 197,4

8. Pomponius Mela III, II.

9. Sé is the fundamental proto-space that precedes creation, its structure is the source of law and defines the rules of existence.

10. B.T. Parra Witte, 'Living the Law of Origin: The Cosmological, Ontological, Epistemological, and Ecological Framework of Kogi Environmental Politics', (unpublished PhD thesis, University of Cambridge, 2018), 4.4.2.

11. E. Julien and M. Fifils, Les Indiens Kogis: La memoire des possibles (PLACE: Actes Sud, 2009), p. 264.

12. J.V. Murra, 'An Aymara Kingdom in 1567', *Ethnohistory*, 15(2) (1968): pp. 115-51.

13. Yanelia Mestre Pacheco, Peter Rawitscher Adams et al., *Shikwakala* (private publication of Organización Gonawindua Tairona, Santa Marta, 2018), p. 225.

14. Alan Ereira, *The Elder Brothers' Warning* (London: Tairona Heritage Trust, 2008), p. 111.

15. *From The Heart of the World: The Elder Brothers' Warning*, BBC TV (1990), 00.39.35-00.40.02 https://www.youtube.com/watch?v=HfSnTUc52C8&t=9s

16. 'El arte de hablar, el placer de escuchar', in *Memorias del Simposio Participación de las lenguas en la construcción de sentidos sociales*. II Congreso de Etnoeducación. Instituto Caro y Cuervo, Fondo de Publicaciones de la Universidad del Atlántico, quoted in M. Trillos Amaya, 'Por Una Educación Para La Diversidad', *Nómadas* 15 (Col) (2001): p. 167.

17. The Kogi *sugui* is usually called a *poporo* in academic literature.

18. G. Reichel-Dolmatoff, 'The Loom of Life: A Kogi Principle of Integration', *Journal of Latin American Lore* 4:1 (1978): 5-27.

19. For the following extracts see *El Mar y Las Lagunas*, a pamphlet in the series *Las Palabras de las Mamas*, La Dirección de Asuntos Indígenas (1992). Much of the text is available in Rocher Vivas, *Antes el Amanacer*.

20. *El Mar y Las Lagunas*.

21. Y. Mestre Pacheco, P. Rawitscher Adams, et al., *Shikwakala* (private publication of Organización Gonawindua Tairona, Santa Marta, 2018), p. 16.

22. Organisación Gonawindúa Tayrona, UAESPNN y Dirección General de Asuntos Indígenas del Ministerio del Interior, 1999, quoted in G. Sánchez Herrera et al., *Plan de Manejo 2005-2009; Parque Nacional Natural Tayrona* (Ministerio de Ambiente, Colombia, 2006), p. 74.

23. Republica de Colombia, Ministerio del Interior, Decreto numero 1500 de 6 Ago 2018, Articulo 4 (b), (c).

Harmonising the Land and Sky in Aboriginal Dreamings

Trevor Leaman

THIS CHAPTER EXPLORES THE WAYS in which Aboriginal and Torres Strait Islander peoples see the realms of Earth, sea and sky as aspects of a unified 'cosmoscape' – in which the skyworld is every bit as real as Earth, complete with rivers and forests inhabited by fish, birds, animals and ancestral beings.[1] Certain important stars and asterisms were seen as the skyworld counterpart of terrestrial animals and their annual appearance and movement through the night sky informed people of the seasonal migrations, lifecycles, abundance and food resource availability of the animals they represented.[2] The examples I deal with here are only a handful of the many Dreamings which harmonise the celestial cycles of the animal constellations in the sky with the lifecycles of their terrestrial counterparts and serve to demonstrate the keen-eyed observations of the natural world by the Indigenous First Australians.

Reflecting the many diverse biogeographic regions and habitats across the continent, and the Dreamings connecting them to the people of that country, a single star or asterism can represent a different animal 'constellation' to each of the many language groups. There is, therefore, no single Aboriginal astronomy – each of the 250 or more language groups has its own Dreamings associated with the land, sea and sky, but they all interconnect through the songlines which criss-cross the land.

The Celestial Emu

Perhaps the best known of the Aboriginal constellations is the Emu in the Sky or Celestial Emu. The Celestial Emu is found in the dark dust lanes of the Milky Way between the Southern Cross (head), Scorpius-Sagittarius (body) and Ophiuchus-Aquila (feet) (Figure 1, left). This constellation is found right across the continent, but the best studied version of the Dreaming narrative associated with it comes from the Wiradjuri, Kamilaroi and Euahlayi peoples of central west New South Wales.[3] His form can also be seen among the extensive rock art sites around the Sydney Basin.

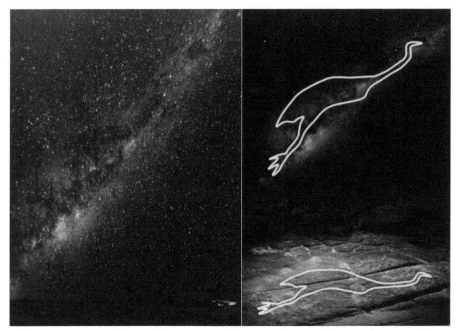

Figure 1: The Celestial Emu, Image: Wikimedia Commons/Barnaby Norris

The Celestial Emu is one of the easiest of all the Aboriginal constellations to find and recognise. Look for the dark dust lanes of the Milky Way between the Southern Cross (the head of the emu, marked by the Coalsack Nebula) and Scorpius (the big bulge of the emu's body). His legs stretch further still across the sky, making it a very large constellation. The emu rock engraving at Elvina Track, Ku Ring Gai National Park, Sydney is thought to be a representation of the Celestial Emu (Figure 1, right).

The orientation of the Celestial Emu in the sky after sunset was used to inform Aboriginal people of the lifecycle of the terrestrial emus, and specifically when it was the best time to collect eggs, an important protein resource.[4] The first appearance of the Celestial Emu in April-May (Figure 2A) signifies the start of the Emu breeding season when the females chase the males before mating. When the Celestial Emu is horizontal in the sky in June (Figure 2B), this is the time when terrestrial Emus are nesting and laying eggs. This is the best time to harvest emu eggs, but when doing so the Aboriginal people take only enough for their needs and leave a couple of eggs behind in each nest to ensure the viability of the breeding population. When the Celestial Emu starts to dip head-down in July (Figure 2C), this signifies that it is too late to harvest eggs as they now contain chicks ready to hatch.

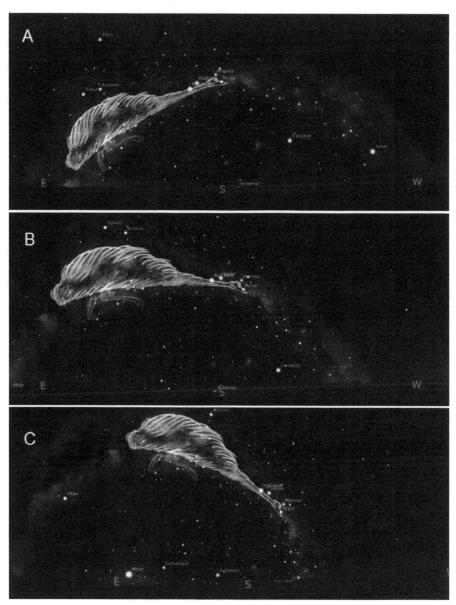

Figure 2: The orientation of the Celestial Emu in the sky.
Image: Stellarium/Robert Fuller.

Figure 3: Neilloan the Mallee Fowl. Image: Stellarium/Wikimedia Commons.

NEILLOAN AND MARPEANKURRK

To the Wergaia people of northern Victoria, the star Vega, in the western constellation of Lyra, the Harp, is *Neilloan*, the Mallee Fowl.[5] The star's first appearance in the east after sunset, or acronychal rise, in late winter-early spring coincides with the males preparing the nest for the breeding season. The females start to lay eggs around the time Vega crosses the meridian at sunset (dusk meridian crossing) in September, and hence when *Neilloan* reaches its highest elevation in the sky (Figure 3). The first chicks of the season start to hatch in November, coinciding with the last appearance of Vega in the north-western sky after sunset (heliacal set).[6]

Also from Wergaia country is an important Dreaming associated with the star Arcturus, in the western constellation of Boötes, the Herdsman.[7] The dreaming tells of a time long ago when the people were suffering under a big drought, *Marpeankurrk* wandered away from camp to die in peace. Whilst awaiting the inevitable, she noticed a trail of ants disappearing down a hole. She dug down to uncover an ant's nest full of larvae (*bitturr*), an important high-protein food source She started eating the *bitturr* and noticed her strength returning. By showing

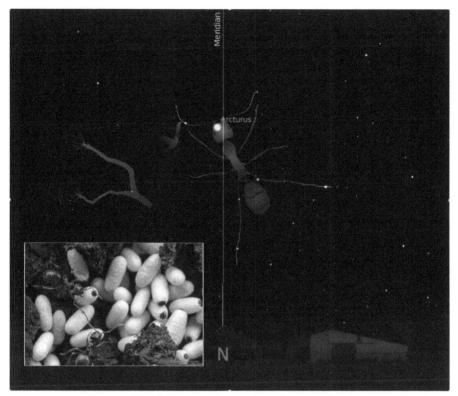

Figure 4: The dusk meridian crossing of the star Arcturus.
Image: Stellarium/Wikimedia Commons

other members of her family how to find more *bitturr* she ensured their survival. To commemorate her deed, she was placed in the sky on her passing to become the star we know as Arcturus, but to the Wergaia it was always *Marpeankurrk*, the 'Wise Woman Star'. Its dusk meridian crossing in August (Figure 4) served as a reminder to her people of her deed, and tells them the time of year when *bitturr* was available as a food source.[8] The orange hue of Arcturus also mimics the colour of the species of ant from that particular region of northern Victoria.

THE PLEIADES OR SEVEN SISTERS

The asterism known as the Pleiades or Seven Sisters (Figure 5), is linked to many important Dreamings across Australia. It is a very young (in astronomical terms: about 115 million years old) and prominent open cluster of stars in the zodiacal constellation of Taurus, the Bull. Along the east coast, the first predawn

Figure 5: The Pleiades. Image: Wikimedia Commons.

appearance (heliacal rise) of the Pleiades in early June signifies the beginning of the northern migration of humpback whales and orca from their summer feeding grounds in Antarctica to their winter breeding grounds in Northern NSW and Southern Queensland, The acronychal setting (last appearance in the west just before sunrise) of the Pleiades in late October to early November coincides with the southerly migration back to Antarctica, with young calves in tow and orca in pursuit.

In the central desert, the heliacal rise of the Pleiades signified the 'official' start of winter and the peak in the dingo breeding cycle.[9] Dingoes were important to the desert Aboriginal people, both as a source of warmth against the cold winter nights and as a source of food when other foods were scarce during droughts. Seeing the Pleiades in the dawn sky told them it was time to look for dingo pups. The Seven Sisters were also totemically-linked to many other plants, insects and animals through interrelated Dreamings, such as bush tomatoes (*kutjera*), honey ants (*tjala*) and thorny devils (*mingari*).[10]

THE TREE GOANNA

According to the Wiradjuri and related language groups of central New South Wales, the bright orange star Antares, and associated stars making up the western

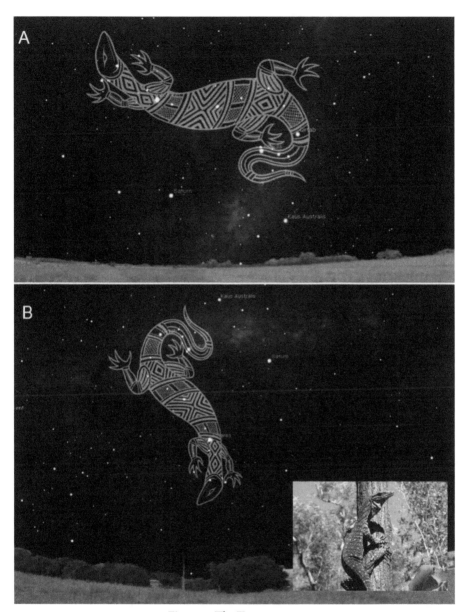

Figure 6: The Tree goanna.
Image: Wikimedia Commons/Stellarium/Scott "Sauce" Towney.

constellation of Scorpius, the Scorpion, was *Guggaa,* the tree Goanna.[11] This constellation informed the Wiradjuri of the best time of year to hunt the tree Goanna. When the *Guggaa* was first seen to rise in the dusk sky in July-August,

Figure 7: Baidam, the Shark. Image: Brian Robinson/Summer Ash

and hence was about to 'climb up' his 'celestial tree' (Figure 6A), it was deemed the wrong time of year to hunt the goanna as it was still lean and thin from lack of food availability over winter. However, when it was seen to be facing downwards in the western sky at dusk in October-November (Figure 6B), it informed the people that it had now 'climbed down' from the 'celestial tree' after feasting on eggs and chicks from nesting birds, and thus its terrestrial kind were likewise full of the nutritious fat that was both an important food source and valuable medicine.

BAIDAM THE SHARK

In the astronomical traditions of the Torres Strait, the shark constellation, *Baidam* (also spelt as *Beizam*), is made up of the stars in the asterism of the Big Dipper, which is part of the constellation of Ursa Major, the Big Bear.[12] When these stars appear in the northern sky in the direction of New Guinea in July-August, Islanders know the mating season of the shark is starting. This is when sharks are more plentiful close to shore, so the Islanders are extra wary and vigilant when wading into the shallow waters close to the shoreline. The appearance of *Baidam* also informs the Islanders of a seasonal change, and that it is the time to plant banana, sugar cane, and sweet potato.

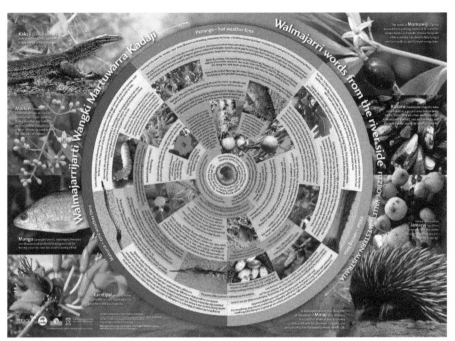

Figure 8: A seasonal resource calendar for the Walmajarri people.
Image: CSIRO

ENCODING THE KNOWLEDGE

Traditionally, Aboriginal and Torres Strait Islander people would encode this seasonal astronomical and ecological knowledge in song, dance and ceremony.[13] Today, there are efforts to also record and preserve this knowledge in the form of colourful and informative regional resource calendars (CSIRO; Figure 8) each specific to a language group and region.[14] Each chart may contain and convey several layers of interrelated information, such as the name for the season and time of year, the seasonal quality (wet/dry, hot/cold, predominant wind direction, etc.) what animal, plant or insect is abundant for each season and other Dreaming relationships including the appearance of important seasonal stars heralding these changes. These colourful charts are being used in the classroom to help pass on this important and vital cultural knowledge to the younger generations, both Indigenous and non-indigenous.

NOTES

1.   P. A. Clarke, 'The Aboriginal cosmic landscape of southern South Australia', *Records of the South Australian Museum*, 29 (1997): pp. 125-145; P.A. Clarke,. An overview of Australian Aboriginal ethnoastronomy', *Archaeoastronomy*, 21 (2007/2008): pp. 39-58; P.A. Clarke, 'The Aboriginal Australian Cosmic Landscape. Part 1: The Ethnobotany of the Skyworld', *Journal of Astronomical History and Heritage*, 17(3) (2014): pp. 307–325; P.A. Clarke, 'The Aboriginal Australian Cosmic Landscape. Part 2: Plant Connections with the Skyworld', *Journal of Astronomical History and Heritage*, 18(1) (2015): pp. 23–37 and P.A. Clarke, 'Australian Aboriginal Astronomy and Cosmology', Ch 214 in C.L.N. Ruggles (ed.), *Handbook of Archaeoastronomy and Ethnoastronomy* (New York, Springer, 2015).

2.   D.W. Hamacher,'Identifying seasonal stars in Kaurna astronomical traditions', *Journal of Astronomical History and Heritage*, 18(1) (2015): pp. 39–52 and T.M. Leaman, D.W. Hamacher and M.T. Carter, 'Aboriginal Astronomical traditions from Ooldea, South Australia, Part 2: Animals in the Ooldean sky', *Journal of Astronomical History and Heritage,* 19(1) (2016): pp. 61-78.

3.   R.S. Fuller; M.G. Anderson; R.P Norris and M. Trudgett, 'The Emu sky knowledge of the Kamilaroi and Euahlayi peoples', *Journal of Astronomical History and Heritage*, 17(2) (2014): pp. 171-179.

4.   Fuller, et al.

5.   J. Morieson, 'The Night Sky of the Boorong: Partial Reconstruction of a Disappeared Culture in North-West Victoria' (MA Thesis, University of Melbourne, 1996) and W.E. Stanbridge, 'On the astronomy and mythology of the Aborigines of Victoria', *Proceedings of the Philosophical Institute of Victoria, Transactions*, 2 (1857): pp. 137-140.

6.   Leaman et al.

7.   Morieson; Stanbridge.

8.   Leaman et al., 2016.

9.   Leaman et al. and N.B. Tindale, 'Celestial lore of some Australian Aboriginal tribes', *Archaeoastronomy*, 12/13 (1983): pp. 258-379.

10.   Leaman et al., 2014; 2016.

11.   Wiradjuri custodians, personal communication.

12.   D.W. Hamacher, 'A Shark in the stars: astronomy and culture in the Torres Strait', *The Conversation*, 10 July 2013.

13.   D. W. Hamacher, 'Dancing with the stars': astronomy and music in the Torres Strait', in N. Campion and C. Impey (eds.) *The Inspiration of Astronomical Phenomena: Proceedings of the INSAP IX Conference* (Lampeter: Sophia Centre Press, 2016); and T.M Leaman and D.W. Hamacher, 'Aboriginal Astronomical traditions from Ooldea, South Australia, Part 1: Nyeeruna and the "Orion Story"', *Journal of Astronomical History and Heritage*, 17(2) (2014): pp. 180-94.

14.   CSIRO, Indigenous Seasons Calendars, https://www.csiro.au/en/Research/Environment/Land-management/Indigenous/Indigenous-calendars [accessed 10 March 2019].

The Farm as an Ecosystem

Helen Browning

FARMING AND FOOD PRODUCTION are some of the vital issues of our time. I want to consider briefly the idea of the farm as an ecosystem, or as a sort of self-contained living organism, which has been one of the fundamental tenets of organic husbandry. Indeed, this was one of the concepts I encountered thirty years ago when I started my farming life. I'm a farmer myself in Wiltshire, as well as Chief Executive of the Soil Association.[1] And I've spent the last thirty years in some ways trying to live up to this expectation of us as organic farmers. I'm going to raise three issues in the hope that we can stimulate some conversation around them. Firstly I want to give a brief sense of the trials and tribulation of being a practical farmer and, secondly, I will ask whether organic farms really are self-contained islands in the way we sometimes think and thirdly, I want to raise a question which is an increasingly big one for me, which is that of the relationship between the farm as an ecosystem and the wider environment.

So, my own farm: briefly, I farm about 1,500 acres, a thin strip of land from the top of the Marlborough Downs into the Vale of The White Horse in southern England. We have lots of different soils and lots of different geographies through that five-mile strip of land. And we have all the things one would expect of any self-respecting organic farm. We have a balance of fertility generating and fertility using crops, along with a balance of livestock of different types, sheep, dairy cows, beef, pigs and a few chickens, so that we have a clean grazing system and largely-disease free herds. We have a balance of people and skills to run all these enterprises, crops and livestock, and we've got pretty much all the machines and the resources to do all the things that we need to do from repairing machines to cutting our hedges within our own farming business. So, we are largely, though not entirely, self-contained.

And over time we've tried to balance things up a bit more by creating some of the habitats that were lacking on the farm. We've planted woodlands and we've recreated the chalk downland pastures that were so appallingly destroyed after the war; something like 97% of our chalk down land biodiverse grasslands have gone.[2] We've got a little bit of it back again. And we've tried over time to feed our local community as far as possible. The way today's business economics works meant that our enterprises didn't develop in the way we'd initially hoped;

we ended up feeding people all over the place including in other countries rather than in our local community. But, over the last ten years we have managed to re-localise some of our food business, feeding people through our own pub, for example. And now we are even working with some Italians making mozzarella on the farm so the whey from the mozzarella which comes from our own milk can go back to feed our pigs, one of those age-old cycles.

This has all been helpful, especially in pushing back against over-specialisation which is a scourge of our time, and thinking more and more about the resources we have already got on our own farm rather than constantly buying them in. But at the same time, the wider system within which we operate is deeply flawed. As one example, it is impossible today to return the waste from the people who are eating our food back to that land, certainly within the organic system. It's also very hard to feed our local community directly when we don't even have the infrastructure such as abattoirs and processing opportunities, without us all having to capitalise our farms beyond sensible levels. And most of us are not yet tackling the energy issue on our farms, either. It's impossibly difficult because as farmers we are not very good at managing the degree of complexity that this kind of system requires. And we still struggle to find people who both have the values and are practically competent to run these kinds of complex systems. But it's also complicated because no farm is an island, and nor should it be. And, then, what is a farm anyway apart from an arbitrary unit of ownership or stewardship? We know that to secure our biodiversity we need to connect up habitats across the wider environment, not divide it into little ecosystems. We must start to think beyond the unit of the farm in a way that is going to allow species to spread and thrive and allow genetic flow through our landscapes, and to manage our water on a catchment scale rather than just thinking about it at the farm level.

And so I hope that in the future some of our new policy incentives will incentivise management of the wider landscape rather than just focusing on the unit of the farm. And, furthermore, I think one of the only ways that our farmers are going to survive in the longer term is for them to start to share and collaborate and work across their boundaries, whether sharing machinery or labour or marketing collaboratively, as well as improving the ecosystem through engaging beyond the farm gate.

But my final question today is whether our obsession with farming in a narrow sense needs to end. What I mean is that we need to start thinking about how we work with and harvest nature rather than constantly intervening, subjecting nature to our idea of how we want it to be. This leads me to an interest in practices such as permaculture where, instead of constantly ploughing and

Helen Browning at Home on her Farm

sowing and reaping, which we do even within organic and rotational farming, we can develop more permanent systems of farming. Recently, I have become very involved in agroforestry, trying to develop an approach which is more productive per acre while supporting wildlife, protecting soils, providing shade and shelter for farm animals, and sequestering carbon. We are using trees like nuts and fruits, integrating them with our pasture land and cropping. What really interests me is how none of this is ever static, and that harmony is not a point we get to and then preserve, but always a dynamic process.

NOTES

    1.   The Soil Association, https://www.soilassociation.org
    2.   Lucy E. Ridding, John W. Redhead, Richard F. Pywell, 'Fate of semi-natural grassland in England between 1960 and 2013: A test of national conservation policy', Global Ecology and Conservation, Volume 4, July 2015, pp. 516-525, https://www.sciencedirect.com/science/article/pii/S2351989415300184

The Permaculture path to Harmony:
a Story of Personal Emergence

Angie Polkey

PERMACULTURE IS A PROFOUNDLY RADICAL APPLICATION of the principles of Harmony: because for most practitioners it is deeply based on such concepts as local democracy, social justice and equality. The term permaculture was coined in 1978 by Bill Mollison and David Holmgren, and defined by them as 'an integrated, evolving system of perennial or self-perpetuating planet and animal species useful to man'.[1] Mollison and Holmgren contrasted the then dominant and inherently unsustainable system of energy-expensive, mechanistic, soil-destroying agribusiness with what they hoped would be a sustainable (the 'permanent' in permaculture) low-energy, high yield model, working with local ecologies. For Mollison and Holmgren permaculture should be small-scale, intensive (rather than extensive), preserve diversity in plant and animal species, adapt to local conditions, plan for the long-term rather than short-term gain, and be applicable to urban environments as well as rural.[2]

There have been many updated definitions of permaculture since then but most emphasise that it's about much more than simply growing food in a sustainable way. The UK Permaculture Association website explains:

> Permaculture is a design process. It helps design intelligent systems which meet human needs whilst enhancing biodiversity, reducing our impact on the planet, and creating a fairer world for us all. People across the globe are creating thriving communities with permaculture.[3]

Permaculture magazine is equally broad in its view, and states that,

> Permaculture is...an innovative framework for creating regenerative ways of living; a practical method for developing ecologically harmonious, ethical, human-scale and productive systems that can be used by anyone anywhere.[4]

I have been involved in permaculture since 1995 and am a founder member of the Lampeter Permaculture Group in West Wales.[5] I am a permaculture teacher and a conservation & sustainability advisor, running regular courses on permaculture, forest gardening and conservation subjects at Denmark Farm Conservation Centre near Lampeter and the Centre for Alternative Technology near Machynlleth, as part of Aberystwyth University's Lifelong Learning programme.[6] I also run my own rural permaculture holding with my husband and fellow ecologist Andy. It's now a gloriously abundant place where we we've been 'practicing what we preach' for over twenty years – however, my ecological awareness began much, much earlier than that, as I'll explain.

I was five years old when I noticed what Harmony isn't – and what it could be – during my first day at school. As I raced outside with my playmates in the break, I was confronted by an endless, uniform sea of green – the playing field. It looked vast to my young eyes, monotonous and uninteresting. Luckily, one side was bounded by a rusty chain link fence – and beyond the fence was a narrow strip, only a few metres wide, of unfettered wildness. Here there was vegetation of different heights and colours,with flowers and seedheads bursting in exuberance. In my imagination, it was also where 'mysterious creatures' lived! I was hooked.

As an older child, escaping from exam revision and family disunity, I took solace in the wild, wet vista of the seascape that bordered my coastal home. Watching fishermen casting their lines, hearing the endless pounding of the waves, browsing the strandline for exciting new finds, the changing nature of the tides – all combined to show me a cycle of life in which humans were integral but not dominant. But this halcyon view was not to endure. As I grew older I became more aware of the damage we were inflicting on each other and the planet so, when I looked out to the watery horizon as a young adult, I dreamt of a better world beyond.

Upon acquiring my very own first home and garden, my overriding memory is of carrots – large, earth-covered and billowing with feathery foliage! It may be a simple thing to grow one's own food but it impacted me hugely. It seemed nothing short of a miracle that I could help nature to grow and share her abundance. I felt gratitude, and quite a bit of pride, for the vibrant, healthy, colourful sustenance that nature and I could co-produce. The natural world, as in my younger days, became my icon, my teacher, the way I saw meaning in life and my escape route from a growing disconnect with mainstream society.

It was obvious that I was going to study ecology at university. From being knee high to a grasshopper I'd loved playing in the mud, with worms and, perhaps most importantly, having quiet time in nature. As the oldest child and only girl in a noisy, bustling household, something in my psyche often preferred the company of creatures over people. My body, mind and soul craved wholeness and, to me, this meant embracing the complexity of ecology. I loved its intricate, dynamically changing patterns and the fact that everything, from the tiniest to the biggest being, is interconnected.

After university, I landed my dream job with a national nature conservation organisation. However, the term 'holistic' was virtually unknown back then and wholeness was definitely not on the agenda. Much as I admired the skills and dedication of those trying to save species and habitats, it seemed that humans weren't included. Narrow-minded organisational reviews and punishing financial targets got to me. I survived by playing relaxation tracks to and from work but I still woke up in the middle of the night, my heart beating fast and anxious thoughts racing through my mind. I asked myself: 'why is it OK to treat other species with awe and wonder but not care for humans too?' It was then that I discovered permaculture and it truly saved my life!

Permaculture is a holistic, systems-based approach to designing sustainable human settlements, communities, businesses and other endeavours. It has a set of twelve guiding principles, modelled on natural systems, that teach us to,

Observe and interact; Catch and store energy; Obtain a yield; Apply self-regulation and respond to feedback; Use and value renewable resources and services; Produce no waste; Design from pattern to details; Integrate rather than segregate; Use small and slow solutions; Use and value diversity; Use edges and value the marginal; Creatively use and respond to change. [7]

These principles help to ensure that harmonious systems emerge, in the knowledge that everything is dynamic and nothing is set in stone. In short, we design as nature does, within an overall pattern or framework that allows individual elements and their responses not only to be expressed but also to teach us what to do next.

On my first permaculture course in 1995 I was taken back to my infant days – looking longingly over the fence at nature's masterpiece, the abandoned strip of land next to my school field. But this time we were treated to a master 'design-scape' in the form of a forest garden visit. Here there were overlapping layers of vegetation, multiple yields, bees buzzing and birds singing and a beauty that I could hardly imagine from the boring, municipal plantings of my home town. I once heard someone say that beauty is sustainability in action – and it can be if we are tuned in to the Harmony of natural systems.

But the biggest revelation of all was that permaculture has three Ethics – a set of values – that underpins all that we design. *Earth Care*, looking after the earth, sits alongside *People Care*, looking after people, and *Fair Shares* – only using our 'fair share' of the earth's resources. This was a light bulb moment for me – and was patently <u>not</u> what my conservation employer was practicing at the time. It all made sense, rationally and emotionally, and I felt the next stage of my own path to Harmony beginning to emerge.

Moving *away from* disharmony and *towards* harmony was important for my process – and is, in fact, a feature of everything that's alive. If they can, living beings avoid stimuli that don't serve them, such as toxic chemicals, adverse environmental conditions, predators, and are attracted to those that assist survival, such as nourishing foods, warmth, shelter, companionship. My emotional survival depended on moving away from a dysfunctional workplace – and so it was that my husband and I used our new permaculture tools to design our move to a small land holding in West Wales in 1997.

By 2000, I had linked up with other local permaculture teachers to teach a Permaculture Design Course, at the suggestion of someone who saw more in me than I did myself at the time! Permaculture ethics guided our team to create a safe, comfortable, inclusive space to teach in, with scrumptious food, plenty of participative activities and time for self-care and celebration. The students bonded like no other group I'd witnessed and, by the end of the course, the Lampeter Permaculture Group (LPG) was born. And so began the next phase of my life!

The Lampeter Permaculture Group has a constitution and practices based on permaculture ethics and principles. Group meetings echo the format of the original course, sharing activities, company, food and business in one seamless operation. We use zones of influence (another permaculture tool) to reach out to the public through events such as Apple Days, Seed Swaps and Plant Swaps and to engage with decision makers in the local community. The Group has also helped to spawn Transition Lampeter (all Transition Towns reference permaculture) and this has led to further community developments in the shape of a community hub, local 'People's Market' and other events for the community.[8]

Back home, our 'design' has been emerging as we *Observe and Interact* to see how the land and associated living things respond to our interventions. The permaculture principle of *Creatively Use and Respond to Change* comes into sharp focus as we we experience the effects of climate change on our weather. After 22 years of living here the changing patterns are clear and we need to consider growing some different crop and tree species as well as other ways to build resilience. Another principle, *Apply Self Regulation and Accept Feedback,* is also poignant in this regard – how long before the human race accepts that its impacts are causing dis-Harmony in the natural systems that serve us all?

We live in a very old Welsh stone farmhouse and have been very mindful of its heat leakage and associated carbon emissions. Applying the principle of *Produce*

No Waste is incredibly difficult but we've managed to *Catch and Store Energy* in several ways. Firstly, we've heavily insulated the windows, floor, roof and one of the interior walls, using materials that are carbon negative over their lifetime (*Obtain a Yield*). Secondly, we installed solar panels so that we can use our own electricity and export to the national grid, thereby helping to decarbonise the national supply. Thirdly, we've planted our own, diverse woodland to create a sustainable harvest of fuel for our incredibly efficient thermal mass stove (made in the UK and fitted by a local Welsh supplier).

One day, we might insulate the walls externally and install solar hot water but, applying *Small and Slow Solutions* and *Designing from Patterns to Details*, shows that our household is a really low energy consumer so we prefer to adjust our behavior rather than install expensive systems at the moment. A passive solar conservatory adds warmth both directly and via thermal mass in the walls and floor and has cut our heating season down by two months. This was made possible because we used permaculture tools to help us choose the right property in the first place, with a favourable aspect. If only all new developments were designed with solar gain in mind. Added to which, the conservatory is a multifunctional space where we can 'eat out' even when it's raining, enjoy gatherings with friends, hang and air the washing, dry crops and grow tender plants. A perfect way to *Integrate rather than Segregate*!

So what has all this got to do with Harmony? On a personal note I can honestly say that, whatever's going on in the world, when I walk around the land that we steward, I feel a deep sense of wellbeing. This is partly to do with a huge sense of satisfaction from having helped the land and wildlife to flourish, but more than this, I very much appreciate what natural systems can accomplish if we allow them to help us. The permaculture ethics of *Earth Care*, *People Care* and *Fair Shares* are all expressed perfectly in nature and will sustain us if we learn to cooperate rather than dominate.

My feeling of connection goes wider though, across to the broad surrounding hills of the Cambrians that are part of my sense of place. *Use Edge and Value the Marginal* applies here. Knowing that towns like Lampeter and Aberystwyth, with their myriad sustainable initiatives and community networks, exist within our bioregion sustains me immensely. Projects such as the Cambrian Mountains initiative, which aims to restore biodiversity to the well-known mountain range in mid-Wales and help our communities to vision a better world in response to the environmental and social challenges ahead, are food for my soul.[9] We are all interconnected, and Harmony, for me, comes from learning about, nurturing and participating in connections that are an expression of the whole.

NOTES

1. Bill Mollison and David Holmgren, Perma-Culture One: A Perennial Agriculture for Human Settlements (Tyalgum: Tagari Publishers, 1990 [1978]), p. 1.

2. Bill Mollison and David Holmgren, Perma-Culture One: A Perennial Agriculture for Human Settlements (Tyalgum: Tagari Publishers, 1990 [1978]), pp. 6-7 and 90-4.

3. For the Permaculture Association see https://www.permaculture.org.uk/ [Accessed 07 June 2019].

4. Permaculture Magazine, Issue 100, Summer 2019 see https://www.permaculture.co.uk/; Maddy Harland, Lush: 'Science shows how permaculture could be the sustainable future of farming', Lush, https://uk.lush.com/article/science-shows-how-permaculture-could-be-sustainable-future-farming

5. For Lampeter Permaculture Group see http://www.lampeterpermaculture.org/ [Accessed 07 June 2019].

6. For Denmark Farm Conservation Centre see https://www.denmarkfarm.org.uk/ [Accessed 31 May 2019]; For the Centre for Alternative Technology see https://www.cat.org.uk/ [Accessed 07 June 2019]; For Aberystwyth University Lifelong Learning courses see https://www.aber.ac.uk/en/lifelong-learning/ecology/ [Accessed 07 June 2019].

7. Permaculture Principles, 'Permaculture Design Principles', https://permacultureprinciples.com/principles/ [Accessed 9 June 2019] And see David Holmgren, *Permaculture: Principles & Pathways Beyond Sustainability* (Hepburn, Vic.: Homgren Design Services, 2002).

8. For Transition Towns see https://transitionnetwork.org/ [Accessed 07 June 2019]; for the People's Market see https://www.facebook.com/thepeoplesmarketlampeter/ [Accessed 23 September 2019].

9. Land Use Consultants, Bangor University and Victoria University of Wellington, NZ, 'Cambrian Mountains Adaptive Landscape Project', 10353_DefraCambriansALPCaseStudyReportFINAL.pdf [Accessed 9 June 2019]; For Tregaron Community Hub see https://www.facebook.com/whilenyporthmyn.droverswheel [Accessed 07 June 2019].

Harmony and the Food System

Gunhild A. Stordalen

From a talk at the Harmony, Food and Farming Conference, organised by the Sustainable Food Trust, Llandovery College, 10 July 2017.

WHAT WE EAT AND HOW WE PRODUCE is currently at the heart of some of our greatest health and environmental challenges, including the double burden of malnutrition and the epidemic of diet-related diseases, along with climate change and wider environmental degradation. One third of all food produced is either lost or wasted.[1] And when talking about a sustainable food system, we cannot forget the more than one billion people who work to produce and serve food every day, including some of the world's poorest and most vulnerable – the 500 million smallholder farmers in the world.

The growing demand for cheap meat is driving up intensive livestock production at the expense of animal health and welfare with multi-resistant superbugs and scandals like swine flu, avian flu and horsemeat representing the top of the iceberg of detrimental consequences caused by factory farming. And all this is set to get worse as we will have another two billion people to feed over the next decades. Yet, with business as usual on food, there is no way we can achieve the targets of the Paris Climate Agreement or the United Nations Sustainable Development Goals (SDGs) as food is linked, whether directly or indirectly, to almost all seventeen of them.[2] On the other hand, getting it right on food could be our greatest opportunity to improve our health and wellbeing while at the same time protecting our planet.

The American food journalist Michael Pollan famously stated, 'Eat real food. Not too much. Mostly plants'.[3] In fact, now there is consistent evidence that he is right: plant-based diets tend to be both healthier and better for the environment. Although this is true for the world as a whole, in some regions, people may actually have to increase consumption of meat and animal products to meet dietary needs. Currently, there is no agreement on how we can create a global food system that can deliver healthy diets to a growing population within the boundaries of our planet.

To alert us to the dangers of climate change, we have the Intergovernmental Panel on Climate Change (IPCC), which has enabled the world to agree on limiting global warming to less than two degrees Celsius, as well as to coordinate action to

417

restrict greenhouse gases. But so far, there is no equivalent to the IPCC to address challenges in the food system and therefore no science-based targets for decision-makers. As an initial attempt to define the two-degree equivalent for the food system, my foundation, EAT, has, in collaboration with the *Lancet*, the leading medical journal set up a commission of twenty leading experts in nutrition and environmental sciences and policy.[4] The 'EAT-Lancet Commission on Food, Planet, Health', published in 2019, provides the first comprehensive assessment of what constitutes healthy diets from sustainable food systems and brings us one step closer towards a scientific consensus and evidence-based targets for a global food system that meet the needs of a 2030 world.[5]

But knowledge alone is not enough to change the world. We need innovation, not only in technology but also in business models. Although a growing number of actors in the food industry are making serious efforts to change, it is so far not nearly radical enough to take us to where we have to be in 2030. For most large producers, change is still about 'less bad', about making their products and production practices less unhealthy and with a smaller environmental footprint. And instead of investing in research and development in order to improve the health of both people and planet, many companies continue to put their research and development money into nutria pharma in order to tackle problems that they themselves are causing in the first place. As an example: only 40% of new products entering the US food and beverage market in 2016 were classified as net positive for health.[6] Therefore, in order to meet the needs of a 2030 world and achieve the SDGs, we need to shift from 'reducing, repairing and less bad' to 'preventing, mitigating and regenerating'! By starting to address the root causes, we can prevent problems from happening in the first place.

Today, there is a total disconnect between production and consumption policies in different government departments. Around the world, agricultural departments tend to subsidise the production of animal feed and junk food, fueling health and environmental crises which then have to be tackled by other departments across the corridor. What we need is coherent food and agricultural policies and aligned incentives, linking what we produce to what we actually can and should consume. And to make healthy and sustainable food available and affordable for all, health and sustainability objectives need to be integrated in all policies at all levels, from the ways that agriculture is subsidised to the ways in which consumers are incentivised.

Last, but not least, we need more collaboration and better alignment between sectors and disciplines and across borders, as no single actor, government or industry can change the system alone. This will require major commitment from science, business and politics as well as civil society, and will have to be a well-coordinated effort, because, in spite of new research projects, government commitments, business solutions and positive consumer trends, progress is still fragmented and too slow. That is why the EAT Foundation works to help turn these many scattered changemakers into an efficient and comprehensive game-changing movement. Through events like the annual EAT Stockholm Food Forum, or EAT Asia-Pacific Food Forum in Jakarta, EAT offers neutral, collaborative platforms for stakeholders to take part, take action and take the lead on finding win-win solutions to food system challenges.[7] And throughout the year, we are working to translate the latest science into action, through partnerships with cities, governments, academic institutions, the food industry, investors and innovators. By working together we can shift food production and consumption from a cause of problems to a cure, thereby creating a healthier and more prosperous world for all.

Radical changes will be needed. But to get there, we need more integrated knowledge and science-based targets. We need more innovation in technology but also new business models, and more collaboration across governments and across sectors. Last, but not least, we need to hurry up as 2030 is only eleven years from now. So let's get to work!

NOTES

1. Food and Agriculture Organisation of the United Nations, 'Key facts on food loss and waste you should know', http://www.fao.org/save-food/resources/keyfindings/en/ [accessed 7 June 2019].

2. For the Paris Climate Agreement see 'Paris Agreement', https://unfccc.int/files/meetings/paris_nov_2015/application/pdf/paris_agreement_english_.pdf [accessed 7 June 2019]; United Nations Sustainable Development Goals, https://www.un.org/sustainabledevelopment/sustainable-development-goals/ [Accessed 7 June 2019].

3. Daniel J. DeNoon. '7 Rules for Eating Choose Food Over Food-Like Substances, Food Writer Michael Pollan Tells CDC', WebMD 23 March 2009, https://www.webmd.com/food-recipes/news/20090323/7-rules-for-eating#1 [accessed 10 July 2017].

4. For EAT see https://eatforum.org/ [accessed 7 June 2019].

5. 'Food in the Anthropocene: the EAT–*Lancet* Commission on healthy diets from sustainable food systems', 16 Jan 2019, https://www.thelancet.com/commissions/EAT [accessed 16 Jan 2019].

6. For wider context see the US Office of Disease Prevention and Health Promotion, 'Dietary Guidelines 2015-2020', https://health.gov/dietaryguidelines/2015/guidelines/table-of-contents/ [accessed 16 Jan 2019].

7. EAT Stockholm Food Forum, https://eatforum.org/event/eat-stockholm-food-forum-2018/ [accessed 16 Jan 2019]; EAT Asia-Pacific Food Forum, https://eatforum.org/event/eat-asia-pacific-food-forum-2017/ [accessed 16 Jan 2019].

MIND THE GAP!
EXPLORING THE GAP BETWEEN HARMONY AND THE WATERY MATERIALITY OF CLIMATE CHANGE(S) IN RURAL KENYA

Luci Attala

Luci Attala

INTRODUCTION

This chapter considers the influence water (as a hyperobject) has on climate change. Drawing particularly on my experience with the Giriama in Kenya, I demonstrate the inherent complications of finding balance or harmony when perspectives on the material world dramatically diverge.

For Timothy Morton, the end of the world has already started – and he knows this because of the hyperobjects. Hyperobjects, he tells us, are extensive, ungraspable entities that exist simultaneously in different states. He lists things like the climate, air, lava, water, tsunamis, viruses, algae, meteors, and even capitalism as hyperobjects.[1] To be clear, hyperobjects are not causally linked to the end of the world. They are simply a mechanism or method through which we can get a glimpse at the 'strange-strange …[ness]' of our situation.[2] The value of recognising hyperobjects is that they threaten the detrimental habit of imagining that we are looking down and in on the rest of the world. They remind us that we are of *this* world as their effects flow through us. The behaviours, therefore, of hyperobjects force us to realise we are always inside and, a part of, the totality of whatever is happening. In short, the learning hyperobjects offer is that humanity is not, and cannot be (other than intellectually), materially separated from everything else.

This chapter is ethnographically rooted in the conceptions of water held by a group of Giriama horticultural-pastoralists from rural Kenya who I have lived and worked with on and off since 2007. My initial contact with this group of people came through the Welsh government, when I was tasked to monitor and evaluate a water project that they had co-financed with a not-for-profit reforestation initiative working in the area. After that project had drawn to a close, I stayed working in collaboration with the locals on various alternative livelihood schemes of their instigating, and worked looking at how they negotiate water shortages in this semi-arid region. As an anthropologist my knowledge of the Giriama is drawn from extensive immersive participant observation and informal interviews, which means living with and doing as. Consequently, I consider some of the people I describe here to not only be my friends but also part of my extended family.

To date the local population have relied exclusively on naturally occurring sources for their supply, but with drought deepening due to the climate changing water sources are becoming unreliable. In 2016, the arrival of a water system – in the shape of a single pipeline served by a series of managed kiosks – dramatically altered the Giriama's perceptions of water and has introduced a previously unheard of economic component to their relationships with water. In the light of these coalescing alterations, a pipeline delivering a clean water supply should perhaps be celebrated as a success; however, although its arrival while a striking positive for some members, it is also dividing communities as a result of different abilities to access the water. This division circulates around two key themes: first, distance from the tap and, second, economic ability to spare what small amounts of cash a family might accumulate on water. Consequently, this novel method of engagement with water is responsible for creating a series of material and economic challenges that are further powerfully altering local practice.

The question I pose is whether the notion of 'harmony' can be used to help us understand the inherently convoluted, overlapping, interconnected complexities of the actions of hyperobjects like the climate or water. I will briefly reflect on some of the challenges underpinning using the notion of 'harmony'. I do this by attending to both its vastness and its specificity. If harmony is understood as an equilibrium or balance between all parts of a whole, then this chapter explores the densities of understanding the enormity of that idea, and offers one ethnographic example to illustrate how problematic it is to achieve accord despite best intentions.

WATER'S PART IN THE END OF THE WORLD?

For Morton, the end of the world started in April 1784 after some bright spark appreciated that steam was powerful.[3] This spark ignited the touch-paper that went on to produce an engine that in turn helped pave the road to The Industrial Revolution. Steam, used in this way, offered savings in time, effort and production, enabled immense financial gain for some, and generated an unprecedented burgeoning of technological progression first in Europe, and then beyond. Indeed, frankly, the idea of utilising the pressure that trapped steam offers is responsible for the mechanisation of tasks that many of us expect and enjoy today. To be more specific, perhaps it is accurate to say that the behaviour (physics) of water – that is that it produces steam – was formative and, therefore, instrumental in enabling and inspiring the Industrial Revolution and its consequences.

It may seem trite to claim that the ability to mechanise in this way was only possible because of how materials behave together, but recognition of the materiality of worldly relationships demonstrates quite clearly that if steam did not behave in the way that it does, we may not be where we are today. In consequence and to ensure that the now-considered-valuable steam remained in good supply, people's daily labours altered significantly – as Marx succinctly illustrated in the following passage:

The economist understands very well that men [sic] make cloth, linen, or silk materials in definite relations of production. But what he has not understood is that these definite social relations are just as much produced by men as linen, flax, etc. Social relations are closely bound up with productive forces. In acquiring new productive forces men change their mode of production; and in changing their mode of production, in changing the way of earning their living, they change all their social relations. The hand-mill gives you society with the feudal lord; the steam-mill, society with the industrial capitalist.[4]

The need for heat prompted people to dig deeper into the earth in the search for coal (more efficient than wood in heat production). The need for water (constrained by its fluid density and seasonal shortages that might befall certain geographies – either through drought or freezing) meant that pipes were constructed to cover great distances and storage containers popped up across landscapes to bring water to where it was needed. Both water and heat sources therefore proved to be complicit, even enablers in the initial stages of this part of the project of modernity. In short, looked at through the lens of materialities, if water behaved differently, refused to cooperate (create steam) – or if steam failed to exert pressure – the world today would look otherwise. Water's ability to turn into steam when heated, therefore, appears to be the initial material cause of what we now know of as 'climate change'.

WATER CHANGES

The significance of water's behaviour is increasingly being recognised. As the expanding body of literature on the social lives of water illustrates, the ubiquity of water and the manners by which it seeps into every arena of life should not be ignored or underestimated.[5] Water is almost defiantly independent. Its transformations defy physical laws; it actively (re)shapes the landscape and through its constant motion, dissolves, cleans, creates, has memory and destroys.[6] Moreover, it troubles the intellectual boundaries people described between nature and culture and problematises ownership as it slips through borders of countries.[7] It tracks its way through everything including the earth, rocks, plants, flesh and the air and depending on the surrounding air temperature without it you will surely die after only a few days. And now, at the time of the end of the world, watching how water is behaving demonstrates not only that *all* activities are interconnected, impact on and are in relationship with each other, but also that making rain is water's business and is out of our control.

CLIMATE

According to climate scientists the surface temperature of the earth is undergoing rapid changes, which will affect the weather. This, of course, is not the first time earth's climate has dramatically shifted. Planetary history demonstrates that the hyperobject 'climate' has been responsible for producing numerous intense alterations to global conditions. Past changes are presented as 'natural' happenings out of any one thing's control, and attributed to a series of other 'natural' events without blame. Change of this kind is typically understood using the Gaia hypothesis of planetary homeostasis. Lovelock and Margulis' 1974 paper on the Gaia hypothesis (and later Lovelock's book) outlines a set of ideas that maintains the planet is able to organise itself through a system that has the ability to rebalance or harmonise when it gets out of equilibrium.[8] Significantly, however, this time, blame *is* apportioned for the current changes to the climate (and the inhospitable consequences we are anticipating as a result of them). It is us – the actions of the 'unnatural animal' (to coin a famous oxymoron) – that are to blame.[9] Thus, the current changes being experienced are deemed anthropogenic – that is: caused by human activities. Consequently, where Ice or Stone were previously formative in provoking change, now it is the human who has taken up the title of influencer. To this end, we have named the age we are entering the Age of the Anthropocene.[10]

Whoever is to blame, the brute watery facts of the situation are that: the south of the planet is drying up while the northern hemisphere is becoming wetter.[11] Cook *et al.* call this the "rich-get-richer/poor-get-poorer' mechanism' suggesting that the traditionally wetter areas will have increased precipitation while the arid regions will become drier.[12] Their choice of phrase also unintentionally hints at the economic consequences that are likely to ensue from the predictions. In areas where populations depend directly and entirely on the climate to ensure a harvest or the survival of livestock less rainfall is essentially a death sentence. To complicate things further, documentation reminds us that despite increases in precipitation in some areas, the increased surface temperatures overall are set to elevate evaporation. Thus, global aridity will mount. Indeed, again, it seems it is water's patterns of behaviour that are set to escalate change.

According to the different agencies (IPCC (Intergovernmental Panel on Climate Change) and NOAA (National Oceanic and Atmospheric Administration), in the last 50 years across the continent of Africa average surface temperatures have risen between 0.5-2.0C. As a result of the levels of emissions produced by nations in the global north – a process started in all earnestness with the developments of the Industrial Revolution.[13] The arid zones in Africa, ironically not culpable for global emission production, can expect a greater rise in temperatures than other locations in the coming years. With reference to Kenya specifically, the NOAA states that since 1990 temperatures have risen by approximately 1oC (2017). These changes will increase the intensity and frequency of certain weather conditions – namely drought and flooding. Predictions claim that this will continue – causing temperatures to rise further – until at least 2025. Indeed, the IPCC offers a staggering 5-6oC rise for some areas if global emissions do not reduce (2014).

For the Giriama, these climatic changes are directly altering their daily practices. The farmers here, like others in east Africa, are not strangers to drought.[14] On the contrary, inhabiting a designated ASAL (arid and semi-arid landscape) region, as they do, has meant that sustained periods with very little water are regularly negotiated. As a result, finding water, understanding how water behaves, where it collects, how to accumulate it, how to draw it into one's life and what is necessary to attract it occupies much of Giriama life.[15] Consequently, as with other groups that live in ASAL regions, seeking for water – and, particularly, the practices that articulate relationships with water – organise much of daily life.[16]

However, the current shifting weather patterns are proving challenging to negotiate. The deepening periods of drought, possibly produced by the warming of the Indian Ocean, mean that harvests are inadequate and unable to meet local needs.[17] Over three years with pitiful yields has placed a tremendous strain on resilience; many families have been left without seed to plant even if the rains do come (time of writing June 2017). The fragility of the water system in this area is well documented and consequently predictions on the current situation conclude that over the next few years subsistence will continue to be dramatically and dangerously impacted upon.[18]

In 2016 Reliefweb.int (an online 'international humanitarian information source for global crises and disasters' stated that

> Parts of the ... coastal marginal agricultural livelihood zones have recently moved to Stressed (IPC Phase 2). Due to poor rains over the last two seasons, and in some instances total crop failure, coupled with reduced labor opportunities, poor households have had to rely on markets for food purchase for most of the year with significantly lower incomes.[19]

In May 2017 Reliefweb.int's report showed that despite some rains the population was still at risk.

> A May mid-season assessment by the Kenya Food Security Steering Group (KFSSG) that included FEWS NET determined that food insecurity is set to increase from late June, with more poor households likely to experience Crisis (IPC Phase 3) outcomes.[20]

Late spring of 2017, the price of maize nearly doubled despite the Kenyan government's best efforts to import maize (Miriri 2017). By May 2017 with stocks depleted shops limited maize purchases to one bag per family (author's experience). Many were arrested as they fought to secure maize for their families. Furthermore, predictions continue to be bleak. UNICEF (citing data from the Kenya Meteorological Department) stated that

> Most parts of the country are expected to remain generally dry, implying that crops will be adversely affected ... Food security, problems related to water scarcity and lack of pasture are expected to further deteriorate perpetuating risk of resource- based conflicts.[21]

SOLUTIONS TO WATER INSECURITIES: PIPELINES AND CRYING

In recognition of the impending gravity of the situation, a handful of globally focused, not-for-profit and non-governmental organisations (NGOs) have mobilised support. Using the directives of the Millennium Development Goals that cite the need to improve water conditions, the NGOs are funding and implementing solutions to bring water security to the area.[22] The solutions, typically rooted in methods that imitate those in the developed North, are stimulating change.[23] External agencies are not alone in their engagement with this area. The Kenyan government can also be seen to have responded to development aspirations and the population's needs with regards water provision.

Consequently, with a view to shift exclusive reliance on naturally occurring water supplies in this rural area, the government has constructed a single pipeline across the region. Prior to the pipeline the community used other methods to bring water to their area in times of drought. For example, every area had at least one ritual practitioner (*mganga*) who could control the weather and cry (read: call or pray) for rain when necessary. Stories of the need to walk miles to collect water in times of stress are regularly counterbalanced by accounts of how the *mgangas* of the past were able to call for rain when it was needed.

Crying for water can only be successfully achieved by trained practitioners and through a highly structured gathering of community members – something that a modern lifestyle has constrained. To cry for rain it is necessary to draw together a community committee with a representative of each of the seven clans in attendance – something that Daniel Mwango Vunya Ndurya explains is not easy these days.[24] Daniel is 83 and is still working despite being almost blind in both eyes as a result of a combination of cataracts and damage after a roofing accident (According to other villagers Daniel was a ritual practitioner. They attribute the misfortune of losing his sight to a punishment from the spirits for stopping working for them.) Daniel states there are very few able practitioners left to cry for the community as was done in the past. This is not only a simple consequence of modernity but also hinges on the systematic rejection and destruction of the practice by those inspired by the Christian message that has been adopted by some people in the area. Daniel describes how practicing groups – derogatorily described as pagans by the Church – were regularly persecuted for their beliefs, and thus became concerned for their safety. Previously, he remembers, crying for rain was a community pursuit that bound the clans in adversity to a common purpose. Moreover, he asserts, it worked and, now, with the climate changing, the elderly like Daniel are concerned about what the young are constantly praying for.

In the past prayers were for problems. It wasn't something that you had to do regularly ... prayers are not working or if this is what people are praying for – then I don't want any of it! (Daniel)

Daniel describes the ritual in great detail, telling me that he has seen many of these performances enacted and the rain following. The process is very structured – as with most things the Giriama do – there is a right way to do things and people should not deviate from the procedure to ensure success. Consequently, rain cannot just be made; a complicated, carefully organised event must be planned.

The first stage of rainmaking requires involvement from all seven clans. Only once the clans are together can a discussion that determines how to proceed begin. Once the clans agreed on a schedule and a date was organised, the people would be informed that the ritual would be going ahead. Participants would be selected for certain roles and the rest of the community was instructed to get ready for rain. Then the kiza (a sacred ritual site, typically in the forest alongside a river or water basin) would be swept using a particular type of grass from the river's edge. When a space was cleared a small house was built and placed in the cleaned ritual space. Once the construction was complete, 'a woman who had only slept with one man' (Daniel) would take *Mbono* seeds (castor) to pound and then cook for oil. The oil she made would be placed in a pot in the kiza. Only after that could a member of the Akiza clan enter the 'house of god'. Before entry the individual must start singing and be covered head to toe in the traditional black cloth. (Black is the colour for water and any ritual that hopes for water must use black) The rest of the group then repeats the chorus, started by the individual covered with cloth, as they circle the house.

ziara mkanga rina malua, ziara mkanga rina malua
(water-basin reed grass has flowers, water-basin reed grass has flowers)

The group then empties seven containers of river water into a small hole in the top of the roof. As the water starts pouring and when the person in the house is drenched, the group ululates in celebration saying 'there is enough rain for both seasons' as they continue to circle. The whole performance is accompanied by drumming from the *mganga*.

Like many others in the community, Daniel is adamant that crying for rain is an effective method of water acquisition. He told me,

After the ceremony the leader would address the people. During the address the people were told that whoever has to pass through a valley on the way home must now leave immediately because the rains are coming and it will flood. This was always true. It worked. Then the people would quickly disperse and the rain would fall heavily that very same day – always. (Daniel)

This is backed up by Kathunga Mare, elderly head of a fairly wealthy family:

A long time ago when the dry season came people would ask around to see where the rain was. If it was in Mombasa, they would pray in the kiza and the rain would come. It is difficult to get the people together who could do it now. No more expertise.

Many of the older people talk today with disdain and regret about the social changes that have taken hold in this area. They lament the fact that the community lacks individuals with the special knowledge necessary to bring rain, and they state that the water practices of modern life are wasteful, inconsiderate and disconnected from the environment that their grandparents continuously worked with.

If the river runs dry, these people are in trouble. In the past because people could have to walk very far to get water, we would use it sparingly. (Kathunga Mare)

When asked if rainfall patterns had altered invariably the answer was yes.

Yes, there is a difference. The climate has changed. The rain isn't as it was. In the past, we had two seasons of harvest, but now getting even one harvest is a problem and you need to buy everything. (Kathunga Mare)

In the recent past, lack of rain was attributed to the actions of specific *pepo* (Champion uses the word *peho* [in Kiswahili] to describe spirits but in Giriama the term is *pepo*), which needed to be appeased. Champion explains in some detail that lack of rain could be attributed to the land being 'soiled' by transgressive actions such as murder. When asked to account for the change most were unable to offer an explanation but some in agreement with Champion's claims above Kathunga stated that lifestyle was the cause.

Because of the lifestyle. People kill each other more ... [there is] a lot of bloodshed and so the m'*pepo* aren't happy. The m'*pepo* are angry because humans are not happy amongst themselves. The m'*pepo* are punishing. People

say these are the end of times; if people believe this they will experience it.
In the past people used to pray to appease the m'*pepo* and rain would come.
(Kathunga Mare).[25]

In an area regularly suffering from water shortages, the ability to draw rain to
the area is of obvious importance and should not be underestimated. According
to the older people, it was not only effective but was also a method that provided
water equally across the landscape as the rain fell for everyone similarly. Thus
when the rain fell it would help everyone without discrimination. A situation that
is quite different from the water system currently developing in the area.

NEW WATER, TAP WATER

In 1974, eleven years after establishing independence in 1963, Kenya's
administration declared their intention to serve the population with clean water
under their 'Water for all in the year 2000' initiative. Despite the best of intentions,
the government records from the 1980s show that getting water to the area where
the Giriama live was proving to be problematic. Nonetheless, with help from the
Finnish and Swedish governments over the years, an ambitious water project
named 'The Baricho Water Scheme' actioned the drilling and construction of a
significant borehole just north of the small settlement of Baricho approximately
20 kilometres from Boré Koromi. According to the *Africa: water, sanitation and
hygiene* website the bulk of the coastal water supply is served by four sources
of which Baricho is one of two that were completed in 1980 with significant
support from various donors. It appears that notwithstanding the location being
proximal to Boré and other local settlements in land from Malindi, Baricho water
first served the population of Mombasa much further south.[26] Despite these
developments on the coast, Nyanchaga maintains that by 2005, apart from some
water condensing plants, water supplies remained the same as in the 1900s.

Probably inspired by the time-sensitive goals of the Millennium Project, World
Vision, a charity whose mission is 'to follow our Lord and Savior Jesus Christ in
working with the poor and oppressed to promote human transformation, seek
justice, and bear witness to the good news of the Kingdom of God' began a 15
year project based in and around the town of Marafa, a few kilometers from
Boré Koromi.[27] World Vision is an enormous operation. It has projects in 98
countries, employs 40,000 staff and manages an annual budget of US$2.6 billion,
and part of their aim in bringing support to this area was to alter water practices

by financially supporting the Kenyan government to bring piped water to this rural location.[28] As a result of World Vision's involvement in the water from the Baricho borehole now provides a pipeline that runs through Boré Koromi to Marafa. To access the water running in the pipeline World Vision designed, financed and stationed a series of kiosks at various locations along the main road.

Each kiosk represents a position on the pipeline where clean water can be bought. A kiosk is simply a small room built over the pipeline with a meter system to measure the amount of water drawn at each point. Some kiosks have been fitted with a 5,000 litre storage tank; some have not. The kiosk offers a lockable room to the proprietor. It is constructed from concrete, not the naturally occurring materials typically used for local construction, opens with a metal door at the side and uses a metal, lockable stall window over three external faucets where passers-by can place their jerrycan to be filled. The design presents as a water fortress. This is reflected in the ability to lock the building. It is not possible to secure a local wattle and daub construction in this way. Thus, the concrete and metal box signifies not only the value of the water but also heralds this new water as a commodity in a market economy.

Individuals who have been contracted by the local coastal water authority manage the kiosks. A series of initial costs are necessary to assume proprietorship of the kiosk. After those costs are disbursed, the manager of the kiosk is able to use the kiosk to both sell the water and, if capital allows, sell other household products such as flour, oil or soap. Kiosk managers earn one shilling for every 20 litres of water sold, on top of the money they could earn from selling other products out of the small space. Water is presented to the public at a cost of two shillings per 20 litres; the water company determines the price and any deviation is frowned upon – with the potential to lose your kiosk if you are found to have raised the price or otherwise. The price, advertised as significantly reduced compared to other areas of Kenya, is said to simply cover cost. Of the price, half goes directly to the manager, thereby potentially enabling individuals to make money from selling water. Thus, the manager of the kiosk does not receive a wage, but is able to make a profit if they sell the water.

While the kiosks offer access to clean water to numerous families, the reach of the kiosks is still very limited as a result of the extensive distances between kiosks. Consequently, the current system leaves significant portions of the local population reliant on traditional methods of water collection, as they do not have ready access to kiosk water.

Management of a public kiosk offers an income generation activity that does not rely directly on horticulture or rainfall to the area. There are very few income generating activities that exclude a reliance on rainfall in the area. Teaching and tailoring services are exceptions, but are only available to trained individuals. Other mechanisms such as temporary roadside cafes (locally these establishments are erroneously called 'hotels' but which only serve food such as chapattis, beans and chai tea) and other retail outlets currently amount to what is available to the population. Thus, kiosks represent not only new places for water acquisition but manifest as novel economic spaces in a landscape where financial exchange mechanisms, that do not concern field produce, are typically limited.

Thus, the introduction of water kiosks presents a series of opportunities to certain members of the communities. In this location, the opportunities have fallen with only one exception to the more wealthy men of the community as a result of their ability to produce the necessary down payment of 5,000 shillings to assume the position of manager. The only exception was a community group of women who banded together to take charge of the kiosk nearest to their homes. These women, active in the local school, had also instigated a tree nursery to increase their incomes. Assuming management of the kiosk not only supported their family incomes but also enabled the women to ensure the young trees in their nurseries survived. Other than this management group, all of the other kiosks are managed men, who after taking responsibility for the operation then handed it over to a female to serve the customers, thereby creating a gendered chain of command controlling the new water.

ENGAGING WITH WATER

The different methods of collection illustrate how relationships with water are altering. Water collection direct from the river or water basin demands that rain has fallen to a significant level to produce the water source in the landscape. If this has occurred the collector must first travel to the source. Then, because of the lay of the land, women must enter the water source, typically up to knee level, to be able to fill the jerrycan. The woman wades into the water, bringing her skin in contact with the water and introducing any matter that she may be carrying with her into the source. Then she cleans out the container, rubbing of the dirt on the sides and swilling out the contents or dregs at the bottom of it. This she throws out into the water in the same area that she then goes on to use to refill her container. Thus, the process of cleaning out the can is illusory as she refills it with the same water that she has cleaned it with. Furthermore, as there are numerous water restrictions that preclude individuals from using different points, each woman collects from the same place in the basin potentially adding and sharing pollutants each time. Depending on the water basin, the filled container may need to be hauled up the basin's slippery, sloping walls to the pathway after it is filled. With very steep slopes women tend to haul the containers up to the pathway before they place the water on their heads for transportation home.

To collect river water means engaging directly with the water by immersing parts of the body into it thereby amplifying hygiene concerns. Furthermore, water borne diseases that strike from physical contact with the water, such as bilharzia, are commonplace, alongside other issues from ingesting contaminated water. However, accessing tap water brings other unexpected consequences; firstly, with regards commodification of water, and, secondly, with regards time management.

Water collection from a kiosk on the other hand requires different means. For many the distance is similar to the water basins, for some it may even mean a reduction in distance, but for all collectors the method of transportation remains the same – 20 litres at a time on the head. Filling the container using a tap comes with new issues. The most significant of which concerns the need to pay. There is no credit at the kiosks; to fill a container means you must have money. To have money means you must have previously earned it. Without jobs, or incomes from other sources such as selling vegetables, honey or from casual labour, there are many people who simply cannot buy the water. The average family manages on approximately 200l per day – 10 x 20l cans (more on wash day). The cost of this water is minimal, 20 shillings per day or 140 shillings per week, but in an area without wage earners, any price is substantial.

In addition to the financial cost of water there is a time cost. Filling up containers from a tap takes longer than submerging a container in the river. Consequently, tap water collection often adds time on to overall haulage budgets. Not least because water collection creates something of a rush hour around it with people hoping to complete this section of their chores before the heat of the sun hits them, or after the sun has reached its apex. Thus, it is typical to see queues of containers representing the women's place in the line while they attempt to complete other chores for the day, or simply wait in some available shade.

The commodification of water, therefore, precludes universal access to it. Prior to the installation of kiosks water shortages were felt similarly regardless of economic status. Now, water relationships reflect affluence and financial ability, which is creating ripples and exacerbating various community tensions as a result of a division in the community between those who can afford to buy water and those who cannot. Thus, if previously, water acted as a leveler that similarly shaped people's lives, now it acts to accentuate and visibilise differences. Indeed, it is even creating judgement and enforcing social hierarchies as certain individuals openly disapprove of those who still rely on river water.

> They just need to learn how to buy like us. Once they do that, they will be fine.
> (Christine Kanini)

Water once acting as a leveler is now acting as a divider. And while tap water can be used for all domestic tasks it cannot be sued to water the fields. Without significant infrastructure implemented only rain can irrigate.

A pipeline of water unavailable to those without the funds to purchase it, could be thought of as worse than late or problematic rains. Furthermore, with aspirations for the trappings of modernity embodied in the pipeline, kiosks and payment, previous methods of water acquisition are being disparaged by the wealthier in the community. This in turn is provoking cultural traditions to be reconsidered and rejected and finally, it means that lives are adjusting to finding money above other activities previously thought necessary.

CONCLUSION

Hyperobjects, Morton explains, are extensive, ungraspable things multiple-sited, multivocal happenings or events. They exist without our gaze or attention and because of their 'strange strangeness' or ungrasp-ability, they compel (even oblige) us to recognise forces that we can do little about – and they do this by troubling both the global and the local simultaneously. What Morton makes us realise is that we think we know hyperobjects because we know of them, but, in fact, their vastness means we can't really know them fully. Morton is drawing them to our attention at this time, not to illustrate patterns or their power, nor yet with any notion of morality about their destructive (or creative) abilities. Rather, his aim is to remind us that we can never be harmonious with the material world we are intrinsically part of if we continue to maintain the erroneous belief that it is distinct from and made of different stuff from ourselves. Viewing existence using Morton's hyperobjects as a lens forces us to realise we are always inside and, part of, the 'everything'. This profound perspectival shift could come as an affront to your beliefs and it should draw you down from the abstract and the metaphysical to the empirical, raw-physicality of your own, and everything's, materiality.

In the Age of the End of the World, the hyperobjects are talking to us and through listening we can learn how to be this planet together. The hyperobject of global warming is not only a series of upset planetary forces; it is a voice and a method by which to show (make visible) our connections to each other and to everything. The hyperoject of capitalism is also talking to us. Using this lens, global warming becomes a force for further change and a mechanism that reveals the invisible ties that bind, while capitalism in a new setting provokes us to further recognise its limitations and problems.

In the example I offer in this chapter, the impacting factors of history, chemistry, geography, tradition, religion and water demonstrate that sustainable solutions are contextual and contingent. Relationships with water do not simply rely on water availability or cleanliness, but are also enabled and articulated in correspondence with the manner by which the hyperobject acts or behaves as a material. Moreover, it is clear that 'rain and indeed drought are inherently political, not just in terms of water as an essential and often scarce 'natural resource' but also because of its place in a powerful symbolic order'.[29] Collectively, the drought, rain-making rituals and the pipeline embody the 'tension between neo-liberal treatment of water as an 'economic good' and an 'internationalist' humanitarian principle'.[30] What needs to be recognised is the 'strange strange…[ness]' of our connections with the hyperobjects, and that there is a different but 'intricate relationship between water, memory and landscape' in each setting.[31] Rain brings water to the community, falling without judgment or prejudice on community members simultaneously. Piped water, on the other hand, whilst seemingly forthcoming supplies an individual at a time and demands pecuniary recompense. This difference means that its new method of arrival has the potential for dramatically altering perceptions of what water is and how one can engage with it. The social and economic consequences of this insinuate trepidation should be associated with the introduction of novel systems. Development strategies, therefore, must avoid mimicking past methods if they are to be culturally effective, harmonious and sustainable *in situ*.

Global warming is a hyperobject. It is everywhere at the same time, wearing a different face depending on temporality and location. Every time you walk down the street and breath in bus fumes global warming seeps into your lungs, every time the rain floods the fields global warming soaks into your shoes. For a group of individuals in rural Kenya walking miles to find water, the same global warming that wets your feet and fills your lungs, dehydrates their lives. Thus, the hyperobject global warming is inside, around and running through all of us in different ways. Seeing the hyperobject for what it is enables the illusion of distance to be removed.

The fact that you chose a plastic lid on your takeaway coffee, that you washed your hair this morning and that you had eggs for breakfast has altered your life and the lives of people you will never meet. Realisations like this should bring you to your knees as you appreciate you are orchestrating not only your own slow death but also the death and suffering of others. Which means, be attentive and mind the gap!

But what is harmony in such a complex, fluid system? Harmony is a grand term that evokes heavenly choirs chiming in unison, it offers a picture of goodness beyond the individual and it insinuates at something mighty to aspire to. It summons up notions of congruence, unity, order, agreement, equilibrium and balance. Using the term goes further to suggest that activities, choices and methods can be harmonious, if enacted in the 'right' way, according to the prescriptions of the method. Indeed, those who use the term go further than that: they suggest that when activities are harmonious they will be the 'right' activities, and that one's job is to seek out, and enact, the innate methods of harmony that run through all planetary endeavours to ensure that one's actions are in balance with the rest of it all.

But as I can see, there is no inherent, absolute method to harmony (or right way) for earthly processes; rather than that, there is a system that humanity has found patterns in. Finding patterns does not indicate or imply that there is a greater good nor yet a correct way to be, but rather demonstrates the power of humanity's pattern recognition and meaning making abilities. Of more import to me is how we interpret these patterns that we *think* we have found – In other words: taking what we think we have seen, what should we do, if anything, with the hyperobjects in the Age of the Anthropocene?

One might be tempted here to re-tread some rather worn paths about natural being good and cultural (that is human activity) being bad but the value in associating 'natural with good' and 'cultural with bad' is a tired discussion that leads us nowhere of any use.[32] Everything is natural – how could it be otherwise? If one attempts to argue that a sunflower knows how to 'sunflower', but that a human has forgotten how to 'human', one has fallen into a dreadful trap whereby a naturally occurring aspect on the planet (that is: humanity) is deemed flawed or deviating. Questions to ask: How could this happen in a harmonious system? How could one aspect of the whole be so bad? (Starting to sound biblical!) And of course if one naturally occurring being can go bad...can/have others?

No, more sensible a discussion would be one in which we recognised not that there is an inherent harmony to find and live by but that we recognise that our actions cause effects and so need to be thought through carefully.

ACKNOWLEDGEMENT

The Wenner Gren Foundation supported the fieldwork for this research under the title *The Role of 'New' Water in Shaping and Regulating Futures in Rural Kenya*.

NOTES

1. Timothy Morton, *Hyperobjects: Philosophy and Ecology after the End of the World* (Minneapolis: University of Minnesota Press, 2013).

2. Timothy Morton, *The Ecological Thought* (Cambridge, MA: Harvard University Press, 2010), p. 61.

3. Morton, *Hyperobjects*.

4. K. Marx, *The Poverty of Philosophy: Answers to the Philosophy of Poverty by M Proudhon* (Progress Publishers, Trans. by Zodiac for Marx/Engels Internet Archive: Marxists.org 1999), p. 49; https://www.marxists.org/archive/marx/works/1847/poverty-philosophy/cho2.htm [accessed 15 December 2018].

5. P. Ball, *H2O: The Biography of Water* (London: Orion Books, 2002); J. R. Wagner, ed., *The Social Life of Water* (New York, Oxford: Berg, 2015).

6. P. Ball, *H2O: The Biography of Water*.

7. V. Strang, *Water: Nature and Culture* (Reaktion Books: London, 2015).

8. J.E. Lovelock and L. Margulis, 'Atmospheric homeostasis by and for the biosphere: the Gaia hypothesis', *Tellus* Vol. 26, Nos. 1-2 (1974) at http://www.jameslovelock.org/page34.html [accessed 15 December 2018]; J. E. Lovelock, *Gaia: A New Look at Life on Earth* (Oxford: Oxford University Press, 1979).

9. H.G. Wells, *A Modern Utopia* (1905; repr. Pennsylvania: Pennsylvania State University, 2004), p.86.

10. C.M. Tobias and J.G. Morrison, *Anthrozoology: Embracing Co-Existence in the Anthropocene* (Switzerland: Springer, 2017); J. Zalasiewicz, n.d., Working Group on the Anthropocene: what is the 'Anthropocene'?, current definition and status at http://quaternary.stratigraphy.org/working-groups/anthropocene/ [accessed 15 December 2018].

11. B.I. Cook, J.E. Smerdon, R. Seager and S Coats, Global warming and 21st century drying, Clim Dyn., 43:2607-27; DOI 10.1007/s00382-014-2075-y (National Aeronautics and Space Administration, Goddard Institute for Space Studies, 2014).

12. Cook, 'Global Warming', p.2608.

13. IPCC, 'The IPCC fifth Assessment Report: What's in it for Africa?' at https://cdkn.org/wp-content/uploads/2014/04/AR5_IPCC_Whats_in_it_for_Africa.pdf [accessed 15 December 2018].

14. R. Cassidy, 'Lives With Others: Climate Change and Human-Animal Relations', *Annual Review of Anthropology*, Vol. 41 (2012): pp. 21-36; H. Cooley, 'Floods and Droughts' in P. H. Gleick, *The World's Water 2006-7: the Biennial Report on Freshwater Resources* (Washington, Covelo, London: Island Press, 2006); B. Derman, R. Odgaard and E. Sjaastad, eds., *Conflicts over Land and Water in Africa* (Suffolk: James Curry, 2007).

15. L. Attala, 'Bodies of Water: Exploring Water Flows in Rural Kenya', in L. Steel and K. Zinn, (eds.), *Exploring the Materiality of Food 'Stuffs': Transformations, Symbolic consumption and Embodiment* (Oxon, New York: Routledge, 2016).

16. E.F. Moran, *Human Adaptability: An Introduction to Environmental Anthropology*

(3rd ed.) (Philadelphia: WestView Press, 2008); Derman, *Conflict over Land and Water.*

17. C. Funk, A. Hoell, S. Shukla, I. Blade, B. Liebman, J. B. Roberts, F.F. Robertson, and G. Husak, 'Predicting East African spring droughts using Pacific and Indian Ocean sea surface temperature indices', *Hydrology and Earth System Sciences*, Vol. 18 (2014): pp. 4965-78.

18. M. Jacobsen, M. Webster and K. Vairavamoorthy, 'The Future of Water in African Cities: Why Waste Water?' at http://water.worldbank.org/sites/water.worldbank.org/files/publication/iuwm-africa.pdf [accessed 12 December 2016]; World Bank. 'Financing Small Piped Water Systems in Rural and Peri-Urban Kenya', *Water and sanitation program working paper* at https://openknowledge.worldbank.org/handle/10986/17316 [accessed 12 December 2016].

19. See Reliefweb.int, 'Kenya Food Security Outlook Update' at http://reliefweb.int/sites/reliefweb.int/files/resources/KE%20August%202016%20FSOU_Final_0.pdf [accessed 12 December 2016].

20. See Reliefweb.int, 'Kenya Key Message Update, May 2017' at http://reliefweb.int/report/kenya/kenya-key-message-update-may-2017 [accessed 20 June 2017].

21. See UNICEF, 'Kenya Humanitarian Situation Report' at https://reliefweb.int/sites/reliefweb.int/files/resources/UNICEF%20Kenya%20Humanitarian%20Situation%20Report%20-%205%20June%202017.pdf [accessed 15 December 2018].

22. UN.org, 'The Millennium Development Goals Report 2015: Summary', at http://www.un.org/millenniumgoals/2015_MDG_Report/pdf/MDG%202015%20Summary%20web_english.pdf [accessed December 2016].

23. R. Cassidy, 'Lives with Others'; J. Fontein, 'The Power of Water: Landscape, Water and the State in Southern and Eastern Africa: An Introduction', *Journal of Southern African Studies*, Vol. 34, No. 4 (2008); P. G. Jones and P. K. Thornton, 'Croppers to livestock keepers: livelihood transitions to 2050 in Africa due to climate change', *Environmental Science & Policy*, Vol. 12, No. 4 (2008): pp. 427-37.

24. The seven clans of this area are Amwabayawaro, Akiza, Amwakithi, Amilulu, Amwamweri, Amwandundu and Amwakombe. According to my informants Amwandundu started the kiza. These clan names are quite different from those documented by Parkin, who worked with Giriama who lived at the southern part of Giriama land on the coastal strip: see D. Parkin, *The Sacred Void: Spatial Images of Work and Ritual among the Giriama of Kenya* (Cambridge: Cambridge University Press, 1991).

25. A.M. Champion, 'The Agiryama of Kenya', *Royal Anthropological Institute Occasional Paper No. 25* (London: Royal Anthropological Institute of Great Britain and Ireland, 1967), p. 32

26. Champion, 'The Agiryama of Kenya', p. 126.

27. WorldVision.org, 'Mission Statement', https://www.worldvision.org/about-us/mission-statement [accessed 13 June 2016].

28. Wvi.org, 'World Vision Annual Review' at http://www.wvi.org/sites/default/files/1_WVI-Annual-Report-2016%20%281%29.pdf [accessed 13 June 2016]; Personal communication with Joffe, a World Vision employee from Nairobi working in Marafa.

29. Fontein, 'The Power of Water', p. 744.

30. Fontein, 'The Power of Water', p. 742.

31. Morton, *The Ecological Thought*, p. 61; Fontein, 'The Power of Water', p. 746.

32. P. Descola, *Beyond Nature and Culture* (Chicago: Chicago University Press, 2013); T. Ingold, *Perceptions of the Environment: Essays on Livelihood, Dwelling and Skill* (London, New York: Routledge, 2000).

Harmony and the Climate Crisis

John Sauven

From a talk at 'The Harmony Debates', conference at the University of Wales Trinity Saint David, Lampeter 2 March 2016.

When we consider the principles of Harmony – such as interdependence; cycles and seasonality; diversity; health and meeting human needs without undermining natural systems or the needs of future generations – we need first to step back and retrace where we have come from.[1] And I want to do that in relation to climate change.

I was listening to the BBC the other week and heard Owen Paterson, the former Tory environment minister, say that he could live with one or two degrees rise in global average temperatures.[2] In fact, he said it would mostly be good news. Even if we discount Owen Paterson's climate scepticism, when scientists talk about climate change, it appears to the general public that they are talking about tiny differences in temperature – one or two degrees Celsius. The significance of such small changes can be challenging for us humans to understand, given the massive daily fluctuations in temperature, say between the northern and southern hemisphere, between winter and summer, or between night and day. But, if we look at global average temperatures from a historical perspective, through studying the history of planet Earth, we can get a better understanding about our climate and why these small temperature changes are so critical for life on Earth.

The recent history of planet Earth, going back over at least 2.6 million years, shows that we oscillate between ice ages and periods we know as interglacials. Basically, there are around one hundred thousand years of glaciation followed by as little as thirty thousand years of warmer interglacial. These changes are part of the natural cycle tied mainly to variations in the Earth's orbit of the Sun. We have a great library of the planet's history in ice cores and ocean sediments which gives us an excellent record of these repeated hot and cold periods going back millions of years: right now we are in the middle of an interglacial period. Some interglacials in the past have been slightly warmer than the one we are currently living in which gives us an understanding of what life on a warmer planet might be like. The current geological phase, which we know as the Holocene, is dated to the end of the last ice age over ten thousand years ago. The Holocene has been called the Garden of Eden on the grounds that it is ideal for supporting life on Earth, including the human species. The important point is that the average global temperature range, within which the planet has remained during the Holocene, is only around plus or minus one degrees Celsius. Such a plus one-degree global change is significant because it takes a vast amount of heat to warm all the oceans, atmosphere and land by that much. But what is key is that we have existed, for most of this period, within a very narrow global average temperature range, plus or minus one degree Celsius. When we last had an interglacial state in the Eemian epoch, ending around 115,000 years ago (before the last ice age) the average temperature on Earth was in the order of two to three degrees Celsius warmer than now. And sea levels were on average six metres higher mainly due to the partial melting of the Greenland ice sheets. Each epoch is not exactly comparable, but it gives us the best available evidence of what life would be like on a slightly warmer planet.

This information from the Earth's geological library gives us an indication of what we may be facing in future. For example, a rise of over three degrees Celsius would be a similar situation to that of around five million years ago, when sea levels were twenty-five metres higher then they are now. Small differences in global average temperature can bring about dramatic differences to our climate, ultimately with a huge impact on our well-being.

Modern homo-sapiens have been on this planet for one hundred thousand years or so and the cognitive revolution, when our ability to acquire knowledge and understanding developed, happened near the beginning of this period. So we've had the same ability, physically and intellectually, to develop societies as we know them over this period. But we had a bumpy ride over last hundred thousand years. Initially we were hunters and gatherers, and perhaps there were a few million people at the most. Then we had an exceptionally cold point during the last ice age, probably triggered by a volcanic eruption that blotted out the sun roughly 75,000 years ago. Most of the fresh water was tied up as ice in the poles and sea levels were almost eighty metres lower than today. Genetic analysis indicates that, most likely, we were down to just a few thousand fertile adults at that point, living somewhere in central Africa, the only place, apparently, where there was food and fresh water. And humans were, to put it simply, virtually extinct as a species. If we had gone extinct the world today would be teeming with life.

This is a remarkable reminder of the fact that we depend on a stable climate for our own human wellbeing. Humans are lucky to have survived to this point. It also shows we are very closely related to each other. When we left the last Ice Age just over ten thousand years ago we moved into a remarkable, not to say almost miraculously stable, interglacial period. The rainy seasons established themselves. If you lived in the northern temperate regions you knew that the temperatures rose above fifteen degrees Celsius roughly in the spring and temperatures remained high until the autumn. Everything we love, everything we depend on, from the grasslands, the rainforests, the marine systems and the coral reefs, settle in the Holocene. The Holocene is almost perfect for us humans.

Now, if the present accumulation of carbon dioxide in the atmosphere continues unchecked, greenhouse warming of four or five degrees Celsius might occur by as soon as the end of this century – this scale and speed of change would be unprecedented. Scientific evidence shows that the dominant cause of the rapid warming we are experiencing today is due to the increase of the amount of so-called greenhouse gases in the atmosphere, mainly carbon dioxide – CO_2 – from burning fossil fuels and cleaning ecosystems like rainforests. These greenhouse gases prevent more of the energy we receive from the Sun from escaping back to space, and this warms the planet.

Today's carbon dioxide levels in the atmosphere vastly exceed those reached at any point during the past 800,000 years. And what's even more startling is that CO_2 is being pumped, by us, into the atmosphere at a rate unprecedented during the past sixty-six million years. It's mainly because of these human activities that scientists are claiming we are entering a new geological epoch, one that has unofficially been called the Anthropocene, or 'human age'. We are waving goodbye, with some considerable risk, to Eden's Garden.

Until recently humans constituted a relatively small world on a big planet. But unsustainable economic development delivered increasing wealth because we had ample atmosphere in which to dump our CO_2 emissions, ample ocean in which to dump our waste, ample plants and animals to consume from the forests and grasslands, and ample biomes like rainforests and marine systems to exploit. And no invoices were being sent back from the natural world to human societies at the planetary scale. But now we have tipped over into what we can today call the era of humans being a relatively big world on a small planet, rather than a small world on a big planet. We are actually, as humanity, hitting the biophysical ceiling of the planet's capacity to support our species. In the context of geological time and the sequence of glacial and interglacial periods, this has occurred in a very short space of time. The process started in the United Kingdom in the mid-eighteenth century with the use of the coal-fired steam engine and the start of the industrial revolution. At that time the global population was just seven hundred million. Industrialisation continued through to what has been termed the 'great acceleration' in the 1950s when production and consumption entered a period of exponential growth. The consequence is the sixth mass extinction of species (the first to be caused by another species, humans), and abrupt changes resulting from climate change, such as the accelerated melting of the polar ice sheets, the retreat of glaciers worldwide and the massive bleaching of coral reefs. So we have, in just seventy years, from the great acceleration in the 1950s until today, managed to start pushing ourselves outside of a state that we've been in for over ten thousand years when we left the last ice age. This is no small 'achievement'!

And now, we are at a point when what we do over the next decade or so will very likely determine the state for humanity over the next ten thousand or more years. So we are at this pivotal point. That's why one or two degrees Celsius is so important. We can live outside the Holocene. We've done it before: we were a few thousand people, hunters and gatherers in equatorial Africa 75,000 years ago, during the last ice age.

But if we take ethical responsibility for 7.4 billion co-citizens today (and this is projected to be 9.7 billion by 2050), the scientific message becomes as simple as it is dramatic, that the climate of the Holocene is the only one that we know for certain that can support the modern world as we know it. We can no longer, in any societal kind of policy making, consider that what is out there in the oceans and the atmosphere are somehow things that can be endlessly exploited, or can endlessly receive all our pollution. It's exactly the reverse. We now need to internalise rather than externalise everything inside our own economic models because we are now truly at a global saturation point.

The economy has grown very successfully on one measurement, but at the expense of human capital and nature, and it's now urgent that we shift into a new situation where we actually see the economy delivering for society within a safe operating space of planetary boundaries. We now need to understand that nature actually is the fundamental reason why we can have prosperity and economic development on planet Earth. And that is the most important message we need to understand – that nature comes first because nature can do much better without us but we cannot survive without nature. If humans are now a big world on a small planet, and if we depend, all of us equally, on nature, then we must recognise that we all have a personal responsibility for nature. Of course, this has profound consequences for us as individuals. If we view this in terms of a budget, we can 'afford' to pump 3,000 billion tonnes of CO_2 into the atmosphere and stay within two degrees Celsius. We have used up about 2,000 billion tonnes of CO_2 since the 1850s, so we have 1,000 billion tonnes of CO_2 left. At current rates of CO_2 emissions (36 billion tonnes a year) we have a budget that will be spent within a few decades. To stay within 1.5 degrees Celsius would mean even faster action. This means leaving most of our reserves of coal, oil and gas in the ground and ending deforestation.

This wouldn't keep us in the Holocene, because we are leaving it. But it would keep us in a manageable interglacial state. To get there we need to end the internal combustion engine and shift to electric vehicles. We need to rapidly develop renewable energy and storage technology. We need to eat a more plant-based diet in order to end agricultural expansion into forests and other vulnerable ecosystems mainly driven by meat production. And we need biomes like rainforests to recover in order to store more carbon and protect biodiversity. Finally, we need to move from a linear to a circular economy, reducing, reusing and recycling. The solutions are largely available but the politics lags far behind. It requires a new mindset in terms of the values between us humans and our planet Earth. We need to recognise not just the interdependence between humanity and nature, but that humanity is part of nature.

NOTES

1.   David Cadman, 'Principles of Harmony', The Harmony Project, https://www. theharmonyproject.org.uk/principles-of-harmony/, 29 January 2019 [Accessed 10 May 2019].
2.   And also see Adam Vaughan, 'Owen Paterson v the science of climate change', The Guardian, 30 September 2013, https://www.theguardian.com/environment/blog/2013/sep/30/owen-paterson-science-climate-change [Accessed 10 May 2019]

REFLECTIONS UPON EDUCATION FOR SUSTAINABILITY: SUPPORTING STUDENTS' KNOWLEDGE, UNDERSTANDING AND PRACTICE

Caroline Lohmann-Hancock and Nichola Welton

INTRODUCTION

GLOBALISATION HAS BROUGHT WITH IT both perceived benefits and challenges to the environment, communities and people. The global economy has brought jobs and homes to many, whilst simultaneously hastening poverty and debt.[1] No longer can society view itself in a vacuum with the luxury of choice and power without responsibility for their own country and peoples, rather, every action now has repercussions not only for those in power but also those who are powerless in the world.[2] Indeed, the connected world is now upon us; no longer are we separated and separate from each other. Rather, we are inextricably linked to each other with or without consent.[3] What does this modern world mean for the individual or the community or indeed the leaders of countries or businesses? Should a minority have the power to decide on behalf of the majority as they are 'educated' and thus intrinsically 'know best'? An alternative position could include the rights of the planet and the rights of future citizens, as yet unborn, where harmony is central.[4] This chapter explores sustainability through the lens of the next generation of thinkers, students on a social studies degree module. From theory to practice through reflection and reflexive diaries, the output from this module are explored within a higher education context.

Initially, there will be an exploration of what constitutes 'sustainability' within a globalised world.[5] The data within this study was collected through assessments completed for this module through a reflective diary.[6] This reflective diary allowed a window on student's progression and attitudes through the module's linked to sustainability theoretical perspectives. The themes from these assessments will be explored in regards to the journey these students undertook and their reflections upon what sustainability meant to them pre-, during and post-module. Finally, conclusions will be drawn as to the lessons learned from such a process and implications for future practice.

LITERATURE REVIEW

The speed of change in the modern world is unsustainable with society consuming resources at an unprecedented level. Resources, physical as well as human, are being consumed, and some would say abused, at an ever faster rate. As with all forms of capital there is a finite limit to the consumption of such resources.[7] For some, the world has already reached a point of no return in regards to the use of the Earth's resources in all its forms; 'It is clear that we are currently using non-renewable resources at rates vastly exceeding the rates of replacement'.[8] With poverty and hunger, floods and droughts, the equilibrium of the nature rhythms of the earth are no longer in harmony. Is this a natural process of the ebb and flow of the earth's natural cycles or is this a human-made disaster? 'Natural disasters may be initiated or accelerated by human induced activities' and as such they are not 'completely unavoidable'.[9] Certainly, whatever the cause, the reality is that humanity stands on the edge of a new world order in which the Earth will no longer be able to regain the rhythms of the past. Not only does this disequilibrium impact environments and animals it also impacts communities and us as individual citizens on a finite resource we call the 'Earth'.

The adoption of an eco-centric/social-ecological perspective on sustainability frames the current research and underpins a critical reflection on the complexity of definitions of sustainability models. Social ecology recognises that nearly all our present ecological problems arise from deep-seated social problems.[10] Within a social studies programme, a social ecology perspective is particularly relevant in developing critical reflective practitioners as well as providing a theoretical framework with which to examine the power dynamics within dominant sustainability models, enabling students to recognise the impacts on contemporary society and for future generations. The process of asking 'questions about relationship, justice, fairness and respect in economic, social and political relationships' are central to the concept of 'Harmony' and indeed to social studies.[11].

If we adopt a social ecological model of sustainability, we can begin to explore the development of a fair and equal society; a society where there are both rights and responsibilities; a society which is not ruled by greed and capitalism. Rather, a world where the ebb and flow of the natural world is emulated in the very fabric of the modern world without which society will decay and destroy not only the forests but society itself. Sustainability is more than recycling; although part of the process. It is more than thinking about how we travel. Sustainable practice is a value base, a way of thinking, an attitude of mind about all resources in the world. As the world is interconnected, one could ask do we, humankind, have the right to buy and sell earth's resources through the stock markets? Are they 'ours' to dispose of as we please? Capitalism, the model of neo-liberalism, puts forward that the market is all and profit is king (engendered and culturally Western) but is this just imperialism and colonialism revisited? A social ecological model of sustainability is the antithesis of capitalism; such a tension is ignored by many. Alternatively, sustainability is seen as just another commodity to be brought into the capitalist stable where profits can be made at the expense of the Earth's resources. Indeed, even the sustainability agenda has become a market place for capitalism:

> Free market environmentalism, which conceives of sustainable development within the dominant free market paradigm. In this paradigm, nature itself is, to some extent, treated as a commodity subject to market principles linked to its ability to supply environmental goods and services.[12]

WHAT IS EDUCATION FOR SUSTAINABILITY?

In order to engender a critical perspective on sustainability the education process is essential. It was recognised in the United Nations' Agenda 21 that there was a need to 'reorient existing education to address sustainable development'.[13] Sterling[14] illustrates how 'Progress towards a more sustainable future critically depends on learning, yet most education and learning take no account of sustainability'. He also suggests, therefore, that current transmissive or modernistic models of education continue to develop non-sustainable practice, and that: 'education for sustainable development/or in' needs to shift towards a transformative educational paradigm.[15] Orr further illustrates the requirement for re-orientating education:

> The destruction of the planet is not the work of ignorant people rather it is largely the result of work by people with BAS, BSCS, LLBS, MBAS and PHDS.[16]

The challenge therefore for educators in higher education is concerned with developing students' ecological/sustainable literacy[17] is to develop a transformative learning experience with students to ensure that the principles of ESD as a process of reorienting education are enabled.

Education by its nature is a progressive and forward looking process which draws on the past to make informed decisions in the present for the future.

Education is thus, potentially, a sustainable process. Beyond knowledge education has the potential to 'enlighten' and 'empower' particularly if at the heart of the process is the development of praxis.[18]

The promotion of education for sustainability has been seen as a priority within Welsh government and evidenced by its published guidance and documentation for education sectors up to further education.[19] It is also central to the Welsh Government's legislative framework its most recent policy demonstrates Wales' commitment to sustainability agenda, the Well-being of Future Generations (Wales) Act 2015 which focuses on 'improving the social, economic, environmental and cultural well-being of Wales'.[20] The premise of its ESDGC approach is one based on adopting a more transformational learning experience through viewing ESDGC as a pedagogical process with a set of values underpinning it rather than just focusing on content and knowledge: the skills, knowledge and understanding and values are illustrated in the following questions and reflect to some extent Sterling's[21] above discussed perspective on Education for Sustainability model:

i. Does the learning offer past, present and future perspectives?

ii. Does the learning address any of the major themes within ESDGC, i.e., wealth and poverty, health, climate change, the natural environment, consumption and waste?

iii. Does it make interconnections between these themes and between people, places and events both locally and globally?

iv. Is the learning relevant to learners' lives?

v. Does it encourage critical examination of issues?

vi. Does the learning address controversial issues and examine conflict resolution?

vii. Does the learning explore values and cultural perspectives?

viii. Does it empower learners to take appropriate action?[22]

The document further emphasises the importance of reflecting on personal values and perspectives so as to ensure a critical thinking approach is adopted when teaching and engaging with ESDGC:

As teachers we need to be aware that our own values are shaped by our experiences and to consider how these will impact on our teaching. We should be facilitators of learning and not be expecting pupils to accept our own values or a particular set of values.[23]

Within higher education there has also been a significant shift towards embedding ESD[24] with the emphasis on the pedagogic approach again evident:

Education for sustainable development is the process of equipping students with the knowledge and understanding, skills and attributes needed to work and live in a way that safeguards environmental, social and economic wellbeing, both in the present and for future generations.[25]

Education for sustainable development means working with students to encourage them to:

i. Consider what the concept of global citizenship means in the context of their own discipline and in their future professional and personal lives;

ii. consider what the concept of environmental stewardship means in the context of their own discipline and in their future professional and personal lives;

iii. think about issues of social justice, ethics and wellbeing, and how these relate to ecological and economic factors;

iv. develop a future-facing outlook; learning to think about the consequences of actions, and how systems and societies can be adapted to ensure sustainable futures'.[26]

The focus as with compulsory education is on a more critical transformational learning approach:

Education for sustainable development encourages students to develop critical thinking and to take a wide-ranging, systemic and self-reflective approach, adapting to novel situations that can arise from complexity. An ability to anticipate and prepare for predictable outcomes and be ready to adapt to unexpected ones is an important goal.[27]

The challenge for us as educators is to engage with both the content of ESD and the underpinning pedagogical approach, including examination of our values if we are to address the reorientation of education, and to design our learning and teaching so as to promote a transformational learning experience for students and not pay lip service alone to the key content areas of ESD. In doing so, however, we also need to engage in critical discussion and reflection including exploring alternative modules of sustainability such as a social ecology perspective.[28] In doing so we may be going some way to address what Giroux highlighted as key issues in teacher education: 'its culpability in promoting teacher political neutrality and disengagement in political discourse related to education'.[29] He further discusses teachers lacking the political language to deconstruct hegemonic systems of power. In adopting a social ecology model, the development of both a transformational educational experience and an underpinning theoretical base to challenge inequality can therefore be achieved.

Adopting such a theoretical framework of ESD and social ecology entails questioning dominant perspectives on sustainability as well as the policy context; this enables engagement in critical reflection through exploring disorientating dilemmas.[30] For example engaging students in evaluation of different models/ paradigms of sustainability to aid their reflection may result in them questioning their own attitudes and behaviour. Ther are different options for understanding sustainability: it may be considered 'technocentric', in which case it maintains the societal status quo, giving prime importance to people and technology above the environment, or it may be understood as 'ecocentric', emphasising the need for system change and recognising the people, society and the environment and intertwined phenomena.[31]

THE NEED FOR A CRITICAL TRANSFORMATION OF CONSCIOUSNESS

Engaging students with different models / paradigms of sustainability and the activity of critical reflection on their own values thus adopts the approach that Sterling, Orr, Giroux and indeed the policy documents of ESD recommend.[32] Therefore, this should result in the outcome of a critical consciousness on sustainability therefore producing a transformational educational experience one that Ferreira, argues will assist towards developing a sustainable future:

If we are to envision and construe actual sustainable futures, we must first understand what brought us here, where the roots of the problems lie, and how the sustainability discourse and framework tackle—or fail to tackle—them. To do this is to politicize sustainability, to build a critical perspective of and about sustainability. It is an act of *conscientização* (or conscientization), to borrow Paulo Freire's seminal term, of cultivating critical consciousness and conscience.[33]

In order to cultivate critical consciousness and conscience as educators we therefore need to engage our students in a critical reflective process.

The aim of the social studies degrees which form the basis of this study is to explore diversity, difference and equality in society with a clear focus on the foundational triad of theory, policy in practice.[34] A main element is understanding the attitudes and values within an anti-oppressive practice paradigm, for those working with vulnerable groups in a globalised context.[35] In addition, students are encouraged to understand inequality in society through exploring how barriers and challenges impact upon individuals at risk of inequality and poverty through personal engagement and reflection. Kroth and Cranton explain Mezirow's position in this regard as a form of 'storytelling' which allows for the 'transformative learning through first-hand accounts of others' transformative experiences'. In turn this exposes a:

Learner to alternative perspectives, a process that is at the heart of critical reflection and critical self-reflection, which is, in turn, central to transformative learning.[36]

To this end, the module explored the social ecological model of sustainability and was also underpinned by such values. In addition, the learning outcomes engaged with the contested nature of sustainable development; the role of sustainable development in building a world based on principles of justice, equity, participation and transparency; strategies and frameworks for promoting sustainability; and social justice in communities through reflecting upon their own value and attitudes in relation to sustainable development.[37]

Students undertaking the sustainability module often come with a range of pre-set ideas which focuses on recycling of plastic bags and the use of public transport.[38] Many students saw the module as not for them with 'low levels of concern about environmental problems'. This demonstrated a lack of insight of why a sustainability module is included in the programme.[39] It is essential to engender a critical perspective on models of sustainability and therefore engage with a transformative education process. This, in turn, allows students to develop philosophical thinking and critical reflection on ones' own practice to allow for meaningful engagement with the global and societal context.[40]

METHODOLOGY

A qualitative constructivist paradigm was used in this study which allowed for a transformation of knowledge and experiences through active engagement with real life tasks.[41] Students were give a qualitative questionnaire alongside reflective diaries to 'tell the story' of their sustainability learning journey from what Johari[42] calls the 'unknown' to what is 'known' by the individual. Applied Thematic Analysis was used to interrogate the data resulting from the questionnaires and reflective diaries alongside the students' written assessments for the module.[43] Applied Thematic Analysis is a 'rigorous, yet inductive, set of procedures designed to identify and examine themes from textual data in a way that is transparent and credible' and draws upon a 'broad range of theoretical and methodological perspectives' with a 'primary concern' on 'presenting the stories and experiences voiced by study participants'; in this study hearing the 'story' of the students' journey is central to the research.[44] Initially a pre-module qualitative questionnaire was given to students in the first lecture to gain an understanding of their initial position in regards to sustainability. During the module students were asked to keep a reflective diary of their journey and this was then fed into their assignments. Finally, a post-module qualitative questionnaire was used to collect data on the distance travelled.

CONTEXT

The module had a strong focus built upon an understanding of the various theoretical perspectives of sustainability alongside visits to eco-villages and forest schools. These 'real life' activities allowed for deeper understanding of policy and theory in practice.[45] Prior to visiting the eco-village little was explained to students about what to expect; this was to ensure that they engaged with the journey in their own way. Initially, we advised students to wear suitable clothing and a group of female students asked:

What footwear should we wear?

The team suggested hiking or walking boots. They indicated that they did not have either. We then suggested wellingtons. They replied:

We'll have to buy or borrow these.

This lack of previous experience within a rural environment also mediated their engagement and understanding of the module content. Through encountering real-life experiences which create a form of 'disorientating dilemma' they begin to:

> Examine, question, and reflect on the their preconceived notion, myopic beliefs, thinking, assumptions, and actions, and as a result, change or broaden their frames of reference.[46]

For example, one of the real challenges for some students was the use of a 'dry compost toilet' which did not have the western water flush but rather the use of sawdust to sprinkle over their human waste; in turn this waste was eventually used on the land.[47] This reality of eco-existence for some students was a real challenge but generated a range of discussions about how society disposes of human waste and the social stigma around toileting; this leads to a critical consciousness and reflection which in turn allows for transformation to occur.[48]

RESULTS

PRE-MODULE QUALITATIVE QUESTIONNAIRES

From the pre-module questionnaires it was evident that students' general perception of sustainability was limited in scope prior to engagement in the lecturers. The data indicated that students generally presented as being techno-centric accommodationists with a fixed mind set. This is in line with society's stereotypical response to the sustainability agenda where the narrative is often focused on 'it's about recycling plastic bags'.[49] They wished to maintain their current lifestyles and saw technology as the future. They also saw ESDGC as content-focused rather than as a process and pedagogical focus; this was also combined with critical reflection on the lack of harmony in the world. Students' comments included:

Upon arriving at the first lecture I felt a lack of interest for the subject of

sustainability.
Vague knowledge about fossil fuels and recycling materials.
Couldn't see the link to people.
It's about global warming, recycling and saving the planet.

At this point of their journey the overwhelming majority in society and in indeed the students in this cohort considered that they wanted an undisturbed environment but also want a nice car, house, and holidays overseas.[50] There were two students who had 'lived in eco-friendly communities' and they felt somewhat marginalised and considered that they should not voice their experiences or opinions as they might be seen as so-called 'yoghurt knitters'; thus they tended not to engage greatly in the class discussions.

REFLECTIVE DIARIES

The reflective diaries allowed students to explore their own stories and journeys through the module and chart how, if at all, their preconceptions changed. It was evident from their diaries that students quickly began to challenge techno-centric policy frameworks and broadened their understanding of ESDGC as an educational process. It was evident that the majority of the cohort moved from a 'techno-centric' to either an 'eco-socialist' or 'eco-feminist' position.[51] Students commented:

> *I am now aware that my initial perspective of sustainability, relating to recycling and renewable and non-renewable resources, was incorrect and likely influenced by a lack of understanding and media communications such as news reports.*

> *However, a successful relationship includes some effective practise that can support principles of sustainability, such as community engagement, government implication, policy makers, environmental and health policies.*

> *I have actually followed an ESDGC pedagogy approach, e.g., through forest schools, eco-schools and school councils but I didn't realise it!*

> *The real challenge is of trying to do this against competing educational policy, e.g., literacy and numeracy project with set targets regardless of ability.*

Once they reflected upon their initial opinion they moved from the 'unknown' to a 'private hidden realisation' within their diaries.[52] This demonstrated critical reflection in practice. This again resonates with the concept of 'Harmony' as the results from these data focuses on issues of 'relationship, justice, fairness and respect in economic, social and political relationships'.[53]

POST-QUALITATIVE QUESTIONNAIRES AND ASSIGNMENTS

The data from the post-qualitative questionnaires and submitted assignments demonstrates a more future orientated position of all students. They began to challenge the ability of the modern society, and themselves, to engage with Agenda 21. In addition, they evidenced in their work that they had begun to challenge their own resource consumption and behaviour as being contributory to the unsustainability of current practice and understood that

> *It's about societal attitudes and understanding of the needs of today set against future generations.*
>
> *It's a social justice issue!*

They also raised concerns that:

- The wealthy and middle classes were equipped to engage in meaningful change.
- Their own behaviours as practitioners mattered.
- This had an impact within their communities through educational practice.

Finally, at the end of their journey through the module, students began to realise that sustainability permeates all aspects of our lives, that everyone has a role to play and that there were going to be challenges to their own lifestyles to put this into practice.

DISCUSSION

The research undertaken with students through reflective diaries illustrated the potential of adopting an ESD pedagogy and enabling the development of critical consciousness through critical reflective/ reflexive activities and experiential learning; the aim being for students to engage in 'praxis' and therefore a truly transformative learning experience. The module content and assessment allowed students to:

i. Reflect upon their own practices and experiences

ii. Examine their own values

iii. Engage them in the development of ESDGC activities applicable to their context

iv. Question current educational models of learning, i.e., transmissive versus transformational

v. Examination of sustainability theories, e.g., techno-centric and eco-centric

vi. Experiential visits and learning

The pedogogicial approach therefore is aligned to current views on developing sustainable literate students and engaged students in future orientated education.[54] Education for Sustainable Development has consequently promoted competencies like critical thinking, imagining future scenarios and making decisions in a collaborative way.[55]

In addition to developing these sustainable literacy skills, the development of a critical conscience is fundamental to the aims of the programme and indeed reflect the core values of 'harmony'.[56] The focus on adopting a social-ecology model to enable critical reflection further enabled students and staff to develop change to their practices and therefore engagement with 'praxis'.[57]

The ultimate aim of the programme is to promote a 'just sustainable paradigm' and to embed a social justice element rather than accept a sustainable paradigm that promotes commodification.[58]

CONCLUSION

The research provided an opportunity for critical reflective/ reflexive practice for both students and lectures, developing praxis and critical consciousness. To this end the model of teaching and assessment could be seen to be transformative in nature. Further it highlighted the need to examine further sustainability theories and consider the importance of a 'just sustainability paradigm'. As educators within higher education we have a responsibility to prepare our students for the future; the challenge or opportunity is very well articulated by Orr:

The plain fact is that the planet does not need more successful people. But it does desperately need more peacemakers, healers, restorers, storytellers, and lovers of every kind. It needs people who live well in their places. It needs people of moral courage willing to join the fight to make the world habitable and humane. And these qualities have little to do with success as we have defined it.[59]

'Harmony' is central to the sustainability agenda at a global, national, local and social level where the concepts and practice of harmonious 'relationships, justice, fairness and respect in economic, social and political relationships' has the potential to benefit all.[60]

NOTES

1. R. Kaplinsky, *Globalization, Poverty and Inequality: Between a Rock and a Hard Place* (Cambridge: Polity Press, 2005), p. 88.
2. V. F. Krapivin, *Globalisation and Sustainable Development: Environmental Agendas* (Chichester: Praxis Publishing, 2007), p. 53.
3. M. Lockwood, C. Kuzemko, C. Mitchell and R. Hoggett, 'Theorising Governance and Innovation in Sustainable Energy Transitions,' Gov Working Paper 1304, Energy Policy Group, University of Exeter, (2013) http://projects.exeter.ac.uk/igov/wp content/uploads/2013/07/WP4-IGov- theory-of-change.pdf [accessed: 20/09/2017]
4. J. Porritt, *Capitalism: As If the World Matters* (London: Earthscan, 2007), p. 8.
5. See S. Harris and D. Throsby 'The ESD Process: Background, Implementation and Aftermath', in *The ESD Process: Evaluating a Policy Experiment*, C. Hamilton and D. Throsby (eds.), (Canberra: Australian Academy of Science, 1998), pp. 1-19; J. MacNeill, 'The Forgotten Imperative of Sustainable Development', *Environmental Policy and Law*, Vol. 36, No.3/4 (2006): pp. 167-70; and WCED, *Our Common Future, World Commission on Environment and Development* (London: Oxford University Press, 1987).
6. L.M. Baran and J. E. Jones, *Mixed Methods Research for Improved Scientific Study* (Hershey, PA: ISF, 2016), p. 129.
7. O. Scharmer and K. Kaufer, *Leading from the Emerging Future: From Ego-System to Eco-System Economies* (San Francisco: BK Currents Book, 2013).
8. C. Hill, *An Introduction to Sustainable Resource Use* (Abingdon, Oxon: Earthscan, 2011), p. 44.
9. D. Satendra and V. K. Sharma, *Sustainable Rural Development for Disaster Mitigation* (New Delhi: Ashok Kumar Mittal, 2004), p. 4.
10. See M. Bookchin, *The Ecology of Freedom: The Emergence and Dissolution of Hierarchy* (Oakland, CA: AK Press, 2005) and M. Zimmerman (ed.), *Environmental*

Philosophy: From Animal Rights to Radical Ecology (Englewood Cliffs, NJ: Prentice Hall, 1993).

11. David Cadman, 'Harmony', https://www.uwtsd.ac.uk/harmony-institute/ [accessed 17 March 2019]

12. F. J. Schuuman, *Globalization and Development Studies: Challenges for the 21st Century* (London: Sage, 2001).

13. UNCED (1992) Agenda 21, *United Nations* DESA, https://sustainabledevelopment. un.org/outcomedocuments/agenda21 [accessed 20 September 2017]; WAG, *Education for Sustainable Development and Global Citizenship A Common Understanding for Schools* (Cardiff: WAG, 2008).

14. S. R. Sterling, *Sustainable Education: Re-visioning Learning and Change* (Cambridge: Green Books for the Schumacher Society, (2001), p. 94.

15. Sterling, *Sustainable Education*, p.10.

16. D. Orr, 'What is Education For?', p. 52.

17. See D. Orr, *Ecological Literacy: Educating Our Children for a Sustainable World* (The Bioneers Series), (San Francisco: Berkeley: Sierra Club Books, 2005); S.R. Sterling, 'The Future Fit Framework: An introductory guide to teaching and learning for sustainability in HE' (Plymouth: Centre for Sustainable Futures, Teaching and Learning Directorate, Plymouth University for the Higher Education Academy, 2011) and QAA, 'Education for sustainable development: Guidance for UK higher education providers June 2014 (QAA, 2014).

18. See M. K. Smith, 'What is praxis?' in *The Encyclopaedia of Informal Education*, (1999, 2011), http://infed.org/mobi/what-is-praxis/ [accessed May 2017].

19. See Welsh Government, 'Education for Sustainable Development and Global Citizenship In the Further Education Sector in Wales' (Cardiff: Wales, 2008); Welsh Government. '*A Common Understanding for Schools*' (Cardiff, WalesL Welsh Government, 2008); Welsh Government, 'Education for Sustainable Development and Global Citizenship Information for teacher trainees and new teachers in Wales' (Cardiff, Wales: Welsh Government, 2008) and Welsh Government, 'A Strategy for Action – Updates January 2008 (Cardiff, Wales: Welsh Government, 2008).

20. Welsh Government, 'Well-being of Future Generations (Wales) Act 2015 The Essentials' (Cardiff: Welsh Government, 2015), p. 3.

21. Sterling, *Sustainable Education*.

22. Welsh Government, 'A Strategy for Action,' p. 12.

23. Welsh Government, 'A Strategy for Action,' p. 14.

24. See Sterling, 'The Future Fit Framework' and QAA, 'Education for sustainable development'.

25. QAA, 'Education for sustainable development', p. 5.

26. QAA, 'Education for sustainable development', p. 5.

27. QAA, 'Education for sustainable development', p. 7.

28. See Bookchin, *The Ecology of Freedom*.

29. Cited in S. M. Tomlinson-Clarke and D. L. Clarke, *Social Justice and Transformative Learning: culture and Identity in the United States and South Africa* (New York: Routledge, 2016), p. 122.

30. Clark cited in Tomlinson-Clark and Clark, *Social Justice and Transformative Learning*, p. xiii.

31. E. Cudworth, *Environment and Society* (London: Routledge, 2003), pp. 37-38.

32. S. Sterling, 'Whole Systems thinking as a basis for paradigm change in education,

explorations in the context of sustainability' (2003), available at: http://www.bath.ac.uk/cree/sterling/index.html [Accessed: 20/09/2017]; D. Orr, 'Love it or Lose it: The Coming of Biophilia Revolution', in S.R. Kellert and E.O. Wilson (eds.), *The Biophila Hypothesis*, pp.415-40, (Washington, DC: Island Press 1993), p. 428; H.A. Giroux, *Teachers as intellectuals* (New York: Bergin and Garvey, 1988).

33. F. Ferreira, 'Critical sustainability studies: A holistic and visionary conception of socio-ecological conscientization', *Journal of Sustainable Education* [http://www.susted.com/wordpress/content/critical-sustainability-studies-a-holisticandvisionary-conception-of-socio-ecological-conscientization_2017_04/ [accessed 1 May 2017], p. 1.

34. UWTSD PROGRAMME DOCUMENT: MSocStud (Hons) Social Studies: Additional Needs; MSocStud (Hons) Social Studies: Communities, Families and Individuals; MSocStud (Hons) Social Studies: Health and Social Care; MSocStud (Hons) Social Studies: Advocacy; BA (Hons) Social Studies: Additional Needs (Carmarthen: UWSTD, 2017).

35. K. van Wormer and F. H. Besthorn, *Human Behaviour and the Social Environment: Groups, Communities and Organisations* (New York: Oxford University Press, 2017), p. 33.

36. M. Kroth and P. Cranton, *Stories of Transformative Learning* (Rotterdam, NL: Sense Publishing, 2014), p. xiv.

37. UWTSD PROGRAMME DOCUMENT

38. Sterling, *Sustainable Education*, p. 297.

39. R. G. Mira, J. M.S. Cameselle and J. R. Martinez (eds.), *Culture, Environmental Action and Sustainability* (Cambridge, MA: Hogrefe and Huber, 2002), p. 28.

40. UNCED Agenda 21.

41. See R. Yin, Case Study Research (Thousand Oaks, CA: Sage, 2003), p. 1 and H. Simons, Case Study Research in Practice (Thousand Oaks, CA: Sage, 2009), p.21.

42. Cited in M. J. Sheehan, *Sustainability and the Small and Medium Enterprise (SME): Becoming More Professional* (Australia: Xlibris Corporation, 2013), p. 96.

43. G. Guest, K. M. MacQueen and E. E. Namey, *Applied Thematic Analysis* (Thousand Oaks, CA: Sage, 2012).

44. Guest et al., *Applied Thematic Analysis*, pp. 15.

45. Yin, *Case Study Research*, p.1 and Simons, *Case Study Research in Practice*, p. 21.

46. Clark cited in Tomlinson-Clarke and Clarke, *Social Justice and Transformative Learning*, p. 122.

47. W. Legrand, P. Sloan and J. S. Chen, *The Hospitality Industry: Principles of Sustainable Operations* (Abingdon, Oxon: Routledge, 2017), p. 138.

48. Clark cited in Tomlinson-Clarke and Clarke, *Social Justice and Transformative Learning*, p. 122.

49. U. Mander, C. A. Brebbia and E. Tiezzi E, *The Sustainable City IV: Urba Regeneration and Sustainability* (Southampton: WIT Press, 2006), pp. 135-136x`.

50. Cudworth, *Environment and Society*.

51. See N. Carter, *The Politics of the Environment: Ideas, Activism, Policy* (Cambridge: Cambridge University Press, 2007) and G. Gaard, *Ecofeminism: Women, Animals, Nature* (Philadelphia, PA: Temple Univesrity Press, 2010).

52. L.M. Baran and J.E. Jones, *Mixed Methods Research for Improved Scientific Study* (Hershey, PA: ISF, 2016).

53. David   Cadman,   'Harmony',   https://www.uwtsd.ac.uk/harmony-institute/ [accessed 17 March 2019]

54. See Sterling, 'The Future Fit Framework'; QAA, 'Education for sustainable

development'; Welsh Government, 'Education for Sustainable Development and Global Citizenship In the Further Education Sector in Wales'; Welsh Government, 'A Common Understanding for Schools'; Welsh Government, 'Education for Sustainable Development and Global Citizenship Information for teacher trainees and new teachers in Wales' and Welsh Government, 'A Strategy for Action.

55. UNESCO, 'Education, Education for Sustainable Development (ESD),' (2014) http://www.unesco.org/new/en/education/themes/leading-the-international-agenda/education-for-sustainable-development/ [accessed 20 September 2017]

56. David Cadman, 'Harmony', https://www.uwtsd.ac.uk/harmony-institute/ [accessed 17 March 2019]

57. Bookchin, *The Ecology of Freedom*.

58. J. Agyeman, *Sustainable Communities and the Challenge of Environmental Justice* (New York: New York Press, 2005), p. 90.

59. Orr, *Ecological Literacy*, p. 242.

60. David Cadman, 'Harmony', https://www.uwtsd.ac.uk/harmony-institute/ [accessed 17 March 2019]

Early Years Education, Education for Sustainable Development and Global Citizenship and the Principles of Harmony

Glenda Tinney

IN THIS CHAPTER I WILL EXAMINE CONCEPTS OF HARMONY in relation to my work in education and sustainability, relying on David Cadman's suggestion that:

> Harmony is an expression of wholeness, a way of looking at ourselves and the world of which we are part. It's about connections and relationships. The emotional, intellectual and physical are all connected. We are connected to our environments, both built and natural; and all the parts of our communities and their environments are connected, too. Harmony asks questions about relationship, justice, fairness and respect in economic, social and political relationships. As an integrative discipline it can be expressed in ideas and practice.[1]

This definition offers several parallels with the concept of sustainable development which also recognises the interconnectedness of the social, the environmental and the economic, together with a need for human communities to acknowledge that their wellbeing is closely interlinked to that of other communities as well as the wider living and non-living environment.[2] There is a growing awareness that young children are the decision-makers of the future and that those working in the early years' sector are well placed to support children see the interconnectedness of the world and support them understand the cycles and patterns of the natural environment indicative of both harmony and sustainable development.[3] In this context there are however also concerns that children are increasingly separated from opportunities to see the interconnectedness of their world, engaging more in a digital world with less time engaging with nature and less opportunities to be involved in essential life skills such as growing food, cooking and building. In response, early years researchers and practitioners have in recent years developed international research exploring how educators and young children can engage with issues linked to sustainability and citizenship.[4] These authors also refer to some of the principles noted by Cadman in terms of acknowledging wholeness, connection, interdependence and diversity and dovetail closely with the principles

461

of both early childhood and Education for Sustainable Development and Global Citizenship (ESDGC) discussed later in this chapter.[5] Globally, links have been made to approaches to education which emanate from the Agenda 21 goals which were developed during the United Nations Conference on Environment and Development UNCED Earth Summit (1992). Agenda 21 noted the need for education to support adults and children to understand and take action to deliver long term environmental, social and economic sustainability.[6] Internationally, this approach to education has been referred to as Education for Sustainability/Education for Sustainable Development (ESD), and different countries have developed their own approach based on the underpinning principles of Agenda 21.[7] In Wales, since devolution sustainable development has been a part of the Welsh Government constitution underpinning work across the different government sectors.[8] In the education sector, Education for Sustainable Development and Global Citizenship (ESDGC) has been developed as a cross-curricular approach to engaging learners and educators in the values and principles of sustainable development and citizenship as a means to support the future wellbeing of human beings, other living things and the wider environment. As noted by the Welsh Government's Department for Children, Education, Lifelong Learning and Skills (DCELLS), ESDGC is:

> About the things that we do every day. It is about the big issues in the world – such as climate change, trade, resource and environmental depletion, human rights, conflict and democracy- and about how they relate to each other and to us. It is about how we treat the earth and how we treat each other, n o matter how far apart we live. It is about how we prepare for the future.[9]

I will explore how ESDGC can be developed for early years children and how some of the underpinning principles and values of early years' education and care is closely related to the principles of both Harmony and ESDGC.

THE DEBATE

The engagement of young children in the discussion and actions required to create a more sustainable society continues to be contentious.[10] Concerns regarding frightening children regarding the significant problems facing human society link closely to the consideration that childhood is innocent and that young children should not be concerned by problems such as global warming, poverty, inequality, habitat destruction and pollution.[11] However, for many children globally, these are real concerns impacting on their everyday lives, whether they have been forced to migrate due to environmental degradation, war or poverty or other factors, or whether their health and wellbeing is impacted by pollution or lack of resources. The development of 24-hour media and sharing of images and digital technology from across the planet means that many young children in Western society are increasingly aware of the inequalities and consequences of living in an unsustainable society, and authors such as Siraj-Blatchford and Huggins and Davis suggest discussing such issues with children may support their understanding and offer possible solutions as opposed to frightening them more.[12]

Other authors have argued that it is not possible for young children to understand some complex issues, such as those linked to poverty, inequality, climate change, overconsumption and pollution.[13] However, a growing body of research has suggested that young children are both aware of, and concerned about, the challenges facing human society.[14] As noted later in this chapter, theorists such as Bruner suggest that children can engage with complex learning. However, this will be at a level appropriate to their understanding, and for young children, understanding the complex chemistry of climate change may therefore not be appropriate. However, exploring the seasons and being outdoors to see how the weather can change and how these changes can affect people would allow them to start considering the significance of the climate.[15]

It could also be argued that, if young children were to engage with the problems facing society, they may still have no impact as they lack a political voice or the independence to take action.[16] However, the international United Nations Convention on the Rights of the Child 1989 has been ratified in most countries of the world in order to uphold children's rights and to underpin responsibilities to them. Article 29 suggests that:

A child's education should develop each child's personality, talents and abilities to the fullest. It should encourage a child to respect others, human rights and their own and other cultures. It should also help a child to learn to live peacefully, protect the environment and respect other people.[17]

There is an expectation that children should be supported in their ability to respect other people and the environment and to learn to live harmoniously with others. Furthermore, authors such as Heft and Chawla and Elliott and Davis have discussed the significance of viewing children as competent and active participants.[18] Mackay suggests that adults should be 'honouring the young child's right to know about social and environmental issues; to be part of conversations and possible solutions; to have their ideas and contributions valued'.[19] This focus on the child's voice, and the recognition of children as competent, is also the focus of approaches to early years education such as that of Reggio Emilia. In the Reggio approach children are afforded choice and a variety of ways of communicating their ideas through several different 'languages' or modes of expression, including through art, dance, drawing and sculpture.[20] In Wales, the Welsh Government, through the Rights of Children and Young Persons (Wales) Measure 2011 has given 'due regard' to children's rights in all decisions which impact on children and young people. In doing so the Welsh Government has reinforced rights such as that of Article 12 which notes 'children have the right to say what they think should happen, when adults are making decisions that affect them and have their opinions taken in account'.[21] In terms of the intergenerational, long term implications of some of the problems that young children will have to respond to in the future, listening to children now and providing them with opportunities to articulate their views and opinions when young could support their meaningful participation as they grow older.[22]

THE PHILOSOPHY OF EDUCATION FOR SUSTAINABLE DEVELOPMENT (ESD/ESDGC)

The writers of Agenda 21 and other authors on Education for Sustainable Development have indicated that it does not constitute merely a set of subjects to be taught but rather a values based curriculum which both informs and allows children to understand the interconnected nature of their world, as well as to develop the skills and knowledge to be able to actively benefit society and the environment for generations to come.[23] In doing this, authors across the ESD research field have identified the key principles of ESD.

It is striking that the principles of ESD globally, and thus ESDGC in Wales (as noted above), are in close parallel with the theory and values underpinning early years education.[24] For example, this can be observed directly in early years activities which incorporate outdoor learning or are linked to approaches, such as Forest School.[25] For example, children playing in a woodland environment trying to move logs from one place to make dens are engaged in experiential practical learning which involves problem solving and creative thinking when working out the best methods to move the logs or build a den. Such play

PRINCIPLES OF ESD

- Interdisciplinary
- Holistic and interconnected
- Critical
- Creative
- Complicated
- Active discovery
- Problem solving
- Reflective
- Whole setting approach
- Futures approach
- Active participation
- Inclusive
- Student led
- Practitioner facilitates rather than instruct

involves cooperation and team-work skills, which, if child-led can also place the adult in a facilitation role supporting peer learning and avoiding didactic teaching. Furthermore, when such activities are planned through a sustainability lens, they can also engage cross-curricular and interdisciplinary study with the outdoors providing a backdrop for all areas of the curriculum from science and mathematics to language development, art, drama, and humanities. Regular outdoor experience may also allow for significant interactions with nature as well as peers and can be an opportunity to learn about the environment, to co-work and show empathy for both the living and non-living world.[26] Natural areas can also stimulate discussion exploring how, for example, woodland was used in the past and how it can be protected for other people to enjoy in the future.

In the context of young children, as Froebel pointed out, 'play at this stage is not trivial; it is highly serious and of deep significance' and as noted previously child-led play can provide the backdrop for children to develop more detailed understanding.[27] Practitioners can also use this context to develop sustained shared thinking by allowing a child to collaborate with other adults or peers to extend his/her understanding in relation to specific concepts, problems or other enquiry. In terms of approaches to pedagogy, Dewey suggested learning should be grounded in real situations and Piaget referred to children as 'young scientists' that gain understanding from interacting with the environment and learning from mistakes.[28] Play offers opportunities for children, therefore, to construct their own knowledge. In terms of understanding sustainability, children can learn away from a rigid classroom environment through building, cooking and in nature which offer a platform to reengage with the patterns and interactions of life.[29]

Socio-constructivism, which holds that human development takes place within a social context, and that knowledge is constructed through interaction with others, points to the significant role of the adult in supporting and facilitating learning. For example, Lev Vygotsky referred to the 'more knowledgeable other', including peers who could support a child to achieve a higher level of understanding by offering support and guidance, and Bruner referred to this support as 'scaffolding', where a child develops deeper knowledge as initial learning is embedded, and new support is required to develop further.[30] Thus, as noted by Edwards and Cutter-Mackenzie, when required the adult will guide play and provide the environment to ensure deeper understanding.[31] With young children this could be developing empathy for living things; for older children this could involve developing an understanding of a natural cycle or the links between environment and wellbeing. Thus practitioners also have a significant role in modelling the positive behaviour, and work by Vygotsky and Bandura suggests that young children are influenced by the practice they see around them.[32] Thus developing ESDGC is not the role of only one enthusiastic 'ecowarrior' or an opportunity for 'do as I say, not as I do' instead as reflected in programmes such as Ecoschools is a whole school approach where all adults working with children support good practice and engage in work with parents and the community to create more sustainable environments.[33]

Early years practice which is informed by this theory may also support the ESDGC agenda and aspects of the Harmony principles in terms of children being able to recognise their interconnectedness with their world, empathise with other living things, and recognise the holistic and interdisciplinary nature of the world they live in.

In Wales, such theory underpinned the creation of the Foundation Phase Framework, the early years curriculum designed for the education of children between three and seven years old.[34] The curriculum was introduced in 2008 as a play curriculum with a focus on experiential learning, a learning environment

Diagram 1: Foundation Phase areas of learning (based on DCELLS, see note 34)

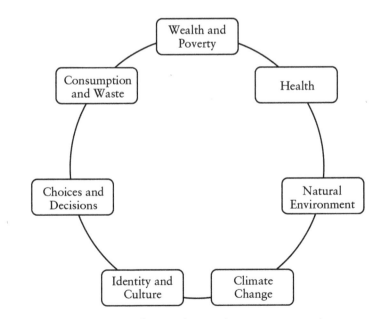

Diagram 2: Themes of ESDGC (DCELLS, see note 9)

which should have both an indoor and outdoor learning ethos, based on seven cross-curricular areas of learning (see Diagram 1).[35]

The ESDGC curriculum includes seven themes which are designed to be cross-curricular (See Diagram 2).

In reality these two curricula interlink where children have opportunities to explore the Foundation Phase areas of learning and ESDGC themes simultaneously.

For example, when cooking with young children, such as making a pizza, there are opportunities to be creative when choosing ingredients, to develop language literacy and communication when working together, listening or reading a recipe and sharing instructions, as well as describing the taste and smells encountered. Weighing ingredients and sharing pieces of pizza develops mathematics skills. In terms of understanding and knowledge of the world, children may grow some of the herbs and other ingredients involved or may discuss how they grow, or explore how the different foods change state when cooking. Making pizza involves rolling and cutting which develop children's fine motor skills (physical development), and in Wales much of the discussion could incorporate incidental Welsh or be through the medium of Welsh depending on the context of the setting. It is also an opportunity to engage other languages used by the children or linked to the food stuffs being used. ESDGC themes could also be incorporated at the same time. For example, pizza could be explored in terms of identity and culture in terms of countries where pizza or similar foods originated and how pizza crosses several countries and cultures. Health could be developed in terms of considering the health benefits of balanced diets and choosing healthier toppings. The natural environment could be explored in terms of where the ingredients were grown or children could be encouraged to grow their own ingredients and thus start understanding plant biology, plant-animal interactions and adaptations to the environment. Choices and decisions would be part of the process of allowing children to choose the best ways to approach the cooking but could also link to choosing ingredients and deciding on, for example, Fair Trade, locally sourced, home-grown foods or options linked to minimal air miles or packaging. This could also link to climate change and consideration of how the choices made could impact on the environment and carbon emissions. Consumption and waste could involve considering where any waste should go in terms of composting, the rubbish bin, or being taken for the school hamster or local pig farm. Wealth and poverty could also be considered in terms of thinking about the amount of resources available to make the pizza and to think if this would be the case in all contexts. All aspects of a child's daily play and activity could support this cross-curricular approach in terms of gardening, building, playing with mud, building dens, role play, water and sand play and thus there would be no expectation to cover everything in one activity. Rather, this would be a long term approach where as children developed they would return to familiar ideas and build on these with different insights as explored in Bruner's spiral curriculum.[36]

As noted above there appear to be significant opportunities to support ESDGC in the early years, and also to support Harmony principles where children are supported to have empathy for their world, to understand the patterns and interconnectedness of the environment and societies in which they live, and to have respect not only for their current communities locally and internationally but to act in ways that will protect generations to come.

However, such opportunities can be limited. For example, Sterling suggests that:

Most mainstream education sustains unsustainability-through uncritically reproducing norms, by fragmenting only a narrow part of the spectrum of human ability and need, by an inability to explore alternatives, by rewarding dependency and conformity, and by servicing the consumerist machine.[37]

Other authors note that practitioners may feel uneasy and lack confidence to engage with sustainability and citizenship due to feeling they do not have the knowledge to respond to the complexities of the issue or because they come from a tradition where teaching is about facts rather than promoting a set of values or action.[38] However, ESDGC fits the wider personal social and emotional development curriculum linked to supporting positive behaviour and inclusive settings and it being promoted in this way may be less daunting. One of my own previous studies noted that practitioners working in the Foundation Phase perceived certain barriers involved in delivering ESDGC, including perceptions of time constraints, a curriculum which is already full, and themes that were too complex for young children.[39] This suggests that practitioners may value training and ideas to support ESDGC in the Foundation Phase and that this should acknowledge that ESDGC is not about being all-knowing, but involves a pedagogy linked to child-led learning, and adults as facilitators who encourage children to be creative and reflective problem solvers. Such skills may allow children the disposition to respond to some of the problems facing humanity as they grow older, as well as developing empathy for others which allows them to become positive role models and citizens in later life.[40]

Others barriers to promoting ESDGC and some of the principles of Harmony are also relevant to an early years context. Practitioners and young children, especially in a Western context, live in a society increasingly removed from the natural environment. Several children may not have access to green space regularly or be involved in growing or preparing food. With the advent of central heating and other technological innovations, children may not be able to see the links between switching on electric appliances and non-renewable resources such as oil, gas and coal energy sources that make this possible. For others, natural cycles such as the carbon or nitrogen cycle may not be recognised in terms of the links between the living and non-living parts of nature. For example, in my own interactions with my own children, linking wood, coal, petrol and gas to carbon stores and thus to carbon dioxide pollution is not straightforward, and thus the expectation that children will understand how 'walk to school days' or switching off lights and electric appliances are designed to lower carbon emissions may need children and adults to revisit the more basic connections between the natural and human world.

Another perspective is provided by Urie Bronfenbrenner's ecological model, which suggests that a child's development is influenced by his/her environment at different levels.[41] Such different levels include the immediate environment (microsystem, such as parents, peers, setting); the interactions between their immediate environment (mesosystem, such as the interaction between parents, setting and peers); the exosystem which are factors such as parents' employment conditions which will indirectly affect the child's wellbeing in terms of how often they will see their parents and the macrosystem which is the cultural and political context of the child's environment (which may include poverty, living in a developed or industrialised country and can evolve with time as successive generations impact on their environment). Another level, the chronosystem explores how socio-historical changes can impact on a child's development. In this context, the lives of children growing up today may be very different from those of the practitioners they work with: for example, the development of digital technology as a significant communication tool amongst the young. However, the context of the practitioners themselves will also be impacted by their own early environments, and in richer nations, as Davis argues, this has been characterised by overconsumption and consumerism which is at times at odds with long term sustainability.[42] However, to ignore this context can lead to ESDGC existing in a vacuum, and there may sometimes need to be a middle ground where practitioners and children can be supported to engage with approaches, which may not be part of their current experience. On early years courses which I am personally involved in delivering, the provision of a safe environment in which to explore possible alternative approaches has been valuable. Such experiences include visits to Down to Earth, an outdoor centred in the Gower, on the Welsh coast, in order to experience learning in nature; visiting a Forest School in order to learn how to light fires, cook food and see the practical links between fuel and natural cycles; joining the John Muir Award in order to learn about a special environment and to support its conservation and learning about more sustainable ways of building can all provide valuable tasters to encourage engagement with ESDGC. In doing this, practitioners and young children can develop innovative approaches to ESGDC that may also support the values of Harmony.

NOTES

1. David Cadman, 'Harmony', The Harmony Institue, https://www.uwtsd.ac.uk/harmony-institute [Accessed 16 May 2019].
2. I. Siraj-Blatchford and V. Huggins, 'Sustainable development in early childhood care and education (SDECCE)', *Early Education Journal*, Vol. 76 (Summer, 2015): pp. 3-5.
3. A. Cutter-Mackenzie, S. Edwards, D. Moore and B. Boyd, *Young Children's Play and Environmental Education in Early Childhood Education* (Heidelberg: Springer, 2014).
4. See I. Pramling-Samuelsson and Y. Kaga (eds.), 'The contribution of early childhood education to sustainable society' (Paris: UNESCO) at: http://unesdoc.unesco.org/images/0015/001593/159355e.pdf [accessed 24 February 2013]; J. M. Davis, 'Revealing the research 'hole' of early childhood education for sustainability: a preliminary survey of the literature', *Environmental Education Research*, Vol. 15, No. 2 (2009): pp. 227-41; J. Davis and S. Elliott, 'Exploring the Resistance: An Australian Perspective on Educating for Sustainability in Early Childhood', *International Journal of Early Childhood*, Vol. 41, No. 2 (2009): pp. 77; E. Pearson and S. Degotardi, 'Education for Sustainable Development in Early Childhood A Global Solution to Local Concerns', *International Journal of Early Childhood*, Vol. 41, No. 2 (2009): pp. 97-111; A. N. Cutter-MacKenzie and S. Edwards, 'Environmentalising early childhood education curriculum through pedagogies of play', *Australasian Journal of Early Childhood*, Vol. 36, No. 1 (2011): pp. 51-59; Cutter-Mackenzie et al., *Young Childrens' Play*; A. Warwick and P. Warwick, 'Towards a pedagogy of love: sustainability education in the early years', *Early Education Journal*, Vol. 76 (Summer 2015), pp. 6-8.
5. D. Cadman, Conference Presentation David Cadman, Harmony Professor of Practice, the University of Wales Trinity St David 2 and 3 March 2017.
6. G. Tinney, 'A all plant ifanc newid y byd? Addysg ar gyfer datblygu cynalidawy a'r Cyfnod Sylfaen', in S.W. Siencyn (ed.), *Y Cyfnod Sylfaen 3-7 oed. Athroniaeth, Ymchwil ac Ymarfer* (Caerfyrddin: Cyhoeddiadau Prifysgol Cymru Y Drindod Dewi Sant, 2010).
7. Davis, 'Revealing the research "hole"'.
8. Tinney, 'A all plant ifanc newid y byd?'
9. DCELLS, 'Education for Sustainable Development and Global Citizenship :

information for teacher trainees and new teachers in Wales' (Cardiff: Welsh Assembly Government: 2008), p. 4.

10. S. Elliott and J. Davis, 'Exploring the Resistance'; G. Tinney, 'Perceptions and understanding of Education for Sustainable Development and Global Citizenship (ESDGC) by teachers and teacher trainees and the opportunities and challenges of integrating ESDGC into early childhood learning within the Foundation Phase' (MA Early Childhood Research Thesis, Carmarthen, UWTSD, 2013).

11. Tinney, 'Perceptions and Understanding of Education'.

12. Siraj-Blatchford and Huggins, 'Sustainable development in early childhood care' and Davis, 'Revealing the research "hole"'.

13. M. Warnock, 'Moral values', In J.M. Halstead, and M.J. Taylor (eds.), *Values in Education and Education in Values* (London: The Falmer Press, 1996), pp. 45-54.

14. See J. Page, *Reframing the Early Childhood Curriculum: Educational Imperatives for the Future* (London: Routledge Farmer, 2000); I. Engdahl. and M. Rabusicova, 'Children's Voices about the State of the Earth and Sustainable Development', *A report for the OMEP World Assembly and World Congress on the OMEP World Project on Education for Sustainable Development 2009-2010* (World Organisation for Early Childhood Education, 2010).

15. J. Bruner, *The Process of Education* (Cambridge: MA: The President and Fellows of Harvard College, 1960).

16. Tinney, 'Perceptions and Understandings of Education'.

17. UNICEF, 'The Convention on the Rights of the Child Guiding principles: general requirements for all rights' at https://www.unicef.org/crc/files/Guiding_Principles.pdf [accessed 30 July 2017].

18. H. Heft and L. Chawla, 'Children as agents in sustainable development: Conditions for competence', in M. Blades and C. Spencer (eds.), *Children and Their Environments* (Cambridge: Cambridge University Press, 2005); and J. Davis and S. Elliott, *Research in Early Childhood Education for Sustainability: International Perspectives and Provocations* (London: Taylor and Francis, 2014).

19. G. Mackay, 'To know, to decide, to act: the young child's right to participate in action for the environment', *Environmental Education Research*, Vol. 18, No. 4 (2012): pp. 473-84.

20. P. C. Edwards, G. Firman and L. Gandini, *The Hundred Languages of Children. The Reggio Emilia Approach. Advanced Reflections* (2nd ed.) (Greenwich, CT: Ablex, 1998).

21. UNICEF, 'The Convention on the Rights of the Child Guiding principles'.

22. G. Symons, 'The Primary Years', in J. Huckle and S. Sterling (eds.), *Education for Sustainability* (London: Earthscan, 1996), pp. 55-71.

23. See J. Fien, 'Education for Sustainability', in R. Gilbert, *Studying Society and Environment: A Guide for Teachers* (Melbourne: Thomson Social Science Press, 2004), pp. 189-200; Davis, 'Revealing the research "hole"'; A. Peterson and P. Warwick, *Global Learning and Education* (Oxon: Routledge, 2015).

24. Davis, 'Revealing the research "hole"'.

25. S. Knight, *Forest School in Practice. For All Ages* (London: Sage, 2016).

26. G. Tinney, 'A all plant ifanc newid y byd?'

27. F. Froebel, *The Education of Man* (1826; repr. New York: Dover, 2005).

28. I. Siraj-Blatchford, K. Sylva, S. Muttock, R. Gilden and D. Bell, 'Researching Effective Pedagogy in the Early Years (REPEY)'. DFES Research Report 365 (London: HMSO

Queen's Printers, 2002); J. Dewey, *Experiences and Education The 60th Anniversary Edition Indiana* (1939; repr. USA: Kappa Delta Pi, 1998); J. Piaget and B. Inhelder, *The Psychology of the Child* (USA: Basic Books, 1972).

29. Knight, *Forest School in Practice.*

30. L. Vygotsky, *Mind in Society: Development of Higher Psychological Processes* (Boston: Harvard University Press, 1978); J. Bruner, *Acts of Meaning* (Boston: Harvard University Press, 1990).

31. Cutter-MacKenzie and Edwards, 'Environmentalising early childhood education curriculum'.

32. Vygotsky, *Mind in Society*; A. Bandura, *Social Learning Theory* (Englewood Cliffs, NJ: Prentice-Hall, 1977).

33. E. Merriman, A. Power-Rekers and G. Tinney and A. MacGarry, *An Evaluation of the Eco-Schools Programme in Wales; Final report*, Natural Resources Wales (Agency), Wales and WAG (Carmarthen: University of Wales Trinity Saint David, 2014).

34. E. Merriman et al., *An Evaluation of the Eco-Schools Programme*; DCELLS, 'Framework for Children's Learning for 3 to 7-year-olds in Wales' (Cardiff: Welsh Assembly Government, 2008); Welsh Government, Curriculum for Wales. Foundation Phase Framework (Cardiff: Welsh Government, 2015).

35. DCELLS, 'Framework for Children's Learning'.

36. Bruner, *The Process of Education.*

37. S. Sterling, *Sustainable Education. Revisioning Learning and Change* (Darlington: Schumacher Briefing Number 6, Green Books, 2001).

38. For knowledge of complexities see Davis and Elliott, 'Exploring the Resistance' and Tinney, 'A all plant ifanc newid y byd?'. For teaching about facts see Tinney, 'A all plant ifanc newid y byd' and Cutter-Mackenzie et al., *Young Children's Play.*

39. Tinney, 'Perceptions and Understanding of Education'.

40. Davis, 'Revealing the research "hole"'.

41. B. Bronfenbrenner, The Ecology of Human Development: Experiments by Nature and Design (Cambridge, MA: Harvard University Press, 1979).

42. Davis, 'Revealing the research "hole"'.

HARMONY IN EDUCATION:
APPLYING THE PRINCIPLES OF NATURAL SYSTEMS TO LEARNING

Richard Dunne and Emilie Martin

The truth is that many things on which your future health and prosperity depend are in dire jeopardy: climate stability, the resilience and productivity of natural systems, the beauty of the natural world, and biological diversity.[1]

THE EARTH IS A COMPLEX, SELF-REGULATING SYSTEM. Like all systems, it is the sum of its parts; parts which are interconnected and interdependent in ways we are only beginning to understand. We know that the currents in our oceans regulate the climate of the entire planet and that the trees in our rainforests keep in check the amount of carbon dioxide present in our atmosphere. The Earth relies on the existence of a natural balance between its parts in order to function as a whole. This balance is one of many ways in which the natural world exists in a state of harmony. The Prince of Wales, Tony Juniper and Ian Skelly described this state of harmony in 2010:

> There is a deep mutual interdependence within the system which is active at all levels, sustaining the individual components so that the great diversity of life can flourish within the controlling limits of the whole. In this way, Nature is rooted in wholeness.[2]

But there is now an overwhelming body of scientific evidence showing that human activity is threatening the natural world's ability to sustain itself – and us. Human activity over the last two centuries, and particularly within the last sixty years, has caused accelerated change that has left its imprint on a global scale, as research from the Intergovernmental Panel on Climate Change makes clear:

> Warming of the climate system is unequivocal, and since the 1950s, many of the observed changes are unprecedented over decades to millennia. The atmosphere and ocean have warmed, the amounts of snow and ice have diminished, and sea level has risen.[3]

Our consumption of the planet's resources and our degradation of the environment are destabilising the systems upon which we depend, compromising the ability of the planet to support human life and human development. And yet we continue to behave in ways that suggest we are unaware of the very fact that we exist within a finite system. Our disconnection from the systems that sustain us is at odds with our fundamental reliance upon them.

A contributing factor to the ecological and environmental problems we now face is our tendency to consider things in isolation with little consideration of cause, effect or provenance. We are disconnected from the reality that lies behind the choices we make; a reality which is already impacting our collective wellbeing.

THE COST OF NATURE DEFICIT

It seems that our disconnection from nature begins at an early age. The fact that children in western societies today spend less time outdoors, in general – and in natural environments, in particular – has been well-documented.[4]

Research commissioned by Natural England, which advises the government on issues relating to the natural environment, shows that in the two years to February 2015, just over ten percent of UK children had not visited a natural setting.[5] The definition of natural settings used for the purposes of the research includes parks, canals and nature areas; coast and beaches; and wider countryside such as farmland, woodland, hills and rivers. This is the phenomenon that Louv, writing in 2005, terms 'nature deficit disorder', noting that one outcome of this lack of interaction and engagement with the natural world is that 'for a new generation, nature is more abstraction than reality'.[6]

Making the case for action to address this disconnection four years later, he referred to the 'psychological, physical and cognitive costs of human alienation from nature' – costs which are particularly damaging for children in their formative years.[7] A study by Cervinka, Röderer and Hefler also shows a strong correlation in adults between a sense of connectedness with nature, and wellbeing (in particular, vitality) and meaningfulness. The authors of the study offer the following definition of meaningfulness:

Meaningfulness, in contrast to depression, is understood as a developmental motive, referring to a human's need of being in the world and experiencing a sense of purpose in life. People scoring high on meaningfulness conceive their lives as fulfilling and relatively free from feelings of powerlessness, helplessness, fear and depression. They feel accepted by others and experience social connectedness and high satisfaction in their lives.[8]

While there are, no doubt, many consequences of a lack of connectedness with nature which affect us as individuals, the impact of nature-deficit extends beyond our selves to threaten the collective wellbeing of humankind. Louv notes: 'Reducing that deficit – healing the broken bond between our young and nature – is in our self-interest, not only because aesthetics or justice demands it, but also because our mental, physical, and spiritual health depends upon it'.[9] This was echoed by research by the Countryside Commission, which identified engagement with the natural environment as a key factor in determining our 'contentment with life' and in achieving 'better social ties and sense of community, improved physical and mental health, strengthened economic prospects, reduced crime rates, and enhanced children's play and learning'.[10]

THE ENVIRONMENTAL CONSEQUENCES OF DISCONNECTION

A further consequence of our disconnection from nature is evident in our attitude to the environment and the esteem in which we hold the natural world; 'The health of the earth is at stake as well. How the young respond to nature, and how they raise their own children, will shape the configurations and conditions of our cities, homes – our daily lives'.[11]

Our view of nature and the beliefs we hold about our dependence upon it inevitably shape the choices we make in our daily lives, and the choices we make as consumers. This is particularly true of the choices we make about the food we eat. A 2017 study of more than 5,000 school children aged 5-16 years by the British Nutrition Foundation (BNF) revealed that some of them held some startling ideas about where our food comes from.[12] The survey found that more than one in ten of the eight to eleven-year-olds in the sample thought pasta comes from an animal, and almost one fifth of five to seven-year-olds believed fish fingers are made from chicken. Of the 11-14-year-olds surveyed, one in ten was unaware that carrots and potatoes grow underground and one in twenty of the oldest children in the sample aged 14-16 years thought cows produce eggs.

On the face of it, such misinformation might seem amusing. A more interesting question to ask – at least in the context of this chapter – would be: Why is it the case that a significant number of children appear to have so little understanding of the provenance of their food? The answer may lie in the study itself, which revealed that fifteen percent of the children surveyed had never visited a farm. Could this be another outcome of nature-deficit? Through the eyes of some children (even some adults), the pre-packaged cuts of meat they see when they visit a supermarket may be entirely disconnected from the animal they came from. In the same way, the bags of pre-sliced carrots may be disconnected from the soil in which they grew.

The final question in the BNF study asked the older children, aged eleven to sixteen years: 'Where would you learn about or go to find reliable information about your food?'. 'School' and 'the internet' were the two most popular answers. It is encouraging to know that children look to their schools and teachers as sources of trusted information about food. However, a second study carried out by the BNF in the same year showed that only a quarter of primary school teachers and half of secondary school teachers had received professional development to deliver food education in the preceding three years.[13]

It is not our intention in this chapter to focus solely on children's ideas about food and food provenance. Rather, we use this example to illustrate one of the many ways in which our understanding of our reliance on the natural world has become divorced from reality. It is also important, however, to recognise the opportunity that schools, teachers and education policy makers have to counter this. David Orr talks of the role of educators in the development of children's 'ecological literacy' and argues for the need for educational policy to acknowledge and respond to this:

> More of the same kind of education will only compound our problems. This is not an argument for ignorance, but rather a statement that the worth of education must now be measured against the standards of decency and human survival – the issues now looming so large before us in the decade of the 1990s and beyond. It is not education that will save us, but education of a certain kind.[14]

AN EDUCATION OF A CERTAIN KIND

Humanity faces complex problems in securing for itself a healthy and sustainable future. Whatever the solutions may be to these problems, what is clear is that progress towards a more sustainable way of living will only be possible if, collectively, we undergo a shift in mindset about how we perceive the world around us and our relationship to it. There is a clear reference to this in the title of the 2010 book by HRH The Prince of Wales, Ian Skelly and Tony Juniper, *Harmony: A new way of looking at our world*, as the use of the phrase 'new way' in its title implies. It is our belief that we must challenge the way in which we currently structure the education of our students and question whether it is fit for purpose.

We must offer our young people an education of a kind that equips them to engage with, and take the lead on, issues around environmental health and sustainability, and which counters thinking that see things in isolation. We need only look to the natural world to learn about the connectedness that exists within any system. By looking to nature for examples, we see how elements within systems are interdependent, how each element has its own purpose and role to play, and how the health of the system is preserved when there exists a dynamic balance between those elements. If we are to learn to replicate such systems, we need to find a way of working that facilitates a paradigm shift towards a mindset rooted in connectedness. This will be no small undertaking as our disconnected ways of thinking are deeply engrained but education has a key role to play in addressing this. We need to design a new model for learning that enables our young people to see connections and relationships. Furthermore, we need to support them in developing a greater appreciation of how essential this joined-up outlook is to understanding the relationship between actions and consequences. In the next section we will look at how the integration of skills and knowledge from different subject disciplines can help support children in developing the thinking skills needed to tackle some of the biggest challenges to planetary and human health.

THE COMPARTMENTALISATION OF LEARNING

Before we look at how we might achieve this kind of education, let us take a moment to reflect on the position that we, as educators, find ourselves in today. This chapter focuses on the education system in England, but the same issues occur in many countries. So far, we have created within English education a system that is organised neatly but artificially into compartmentalised subjects. This is engrained in the fabric of the National Curriculum here, in England, which 'is organised on the basis of four key stages and 12 subjects, classified in legal terms as 'core' and 'other foundation' subjects'[15]. Programmes of study published by the Secretary of State for Education, set out the 'matters, skills and processes'[16] to be taught for each subject at each key stage. By structuring the curriculum in this way, we have created rigid subject silos which do not accommodate the natural links between knowledge, skills and understanding across different areas of the curriculum. In our systems of assessment, too, we test subject-specific knowledge and skills in isolation from one another. We are perpetuating, through the structure of the education in our schools, disconnected ways of thinking and looking at the world. This is a flawed approach if we are to live more sustainably.

There are certainly some schools and teachers who encourage students to make links between different areas of their learning and actively plan for teaching and learning to occur in a more coherent, cross-curricular format. But, as this is not a statutory requirement, whether or not a student is exposed to a more holistic model of learning is largely a matter of luck. In short, we have created a system in which separation – rather than connection and integration – is the norm. It is our belief that at its best, primary education – indeed all education – should be about helping young people to understand the world around them, and to make sense of it. Taking the child as its starting point, a more 'joined-up' approach to learning acts as a bridge between the child's natural inquisitiveness and the wider world. When we teach – and when we encourage young people to learn – in a more integrated, cross-curricular way, we support them in applying their learning to matters outside the confines of a particular subject, and in engaging with the world around them. Beane argues that this approach has two beneficial outcomes for students:

> First, young people are encouraged to integrate learning experiences into their schemes of meaning so as to broaden and deepen their understanding of themselves and their world. Second, they are engaged in seeking, acquiring, and using knowledge in an organic – not an artificial – way. That is, knowledge is called forth in the context of problems, interests, issues, and concerns at hand. And since life itself does not know the boundaries or compartments of what we call disciplines of knowledge, such a context uses knowledge in ways that are integrated.[17]

One chapter of the National Curriculum in England acknowledges that learning can be structured in ways other than distinct academic subjects. The Early Years Foundation Stage (EYFS) curriculum, introduced in 2008 and revised in 2012, sets out standards for the learning, development and care of children from birth to 5 years of age. It is applicable to all early years settings, extending through nursery school and into the first year of formal schooling. The EYFS curriculum is structured not in terms of subjects but in terms of areas of learning and development. These are: communication and language; physical development; personal, social and emotional development; literacy; mathematics; understanding the world; and expressive arts and design.[18]

The introduction of the EYFS curriculum has created a learning environment which recognises the importance of child-initiated learning and values fluidity over rigidity. The adult may shape the learning by selecting the activities to which the child has access, but their role is then to support the child in taking their learning forward. The child is involved in influencing the content of learning, bringing their own interests and prior knowledge to an activity and determining the way in which that activity will develop. The child is equally central to the structure of the learning – he or she is the unifying thread running through his or her interaction with each element of the curriculum. Two years after the introduction of the EYFS, Ofsted noted an increase in the proportion of children meeting expectation in personal, social and emotional development at age five, which rose from 72% in 2008 to 77% in 2010. Over the same period, increased attainment was also noted in communication, language and literacy, with 53% meeting expectation in 2008 compared to 59% in 2010, and in children reaching a good level of development by the age of five (49% in 2008 and 56% in 2010).[19]

Despite the perceived success of the EYFS, this acknowledgement that young people learn well when they are free to explore and to make connections across traditional subject boundaries is not reflected in the curriculum beyond the child's first year at primary school. From Year 1 onwards, when children are five or six years of age, the curriculum taught in our schools is structured around defined and separate subjects.

In the modern curriculum we have fragmented the world into bits and pieces called disciplines and subdisciplines. As a result, after 12 or 16 or 20 years of education, most students graduate without any broad integrated sense of the unity of things. The consequences for their personhood and for the planet are large. For example, we routinely produce economists who lack the most rudimentary knowledge of ecology. This explains why our national accounting systems do not subtract the costs of biotic impoverishment, soil erosion, poisons in the air or water, and resource depletion from gross national product... As a result of incomplete education, we've fooled ourselves into thinking that we are much richer than we are.[20]

At the level of primary education, students may learn about materials in Science and about our oceans in Geography. But unless the learning is structured in a way that encourages them to bring together their knowledge from both subjects, students may not develop an understanding of the ways in which these two areas of learning are linked. They may not, for instance, see the consequences of our use of materials such as plastic on the health of the ecosystems in our seas and on the sustainability of the oceans' resources. On the other hand, if we make what is learnt in the classroom relevant to the world in which our students live, if we support them in engaging with the issues that affect their future wellbeing and encourage them to act purposefully to resolve these issues, the outcome of their learning becomes much more purposeful.

The effect of assessment systems in perpetuating separateness in learning

The compartmentalisation of knowledge and skills within mainstream education in England today is evident in teaching and learning, but also permeates the systems of assessment that we use to judge student outcomes. Whether through formal examination or teacher assessment, the judgements made about a student's educational success are subject specific: grades are awarded and decisions about attainment are made in separate areas of the curriculum. The systems of assessment used in our schools follow a linear model of academic progress measured against criteria of increasing complexity that lead towards the end goal of a final, summative grade.

The selection of content to be assessed – the domains that form the basis for testing and against which a student's abilities are judged – represents a further separation of perceived high-value knowledge and skills from the rest of the content within a programme of study. By creaming off certain knowledge and skills from the rest of the curriculum to form the focus for testing, this tested content takes on a higher status. In this way, 'higher value' content is set apart from the rest of the content taught in the classroom: it becomes disconnected. Whilst there is clearly a need to measure learning outcomes, it is not hard to see how inevitably, (although unintentionally), what we choose to assess begins to dominate the content taught in the classroom.

In recent years, a growing number of voices within the teaching profession and within the wider education community have expressed concern over the narrowing of the curriculum as a result of the dominance of the 'standards agenda' and high-stakes testing. More recently, Ofsted has joined the debate on the narrowing of the curriculum, with Her Majesty's Chief Inspector, Amanda Spielman, noting:

> Across the whole education sector, a mentality of 'what's measured is what gets done' trumps the true purpose of education, and curriculum thinking – the consideration of what needs to be taught and learned for a full education – has been eroded.[21]

These are observations that have also become increasingly apparent to those outside the education sector. Ofsted's 2017 interviews with parents of primary school-aged children revealed that around half of those surveyed 'believed that test preparation had reduced the teaching time available for the other foundation subjects or for reading for pleasure'.[22] The Ofsted Chief Inspector also notes a corresponding narrowing of the curriculum in secondary education, with many students selecting the subjects they will study to GSCE level a year earlier than was previously the case, in order to maximise the time spent studying towards the Year 11 exams.[23] The impact of high-stakes testing on the breadth of content taught in schools has implications that reach further than the educational experience of individual students:

> The dominance of testing as part of American and British school reform policies insures that many of the skills thought to be most useful in the twenty-first century will not be taught. Thus, students and their national economies will suffer when nations rely too heavily on high stakes testing to improve their schools.[24]

In the next part of this chapter we will look in more detail at how one school has responded to these issues, developing cross-curricular learning and a broad and distinctive curriculum.

The development of a connected, nature-inspired curriculum

The school in question is Ashley Church of England Primary School, located in Walton-on-Thames, near London. For the last five years, it has been pioneering an enquiry-based curriculum based on nature's principles of Harmony. Even before the school began to develop its nature-inspired curriculum, the everyday practices of the school community had for many years been shaped by a concern for sustainability:

> The school hadn't just created a garden where children could learn about growing vegetables. Its allotment was supplementing the school kitchen's procurement program. The school wasn't simply aiming to reduce food waste. The children were weighing and monitoring the food discarded at the end of every lunch sitting and using the data to help the kitchen staff revise menus and refine portion sizes.[25]

Figure 1. Young beekeepers learn how to maintain a hive and how bees work together

The school decided to take this further, introducing Harmony principles as the frame for learning across the curriculum. The Harmony curriculum developed at the school has three main aims:

Firstly, to reconnect children with nature. As you might expect, the school emphasises the importance of learning *in* nature and learning *about* nature, but also supports children in learning *from* nature. To take an example from the Year 2 curriculum, in the summer term the children in this year group explore the enquiry question: *Why are bees so brilliant?* As part of this learning enquiry, they identify the roles of different types of honey bee within a hive – an example of diversity at work in the natural world. They carry out beekeeping activities to maintain the school's hives (learning *in* nature), learn about the bees they identify in the hive and how they work together (learning *about* nature), then apply what they have learnt to their own lives, looking at how diversity is beneficial to the different social groups they are part of (learning *from* nature).

Secondly, to engage the children in problem solving, drawing on their knowledge and understanding of content from different curriculum subjects. The focus for a half term's learning is framed for the children as a question, often with an environmental or ecological focus. This might be 'What are the cycles of our solar system?' or 'How can we build community in our town?'. At the end of each enquiry, each class is asked to reflect on the enquiry question, based on their learning and experiences across all subjects over the half term. By structuring learning through cross-curricular enquiry, students experience connectedness in their learning first-hand. As Beane notes: 'Curriculum integration begins with the idea that the sources of curriculum ought to be problems, issues and concerns posed by life itself'. These may be categorised as 'self- or personal- concerns' and 'issues and problems posed by the larger world'[26].

Thirdly, to make learning purposeful. At the end of each learning enquiry, the children work together to create a Great Work that represents a culmination of their learning in all subject areas. This could be an art exhibition, a meal to share with their parents, a puppet show, the planting of a fruit orchard or the creation of a soundscape – to name but a few. This approach embodies a valued system at work.

When we structure the education of our young people in a way that allows them to draw upon and link together knowledge and skills gained in different areas of the curriculum, we create a very different learning experience – one that highlights and accentuates the value of connections in what we learn and the importance of good working relationships in achieving successful outcomes. These are essential life skills. As educators, we need to recognise that in order to help students experience a sense of connectedness, we must as educators develop practices that model this connectedness:

all education is environmental education. By what is included or excluded we teach students that they are part of or apart from the natural world. To teach economics, for example, without reference to the laws of thermodynamics or those of ecology is to teach a fundamentally important ecological lesson: that physics and ecology have nothing to do with the economy. That just happens to be dead wrong. The same is true throughout all of the curriculum. [27]

When children learn from the principles that maintain wellbeing and balance in the natural world, they can learn how to live in a more sustainable and balanced way. By presenting learning to the children in this way, the school hopes to counter the effects of the 'crisis of perception'[28] which is distorting our understanding of our relationship with nature.

PRINCIPLES OF HARMONY AS A LENS FOR LEARNING

Ashley Primary school works with seven principles of Harmony, inspired by HRH The Prince of Wales, Tony Juniper and Ian Skelly's 2010 book. These are: the principle of Geometry; the principle of the Cycle; the principle of Interdependence; the principle of Diversity; the principle of Adaptation; the principle of Health, and the principle of Oneness.

The principle of Geometry – which allows children to explore and respond to the geometry that exists in the natural world – cuts across all learning throughout the academic year. Of the remaining six principles, one is explored by the children in each year group as the focus of their learning enquiry each half term. In this way, by the end of their time at the primary school, a child who has attended since the start of Reception will have explored each of the principles in seven different ways.

If we take the principle of the Cycle as an example, we can 'map' the learning enquiries through which a child will have explored this principle by the end of their time at the school:

- What can we learn about farm animals? (Reception)
  Including the life cycle of chickens and other farm animals
- Which is my favourite wildflower and why? (Year 1)
  Including the life cycle of flowering plants, seed collection and sowing a wildflower meadow
- How can we bring traditional tales to life? (Year 2)
  Including looking at how traditional stories are passed on from one generation to the next
- How can we learn to identify native trees through the seasons? (Year 3)
  Including a focus on how trees change through the cycle of the seasons
- What are the cycles of our solar system? (Year 4)
  Including an exploration of the orbit of the Moon around the Earth, and the Earth around the Sun and how this affects the cycles of day and night, months and years.
- What journey does a river take? (Year 5)
  Including a focus on the water cycle
- How do the Inuit of the Arctic live with nature? (Year 6)
  Including an exploration of how the Inuit live in harmony with the cycle of the seasons

Figure 2. Children collect and sort seeds from the school's wildflower meadows, ready to be sown in new areas

USING HARMONY PRINCIPLES TO FRAME LEARNING

As an example of learning in Year 3, in the autumn term children are posed the enquiry question: *How can we identify native trees through the seasons?* This enquiry is linked to the Harmony principle of the Cycle. To explore this question, the children first use identification keys to name different trees growing in the school grounds, then focus on one tree to create their own identification key for others to use. Alongside this learning, they measure and order the heights of different trees in maths and reflect upon the variety of tree heights. In their Geometry learning, the children explore symmetry in the leaves of different trees and use this knowledge to help them in their identification of native trees. In Science, they learn about why the leaves of deciduous trees change colour in the autumn and why these trees shed their leaves. This links to other Science learning about the process of leaf decay, which contributes to the creation of healthy soil. In English, they write non-fiction texts about the importance of healthy soil to tree growth and poetry about the colours and atmosphere of autumn. Also, in Science, children learn about the parts that make up a tree's seed, and how through the process of germination, a shoot and root grow. They also learn about the function of fruits and how trees disperse their seeds in a variety of ways.

Learning in other subjects is also linked to the central enquiry focus. The children work with watercolours and resist techniques to create artwork depicting a silver birch forest, they taste and comment on native apple varieties and cook with apples in Design Technology. The children return to this enquiry throughout the school year, to observe changes to the trees in different seasons, producing a booklet to record what they have found out. For their Great Work, the children plant native fruit trees in the school's orchard. The fruit from these trees will contribute to the learning of future cohorts of children in the school; in this way the children themselves contribute to cycles in learning at the school.

EXPLORING NATURE'S GEOMETRY IN THE CLASSROOM

Everywhere we look, we see patterns, shapes and proportions that are repeated in micro and macro form throughout the natural world and in ourselves – even throughout the universe. Exploring these patterns is a great starting point for learning and helps children experience the awe and wonder of nature – and develop a reverence for it. The principle of Geometry is a key part of every learning enquiry at Ashley School and each week's learning begins with a Geometry activity.

The practice of geometry provides a different lens for learning. It helps young people to see how things are. We teach geometry every week in our school and it has produced some stunning results. As the students learn the proportions and ratios of Nature's patterns, they start to understand that there is an order to life that gives it balance and harmony. They see the world from a different perspective, and they begin to develop a much deeper insight into what harmony means. Geometry is a mindful art. [29]

The school has worked closely with the Prince's Foundation School of Traditional Arts (www.psta.org.uk) to bring the principle of Geometry to the curriculum.

The children themselves have a lot to say about their Geometry learning. They appreciate the fact that it helps them develop a 'new way of looking at the world', giving them a new perspective on familiar things as well as providing a route in to learning about aspects of the natural world that are new to them. The following quotes from conversations with the children at Ashley School, gathered in the course of research into the school's curriculum development work, clearly show this.[30]

Figure 3. Using the vesica piscis created by overlapping circles, a Year 5 child works on a fish painting

Figure 4. A Year 6 child explores the Geometry of snowflakes
as part of their learning about the Antarctic

The research is available in full on The Harmony Project website.

www.theharmonyproject.org.uk.

'You start to notice the patterns in different things. You pay attention to the detail, to what things really look like. Like a Jaguar has rosettes in its fur. It makes you want to learn more about the animal.' Charlie, Year 3.

'The Geometry is really interesting because you might look at something you've known about for a long time but it makes you see it in a different way and really concentrate on that. It's more interesting to actually do something like making a five-petaled flower from the orbit of Venus and Earth than just being told about it. After that you might decide to look at the orbits of other planets, too.' Charlie, Year 4.

'You can take something small as a starting point then the learning leads on from there. There are big ideas and beauty in the small things. You learn to see beauty in the whole world.' Eden, Year 5.

There is also reference to a sense of wellbeing in some of the children's responses to their Geometry learning, as the following quote from one Year 6 child reveals:

'The Geometry learning makes me feel calm. There's not one answer so no-one's ever going to tell you that's wrong or right. You just keep trying and changing things until you have something you're happy with.' Hannah, Year 6.

A MORE HOLISTIC APPROACH TO PLANNING

As might be expected, this kind of organisation of National Curriculum content into learning enquiries, and into the school's wider distinctive curriculum, requires a shift in working practices.

Usually the process of planning and delivering a curriculum of learning starts by referencing the National Curriculum and pulling curriculum objectives into a planning document. This is a very logical thing to do, but it means that the National Curriculum objectives drive the learning and, if planned in subject-separate ways, the learning quickly becomes very disjointed and disconnected. It can all feel rather piecemeal and even pointless from a student perspective.[31]

Instead, teachers at Ashley School take as the starting point for their planning a principle of Harmony and from this plan an enquiry to frame learning across all subjects. In Year 2, for example, the students explore the principle of Health through the enquiry question, 'What does it mean to be healthy?'. In PE, they learn about the effects of exercise and experience different forms of activity, including yoga and karate, that contribute to different aspects of our physical and mental wellbeing. In Science, they learn about the importance of hygiene and the role of different food groups in contributing to good health. In History they learn about pioneers in the field of healthcare, and in Maths they investigate the symmetry and asymmetry of the human body and learn to measure the body accurately using hand spans, as well as metric units of measurements. In English they write poems about wellbeing; instructions about effective hand washing; and recipes for making healthy meals. At the end of the half term, the students research and prepare their own healthy meal, which their parents are invited to share.

This web of interconnected learning allows children to reflect on different ways in which the question at the centre of their learning enquiry could be answered. They are provided with opportunities to participate in debate about the challenges we face and to design projects to secure a healthier, more sustainable world in which we can all thrive.

The students' own reflections give perhaps the best indication of the impact of teaching and learning based on principles of Harmony. In the younger years, the children's responses to their learning enquiries may be simply articulated but show an emerging connectedness in their thinking. Olivia, age 6, responded to the question, 'What does it mean to be healthy?' by saying: 'Smiling makes you healthy'. Her exploration of the principle of Health had allowed her to start making a connection between emotional wellbeing and overall health. As we might expect, the reflections of older students show greater sophistication. Sam, age 10, was asked to reflect on his enquiry question, 'How can we ensure our oceans stay amazing?' and responded: 'Plastic rubbish is killing the creatures of the ocean. Because all the creatures are connected to each other, if we kill the creatures, the food chain will be destroyed. We need to work out how to stop this'. Sam shows a more developed understanding of the interdependence of elements within a system and is beginning to articulate the need for humans to change the way they work to maintain the wellbeing of that system.

OUTCOMES OF A HARMONY CURRICULUM

Research into the outcomes of the introduction of a Harmony curriculum at Ashley School is still at an early stage.

In the school year 2017-18, researchers at the University of Canterbury Christ Church included the school in a study of three schools developing their own curricula around concepts of sustainability. The study revealed two key outcomes of this approach to learning and to building a school community around principles of Harmony. Firstly, the children were aware that their education at the school was preparing them for life outside and beyond their time in school. Secondly, the children knew that they were being supported in developing leadership skills and had a sense of their own agency as a result. The researchers noted that these outcomes are not attributable to any single initiative at the school but to a 'holistic and multi-layered approach' that relies on the involvement of the entire school community.[32]

Standards in the core subjects of English and Maths remain a priority at the school and attainment is high. The balance between securing core skills and knowledge, and purpose in learning is constantly reviewed and refined to ensure it is right. We can check outcomes by examining the results of SATs. These are national tests that children take twice during their primary school life, first at the end of Key Stage 1 (KS1) in Year 2, and then at the end of Key Stage 2 (KS2) in Year 6. For the academic year ending 2018, Key Stage 2 (KS2) SATs results were significantly above the national average:

Children working at, or exceeding, expectation at the end of KS2 (%)

	Ashley C of E School	National Average
Reading	98	75
Writing	93	78
GPS*	98	78
Maths	93	76

*Grammar, Punctuation and Spelling

Hand in hand with this high level of attainment, the students are supported in developing less easily quantifiable skills that we increasingly see are important to other aspects of their learning and personal achievement. For example, through engagement in activities linked to the principle of Geometry, both teachers and students have noticed an improvement in fine motor skills; greater attention to detail, better concentration and mindfulness; and high levels of self-esteem through the realisation of challenging, high-quality geometry work.

THE CHALLENGE FOR EDUCATORS

If we are to secure for ourselves a healthy and sustainable future, it is clear that we need to change the way we think about the world and our place in it. And if we believe this to be necessary, we must face another fact: we will not succeed in shifting the paradigm in a reductive education system that teaches for separateness.

'We need to teach our children, our students and our corporate and political leaders the fundamental facts of life: that one species' waste is another species' food; that matter cycles continuously through the web of life; that the energy driving the ecological cycles flows from the sun; that diversity assures resilience; that life, from its beginning more than 3 billion years ago, did not take over the planet by combat but by networking.'[33]

For educators, this poses a new challenge: How do we develop thematic and integrated ecosystems of learning, relevant to the age in which we live and appropriate to the age of our students? What is a certainty is that we cannot afford to continue with an education system which does not prepare our young people to bring about the revolution in thinking and practice that will address the challenges we face now and in the future.

NOTES

1. David Orr, 'What is Education For?', https://www.eeob.iastate.edu/classes/EEOB-590A/marshcourse/V.5/V.5a%20What%20Is%20Education%20For.htm [accessed 29 March 2019].

2. HRH The Prince of Wales, Tony Juniper and Ian Skelly, *Harmony: A New Way of Looking at Our World* (London: Harper Collins, 2010) p.17.

3. IPCC, 2014: Climate Change 2014: Synthesis Report. Contribution of Working Groups I, II and III to the Fifth Assessment Report of the Intergovernmental Panel on Climate Change [Core Writing Team, R.K. Pachauri and L.A. Meyer (eds.)]. IPCC, Geneva, Switzerland, p. 2, https://www.ipcc.ch/site/assets/uploads/2018/02/AR5_SYR_FINAL_SPM.pdf.

4. Richard Louv, *Last Child in the Woods* (New York: Atlantic Books, 2013); A. Hunt, D. Stewart, J. Burt and J. Dillon, 'Monitor of Engagement with the Natural Environment: a pilot to develop an indicator of visits to the natural environment by children – Results from years 1 and 2 (March 2013 to February 2015)', (Natural England Commissioned Reports, Number 208, 2016).

5. H. Woolley, L. Pattacini and A. Somerset-ward, *Children and the natural environment: experiences, influences and interventions – summary*, (Natural England commission reports, 24 March 2011).

6. Louv, *Last Child in the Woods*, p. 2.

7. Richard Louv, R., 'No More "Nature-Deficit Disorder"' (2009) at https://www.psychologytoday.com/gb/blog/people-in-nature/200901/no-more-nature-deficit-disorder [accessed 29 March 2019].

8. Renate Cervinka, Kathrin Röderer and Elisabeth Hefler, 'Are nature lovers happy?' in *Journal of Health Psychology*, 17(3), 2011, p. 384

9. Louv, *Last Child in the Woods*, p. 3.

10. Countryside Commission (1997), cited in Willy, T. and Catling, S., *Understanding and Teaching Primary Geography* 2nd *Edition*, (London: Sage, 2018), p. 211.

11. Louv, *Last Child in the Woods*, p. 3.

12. British Nutrition Foundation, 'National Pupil Survey 2017: UK Survey Results', 2017, https://www.nutrition.org.uk/attachments/698_UK%20Pupil%20Survey%20Results%202014.pdf.

13. British Nutrition Foundation and the Food Teachers Centre, 'Food education learning landscape: Teacher research', 2017, https://www.nutrition.org.uk/attachments/article/1085/FELL%20Appendix%2011.2%20Teacher%20Survey%20Research.pdf.

14. Orr, 'What is Education For?'.

15. Department for Education, 'National Curriculum in England: framework for Key Stages 1-4', 2014, https://www.gov.uk/government/publications/national-curriculum-in-england-framework-for-key-stages-1-to-4/the-national-curriculum-in-england-framework-for-key-stages-1-to-4 [accessed 29 March 2019].

16. Department for Education, 'National Curriculum in England: framework for Key Stages 1-4', 2014, https://www.gov.uk/government/publications/national-curriculum-in-england-framework-for-key-stages-1-to-4/the-national-curriculum-in-england-framework-for-key-stages-1-to-4 [accessed 29 March 2019].

17. J. A. Beane, 'Curriculum Integration and the Disciplines of Knowledge', *The Phi Delta Kappan*, Vol. 76, No. 8, 1995, p. 616.

18. Department for Education, 'Early Years Foundation Stage', https://www.gov.uk/early-years-foundation-stage [accessed 29 March 2019].

19. Ofsted, 'The impact of the Early Years Foundation Stage', 10th February 2011, https://www.gov.uk/government/publications/the-impact-of-the-early-years-foundation-stage [accessed 29 March 2019].

20. Orr, 'What is Education For?' (1991)

21. Amanda Spielman, 'Amanda Spielman launches Ofsted's Annual Report 2017/18' (2018), https://www.gov.uk/government/speeches/amanda-spielman-launches-ofsteds-annual-report-201718 [accessed 29 March 2019].

22. Amanda Spielman, 'HMCI's commentary: recent primary and secondary curriculum research', (2017),https://www.gov.uk/government/speeches/hmcis-commentary-october-2017 [accessed 29 March 2019].

23. Spielman, 'HMCI's commentary: recent primary and secondary curriculum research'.

24. David Berliner, 'Rational responses to high-stakes testing: the case of curriculum narrowing and the harm that follows', *Cambridge Journal of Education*, Volume 41, Issue 3, 2011, p. 287, https://eric.ed.gov/?id=EJ950193.

25. Emilie Martin, 'Is it time to rethink our perspective on holistic education?', *Learning for Well-being Magazine*, Issue 6, 2019, https://www.l4wb-magazine.org/mag06-vpto1 [Accessed 29 March 2019].

26. Beane, 'Curriculum Integration and the Disciplines of Knowledge', p. 616.

27. Orr, 'What is Education For?'.

28. HRH The Prince of Wales, Juniper and Skelly, *Harmony*, p. 6.

29. Richard Dunne, *Harmony: A new way of looking at and learning about our world. A teacher's guide,* (London and Bristol: The Harmony Project/Sustainable Food Trust, 2019), p. 51.

30. The research is available in full on The Harmony Project website: https://www.theharmonyproject.org.uk/tag/ashley-school/ [Accessed 11 November 2019].

31. Richard Dunne, *Harmony*, p. 46.

32. Nicola Kemp and Alan Pagden, 'Doing it Differently: The Place of Education 'Alternatives in the Mainstream', Unpublished Report, University of Canterbury Christ Church, 2018.

33. Fritof Capra, 'Pedagogy of Sustainability', *Resurgence*, March/April 2014, Issue No. 283, p. 28.

An Investigation into Well-Being in Wales

Tania Davies

THE RELATIONSHIP BETWEEN HARMONY AND HEALTH is increasingly recognised. It has been the focus of my research over the last few years, and although I am dealng specifically with the situation in Wales, the discussion can be applicable to the wider global contect. I am also personally engaged as I am currently one of the two programme directors for the Health Portfolio at the University of Wales Trinity Saint David (UWTSD), as well as a PhD student. My research at the University has involved an in-depth examination of well-being in line with the aim, statutory duty and impact of the Well-Being of Future Generations (Wales) Act (WFG), which was passed in 2015. This is an innovatory and world-leading piece of legislation to which all public bodies in Wales must conform: its radical nature is based in the requirement it established for all policy decisions and implementation to take into account the well-being of future generations. The problem is how, exactly, well-being can be measured. My aim is therefore to create a Wellbeing Assessment Framework, or WAF for short, which aims to measure the collective wellbeing of a small population, which in this instance consists of the University's staff, both academic and non-academic. This framework should assist both the University and any public body or interested organisation in meeting any well-being goals and self-regulated objectives which are linked to the Act, and to implement and monitor improvement in their implementation.

Both the WAF and the conclusions resulting from my research could be applicabile to all forty-four public bodies and affiliated organisations in Wales, as well as all over the world. My research should also help to situate well-being in relation to Harmony. David Cadman describes Harmony as consisting of 'broad and all-embracing principles about the nature of the cosmos and [...] our relationships within it, because we are a part *of* not apart *from*' with sustainability defined as 'the relationships between environment, society and economy that can be sustained and nourished over long periods of time for mutual benefit'.[1] In carrying this theme of cyclical relationships and interconnectivity, perhaps well-being could be defined as based in the individual relationship with self, and how health, society and environment impacts upon this.

Introduction

The Well-Being of Future Generations Act is currently the only piece of legislation in the world which places a statutory (albeit self-regulatory) duty on public bodies to carry out sustainable developmentsustainable development.[2] It is centred on traditional definitions of sustainable development but also goes on to specify sustainable development as a process of improvement. The Act introduces seven well-being goals which all must be met in order to work toward the enhancement of the economic, social, environmental and cultural well-being of Wales.[3] These goals are:

1. A prosperous Wales

2. A resilient Wales

3. A healthier Wales

4. A more equal Wales

5. A Wales of cohesive communities

6. A Wales of vibrant culture and thriving Welsh language

7. A globally responsive Wales[4]

Public bodies must first set objectives which meet these goals in accordance with the sustainable development principle and then take all 'reasonable' steps to meet them. A position of Future Generations Commissioner for Wales has also been created, to which all public bodies are to be held accountable.[5] Currently, only the civil service appears to regularly monitor the well-being of staff, and no research has yet been undertaken to evaluate the implantation of the Well-Being of Future Generations Act in a higher education setting in Wales.[6] My research will therefore be the first body of work to focus on the implementation of the Act within a public body organisation. It focuses directly upon one area of the Act, that of a healthier Wales. As the Act is somewhat in its infancy this is a timely and topical area of research.

Higher Education

To deal with the facts and figures of Higher Education (HE) in Wales, one of the public bodies included in the Act is the Higher Education Funding Council for Wales (HEFCW) which regulates and scrutinises the quality, service and performance of all Welsh universities.[7] Throughout the UK in 2014-15 there were 403,835 staff classed as 'academic' and 'non-academic' in the Higher Education sector, with almost half of these employed on academic contracts.[8] Within Wales there are just over 10,000 academic and 12,000 non-academic staff employed in Higher Education Institutes. UWTSD is part of HEFCW's scrutiny as one of the nine universities in Wales.

Within UWTSD there are 700 academic staff (60% of these are part-time) and an additional 855 non-academic staff.[9] Within the university there are nearly 27,000 students from over 30 different countries attending a range of campuses.[10] The university itself boasts a strong and award-winning sustainability theme as a core principle, and possesses an institutional approach to delivering this through university culture, curriculum, campuses and community.[11] It has set up an award-winning institute – the Institute of Sustainable Practice, Innovation and Resource Effectiveness, or INSPIRE – in order to champion and embed this principle as well as support the link between systemic change at university and national level.[12] Therefore, UWTSD is situated as an ideal and sector-leading HEI to provide rich and detailed data and context within which to study the implementation of this new legislation.

While there is a sizeable amount of research relating to well-being in general, it tends to focus mainly upon children and students.[13] Student well-being, anxiety and stress are well-documented and measured.[14] There are also examples of student health linked to student well-being in research.[15] Research relating to HE staff tends to focus only upon areas such as the responsibility of the teacher in health education and student well-being or teacher socialisation, the impact the gender of the tutor has upon student achievement, or the importance of teaching quality upon student achievement.[16] The small amount of focus upon educational staff in terms of health is largely depicted in terms of occupational stress, and tends to be somewhat dated or describes stress as an environmental and individual matter in teaching.[17] Traditionally, teaching staff are relied upon to identify their own needs, and the importance of self-regulation is still highlighted in occupational well-being for teachers.[18] Although a small amount of research on the well-being of staff can be found situated in compulsory education, the majority measures staff well-being in relation to benefits for students and higher marks only.[19] Speller *et al.*, for example, reported that a majority of teachers think that schools play an important part in health and well-being promotion, although how they are trained to achieve this is still in question.[20] Additionally, teacher well-being in relation to student well-being is argued as 'two sides of the same coin' by Roffey.[21] The Education Support Partnership collect Health Surveys on educational staff and focus upon staff well-being but, again, largely frame staff well-being in relation to student well-being and achievement.[22] Teaching unions such as the NASUWT have collected data on well-being and published advice on managing mental health and well-being for teachers and school leaders, but this data has not been utilised effectively or measured consistently.[23] It is clear that whilst there is evidence-based research and data available related to well-being in general, there is a distinct gap in our knowledge and disparity in published work relating to not only higher education staff and well-being but also the implementation of policies related to safeguarding the mental health and well-being of academics.

MEASURING WELL-BEING

As I said, my aim is to provide an in-depth examination and assessment of well-being in relation to one small population, staff at the University of Wales Trinity Saint David. On a micro scale, this should assist in measuring the impact of a new national policy within a Welsh University context. But on a macro scale, its could provide a way of measuring and assessing well-being in any organisation, public body, or group. My Wellbeing Assessment Framework (WAF) aims to measure well-being and attribute a well-being score to the sampled group, and the results should then be analysed in order to point to possible improvements. There are four main components of the WAF. These are:

1. Health and Safety Executive (HSE) stress indicators questionnaire

2. Well-Being of Future Generations national well-being indicators

3. Higher Education Statistics Authority (HESA) profiling categorisation

4. The capabilities approach.

Elements and aspects of each will be entwined and carefully utilised in harmony with the other. For each, the theoretical basis, underlying philosophical concepts and policy documentation shall be taken into account.

The concept underpinning the measurement of well-being is adapted from the *capability approach*, a philosophical ideology created by Amartya Sen.[24] The capability approach argues that well-being should be measured regarding the capabilities of an individual and that this should reflect who and how *they* wish to be.[25] The idea is that expansion of such individual capabilities should therefore increase well-being. The approach itself is now primarily utilised as an alternative to utilitarian (or resource or income-based) approaches to human welfare; rather than 'necessities' being used to measure quality of life or well-being in terms of satisfaction or acquisition of 'primary goods'. Sen proposes that the things people are able to do or to be as a result of having resources – simply, the capabilities they command – should instead be valued and measured in terms of well-being. Well-being, then, depends upon the opportunities afforded by goods rather than the goods themselves. The capability approach is a normative paradigm which is now used to evaluate individual well-being and social arrangements, as well as the design of social policy and proposals for social change.[26] Arguably, the hallmark of the capabilities approach is the idea of taking human agency seriously.

Therefore, the central idea of well-being as measured in regard to the capabilities of an individual as a reflection of who and how they wish to be is the underpinning theory of the WAF.[27] The desired outcome, then, should evidence that expansion of such capability impacts positively upon levels of well-being.

While this approach is by no means a perfect measure of well-being, the approach has been classified as one of the most significant theoretical contributions to welfare analysis across multi-disciplinary fields and the analysis of social policy within the developed world.[28] It has also been hugely influential in terms of the creation of the Human Development Index, which is a statistical tool used to measure a country's overall achievement in its social and economic dimensions.[29] For the purposes of my research, however, rather than duplicate how the approach has been utilised directly, the underlying principle of personal freedom and autonomy shall be a key factor in assessing how social policy (in line with the Well-Being of Future Generations Act) impacts upon individual autonomy – a paramount factor in all policy analysis.[30] Revision is also an important aspect of the capability approach: Sen states that the framework can be adapted but 'clarity of theory has to be combined with the practical need to make do with whatever information we can feasibly obtain for our actual empirical analyses'.[31] This perspective is particularly useful as it will help to ensure a well-rounded and bespoke measurement framework which produces the most accurate results possible.

METHODS

My study is located within a relativist ontological position and a pragmatist epistemology. In practice this means that the world can be viewed from the subjective perspective of the participant, and that one single objective reality cannot be assumed.[32] The main methods of data collection are an online questionnaire and a semi-structured interview. The main methods used for data analysis will be the online tool, SPSS and thematic analysis. These methods have not been finalised as yet as they will be determined by the data.

There are four stages to the data collection:

1. *Informal pilot stage*: In this stage I will gain as much input and direction from peers and colleagues as possible and construct a rudimental version of the WAF. I will then ask close colleagues to take the WAF and give feedback and opinions. This will then help me to refine the final version, which will be created using the data collection and analysis software Sensemaker.

2. *WAF rollout stage*: In this stage the final version of the WAF will be sent to all academic and non-academic staff in order to collect as much data as possible. This will be an online questionnaire. The aim is to collect one well-being score for the whole group of participants (the sample group), as well as examine individual scores.

3. *Interview stage*: Semi-structured interviews will then be undertaken with a purposive sample group selected. The aim will be to gain as broad a picture as possible of each participant's experience of well-being (as defined within both the WbFGA and Capabilities Approach), and obtain a self-determined well-being score.

4. *Data analysis stage*: There will be three sets of data: individual WAF scores, sample group score, and self-determined individual scores. These will utilise the same results scale, but the collection of data will differ. The WAF results table will purposely be revisable, dynamic and adaptable. This is so any organisation can adapt the WAF to their own targets, ideologies and aims in regard to measuring well-being. Put simply, this means that the weighting system can be made fluid so that importance can be placed upon different sections of the WAF in order to create a dynamic results scoring system. For example, the HE sector may place more importance on the capabilities approach sections as the important aspect of measuring well-being, whereas the health sector may place importance on the stress section as the most important aspect.

CONCLUSION

At present, my research remains in the initial stages, and I am inviting ideas, input and feedback from academic and non-academic colleagues from within UWTSD, HE as a sector, or any external interested party. The Harmony conference – as well as colleagues working within wellbeing, health, sustainability and happiness across a spectrum of schools and Faculties – have a massively important part to play in this research, as their input is crucial in the design and implementation of the WAF.

In terms of how my research relates to Harmony, and in particular David Cadman's definition, I would particularly like to invite discussion on where the concept of wellbeing fits into Harmony. Is it, as I suggested in the introduction, a more personal relationship between self and health? Can it be typified as a broader concept, no more or less than any other aspect of Harmony and balance as defined by Cadman himself? Could it be classified as the most important aspect as it is so closely entwined with the self and health? Well-being and Harmony may be far too subjective to ever conclusively define, for each of us has such a personal relationship to our own wellness, and our ideas of what being well ought to mean. Is there, then, a way to holistically broaden well-being or Harmony into an umbrella or parent term in a way that resonates with all of us? For if we can view Harmony as our relationship with the world and each other, and well-being as our relationship with oneself, then surely we should not discuss one without including the impact of the other.

NOTES

1. David Cadman, 'Harmony' at https://www.uwtsd.ac.uk/harmony-institute/ [Accessed 19 March 2019].
2. H. Davies, 'The Well-being of Future Generations (Wales) Act 2015: Duties or aspirations?' *Environmental Law Review* (2016): pp. 41-56; Welsh Government, 'Well-being of Future Generations (Wales) Act 2015' (March, 2016) at www.thewaleswewant. co.uk: http://thewaleswewant.co.uk/sites/default/files/Guide%20to%20the%20WFGAct. pdf [Accessed 19 March 2019].
3. Welsh Government, 'Well-being of Future Generations (Wales) Act 2016: The Essentials' (2016) at http://thewaleswewant.co.uk/sites/default/files/Guide%20to%20 the%20WFGAct.pdf [Accessed 19 March 2019].
4. Welsh Government, 'Well-being of Future Generations (Wales) Act 2016: The Essentials' [Accessed 3 April 2019].
5. Welsh Government, 'Well-being of Future Generations (Wales) Act 2016: The Essentials' [Accessed 3 April 2019].
6. Gov.uk, 'Engagement and wellbeing' (2 July 2015) at https://www.gov.uk/ government/collections/engagement-and-wellbeing-civil-service-success-stories.

7. HEFCW, 'About Us' (2016) at https://www.hefcw.ac.uk/about_us/about_us.aspx [Accessed October 2016].

8. HESA, 'Staff at Higher Education Providers in the United Kingdom 2014/15' (2016) at www.hesa.ac.uk: https://www.hesa.ac.uk/sfr225.

9. HESA, 'Publications and products' (January 2016) at https://www.hesa.ac.uk/component/pubs/?Itemid=&task=show_year&pubId=1717&versionId=27&yearId=326 and HESA, 'Free Online Statistics – Staff (1 December 2015) at https://www.hesa.ac.uk/index.php?option=com_content&view=article&id=1898&Itemid=634 [accessed 3 April 2019]

10. University of Wales, Trinity. 'Study in Wales' (2016) at http://www.studyinwales.com/study/universities-wales [Accessed March 2016].

11. Sustainability Exchange, 'Green Gown Awards' (2016) at http://www.sustainabilityexchange.ac.uk/green_gown_awards_2015_best_newcomer_university and UWTSD. 'About INSPIRE (2016) at http://www.uwtsd.ac.uk/inspire/about-inspire/ [Accessed March 2016].

12. UWTSD, 'About INSPIRE' and J. Davidson, 'Systemic Change is the only change in town' in A. Nicholl and J. Osmond, *Wales' Central Organising Principle: Legislating for Sustainable Development* (Cardiff: Institute of Welsh Affairs, 2013), pp. 21-33.

13. For children see T. Ridge, 'Supporting Children? The Impact of Child Support Policies on Children's Wellbeing in the UK and Australia', *Journal of Social Policy* (2005): pp. 121-142 and for students see S. Roffey, 'Becoming an Agent of Change for School and Student Well-being', *Educational & Child Psychology*, Vol. 32 (2015), .

14. W. E. Ansari; C. Stock; S. Snelgrove; X. Hu; S. Parke; S. Davies and A. Mabhala, 'Feeling Healthy? A Survey of Physical and Psychological Wellbeing of Students from Seven Universities in the UK', *Environmental Research and Public Health* (2011): pp. 1308-1323; K. Van Petegem; A. Aelterman; Y. Rosseel; and B. Creemers, 'Student Perception as a Moderator for Student Wellbeing', *Social Indicators Research* (2006): pp. 447-463.

15. Public Health England, *The link between pupil health and wellbeing and attainment* (London: NAHT, 2014) and A. Greig; T. MacKay; S. Roffey and A. Williams, A, 'The Changing Context for Mental Health', *Educational & Child Psychology*, Vol. 33 (2016), pp. 21-30

16. J.M. Gore & K.M. Zeichner, Action Research and Reflective Teaching in Preservice Teacher Education, *Teaching and Teaching Education*, 7 (1991): pp. 119-136; J.M. Lindo, N.J. Saunders & P. Oreopoulos Ability, Gender, and Performance Standards, *American Economic Journal* (2009): pp. 95-117. For the responsibility of the teacher in health education and student well-being see D. Jourdan; P. Mannix McNamara; C. Simar; T. Geary and J. Pommier, 'Factors Influencing the Contribution of Staff to Health Education in Schools', *Health Education Research* (2010): pp. 519-530 and V. Speller; J. Byrne; S. Dewhirst; P. Almond; L. Mohebati; M. Norman and P. Roderick, 'Developing trainee school teachers' expertise as health promoters', *Health Education* (2010): pp. 490-507. For teacher socialisation see K. Z. Gore, 'Teacher Socialization' in W. R. Houston, *Handbook of Research on Teacher Education* (New York: Macmillan, n.d.); for the impact gender of the tutor has upon student achievement see H. Antecol, O. Eren, and S. Ozbeklik, 'The Effect of Teacher Gender on Student Achievement in Primary School (2015) *Journal of Labor Economics,* 63 – 89 and for the importance of teaching quality upon student achievement see OECD, *PISA 2012 Results:Students' Engagement, Drive and Self Beliefs* (OECD, 2013).

17. For in terms of occupational stress see M. Y. Tytherleigh; C. Webb: C. L. Cooper and C. Ricketts, 'Occupational Stress in UK Higher Education Institutions: A Comparative Study of All Staff Categories', *Higher Education Research & Development* (2005): pp. 41-

61; for dated see G. Kinman, *A survey into the causes and consequences of occupational stress in UK academic and related staff* (London: Association of University: Teachers, 1998) and for stress as an environmental and individual matter in teaching see P. T. Knight, *Being a Teacher in Higher Education* (Buckingham: SRHE and Open University Press, 2002).

18. For staff reliance on their own needs see J. Bradley, R. Chesson, and J. Silverleaf, *Inside Staff Development* (Berks: NFER-Nelson, 1983), and for the importance of self regulation see U. Klusmann; M. Junter, U. Trautwein; O. Ludtke and J. Baumert, 'Teachers' Occupational Well-Being and Quality of Instruction', *Journal of Educational Psychology* (2008): pp: 702-715.

19. For research on well-being of staff situated in compulsory education see A. Konu; E. Viitanen and T. Lintonen, 'Teachers' wellbeing and perceptions of leadership practices', *International Journal of Workplace Health Management* (2010), pp. 42-57, and for staff well-being measured in relation to benefits for students and higher marks only see Z. Bajorek; J. Gulliford and T. Taskila, *Healthy teachers, higher marks? Establishing a link between teacher health & wellbeing* (Lancaster: The Work Foundation, 2014).

20. Speller *et al.*, 'Developing trainee school teachers' expertise as health promoters'.

21. S. Roffey, 'Pupil wellbeing – Teacher wellbeing: Two sides of the same coin?' *Educational & Child Psychology*, Vol. 29 (2012).

22. Education Support Partnership, 'Looking after Teacher Wellbeing' (2015) at https://www.educationsupportpartnership.org.uk/looking-after-teacher-wellbeing [Accessed June 2].

23. For data see NASUWT, *Teachers' Mental Health* (Birmingham: NASUWT, 2010) and for published advice see NASUWT, 'Managing your Mental Health and Wellbeing' (2016) at http://www.nasuwt.org.uk/consum/groups/public/@journalist/documents/nas_download/nasuwt_007706.pdf [Accessed August 2016].

24. M. C. Nussbaum, *Women and Human Development – The Capabilities Approach* (Cambridge: Cambridge University Press, 2001); M. C. Nussbaum, 'Introduction: Aspiration and the Capabilities List', *Journal of Development and Capabilities*, (2016): pp. 301-308; M. C. Nussbaum, 'Education and Democratic Citizenship: Capabilities and Quality Education', *Journal of Human Development* (2006): pp. 387- 394; A. K. Sen, *Commodities and Capabilities* (Amsterdam: North Holland, 1985): A. K. Sen, 'Development as Capability Expansion' in S. Fukuda-Parr, *Readings in Human Development* (New York: Oxford University Press, 2003).

25. S. Alkire, *The Capability Approach and Well-Being Measurement for Public Policy* (Oxford Poverty & Human Development Initiative, 2015).

26. I. Robeyns, 'The Capability Approach in Practice', *The Journal of Political Philosophy* (2006): pp. 351-376; C. Harnacke, 'Disability and Capability: Exploring the Usefulness of Martha Nussbaum's Capabilities Approach for the UN Disability Rights Convention', *Journal of Law, Medicine and Ethics* (2013): pp. 768-780; R. Smith; G. Hunter and P. Anand, 'Capabilities and Well-Being: Evidence Based on the Sen-Nussbaum Approach to Welfare', *Social Indicators Research* (2005): pp. 9-55.

27. Alkire, *The Capability Approach*.

28. As a measure of well-being see Alkire, *The Capability Approach* and D. A. Clark, *The Capability Approach: Its Development, Critiques and Recent Advances* (ERSC Global Poverty Research Group, 2006); as a significant social contribution see R. Smith et al., 'Capabilities and Well-Being' and for the analysis of social policy see A. Goerne, 'The Capability Approach in Social Policy Analysis – Yet Another Concept?', *REC-WP Working Papers on the Reconciliation of Work and Welfare in Europe*, 3 (2010).

29. The social and economic dimensions of a country are measured using health,

educational level, and standard of living: see 'Definition of "Human Development Index"', The Economic Times, https://economictimes.indiatimes.com/definition/human-development-index [Accessed 3 April 2019].

30. S. Deneulin and J. A. McGregor, 'The Capability Approach and the Politics of a Social Conception of Wellbeing. Wellbeing in Developing Countries; *European Journal of Social Theory*, 13, 4 (2009): pp. 501-519.

31. Sen, *Commodities and Capabilities*, p. 49.

32. J. A. Smith, *Qualitative Psychology A Practical Guide to Research Methods* (London: Sage Publications, 2011).

HARMONY AND SOCIETAL CHALLENGES: EMPOWERING COMMUNITIES

Rachel Parker

We have to look at the whole picture to understand the problems we face ...
to [understand] the principles that produce the active state of balance which
is just as vital to the health of the natural world as it is for human society.[1]

INTRODUCTION

THIS CHAPTER REFLECTS UPON THE PRINCIPLES OF HARMONY outlined by H.R.H. the Prince of Wales as a method of responding to the challenges facing humanity. At the centre of this approach is the need to empower communities to address societal issues for sustainability and resilience. This chapter suggests a pragmatic response to building community capacity to achieve these goals. It outlines the community-based empowerment method of participatory appraisal (PA) which is embedded in the whole-systems perspective of the socio-ecological model. It describes both the quality and ethical role research can play in supporting this process. It is hoped this work will help to empower communities to respond to the call to action from H.R.H. the Prince of Wales to find sustainable solutions and quality partnerships for building community resilience to overcome societal challenges. This call from the Prince of Wales has arisen from his lifelong work in exploring solutions to tackle environmental and social problems. He defines harmony as both an active and balanced state, a dynamic equilibrium of tensions and forces, embedded within a socio-ecological systems perspective of interconnectivity. A key ingredient within this approach is the need to empower communities through the development of public and private-sector partnerships that meaningfully engage with the community using effective participation.[2]

THE ENTRENCHED NATURE OF SOCIETAL PROBLEMS

The Prince of Wales explains that 'For more than thirty years I have been working to identify the best solutions to the array of deeply entrenched problems we now face'.[3] He highlights the ingrained nature of societal problems which means their key features are that they are challenging, long-standing, difficult to change and multifaceted. These characteristics can act as significant barriers to finding solutions or designing effective support interventions. For example, within the initial analysis of any societal problem, notions of linear pathways of cause and effect are not usually pertinent, due to the complex interactions from numerous components across multiple system dimensions.[4] At a later stage such components may be broken down and targeted, but at the start a linear model is too limited to understand the full picture. This is why a socio-ecological systems model is a useful starting point for gathering a holistic understanding of the whole system that the social issue resides within.

THE SOCIO-ECOLOGICAL SYSTEMS MODEL AND COMPLEX SYSTEMS ANALYSIS

In the socio-ecological systems model, individual and intra/inter-personal relationships are interconnected to the local community as well as to broader societal factors. So there is a complex and dynamic interplay of multiple factors throughout a whole system, ranging from the individual to family and communities and including local and national influences from the wider society such as prevailing societal norms.[5] Mapping the full contextual detail including the range and impact of a problem from this systems perspective in order to understand the issue, and then from this work to also begin to find acceptable and feasible solutions, is complex and resource intensive. Challenges involve non-linear pathways, manifold influences and dynamic properties in the social systems that social issues both stem from and are embedded within.

One solution is to apply a complex systems lens which can accommodate these issues both for analysis purposes and social intervention design.[6] Furthermore, embedding an analytical dimension of sustainability helps in the exploration of socio-ecological system resilience and adaptation.[7] These concepts are informed by complexity science. Understanding the context for this work is therefore paramount, and requires input from multiple perspectives to generate a quality evidence base so that complex system intervention modelling provides effective solutions to the problem.[8]

USING RESEARCH TO ENSURE QUALITY AND ETHICS

Research has a key role to play in this complex socio-ecological systems informed process to ensure that the quality of the evidence-base, analysis, process evaluation and outcomes are fit-for-purpose for real-world social solutions. Another important factor that research brings to enhance this societally-focussed work is ethics. This enables ethical principles and practice to be at the forefront as well as being embedded throughout the lifespan of a project. For example, through utilising the quality framework of social research, the core ethical principles from the Economic and Social Research Council (ESRC) can be clearly activated. The focus of the ESRC is upon shaping society. Embedding its ethical principles in research activities is a strong and pragmatic way of both bringing and consolidating ethical values into society. This is why ethics in practice is a core strand of the ESRC's work which is specifically contextualised in each research project.[9]

A TRANSDISCIPLINARY RESEARCH APPROACH

For research planning, harnessing the conceptual developments surrounding a transdisciplinary approach to complex societal challenges is one way forward.[10] This is when research is designed from across the boundaries of multiple disciplines, creating innovation and new conceptual frameworks.[11] It uses an array of methods for a collective approach to finding solutions to complex real-world issues. This stems from a collaborative dialogue involving researchers and other sectors of society.[12] It includes important and rich detail being shared by lay experts who have lived experience and knowledge of the context.

In this way, a working group of community stakeholders is brought together to facilitate multiple perspectives, knowledge and skills. Such a transdisciplinary approach has the potential to deliver a working model for the collaboration between public and private-sector partnerships, including non-governmental organisations and the community, in order to tackle societal issues. Managing inherent tensions and challenges within the transdisciplinary approach is necessary to this process.[13] One potential barrier is the structural inequalities that are present, with the potential for some stakeholders to have unequal status. There may also be challenges in interdisciplinary communication and knowledge hierarchies. Another barrier centres around the nature of shared knowledge generation and its validity, stemming from the realm of empirical epistemology and ontology. A further issue is in regards to having the knowledge and expertise in facilitating meaningful community participation. There is also the need to offer multiple and inclusive methods to facilitate community communication and ensure equality of access. Additional barriers reside in having the capacity to build the conceptual frameworks for this work for quality research purposes. From the outset these types of potential barriers should be rendered visible by being brought to the stakeholder discussion table so they are planned for, delineated and actively managed.

PARTICIPATORY APPRAISAL

A research method that is suitable for transdisciplinary research, and which can address some of its potential barriers, is *participatory appraisal* (PA). This can incorporate multiple stakeholder perspectives and shared knowledge generation, facilitating equality of access. It is traditionally used within a community-based and localised problem-solving approach to community development.[14] PA creates knowledge in a way that aligns with many of the elements of a transdisciplinary approach to knowledge generation for real-world solutions. For example, PA enables dialogue and input from community laypeople stakeholders, other community organisational stakeholders and community research stakeholders, centred upon the context and real world of where the question/problem resides. This incorporates analysis on a number of levels (for example, from the laypeople's analysis of the current situation, through to researchers' post analysis of the research data within academic research validity, integrity and ethical protocols) which can lead to shared and real-world solutions.[15] Furthermore, a proactive stance is taken at the outset with measures in place to actively address the potential power imbalances that may arise within the process.

PA stems from *participatory rural appraisal*.[16] This was a participatory approach designed for use in rural development agency work from the 1970s onwards. It empowered local people within their communities to unlock their own wealth of expertise from locally situated information to find local solutions to solve local issues. There is a rich history of PA being used to help deliver meaningful participation in democracy, decision-making and governance.[17] PA facilitates perspectives, enabling shared learning and appraisal by the community upon the specific topic under investigation.[18] This is achieved through the use of a highly trained and quality facilitator who enables these goals to be met.[19]

Key facilitation methods include group cohesion activities and rapport building; the use of secondary sources; semi-structured interviewing and creative and visual-based participatory exercises such as participatory diagramming; and focus groups.[20] This way of working is undertaken to draw upon local expert knowledge about a community-based problem and to help provide community solutions to community issues. Through its key methods, PA engenders intensive and rich quality data and encompasses the complexity of the lived social reality under investigation.[21]

The participatory process that is produced by PA may also mean that the knowledge is acceptable to the community setting it is created within, in that it is based on the community's perspectives of the issue/problem. This may help lead the new knowledge to be implemented within the real world setting, as the knowledge is accepted by the community due to the participatory way it has been generated.[22] The community stakeholder partnership relationships that have developed over time within the initial community research enterprise can also aid this work.[23] For example, the research institution that has established the research stakeholder partnerships in the project will have the expertise in facilitating knowledge exchange and its impact, which can be drawn upon and activated. Applying this additional pragmatic dimension, which includes whether a solution to a societal issue is both acceptable and feasible, to the community, and also the success or failure of the uptake of this new knowledge within the community, is part of implementation science. A model design that uses a 'top down' expert approach with little community involvement or no community partnerships risks the construction of systemic barriers to tackling complex social issues. Some of these barriers will include uptake failure and the work not being fit-for-purpose as its causal assumptions will be flawed due to the lack of community-based stakeholders, perspectives and input.[24]

LAW AND COMMUNITY EMPOWERMENT IN WALES

A final brief reflection within this article is to draw attention to the point that in Wales there is an increasingly strong current within education, health and social care legislative frameworks to give a mandate to the use of community participation and empowerment for tackling societal challenges. Two examples are those of the *Social Services and Well-being (Wales) Act 2014*, and the *Well-being of Future Generations (Wales) Act 2015*, where a collaborative community empowerment approach is a key objective, both for working preventatively as well as in finding solutions. All the public bodies in Wales named in both these Acts are directed to work in this way.

In conclusion, engaging in community empowerment is now legally mandated as the way forward for our society. This support from legislation gives legal weight and purpose to revisiting any challenges stemming from ontology and epistemology.[25] As is already acknowledged within transdisciplinary research, we must now continue to extend the frontiers of our knowledge and generate new paradigms that are fit-for-purpose to accommodate these new developments.

As a starting point in beginning this work, in order to find answers that deliver sustainable solutions, we must first create quality community partnerships that can help to build capacity in community empowerment. This is work in which universities can and should play their full part as higher-level public research and education institutions.

NOTES

    1.   H.R.H. Prince of Wales, T. Juniper and I. Skelly, *Harmony: A New Way of Looking at Our World* (New York, NY: Harper Collins, 2010), p. 5.

    2.   H.R.H. Prince of Wales, T. Juniper and I. Skelly, Harmony; H.R.H. Prince of Wales, *A speech by HRH the Prince of Wales on Health and the Environment delivered at The Cathedral of the Assumption, Louisville, Kentucky* (2015) at https://www.princeofwales. gov.uk/speech/speech-hrh-prince-wales-health-and-environment-delivered-cathedral-assumption-louisville [accessed 22 March 2019].

    3.   H.R.H. Prince of Wales, T. Juniper, and I. Skelly, *Harmony*, p.3.

    4.   Medical Research Centre, *Developing and Evaluating Complex Interventions* (London: Medical Research Centre, 2008) and M. Petticrew, 'When are complex interventions "complex"? When are simple interventions "simple"?', *European Journal of Public Health*, 21, 4 (2011): pp. 397-398.

    5.   See U. Bronfenbrenner, *The Ecology of Human Development. Experiments by nature and design.* (Cambridge, MA: Harvard University Press, 1979) and U. Bronfenbrenner, 'Environments in developmental perspective. Theoretical and operational models' in S. L. Friedman and T.D. Wachs (eds.) *Measuring environment across the lifespan. Emerging methods and concepts* (Washington DC: American Psychological Association, 1999), pp. 3-28.

    6.   See M. Schoon and S. van der Leeuw, 'The shift toward social-ecological systems perspectives: insights into the human-nature relationship', *Natures Sciences Sociétés*, 23 (2015): pp. 166-174 and G.F. Moore, R.E. Evans, J. Hawkins, H. Littlecott, G.J. Melendez-Torres, C. Bonell, and S. Murphy, 'From complex social interventions to interventions in complex social systems: Future directions and unresolved questions for intervention development and evaluation', *Evaluation*, 25, 1 (2019): pp. 23-45.

    7.   See E. Ostrom, 'A general framework for analyzing sustainability of social-ecological systems', *Science*, 325, 5939 (2009): pp. 419-422; M. Ungar, 'Social ecologies and their contribution to resilience', in M. Ungar, (ed.) *The Social Ecology of Resilience. A Handbook of Theory and Practice* (London: Springer, 2013), pp. 13-32; M. Ungar, 'Systemic resilience: principles and processes for a science of change in contexts of adversity', *Ecology and Society*, 23, 4:34 (2018): https://doi.org/10.5751/ES-10385-230434; M. Rutter, 'Resilience. Causal pathways and social ecology' in M. Ungar, (ed.) *The Social Ecology of Resilience. A Handbook of Theory and Practice* (London: Springer, 2013), pp. 33-42 and J. Hinkel, P.W.G. Bots, and M. Schlüter, 'Enhancing the Ostrom social-ecological system framework through formalization', *Ecology and Society*, 19, 3, 51 (2014): http://dx.doi.org/10.5751/ES-06475-190351.

    8.   P. Craig, E. Di Ruggiero, K.L. Frohlich, E. Mykhalovskiy and M. White, on behalf of the Canadian Institutes of Health Research (CIHR) – National Institute for Health Research (NIHR) Context Guidance Authors Group. *Taking account of context in population health intervention research: guidance for producers, users and funders of research* (Southampton: NIHR Evaluation, Trials and Studies Coordinating Centre, 2018).

    9.   Economic and Social Research Council, 'Impact Toolkit', (2019) at https://esrc. ukri.org/research/impact-toolkit/ [Accessed 25 March 2019].

    10.   See M. Stauffacher; A.I. Walter; D.J. Lang; A. Wiek and R.W. Scholz, 'Learning to Research Environmental Problems from a Functional Socio-cultural Constructivism Perspective. The Transdisciplinary Case Study Approach', *International Journal of Sustainability in Higher Education*, 7, 3 (2006): pp. 252-275; Schoon and van der Leeuw, 'The shift toward social-ecological systems perspectives' and J. Weichselgartner and B.

Truffer, 'From knowledge co-production to transdisciplinary research: lessons from the question to produce socially robust knowledge', in B. Werlen, (ed.), *Global Sustainability: Cultural Perspectives and Challenges for Transdisciplinary Integrated Research*, (Cham, Switzerland: Springer International Publishing, 2015), pp. 89-106.

11. Harvard School of Public Health 'Definitions. Transdisciplinary Research. (2019) at https://www.hsph.harvard.edu/trec/about-us/definitions/ [accessed 26 March 2019].

12. See C. Pohl, 'Transdisciplinary collaboration in environmental research', *Futures*, 37 (2005): pp. 1159-1178; C. Pohl, 'What is progress in transdisciplinary research? *Futures*, 43 (2011): pp. 618–626 and S.L.T. McGregor, 'Transdisciplinary knowledge creation' in P. T. Gibbs (ed.) *Transdisciplinary Professional Learning and Practice* (New York, NY: Springer, 2015), pp. 9-24.

13. See Schoon and van der Leeuw, 'The shift toward social-ecological systems perspectives' and M.A. Thompson; S. Owen; J.M. Lindsay; G.S. Leonard and S.J. Cronin, 'Scientist and stakeholder perspectives of transdisciplinary research. Early attitudes, expectations, and tensions', *Environmental Science and Policy*, 74 (2017): pp. 30-39.

14. J. Theis, and H.M. Grady, *Participatory Rapid Appraisal. A Training Manual Based on Experiences in the Middle East and North Africa* (London: International Institute for Environment and Development; Save the Children Federation, 1991).

15. See Pohl, 'Transdisciplinary collaboration in environmental research' and D. Boyd, M. Buizer, R. Schibeci and C. Baudains, 'Prompting transdisciplinary research. Promising futures for using the performance metaphor in research', *Futures*, 65 (2014): pp. 175-184.

16. R. Chambers, *Rural Appraisal: rapid, relaxed and participatory*. Discussion Paper 311 (Brighton: Institute of Development Studies, 1992).

17. See IIED, *PLA Notes 40. Deliberative Democracy and Citizen Empowerment* (London: IIED, 2001); G. Mohan, 'Participatory Development: From Epistemological Reversals to Active Citizenship', *Geography Compass*, 1 (2007): pp. 779–796. doi:10.1111/j.1749-8198.2007.00038.x and IIED, 'PLA 44: Local Government and Participation' (2018) at https://www.iied.org/pla-44-local-government-participation [accessed 25 March 2019].

18. Chambers, *Rural Appraisal*.

19. See Chambers, *Rural Appraisal*; R.Chambers, 'Participatory rural appraisal (PRA): analysis of experience', *World Development*, 22 (1994): pp. 1253-68; R. Chambers, 'Participatory rural appraisal (PRA): challenges, potentials and paradigm', *World Development*, 22 (1994): pp. 1437-54; J.N. Pretty, I. Guijt, I. Scoones and J. Thompson, *IIED Participatory Methodology Series. Participatory Learning and Action: A Trainer's Guide* (London: International Institute for Environment and Development, 1995) and R. Chambers, *Whose reality counts? Putting the last first* (London: Intermediate Technology Publications, 1997).

20. See Theis and Grady, *Participatory Rapid Appraisal*; Chambers, *Rural Appraisal*; Pretty et al., *Participatory Learning and Action* and P. Townsley, *Rapid Rural Appraisal, Participatory Rural Appraisal and Aquaculture* (Rome: Food and Agriculture Organization of the United Nations, 1996).

21. For data see Chambers, *Rural Appraisal*. Also see J. Bergold, and S. Thomas, 'Participatory research methods: A methodological approach in motion', *Forum: Qualitative Social Research*, 13, 1, Art 30 (2010); R. Chambers, 'From rapid to reflective: 25 years of Participatory Learning and Action', *Participatory Learning and Action*, 66 (2013): pp. 12-15 and S. Laws; C. Harper; N. Jones and R. Marcus, *Research for Development. A Practical Guide* (London: SAGE Publications, 2013).

22. Stauffacher et al., 'Learning to Research Environmental Problems'.

23. J. Rycroft-Malone; C. Burton; J. Wilkinson; G, Harvey; B. McCormack; R. Baker; S. Dopson; I. Graham; S. Staniszewska; C. Thompson; S. Ariss; L. Melville-Richards and L. Williams, 'Collective action for knowledge mobilisation: a realist evaluation of the Collaborations for Leadership in Applied Health Research and Care', *Health Services Delivery Research*, 3, 44 (2015): doi: 10.3310/hsdr03440.

24. G. Harvey and A. Kitson, 'PARIHS revisited: from heuristic to integrated framework for the successful implementation of knowledge into practice', *Implementation Science*, 11, 13 (2016): doi: 10.1186/s13012-016-0398-2.

25. Schoon and van der Leeuw, 'The shift toward social-ecological systems perspectives'.

From Despair to Hope:
Building Harmony in a Challenged Community

Mike Durke

IN THE PAST, CONVERSATIONS ABOUT HARMONY in relation to the urban environment have emphasised the cosmically-aligned utopian city but a challenge we now face is much less philosophical and far more practical: how to restore harmony in disadvantaged communities? This aspiration sits comfortably in the context of modern local government reform and the diverse challenges of building community. Creating positive, supportive, cohesive communities is all about equality, balance, wholeness, social justice, and integration. Consideration of how we might develop communities which are more sustainable, democratic, and autonomous, with less need for costly and consistent state intervention, places Swansea and the Townhill housing estate centre stage. This exploration of one part of one city in south Wales provides a road map which is relevant to disadvantaged communities wherever we might find them.

By the beginning of the 1990s, there were the better part of four-hundred empty Council houses in Townhill: approximately 10% of the housing stock uninhabited, boarded-up, and used as little more than dangerous playgrounds by bored teenagers. The area had developed a reputation in the media as an unwelcoming, socially-disadvantaged community, particularly due to the high levels of car crime.[1] In her response to the 2001 crime plan for Swansea, Council Chief Executive, Vivienne Sugar, explained that crime continued to blight parts of the city identified as community regeneration areas, 'We know from information collected by our housing department, for example, that for around half of all council tenancies ended in Townhill, Clase, Blaenymaes and Penlan, fear of crime and harassment was given as a key factor'.[2]

Vehicle crime remained a major problem, alongside burglary, issues surrounding public houses, street drinking, aggressive begging, and youth annoyance.[3] Arson was twice the problem in Swansea than it was in surrounding areas. Between April and June 2001, there were eighty-four burglaries in Townhill. Members of the public and the Councillors who represented them were alarmed by poor police response times and a growing attitude that crime in the area was not worth reporting. However, there was a level of enthusiasm to find a way forward and Mrs. Sugar noted some 'excellent working arrangements' in place between the

CRISIS ON THE HILL

By Chief Reporter
Susan Buchanan

● Townhill and Mayhill 'are in chaos'

● We have no control on estates, councillor

● Family life 'has broken down'

SWANSEA's oldest and largest housing estates are in total chaos, according to a former Lord Mayor and leader of the city council.

Councillor Tyssul Lewis says he is in despair about Townhill and Mayhill and has launched a scathing attack on the estates.

"The whole fabric of society has broken down. The city council has no control there any more," said Mr Lewis, who has been a councillor for 24 years.

"I wish I was a councillor for the Mayals, not Mayhill. It has never been so hard to do the job."

Burgling

Mr Lewis, traditionally a staunch defender of life on the hill, says crime — especially amongst youngsters — is spiralling out of control and parental responsibility has hit an all-time low.

Councillor Lewis' was speaking after the Post approached him for reaction to the final part of its Young and Poor series, which deals with the collapse of family life for many young people.

He says youngsters are being allowed to run riot,

Special report
— page 11

with their parents too busy smoking and drinking to do anything about it.

Children are stealing from neighbours and burgling homes to get goods to sell for cash.

girls couldn't care less," he said.

"I have seen some with two or three children, all by different fathers, knocking on my door for a house. All we are doing on the hill is housing single mothers.

"There is no homeless problem. It's something they've created themselves."

And he revealed that he is currently investigating a case of a brothel apparently being run from a council house in Mayhill.

"It was allocated to a homeless man, but he has never lived there. Every night a taxi pulls up and women get out. They entertain men all night and leave again in the morning."

The council had practically lost control over tenancy of its stock, he claimed, with tenants sub-letting and moving friends in willy-nilly.

"There are so many empty houses which tenants have abandoned to go and live with people. It's total chaos, and I admit I don't know what to do."

"Social deprivation, unemployment and a terrible lack of parental control is the core of the problem," he said.

"We are in crisis. Family life has broken down and there are no firm hands there to guide youngsters."

And single-parent families have become a major housing problem.

"Years ago being single and having a baby was a big thing. Now many of these

In despair . . . Councillor Tyssul Lewis

South Wales Evening Post

29th July 1992

Figure 1: Councillor Tyssul Lewis was not alone in feeling exasperated by the many issues the Hill communities were facing. *South Wales Evening Post*, 29th July 1992.

police and new community partnerships. Townhill was at the vanguard of this new way of working thanks to the pump-priming of European funding.

For those engaged with the implementation of the only European Union (EU) URBAN I Community Initiative in Wales (URBAN), the 2nd March 2001 was

Figure 2: Prince Charles, opening the Prince's Trust Office at the Phoenix in March 2002.

the best and worst of days. Prime Minister Tony Blair had visited Townhill and opened the landmark £1.3 million Phoenix Community Enterprise Centre – a first of its kind in Wales.[4] Later that day, Mr. Ian Spratling, O.B.E., chairperson of the management committee which had steered this innovative programme to fruition, passed away after suffering a tragic accident at home.[5] With the best of intentions, local people were engaged with the process of regenerating this downtrodden estate through four area committees, but progress had ground to a halt, there were disagreements and some hostilities.[6] Mr. Spratling was asked to defibrillate the process as an independent chairman. His drive and business acumen ensured that projects were delivered on time and within budget: the Phoenix would rise from the ashes of the burned-out vehicles which had been such a part of the Townhill landscape for so many years.

Over time, statistics had shown that the Townhill and Mayhill communities, 3,850 households collectively known locally as 'The Hill', officially known as the Townhill Electoral Division, were suffering from unusually high levels of social and economic deprivation: high unemployment, high crime, poor health, and low aspirations. Local people shared with their European neighbours dismay at the constant criticism in the press. These were deprived, damaged, disaffected

MIKE DURKE

communities; unsafe and unwelcoming, crime-ridden and dangerous. URBAN
was an ambitious remedy. This pilot European structural programme aimed
to take a bottom-up, grassroots approach to the tackling of very long-standing
community issues in the most deprived parts of the continent. Altogether 165
cities, including Swansea and Townhill, participated in the two phases of the
URBAN programme, the first of which unfolded between 1994 and 1999.[7] The
second ran between 2000 and 2006 and included a further initiative for Wales
based in west Wrexham.[8] The EU evaluation of the first phase acknowledged
teething difficulties and the amount of time taken to build capacity and secure
the support of partner agencies. The council's foresight in spreading a robust
community partnership ethos in other parts of the city was praised.[9]

Exponential growth in the population of Swansea saw the number of residents
rocket from 17,000 at the start of the nineteenth century to over 134,000 as
the clock ticked into 1900.[10] Fortunes were made, great houses were built and
dynastic families like the Dillwyns and the Vivians helped to ensure that the town
could no longer 'be dismissed as a cultural desert hundreds of miles off the beaten
track'.[11] Life in the slums, particularly in Little Ireland as the Greenhill area was
known, was harsh and uncompromising. This was a time when the production of
copper dominated the communities of Morfa and Hafod, with terraced housing
for the workers lining Llangyfelach Street, Neath, and Carmarthen Roads. These
streets converged at the entrance to the expanding town, where we now have the
cross-roads at Dyfatty, and there was a proliferation of poor quality housing.
Little Ireland was, 'wedged between the districts of Waun Wen and Hafod, which
were overwhelmingly non-Irish'.[12] Swansea's Irish population had swelled after
the devastating potato famine caused an exodus from their homeland. There
was tension, conflict, and even murder.[13] Typhus and cholera blighted this
community.[14] Housing was unimaginably poor, squalid, and unsanitary, with
health reports finding as many as sixteen people sharing a single room,

> Irish, Scotch and Welsh, consisting of wives, husbands, children and single
> people, all in the same room...there must be 250-300 of the commonest
> prostitutes at Swansea...very debauched in their habits as regards drink;
> many of them sleep on straw in the corner of the room, whilst they allow
> ordure [excrement] to cover the floor, or throw it with the ashes; so dirty are
> their domestic habits.[15]

For the Health of the Towns Commission, in 1844, Sir Henry De la Beche showed that one could expect to die far more quickly and younger in the slum areas between Greenhill, High Street and the Strand, than in other communities nearby.[16] Over a three-year period, he found nineteen deaths reported in Mount Pleasant, compared with 450 in the centre of the town and 524 in Greenhill.[17] Swansea Local Board of Health Medical Officer, W. H. Michael, predicted that fatal diseases would return. Once again, they would strike in the most crowded, unsanitary parts of the town like Greenhill and the Strand where there were 'dirty, ill-drained and close habitations...where the entire absence of water renders cleanliness, comfort, or health, almost untenable'.[18]

Professionals, charitable organisations, politicians and some of the rich, powerful land-owners of Swansea society turned their thoughts towards the complexities of social welfare. The working masses were provided with recreational spaces where they could walk, breathe clean air, play and have fun. William Thomas's speech, on 9th July 1874, arguing for open spaces for recreational use, was ground-breaking in Wales. Convinced by this compelling rhetoric, John Talbot Dillwyn Llewelyn donated land at Cnap Llywd Farm and a cash donation to work with the council on the creation of a new park.[19] Parc Llewelyn was transferred into public ownership in a manner which would make the authors of the 2015 Well-being of Future Generations Act proud: to be forever utilised as a park, and for no other purposes whatsoever, and properly maintained for public benefit.[20] Swansea was by this time established as an economically powerful regional hub and there was increasing national and local interest in the lives of the working poor.

The great reforming Liberal governments of 1905 and 1914 were building on the foundation laid by their Victorian counterparts with an expansion of the role of the State in the provision of social welfare. This would transform the lives of so many of the town's less affluent residents. David Lloyd George's 'People's Budget' of 1909 proposed a national insurance scheme, income tax revision, support for the unemployed and the Old Age Pensions Act. After the war, his determination to tackle the hidden miseries of the lower classes was clear,

Those of you who have been at the front have seen the star shells, how they light up the darkness and illuminate the obscure places. The Great War has been like a gigantic star shell, flashing over the land, illuminating the country and showing up the dark, deep places. We have seen places that we have never noticed before, and we mean to put these things right.[21]

Housing was a national priority so that the returning servicemen could be provided with homes fit for heroes. As early as December 1906, Swansea Councillors, C. T. Ruthen and H. G. Solomon, attended a housing reform conference organised by the National Housing and Town Planning Council. They forged links with visionary architect and town planner, Raymond Unwin, and in 1910, the South Wales Cottage Exhibition was held in Mayhill.[22] Prior to the First World War, an eager Swansea Council secured a government subsidy of £2,250 from the Local Government Board (LGB) to build eight show houses on land adjacent to the exhibition site. Garden City principles were followed (cul-de-sacs and gardens rather than monotonous terraced housing) and works were carried out under 'direct administration'. The Council built the houses without use of private contractors, completing the project £300 (approximately £24,000 today) below the cost of the lowest private tender. The housing campaign was but one expression of the 'desire to build a genuine new world in which the disadvantaged could share'.[23] When the post-war call to pursue major construction programmes came, Swansea was quick to respond and the Council wrote to the LGB to request a loan of £278,353 (in the region of £10m in 2019) to erect the first 500 houses on the Hill.[24] Nationwide, the target of 500,000 council houses might not have been achieved but, within five years, 213,000 family homes were built and new communities were taking root.[25]

In 1957, when two teenage Teddy boys were fined for carrying dangerous weapons to a showing of the film *Rock Around the Clock* at the Odeon in Sketty, the magistrate protested that the use of knives and razors was 'un-British and unworthy of Swansea inhabitants'.[26] Disorder and delinquency had become so problematic that by 1961, the Home Secretary had cause to approve a new by-law against unruly behaviour in Swansea's places of public entertainment. Magistrates railed against disruptive youths who populated over-crowded dance halls and indulged in under-age drinking, fighting, gang-warfare and stabbings, with named offenders tending to reside in the communities of Townhill, Mayhill, and nearby Mount Pleasant.[27] Residents of Uplands, Cwmbwrla, Clase, and St Thomas were singled out for stealing from businesses through breaking and entering, with young girls and middle-aged women from across the River Tawe and up the Swansea Valley being responsible for the retail crime of shoplifting. In 1966, the 110 bicycles reported as stolen were overshadowed by the 761 motor vehicles illegally taken. Inattentive parents, who lacked discipline and were too ready to take their children's side against the authorities, were blamed for the increase in crime and moral decline, along with unemployment, the attractions of popular culture, and bingo.[28]

The URBAN programme for Townhill was approved by the European Commission on the 6th November 1996. The net was cast as wide as possible to engage a full range of public, private, and voluntary agency partners. Residents were expected to play a key role in the delivery of projects which were designed to make a tangible difference. Research from a broad range of sources established that local problems were characterised by a high benefit-dependency culture; low self-esteem and a lack of community spirit; poor health outcomes and low levels of health awareness; community facilities and services which were, in some cases, outdated or inappropriate; and, high levels of crime and vandalism, along with an associated and pervasive fear of crime.[29]

The £6.3 million total investment through this initiative, match-funded by Swansea Council to the tune of £2.7 million, was spent in accordance with four measures: community revitalisation (improving facilities, reducing crime, providing childcare); vocational education and training (enhancing local people's readiness for work); economic development (boosting the local economy through business and entrepreneurialism); and, environmental improvements ('greening the hill' to tackle eyesore areas, reconstruct pathways and remove graffiti). Each of these measures would help to build on an already strong sense of community spirit. The Community Development Foundation (CDF) was commissioned to explore the possible creation of a charitable company as the vehicle to keep the momentum going after the grants dried up. The recommendation that the Council 'should support the establishment of a development trust...in order to take the regeneration initiative forward' was accepted.[30] Alan Twelvetrees, the author of the report, made some important observations:

- the European Commission emphasised three criteria for the regeneration of disadvantaged areas: a multi-agency partnership; strong community involvement; and, a forward strategy for long term sustainability;
- such initiatives should not overburden local authorities financially;
- without community engagement, such approaches are not effective or sustainable in the longer term;
- funds need to be drawn from a variety of sources, including trading, to lessen the dependency on public funding; and,
- the lead officer would need to have the right combination of qualifications, knowledge, and personal attributes, including excellent communication skills and a genuine interest in people.[31]

The Hill Community Development Trust Ltd (HCDT – originally called the Phoenix CDT until a more apt name change in 2008) was incorporated in 2001 as a company limited by guarantee with charitable aims. The Board of Directors would be populated by volunteers selected from the community, from public agencies or from elsewhere for their specialist knowledge and skills. This multi-faceted partnership would see a diverse range of perspectives shared around a single table with a focus on gathering information, identifying the issues, and exploring the options for bringing about improvements. The company would control its own affairs, sets its own direction in accordance with a three-year rolling business plan, employ its own staff, create its own policies and procedures, generate its own income through trading activities and secure government grants; all with minimal bureaucracy and delay.

At the official launch of the Phoenix Centre in March 2001, a few important statements were made which stayed with me in my twelve years as Chief Executive of the Development Trust. For example, Mr. Spratling, a successful independent businessperson, acknowledged the importance of assets being transferred to the trust debt-free with the potential for financial sustainability. This was an example of EU funds being used through true partnership, dedicated to the purpose of bringing pride back to the Hill and directly improving people's quality of life. The late Townhill Councillor, Tyssul Lewis, said, 'In all my many years as councillor for this area, I can truly say that the Phoenix is the most significant development in the estate's history'.[32]

Swansea was once again in the 'vanguard of planning and design'.[33] Mr. Spratling was supported by colleagues like Housing Director, Arnold Phillips, a true ambassador for innovative community regeneration. Arnold was Chair-Designate from the date of Mr. Spratling's untimely demise until 2003. The Phoenix was created to build on the URBAN Initiative Foundation and to work towards financial self-sustainability when the grants ran out. This would be achieved by utilising the space at the Centre to provide services by and for the local community: business units to be let at a reasonable rent; a community café; a fully registered children's nursery and playground; a floodlit all-weather sports pitch; a fully-functioning modern library and information centre; and, a flexible conference space. The cynics came out in force: who in their right mind would rent a business unit on the Hill ... where was the demand for a community café and a children's nursery in the middle of this estate ... what a waste of time this venture would prove to be ... yet another case of public money wasted!

Phrases like 'building towards financial self-sustainability' are easy to say but

Figure 3: Prime Minister Tony Blair made a special trip to open the Phoenix Centre on 2nd March 2001.

much more elusive to fully understand. Townhill needed a person with higher-order commercial expertise to steer the ship into the future and we were lucky to find one. At the board meeting of 26th September 2003, when Mr. Roy Phelps was confirmed as the new chairperson, he explained that it was a real honour to have been offered this voluntary position. As a friend of Mr. Spratling, Mr. Phelps had a good awareness of our background and success:

> I will try to bring additional, new skills to the board and will look forward to working with the board in moving the business forward alongside its plans to achieve our goals. I will ask that in everything we do that we at all times project integrity whilst always remaining open and honest with each other.[34]

Mr. Phelp's role was pivotal at a crucial time for the new company. He was instrumental in auditing the position of the Trust, opening discussions with key partners, and setting direction towards a solvent and sustainable future. He strengthened the Board with colleagues who had the necessary skills, knowledge, and experience needed to make progress. Content to give up some of their time, for a new way of working in which they believed, these were people with high levels of commercial expertise. From day one, Mr. Phelps was crystal-clear about a number of necessary critical success factors which remained unchanged over the years we worked closely together, including,

- the line which separates the role of Chairperson and CEO which should never be crossed. The Chair oversees the Board and deals with the higher-level strategic matters and line-manages the CEO. The CEO has complete responsibility for day-to-day operational matters;
- the highest levels of public governance must always be adhered to;
- openness and transparency are everything and the principle of 'no surprises' must always be respected; and,
- if the community does not want it, then the Trust should not be doing it.

The Trust took on responsibility as the local branch of the Welsh Government's Communities First programme and board members, staff, and key partners got on with the process of addressing local needs and driving up income levels. It is fair to say that, slowly, we earned our spurs. The head teacher of Townhill Community School, John Brown, valued the way that the Trust had brought together so many different agencies with an active interest in servicing the Hill communities so that people could pool plans and efforts at a time of increasing austerity. Chief Superintendent Mark Mathias, who had experienced first-hand the intense frustrations of car crime in the 1990s as an Inspector in the South Wales Police, was an enthusiastic and committed supporter,

> I have seen the Townhill Ward move from a community of despair to a community of hope and aspiration. HCDT [Hill Community Development Trust], the Phoenix Centre and Communities First have been important elements in developing a strong partnership working approach which continues to bring very significant community benefit.[35]

The community partnership ethos brought a bright-eyed, solution-focused mind-set. The 'No More Repeats Anti-Burglary Initiative' was lauded for its ingenuity and success.[36] A scientific analysis of the data ensured that officers concentrated on clear targets: victim analysis identified repeat victims and the most vulnerable; offender analysis identified those with criminal records for similar offences; and, there was a pro-active drive on crime prevention and community engagement. Repair time for burgled properties was reduced from twelve weeks to two. Crime

Prevention Officers provided 'secured by design' knowledge so that risks could be controlled and avoided and new windows and doors were installed where most needed. Dimly lit areas were illuminated and high trees and hedges trimmed to improve visibility. Eighty-four burglaries between April and June 2001 might have translated into 336 in a full year. No wonder people were leaving in droves due to the fear or experience of crime. 'No More Repeats' resulted in a reduction in burglaries by an average of 50% for the entire community, with a decrease of almost 70% in some parts of the Hill.[37] There were consistent reports of improved communication between personnel from different agencies.

Two-thirds of the Council-built housing stock on the Hill remained in the possession of the local authority while one-third was sold to those taking advantage of the government's right-to-buy scheme. As a result, the Housing Office experienced an increase in rental income as burglaries fell and the exodus slowed. The decrease in void properties fell from its high point and, over the last ten years, has evened out at between forty to fifty houses at any given time.[38]

Despite what the cynics said, every aspect of the Phoenix Centre proved far more popular than expected. Later described as a world-class example of sustainability by UNESCO (United Nations Educational, Scientific and Cultural Organization), the Centre was just one of six international projects showcased at a major United Nations exhibition in Barcelona to highlight best practices worldwide in cultural diversity, sustainable development, and peace resolution.[39] The café and children's nursery were instant hits, both being run to professional standards by local people. Sports clubs clamoured to book the new floodlit all-weather pitch. The business units were snapped up and have remained fully occupied since their inception. Bookings for the conference space came in thick and fast and the target of 12,000 users of the library within the first year was trebled, with not a single book lost to vandalism or damage.

Generating a new independent income stream was the greatest challenge for the Trust. From a standing start in 2001, trading income grew steadily to reach £100,000 in 2007. The Trust took on direct responsibility for the nursery and community café and levels of self-generated income reached £248,000 in the 2009 financial year. In the last ten years, income has been maintained at similar levels and, remembering the old adage 'turnover is vanity, profit is sanity', costs have been tightly controlled. Since 2002, a total income in the region of £5 million, half from government grants and half generated through trading, has been reinvested in staff and operational costs. Given the number of large companies that managed to secure many millions of pounds sterling in public contracts but

526   MIKE DURKE

Figure 4: Phoenix Manager, Leanne Dower, with her 'Breakfast for £1' campaign.

still find themselves bankrupt, such as Carillion, Interserve, and Dawnus, this community-based social enterprise approach does pose an interesting alternative way to get things done efficiently whilst adhering to professional standards of governance.[40]

Perhaps the grassroots partnership approach, best coined by the term 'community practice', does not hold the keys to unlock all of society's problems, but it does take us some way along this path. We can certainly learn the lessons, explore the map to the pot-holes in the road and apply key principles in any interested community, 'deprived' or otherwise. Gabriel Chanan and Colin Miller offer a compelling rationale for community practice.[41] The breeding ground for poverty and social disadvantage consists of low social capital, a lack of community activity, conflict, and the absence of productive dialogue within communities and with public services. Conversely, high social capital, vigorous community activity and harmonious community relations, combined with productive dialogue and collaboration with public services, all help to improve the quality of life for residents and contribute to a reduction in poverty and disadvantage. Our most challenged communities place a disproportionate pressure on public services. They can be perceived as black-holes for public funding where resources are sucked in without any sign of positive impact. Community practice mobilises communities. New opportunities for participation increase the number of volunteers which fosters new communities of interest and local groups. We see an improvement in dialogue with public services, energy and enthusiasm spread, and aspirations rise. Community practice has two kinds of value: a) *intrinsic* wherein relationships, cooperation and mutual support are cultivated within the community; and, b) *extrinsic,* as local conditions and facilities are improved through closer contact with public service providers. Both improvements help to reduce excessive pressures on overburdened public services, leading to less crime, more employment, better health and education.[42]

In 2019, the statistics continue to suggest that we have a concentration of issues on the Hill that we do not find in most of the other thirty-six electoral divisions in Swansea.[43] 13.7% of the working age population have never worked or have been unemployed for extended, long-term periods compared to 5.2% for the city as a whole. Net household income stands at £20,000, 26% below the city average. Smaller houses can be purchased for £25,000 below the average of £109,000 for Swansea, with savings in the region of £60,000 if you set your sights on a more substantial semi-detached home. However, the improvements continue: the council continues to invest in the housing infrastructure, in the three community centres which provide social and recreational activities, in the three primary schools and two comprehensive schools which educate local children, in the children's social services which remain essential for our most vulnerable young people and in adult social services for our elderly residents. Volunteers continue to run so many community groups and sports clubs. Faith communities look after the spiritual needs of believers and the West End Social Club provides a lively place to socialise. Last year, the Development Trust's fourteen-year campaign for major improvements at Mayhill Park came to fruition with the opening of the £2 million Mountain View Primary Care and Family Centre.[44] The original aspiration for the Community Development Trust to run the Centre was not realised, but this major investment in the community is boosting the health and well-being of residents. The Phoenix has received an uplift thanks to several grants, totalling £285,000, from the Welsh Government.[45] The new library is spotless and as busy as ever and the expanded children's nursery will be opening shortly. In addition, when we consider data covering a range of crimes on the Hill, we see significant improvements over time including burglary rates which fell to thirty-nine in 2018, just over 10% of where they were heading in 2001.

Julie James is the Assembly Member (AM) at the National Assembly for Wales who represents the interests of the 83,000 people who live in her Swansea

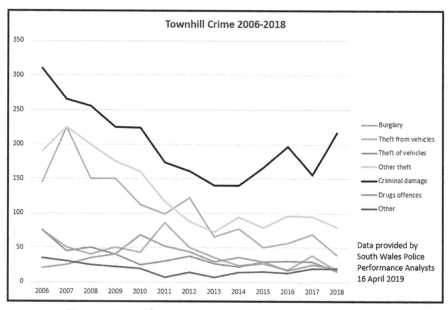

Figure 5: Crime figures suggest a much more settled community.

West constituency.[46] She is energetic in her role and maintains a high level of visibility in the many communities she serves, including Townhill and Mayhill, but her familiarity with the Hill runs much deeper than that. When Vivienne Sugar shared such serious concerns about particular communities across the city, Julie was head of legal services at the City and County of Swansea. She has seen the community evolve over decades and her perspective is clear:

> I have been involved with the Hill community and the Trust for many years and in many roles – from working on the original legal documentation to representing the community as its Assembly Member. In all my interactions with the community it is obvious that given the right enabling support the community can both identify and deliver the solutions and services it needs for itself and this history of the community and its efforts illustrates that emphatically.

In recent years, as an academic and a city councillor, I have had the opportunity to pause and reflect carefully on my thirty years in public services – in policing, youth work, child and family social services and community development, working at a local authority and on the national level, on the frontline and in management roles. It seems clear to me that, as we look ahead to the radical reform of local government in Wales, we must pause and reflect carefully on the lessons learnt, not just by the Townhill experience, but by so many other approaches we have seen from so many different key partners and organisations: from national government to local, from successful corporations to grassroots social enterprises, from major charitable companies to local volunteers who shape their own communities on a daily basis. Face-to-face skilled, supportive, open and honest communication is everything – that is where the magic happens. In his high-level report to the Welsh Government, Joe Simpson was perfectly clear in his appreciation of the need for up-front, close and personal, community-facing public services:

It is in local communities where the connection with citizens can be most fully developed, where community leadership can be exercised and where the opportunities for service integration can be best grasped.[47]

Wales is not alone in seeing the role of elected members evolve towards more collaborative forms of democratic engagement which can only be achieved in the heart of communities themselves.[48] Dynamic community organisations, rooted in the life of the communities they serve, must have a pivotal role if we have any chance of shaping communities from within, as the First Minister urged in a previous ministerial role, by spinning a golden thread to link community aspirations with national goals.[49] Saul Alinsky was right: treating people as 'passive, puppet-like recipients' of services we design and deliver for them is a denial of human dignity and democracy and will never work.[50] If we apply what Barack Obama called Lincoln's simple maxim – that government should only do for communities what they cannot do for themselves – we could well see an upsurge in innovation from frontline professionals like teachers, doctors, social workers, social carers and police officers working with local residents and businesses to search for the solutions to their own problems.[51] The Townhill experience stands as an intriguing example of what can be achieved, even in the most challenging circumstances.

NOTES

1.   Steffan Thomas, '"12-year-old joyriders, police chases and vehicles set alight" When Swansea was the UK car crime capital', *Wales Online*, (18 May 2017), https://www.walesonline.co.uk/news/wales-news/12-year-old-cushion-who-13005163 [Accessed May

2019].

2.   V. Sugar, 'Fear of crime a key issue – Swansea: Why residents leave council estates', 20 November 2001, *Western Mail*, p. 7.

3.   Sgt Bryan Heard, *Safer Swansea Anti Social Behaviour Structure 'Focused on the Causes'* (Swansea: South Wales Police, 2004), https://popcenter.asu.edu/sites/default/files/library/awards/tilley/2004/04-21.pdf [Accessed May 2019].

4.   The Hill Community Development Trust, 'The Phoenix is rising', *View from the Hill Community Newsletter*, Summer 2001.

5.   BBC News, 'Freak fall kills community worker', (3 March 2001), BBC News UK: Wales, (2003), http://news.bbc.co.uk/1/hi/wales/1200643.stm [Accessed 19 March 2019].

6.   Stephan Lucas and DTZ Pieda Consulting, *The Wrexham Urban Partnership Strategy, 2000-2006*, (Manchester, 2000), http://www.wrexham.gov.uk/assets/pdfs/keydocs/afinalreport8-7-00.pdf [Accessed May 2019].

7.   European Union, *Ex-Post Evaluation of the URBAN community Initiative (1994-1999) Final Report* (Brussels and London: GHK, 2003), https://ec.europa.eu/regional_policy/sources/docgener/evaluation/urban/urban_expost_evaluation_9499_en.pdf [Accessed May 2019].

8.   Sonja Haertel and Daniel Mouqué, eds., *Partnership with the Cities: the URBAN Community Initiative*, European Commission (Luxembourg: Office for Official Publications of the European Communities, 2003), https://ec.europa.eu/regional_policy/sources/docgener/presenta/cities/cities_en.pdf [Accessed May 2019].

9.   European Union, *Ex-Post Evaluation*, p. 37 and pp. 53-60.

10.   GB Historical GIS/University of Portsmouth, 'Swansea District through time', Population Statistics, Total Population, *A Vision of Britain through Time*, (2017), http://www.visionofbritain.org.uk/unit/10039708/cube/TOT_POP [Accessed May 2019].

11.   David Painting, 'The Dillwyn Dynasty', *The Dillwyn Day: Science, Culture, Society*, 22 June 2012, (Swansea: Learned Society for Wales and CREW, Swansea University, n/d) https://www.swansea.ac.uk/crew/research-projects/dillwyn/dillwyn-day/dillwyn-dynasty/ [Accessed May 2019].

12.   R.T. Price, *Little Ireland: Aspects of the Irish and Greenhill, Swansea, Studies in Swansea History* (Swansea: City Archives Office, Central Services Department, 1992), p. 31.

13.   R. T. Price, *Little Ireland*, p. 44.

14.   G. Penrhyn Jones, 'Cholera in Wales', *National Library of Wales Journal* X, no. 3 Summer (1958); for online version, see also, https://www.genuki.org.uk/big/wal/Archives/NLWjournals/Cholera (2018) [Accessed May 2018].

15.   Price, *Little Ireland*, p. 35.

16.   Jones, 'Cholera'.

17.   Sir Henry De la Beche, *Health of Towns Commission Report* (1854) cited in Price, *Little Ireland*, p. 35; see also, Sir Henry De la Beche, '"Report on the Sanatory Condition of Swansea", Report on the State of Bristol and Other Large Towns', *Health of Towns Commission* (London: W. Clowes and Sons, 1845), pp. 57-76, http://digital.slv.vic.gov.au/view/action/singleViewer.do?dvs=1557854833569~787&locale=en_US&metadata_object_ratio=10&show_metadata=true&VIEWER_URL=/view/action/singleViewer.do?&preferred_usage_type=VIEW_MAIN&DELIVERY_RULE_ID=10&frameId=1&usePid1=true&usePid2=true [Accessed May 2018].

18.   Price, *Little Ireland*, p. 38.

19.   D. Farmer and B. Lile, *The Remarkable James Livingston: An inspiring leader in the service of Victorian Swansea, its port and its people* (Swansea: Royal Institution of

South Wales, 2002), p. 15.

20. Friends of Parc Llewelyn, 'Friends of Parc Llewelyn', *History of Parc Llewelyn* (n.d.), http://www.friendsofparcllewelyn.co.uk/history.html [Accessed 24 March 2016].

21. D. Lloyd George, 'Classic Podium: A humble recognition of heroes', *Independent*, [7 November 1998 (24 November 1918)], https://www.independent.co.uk/arts-entertainment/classic-podium-a-humble-recognition-of-heroes-1183202.html [Accessed 28 March 2019].

22. N. Robins, *Homes for Heroes: Early twentieth century housing in the County Borough of Swansea* (Swansea: City and County of Swansea, 1992), pp. 1-18.

23. Robins, *Homes*, p. xxvii.

24. Robins, *Homes*, p. 35.

25. L. Hanley, *Estates: An intimate history* (London: Granta, 2007), p. 61.

26. D. J. V. Jones, 'Where did it all go wrong? Crime in Swansea, 1938-68', *Welsh History Review*, Vol 15, Issue 1 (1990): pp. 241-271.

27. Jones, *Where did it all go wrong?*, p. 246.

28. Jones, *Where did it all go wrong?*, pp. 247-271.

29. Townhill/Mayhill URBAN Initiative (1997) URBAN *Action Plan 1997-1999*, URBAN Management Committee March 1997; p15; see also Mike Durke, 'Trust and Partnership: As we look ahead at local government reform, what lessons could be learned from Wales's only URBAN I Community Initiative?', (2018), https://www.mikedurke.co.uk/documents/537_Trust_and_Partnership_2018.pdf [Accessed May 2018].

30. A. Twelvetrees, *Options for Sustainability: A Development Trust for Townhill?* (Cardiff: Community Development Foundation, 1998).

31. Twelvetrees, *Options*.

32. The Hill Community Development Trust, 'The Phoenix is rising', *View from the Hill Community Newsletter*, April 2001.

33. Robins, *Homes*, p. 11.

34. The Hill Community Development Trust Ltd (HCDT), *Minutes of Board Meeting* 26 September 2003.

35. M. Durke, *Trust and Partnership: Why community ownership and close collaboration are the keys to future success* (Swansea: Hill Community Development Trust Ltd, 2012), p. 5.

36. South Wales Police, *Tilley Award: No More Repeats Anti-Burglary Initiative* (Swansea: South Wales Police, 2003).

37. R. Turner, 'Burglaries halved at impoverished estate', *Western Mail*, 14 October 2002, (Cardiff: Reach PLC, 2002), p. 5.

38. Townhill District Housing Office, *Housing Stock Data* (Swansea: City and County of Swansea, 2019).

39. BBC News, 'UN heaps praise on centre', BBC News (Friday, 21 November 2003) http://news.bbc.co.uk/1/hi/wales/south_west/3227066.stm [Accessed 15 March 2019].

40. House of Commons Library, 'The collapse of Carillion', 14 March, 2018 https://researchbriefings.parliament.uk/ResearchBriefing/Summary/CBP-8206 [Accessed 29 March 2019].

41. G. Chanan and C. Miller, *Rethinking Community Practice: Developing transformative neighbourhoods* (Bristol: Policy Press, 2013).

42. Chanan and Miller, *Practice*, pp. 149-50.

43. City and County of Swansea, *Townhill Ward Area Profile*, https://www.swansea.gov.uk/wardprofiles [Accessed 20 March 2019].

44. Ashleigh House, 'Mountain View, Primary Care and Family Centre, Swansea', (2018), http://www.ashleyhouseplc.com/projects/mountain-view-health-and-childrens-centre/ [Accessed 20 March 2019].

45. N. Williams, 'Library's booked in for a £125k facelift', *South Wales Evening Post*, 23 May 2017, p. 25.

46. City and County of Swansea, *Swansea West Constituency Area Profile* non, *Swansea West Constituency Area Profile*, (2019), https://www.swansea.gov.uk/pcaprofiles [Accessed 12 March 2019].

47. J. Simpson, *Local, Regional, National: What services are best delivered where?* (Cardiff: Crown Copyright, 2011), p. 4; see also, http://www.wales.nhs.uk/sitesplus/documents/829/Local%2C%20regional%2C%20national-%20what%20services%20are%20best%20delivered%20where%202011.pdf p. 4 [Accessed May 2019].

48. Sarah Titcombe, *A Guide for New Councillors in Wales* (Cardiff: Welsh Local Government Association, 2017); see also, https://www.wlga.wales/SharedFiles/Download.aspx?pageid=62&mid=665&fileid=976 [Accessed May 2019].

49. M. Drakeford, *Resilient and Renewed: Reforming local government*, (Cardiff: Crown Copyright, 2017), p. 40, https://gov.wales/sites/default/files/consultations/2018-02/170130-white-paper-en.pdf.

50. S. Alinsky, *Rules for Radicals: A Pragmatic Primer for Realistic Radicals* (New York: Vantage Books, 1971), p. 50.

51. B. Obama, *The Audacity of Hope: Thoughts on Reclaiming the American Dream* (Edinburgh: Canongate Press, 2007).

HARMONY AND NATURAL CAPITAL

Tony Juniper

*From a talk at the Harmony, Food and Farming Conference, organised by the
Sustainable Food Trust, Llandovery College, 10 July 2017.*

IT'S NINE YEARS TO THE WEEK since Ian Skelly and I began working with the
Prince of Wales on his book, *Harmony*, but we're still seeing such momentum
being carried forward by this piece of work.[1] It is truly visionary and, I have to
say, working on this book was one of the most important learning experiences
of my life. It really opened my eyes to many things that were previously invisible
to me and, having spent so many years dwelling in the world of conservation,
it was really quite a shock, actually, to engage with the big elements of what I'd
been doing. I'd been neglecting really quite major dimensions of the whole piece.

Patrick Holden has mentioned the separation between the worldly and
the spiritual as being one of the main themes that we dealt with in *Harmony*.
Another is the separation between people and the natural world. It's on that
subject that I'd like to share a few remarks because, following the publication
of *Harmony*, I've spent the last seven years working very intensively on the idea
of natural capital, a concept that flowed directly out of this book.[2] Now, natural
capital is slightly controversial amongst the environmental community because
it talks about a parallel between financial capital and the natural world: if we
look at biodiversity, or a set of ecosystems, we realise that if we husband them
and look after them, we can get dividends long into the future. It's rather like
making a financial investment: if we blow our financial capital, we no longer get
dividends or interest. And the parallel we try to draw in *Harmony* is between the
way we look at the natural world as something that can be endlessly abused and
liquidated in order to meet short-term human needs and the extent to which we
can engage with and manage it in productive and beneficial ways: if we preserve
natural capital we get dividends, if we blow it then we don't.

One of the problems of the modern world is that we have become so
disconnected from nature. The idea of natural capital is a way of reconnecting
people with the reality that that we are one hundred per cent dependent on
healthy natural systems for our wellbeing. When it comes to food and farming,
natural capital is an important factor in terms of how we need to plan for the

future. There are many levels to this, of which perhaps the most fundamental of all is soil.

When we walk across a piece of grass, many of us probably don't think about the ecosystem beneath our feet, but it is one of the most complicated and profound systems on planet Earth. Take a tablespoon of soil, take it to the lab, sit down with an electron microscope and you might count six billion individual organisms: nematodes, bacteria, viruses, micro-fungi, all operating in a complex set of relationships that is enabling the soil to discharge a series of critical functions, including those which support human well-being. In addition to water retention and holding billions of tons of carbon, of course, soil recycles nutrients, enabling plant growth, which is then the basis of agriculture.

These ecological relationships are fundamental and profound, yet we utterly disregard them. It's common for people to use the cultural label of 'dirt' when referring to soil, this particular miracle of nature. And with that in mind, we abuse it, we concrete over it and, of course, we currently farm it in ways which are utterly unsustainable.[3] This lack of sustainability is seen in a range of really alarming trends. I recently produced some material looking at soil damage across the world and the conclusion was that about one-third of agricultural soils are being either highly or disastrously degraded.[4] Organic material is destroyed and not replaced, and soil is compacted, eroded, and washed into the sea.

I guess sometimes we're inclined to think of the problem of soil erosion as being particular to the tropical countries where so much deforestation is taking place but, actually, it's happening right here on our doorstep. One remarkable reminder of that for me, a couple of years ago, was when the International Space Station passed over southern England in March 2014, following a very wet winter. The astronauts, looking down from 300 kilometres up, remarked on the brown fringe lying around the British coast, especially around the big river estuaries. One of the astronauts described this brown fringe as 'runoff'. Actually, what he might equally have said was 'soil' because what he saw was millions of tons of topsoil which had departed the land, travelled across fields, into streams and then, into rivers and finally, into the ocean.

If you think about the way in which we farm, then just look out of the train or car window and you'll see plough lines going straight down the side of hills into water courses: when it rains, the soil leaves the fields. This is utterly insane and yet it is one of the consequences of the ways in which we meet the challenge of a rising population through industrial farming methods. These practices are grossly destructive.

Aside from the impact of farming practices on the soils, and the way in which we are ploughing and leaving ground bare and exposed over the winter, we have moved from mixed farming to industrial monoculture. We have removed many features of local ecosystems: wetlands have been drained, grasslands ploughed, hedges removed and woodlands grubbed up, all in order to make bigger and bigger fields, exacerbating the problems of soil depletion. At the same time as we've done that, we have been depleting another bit of natural capital: biodiversity, above-ground. We have not only been simplifying the soils, but also the ecosystems on top. Of course, those ecosystems are also critical for food production, not least through the activity of pollinators. Most of the flowering plants on Earth today are insect-pollinated, including two-thirds of the world's crop plant varieties.

As a result of the loss of habitat and the use of toxic chemicals, those insect populations have been declining and, as a result of that, food security is being imperilled. We have calculated the value of these kinds of natural services, and when it comes to pollinators, a technical study published in 2010 estimated that we are receiving something like 190 billion dollars of value per year from insect pollination.[5] Now, these kinds of figures are sometimes controversial and they are quite hard to work out in very specific ways. However, we can get an insight as to how those numbers might actually stack up through specific examples. One that really struck me is photographs taken in south-western China where fruit farmers in the springtime climb up apple trees with paintbrushes and move the pollen between the blossoms by hand. They have to do this because all the insects, which would otherwise have moved the pollen, have been killed by industrial quantities of chemicals that have been deliberately deployed in order to control pests. They did control pests but, obviously, at the same time, they killed many of the pollinators. And while we are killing pollinators in an attempt to reduce pest attacks, we are also removing beneficial animals from the landscape.

Many recent studies tell us how we derive enormous benefits from the many natural predators of pest species. One study that struck me as being particularly illustrative was an experiment conducted in a Dutch apple orchard whereby half of the orchard was covered with mist nets, a very fine netting designed to prevent penetration by songbirds, but still allowing moths to lay their eggs and caterpillars to hatch – which damage the fruit.[6] Then, in the other half of the orchard, nest boxes were fitted to encourage great tits to come and nest in there. To cut a long story short, in the half of the orchard where the birds were present there was a fifty per cent higher, high-quality apple harvest compared to where they were excluded.

In these kinds of instances, where we actually begin to gather data, we can reveal hidden values in natural capital that are worth many billions of pounds when translated into financial capital. In the case of those birds in the Dutch orchard, it seems to me that we still lack the proper tools to be able to calculate the kinds of values that are being provided, whereas we can be very adept at putting cash values on the impact of pesticide sales and use on GDP and competitiveness. When we start to look at the other, less obvious values of natural capital, we find enormous negative impact coming from the way in which we're farming in order to make profit in terms of financial capital. As a result we are removing natural capital and, in the end, we will begin to undermine our food security.

Soil health, the recycling of nutrients, the activities of pollinators, the activities of natural pest controllers, all contribute to agriculture in ways that have real economic value. Then, on top of all of that, of course, is the ultimate limiter of agriculture and land: the availability of fresh water. In our modern society we're inclined to think that water comes out of taps or, increasingly, out of bottles, including plastic ones. Yet the provision of water is one hundred per cent governed by the activity of natural systems, ranging from the seeding of rain clouds by plankton in the oceans, through to forests and wetlands which store purified water, and then to the recycling of water on land by ecosystems. When we begin to damage these systems on a large scale, it can have an impact on water security. Over recent years, we have seen several examples of this, including a very long period of drought that affected food production in the La Plata Basin of Northern Argentina, Paraguay, and southern Brazil in 2014-15. And the drought, in turn, was linked with the destruction of tropical forests in the Amazon.[7]

Recently, I travelled to West Africa in order to understand the sustainability of cocoa production there. I was very struck by conversations I had with local farmers, government officials and major cocoa and chocolate companies, all telling me the same story about the impact on yields over recent years of prolonged and severe drought.[8] And, indeed, droughts of unprecedented severity and length. And this, they all put down to the loss of the tropical forests in that part of the world. It was striking to note the highest cocoa yields that remained on the Ivory Coast are now clustered around the Taï Forest National Park, which is the last piece of intact tropical rainforest in that part of the world.

We are also beginning to understand the idea of so-called sky rivers, which transport water from tropical forest ecosystems, which themselves act as a kind of a biological pump, over long-distances. As water is dragged off the ocean, over to the interiors of these continental areas covered in tropical forest, it falls as rain which is then re-evaporated back into the atmosphere. As the water rises as vapour, it condenses and its volume decreases, and, as it does so, it pulls up more air from below, creating long-distance water transfers over thousands of miles.

Indeed, if we travel to the grain belt of North America, the Great Plains of South Dakota and North Dakota, we find water which has travelled north from the Amazon and Central America on these very long-distance air currents, powered as they are by the tropical rainforests. These are fundamental dependencies that we ignore at our peril and, as we look forward to the future of farming, it's going to be essential that we move beyond the idea that technology can solve all of our problems. We must understand that the reality that we inhabit, including the entirety of our financial economy, including farming, is dependent on nature. To ignore this involves us in an unsustainable and destructive crisis of perception. If we destroy the soil, remove pollinators, take away natural pest control and erode the natural systems that purify and renew water, our farming systems cannot go on. We need to change the way in which we look at all of this, not only by developing sustainable food production, but in order to be able to limit and cope with the consequences of climate change. To accomplish this we need to increase organic matter in the soil, put more carbon in to the soil and less into the atmosphere, and conserve the forests, especially the tropical rainforests, that are so important for the planet's hydrology.

Just to conclude with one final thought, aside from the crisis of perception in which humanity sees itself as outside nature, rather than as a part of it, there is also a crisis of economics. If we are going to look forward to a more sustainable and durable future for farming, we need to begin to measure things differently. Cheap food has become the totem for policy-makers and food companies across the world but, actually, it's not cheap at all, not once we factor in the multi-billion pound damage being done, day in and day out, to the systems that sustain agriculture. We therefore need to start measuring natural capital and coming up with a more rational assessment of economic success. If we do that, I think we can go a long way towards sustaining both farming and life on Earth.

NOTES

1. HRH the Prince of Wales, Tony Juniper and Ian Skelly, *Harmony: A New Way of Looking at our World* (London: Harper Collins, 2010).
2. Also see the Natural Capital Coalition at https://naturalcapitalcoalition.org/.
3. Tony Juniper, *What has Nature ever Done for Us?* (London: Profile Books 2013), pp. 35-41.
4. See Jonathan Watts, 'Third of Earth's soil is acutely degraded due to agriculture', *The Guardian*, 12 September 2017, https://www.theguardian.com/environment/2017/sep/12/third-of-earths-soil-acutely-degraded-due-to-agriculture-study (Accessed 25 September 2019), and United Nations Convention to Combat Desertification, *Global Land Outlook*, 1st edition, 2017, https://knowledge.unccd.int/glo/GLO_first_edition.

5.   Joshua Bishop, Nicolas Bertrand, William Evison, Sean Gilbert, Annelisa Grigg, Linda Hwang, Mikkel Kallesoe, Vakrou, Cornis van der Lugt, and Francis Vorhies, 'The Economics of Ecosystems and Biodiversity Report for Business', TEEB – The Economics of Ecosystems and Biodiversity Report for Business – Executive Summary 2010 (Malta: Progress Press, 2010) at https://www.unepfi.org/fileadmin/biodiversity/TEEBforBusiness_summary.pdf [Accessed 10 July 2017]; Tony Juniper, 'Why the economy needs nature', *The Guardian*, 9 January 2013, https://www.theguardian.com/environment/blog/2013/jan/09/economy-nature [Accessed 10 July 2017]; Also see, Tilo Arnhold, 'Economic value of insect pollination worldwide estimated at 153 billion euros', Helmholtz Centre for Environmental Research – UFZ, 15 September 2008, https://www.ufz.de/index.php?en=35639 [Accessed 10 July 2017]: The value of pollination was €153 billion in 2005 for the main crops that feed the world and pollinator disappearance would translate into a consumer surplus loss estimated between €190 to €310 billion.

6.   Tony Juniper, *Nature?*, pp. 139-40.

7.   Jonathan Watts, 'The Amazon effect: how deforestation is starving São Paulo of water', *The Guardian*, 28 November 2017, https://www.theguardian.com/cities/2017/nov/28/sao-paulo-water-amazon-deforestation (Accessed 25 September 2019).

8.   Tony Juniper, *Nature?*, pp. 167-8.

REDISCOVERING THE HUMAN PURPOSES OF BUSINESS

Mark Goyder

Evolution as survival of the fittest has inhibited our observation of coevolution. There is no hostile world out there planning our demise. We are utterly intertwined.
—Margaret Wheatley and M Kener Rogers[1]

Harmony speaks of balance, order and relationship. It is concerned with parts within, and only within, a whole, and of wholes within wholeness. The whole and its parts are always interconnected and related. One affects the other. Neither stands alone. And both are always in flow.
—David Cadman[2]

INTRODUCTION

I HAVE SPENT THE LAST THIRTY YEARS encouraging people in business to reflect on the answer to this question: how do we create the conditions in which society thrives as capitalism thrives? In this chapter I build on this experience in offering my observations on the concept of harmony and its importance to the way companies are owned, led and governed.

If we want a society enriched with freedom, diversity and innovation, then in my view there is no substitute for a vigorous market economy. Markets are good servants, but bad masters. Citizens and future generations will be best served by a market economy in which entrepreneurs can flourish and innovate, provided there is a sufficiently strong framework of external and self-regulation. If regulation loses control of markets, or if the ethics and self-discipline of market participants erodes, greed and short-termism take over. I believe we are now at a moment of choice. We face a choice between two views of capitalism.

I would describe one as extractive capitalism. This treats companies as assets to be sweated. It measures success in terms of the cash returned to shareholders, not in terms of the investment made in future growth or benefits delivered to customers, employees or society, now and in future generations. One of its symptoms is the electronic ticker screen which you can find in the reception area of major companies, relaying real-time information about share prices. Another is the tone of admiration in reports by the financial press of the rewards paid to investment banks and other advisory firms for their part in securing a merger or acquisition even though those advisors' rewards are not in any way linked to the future success of the merged enterprise.

The alternative to extractive capitalism – you might call it attractive capitalism! – is what I would describe as the stewardship view of capitalism. This sees companies as human organisms, entities created by human beings to work with other human beings and apply human skill and technology to meet human needs.

This view is best seen at work in enduring family businesses. In these the current generation are naturally concerned with immediate financial returns but are also thinking of their role as stewards of the assets that they have inherited for generations to follow. They are not pre-occupied with sweating today's assets because they don't intend to sell their shares. They acknowledge the importance of capital as a means to finance the business to achieve its purpose. Their primary focus is on the health of the enterprise and its relationships and reputation which are the foundation for future success. They see ownership of the shares of a company as involving obligations as well as rights.

Yet the stewardship view of capitalism is not confined to family businesses. Indeed, we can witness around us the burgeoning of new organisations that effectively compete in markets while committing themselves to serve human purpose.

Buurtzorg is a Dutch company that has caused a revolution in neighbourhood nursing and is starting to make a difference to the care system in the UK. It was founded in late 2006 by Jos de Blok. It grew from ten to seven thousand nurses in seven years. Nurses work in decentralised self-managed teams of ten to twelve, with each team serving an around fifty patients in a small, well-defined neighbourhood. The goal is for patients to recover the ability to take care of themselves. Neighbours are also invited to play their part. A third of emergency hospital admissions are avoided. An EY study in Holland found that Buurtzorg requires, on average, forty per cent fewer hours of care than other nursing organisations and that patients stay in care half as long, heal faster, and become more self-reliant. EY estimated that the Dutch taxpayer would save to two billion euros if all Dutch patients were served on this basis. At the time of the study absenteeism for sickness was thirty per cent less than the Dutch average among Buurtzorg employees and staff turnover was thirty-three per cent lower.

This is a real-life illustration of harmonious business.

In his 2017 UNESCO paper, David Cadman defined harmony in terms of balance, order and relationship. He said that harmony is concerned with parts within, and only within, a whole, and of wholes within wholeness.[3] Applied to the world of business, that means considering individuals as part of enterprises, and enterprises as part of industries and value chains, and all of these as part of society.

There is, in fact, a strong European tradition of writing and analysis which explores the conditions in which people find balance, order and relationship through their economic activity. There are strong common themes to be found in these Western writings which are suggestive of harmony. Add to this some of the wisdom of the East, and five conditions for harmonious business can be derived.

Five Conditions for Harmonious Business

Karl Marx is best known for his analysis of the causes of class struggle, and for the surplus theory of labour. Earlier on, however, Marx described his vision of the rounded life. He then developed the concept of alienation to describe how human beings allowed themselves to become separated from this ideal. Late Marx may have been the champion of class conflict, but the early Marx was drawn to harmony. He looked forward to the day:

> When labour is no longer merely a means of life but has become life's principal need.[4]

and

> When the productive forces have also increased with the all-round development of the individual, and the springs of co-operative wealth flow more abundantly.[5]

A broader nineteenth-century vision of harmony comes from William Morris who dedicated his life to craftsmanship and wanted to see fulfilling work become available to everyone, but agreed with Marx that class struggle and conflict would be necessary to achieve the conditions in which harmony could prevail. In *How we Live and How We Might Live*, Morris lists four 'claims for a decent life'.

First, a healthy body; second, an active mind in sympathy with the past, the present and the future; thirdly, occupation fit for a healthy body and an active mind; and fourthly, a beautiful world to live in.[6]

John Ruskin, born two hundred years ago in February 1819, wrote about economic justice and celebrated all forms of wealth creation in which the purpose was to enhance human well-being. He said 'there is no wealth but life' and coined the term *illth* to describe wealth created to no useful purpose that causes 'various devastation and trouble'.[7]

A century later there is Ivan Illich, who wrote *Tools for Conviviality*. In that compressed, eloquent and sadly overlooked book written in the 1970s, Illich argues:

I here submit the concept of a multidimensional balance of human life which can serve as a framework for evaluating man's relation to his tools. In each of several dimensions it is possible to identify a natural scale. When an enterprise grows beyond a certain point on this scale it first frustrates the end for which it was originally designed, and then rapidly becomes a threat to society itself.[8]

And

Scientific discoveries can be used in at least two opposite ways. The first leads to specialisation of functions, institutionalisation of values and centralisation of power and turns people into accessories of bureaucracies or machines. The second enlarges the range of each person's competence, control and initiative, limited only by other individuals' claims to an equal range of power and freedom.[9]

So, for Illich, two vital conditions for enterprises to serve society harmoniously are the limitation of scale and the empowerment of individuals.

This leads us naturally to Fritz Schumacher whose *Small is Beautiful* was also published in 1973.

What is it that we really require from the scientists and technologists? We need methods and equipment that are

• Cheap enough so that they are accessible to virtually everyone

- Suitable for small-scale application
- Compatible with man's need for creativity.

Out of these three characteristics is born non-violence and a relationship of man to nature which guarantees permanence.[10]

Here, again, was a writer who was exploring the pre-conditions for harmony in society. Scale, accessibility and room for individual creativity were seen as by Schumacher as the three essentials.

It is interesting to test the criteria developed by these two 1970s visionaries against a contemporary large-scale business such as Facebook. Facebook has been very successful in promoting the accessibility of its inventions but it has been the victim of its own success. By putting revenue growth and scale ahead of client need, it lost sight of the ways in which its activities, and its very dominance might undermine the creativity and self-esteem of many while enhancing that of a few. The accessibility which started as its very raison d'etre (facilitating sharing) became as much a tool for manipulation as for conviviality.

There are strong common themes to be found in all these Western writings which point towards concepts of harmony. Firstly, harmony in society starts with health, well-being or wholeness in the individual. This has implications for how products and services are made and sold and for the importance of relationships. Secondly, scale matters. If you hold everything else constant (human nature, education; economic opportunity) but you increase the numbers tenfold, impersonality and fear are more likely to replace a sense of relationship and interdependence. Thirdly, they all recognise the importance to human fulfilment and therefore harmony of the *right kind of work* – work in which people can find ways of expressing themselves, whether it is the kind of work that is found in the labour market or work in the home. This in turn leads to the importance of overthrowing restrictions based on gender or any other difference and valuing unpaid work in the home or neighbourhood as well as paid work. Fourthly, as is to be expected in the aftermath of more than one industrial revolution, there is a concern with bending technology and markets to human purposes rather than allowing the two combined to become monsters without limits.

In an age of warming oceans, dying coral reefs and increasingly extreme weather there is perhaps one harmony condition missing from all these Western writings. They say little about harmony between generations and the importance of the long term or anything about the dimension of time and the need to assess progress in the light of the needs of future generations. The native American proverb says it well: we do not inherit the world from our ancestors; we borrow it from our children.

In *Small is Beautiful* Schumacher does, implicitly, acknowledge this harmony condition. He quotes with approval the words of ecologists Tom Dale and Vernon Gill Carter:

> Civilised man was nearly always able to become master of his environment temporarily. His chief troubles came from his delusions that his temporary mastership was permanent ... Man, whether civilised or savage, is a child of nature – he is not the master of nature ...When he tries to circumvent the laws of nature, he usually destroys the natural environment that sustains him. And when his environment deteriorates rapidly, his civilisation declines.[11]

Carelessness about the needs of future generations seems to be a particular Western weakness. It can be traced back to a different mental picture of time. In *The Seven Cultures of Capitalism* Charles Hampden Turner and Fons Trompenaars used comparative international surveys of business managers to reveal differences in culture. Their 1993 description of very different attitudes to time was particularly telling.

> The managers were asked to think of the past, the present and the future as being in the shape of circles. The results showed UK managers taking a 'sequential view', drawing three separate circles with a slight intersection of the past and the present, while US managers saw the past as unconnected with the present and the present just touching the future. By contrast, the Japanese took a 'synchronous' view, seeing past, present and future as largely overlapping circles.[12]

Together that makes five conditions for harmony in business:

- Health, well-being and *individual wholeness*
- Appropriate and *accessible scale* of organisation
- *Fulfilling work* open to all talents
- Technology, tools, organisations and markets being *servants and not masters* of human beings
- The *right generational balance* between the needs of today and tomorrow's

A LIVING PICTURE OF HARMONIOUS AND CONNECTED BUSINESS

A few years ago, I spent a week with my wife and family in Myanmar. While I was there, I read Pascal Khoo Thwe's book *The Land of Green Ghosts*, which starts with an account of his growing up as a member of a remote Burmese tribe. This is how he talks about the harvest:

> We grew maize, sorghum for the domestic animals, groundnuts for oil, beans, yams, potatoes, cucumbers, pumpkins, melons, gourds, watermelons, chillies, and other vegetables in the paddy fields. But the most important crop was rice., which was not only a staple food but was used to make the rice wine which we drank every day of our lives.

> I loved the cold season best. It brought the rich fragrance of the rice and other crops as they ripened after the heavy rotting smell of the monsoon ... The monsoon had its own special character of sound, the combination of the bird sounds, the rain, thunder and wind.

> The harvest would start in late December with groundnuts, maize, beans and other crops. As it was the most important, the rice harvest was attended with much ceremony. Friends and relatives came to help with the reaping, threshing, winnowing and storing of the rice with much ceremony.[13]

Sowing, growing, reaping, threshing, winnowing, storing.

Until I read this I had never really thought of these separate activities. I just knew they amounted to something called the harvest.

Pascal Khoo Thwe gives us a living account of economic life. It describes economic activity in the context of human relationships. The purpose of the economic activity is clear. There is a true understanding of the connectedness of the different tasks within the harvest and how the parts relate to the whole: sowing, growing, reaping, threshing, winnowing and storing.

This may sound sentimental, longing for a lost and (far-from) ideal rural existence.

On the contrary, I am saying we have to accept that our economic life is complicated and global and highly specialised and yet we can still demand a better connection between the parts and the whole – between the modern equivalents of sowing, growing, reaping, threshing, winnowing and storing.

This translates as:

- Saving (storing)
- Investing (sowing)
- Fertile soil for starting, growing and financing businesses (growing)
- Stewarding invested assets in line with human values (growing)
- Developing appropriate models of ownership and shareholding (reaping and storing)
- Managing funds with present and future generations in mind and with human well-being as our focus (reaping and winnowing and storing)
- A regime for tending and pruning those businesses so they cannot grow out of control (winnowing)
- And, completing the cycle of the economic harvest, channelling and distributing its fruits in ways that enrich society and take care of future generations, (storing and then sowing again)

CONNECTEDNESS IN ECONOMIC LIFE

The religions of the world share many insights about the right way to live and the best way to make sense of the world. One of these insights is this: everything is connected. Body and soul, mind and spirit, present and future. So if we know from the wisdom of old, that everything is connected, how have those who dominate our economic life come to ignore this vital wisdom? When did they allow the criteria of financial and economic success to become disconnected from human purposes? And how do we reconnect the two in ways that satisfy the conditions for harmony – individual well-being; appropriate scale; the right kind of work and markets and technology that are aligned to these, and the right intergenerational balance? The answer starts with the way we think. That in turn influences the way we act in business, the way investors act and the rules and frameworks that are set by whole societies.

HOW WE THINK

If we want to see the right ideas prosper in the marketplace, we first have to win the battle of ideas. In the course of my own study of economics and philosophy, together with my countless conversations with CEOs, chairmen and institutional investors, I have come to the conclusion that there are five different, yet linked, key variables in the way people think about business. The assumptions people make under these five headings combine to form the 'taken for granted' attitudes for business along a spectrum that runs from the exclusive to the inclusive.

MOTIVATION: MONEY OR MORE

The first is about what motivates human beings in business. To make any kind of generalisation in the economic laboratory, economists have to simplify and concentrate on those parts of economic behaviour that are predictable. So, they invented the caricature or hypothesis of rational economic man. 'Economic man' is characterised by self-interested goals and a rational choice of means.

This is very useful for, say, calculating the response of consumers to changes in the price. But it tells us nothing about how people respond to leadership or how businesses generate loyalty because of their purpose and values – factors which research tells us are central to enduring profitability, let alone harmonious business. It is an exclusive and narrow definition of human nature. It excludes from its calculations a large number of values and emotions which influence human behaviour.

MARKETS: MASTER OR SERVANT

The second question is about how far people trust markets to deliver the right answer. On the exclusive side in this battle of ideas stand those market disciples who believe that markets mechanisms are supreme. A particularly extreme version of faith in markets is the 'efficient markets hypothesis'. Some market disciples (although perhaps fewer now than before 2008) believe that the price of shares reflects all known information about a company. On the other, inclusive, side stand those who believe that this is a naïve belief, which, among many other things, ignores the effect of herding behaviours which have recurred in every market since the Dutch went mad about tulips in the seventeenth century. Markets are flawed human inventions, an important but not sacred moving part in a delicate system which needs constant adjustment. From this perspective, it is important to harness market forces to society's priorities. The difference between the two perspectives was acknowledged all too late by Alan Greenspan, who was head of the US Federal Reserve at the time of the financial crisis. When asked how such a devastating crash happened, his answer was 'Our models didn't work'. He explained that the two hundred and fifty PhDs he had working on the models had ignored two things – the nature and speed of market dynamics and 'people and their unpredictable emotions'! [14] That was quite some omission!

The third question concerns how people view the relationship between business and society. In the exclusive view, business and society sit in separate compartments. At its weakest, the milder form of this argument says that businesses, and those who run or own them, do not need to worry about the needs of society while pursuing profits. The stronger form goes further: it is bad for business and for society if business 'takes its eye off the ball' and thinks about the needs of society.

The inclusive approach says with J. R. D. Tata (of whom more later) that 'What came from the people must go back to the people' – that business is a servant of society.[15]

The fourth question is about how people in business see value. Do they think in terms of abundance or scarcity? Just as some people see motivation in financial terms, some people see value in purely financial terms, and see business value as a zero-sum game. Spend more on your workers' wages or give more to charity or the arts and you will have less left for your shareholders.

Others see value as a positive sum game. Their philosophy is that the more you put in, the more you will get out. Put money back in the neighbourhoods you operate; invest in people and in carbon reduction and somehow it will come back to you. The inclusive approach sees the best companies creating value in ways that are a win-win for shareholders, staff and society.

The contrast between the two is well illustrated by the remarks made by the then CEO of Unilever, Paul Polman, after the food giant Kraft Heinz had attempted, but failed, to take over Unilever.

> Without naming Kraft Heinz, the food group backed by Warren Buffett and the Brazilian billionaires behind 3G Capital, Mr Polman described the failed bid attempt as 'a clash between people who think about billions of people in the world and some people that think about a few billionaires'. Speaking to portfolio managers representing an estimated $25trillion of assets under management, Mr Polman told them to stop asking companies like his why they heeded environmental, social and governance concerns and start pressing those that did not do so to explain 'why [they] have the courage to destroy ...this wonderful planet'. Investors with a responsibility to generate long-term returns to match their pension liabilities had the same responsibility to

ensure that their members 'are retiring in a world they can live in'.[16]

People: Whole or in Part

The fifth question is about how people think about shareholders and other stakeholders. Do they see shareholders, customers, suppliers, workers, and citizens as distinct and separate groups of people? Or do they see terms like 'shareholder' and 'stakeholder' as descriptions of a particular role that people may have to wear in one particular situation?

To a social scientist or anthropologist, it is obvious that human beings all have multiple roles, and this, of course, helps to explain why most of them don't isolate financial considerations from other aspects of human life. There was a tragic and extreme example in the United States in 2012 following the terrible multiple shootings in Newton, Connecticut. In the aftermath, newspapers reported that Cerberus, the owners of the company whose gun was used in the shooting, was considering pulling out of its investment in that company. The reports also mentioned that the father of Cerberus' CEO lived in the town where the shootings occurred. It would be hard to find a clearer example of the truth that no human being can entirely separate into different compartments the way that they live their life.

Assumptions behind the exclusive, compartmentalised, approach have been spreading over my lifetime. Many of them can be found at the heart of the banking crisis. Their prevalence helps to explain the growing public mistrust of business.

We need a holistic, inclusive, connected and harmonious approach. Whatever the labels, and whatever our cultural origins, business attitudes in East and West alike have the potential to move beyond the narrow reductionism of *homo economicus* and the exclusive focus on financial measurement to rediscover the human purposes of business.

How We Act

How we think influences how we act. The practical agenda for a more harmonious approach is really common sense, at least to those who accept that business is a human activity, conducted by human beings in relationships with one another, using the mechanisms and disciplines of the market to help them achieve human purposes. In other words, that business activity is impossible without making a profit but that moneymaking is rarely the purpose.

There is abundant evidence that businesses which over long periods of time seek to serve a purpose beyond profit outperform those who look for money alone. Equally, the literature of entrepreneurship is full of examples to support the idea that financial gain is not the sole or major motivation for entrepreneurs.[17] And it's hardly surprising (except for people who cannot see outside the economist's laboratory).

We do business with those we trust. We are less likely to trust businesses whose only motivation is making money for themselves or for their shareholders unless we have some personal or family connection with those shareholders. We are much more likely to do repeat business with those whom we trust. Getting a new customer is more expensive for a business that building a relationship with an existing customer. Regulation is going to punish businesses which prove untrustworthy. Trust enhances the value of the brand. The best people want to work for organisations that give their work meaning and purpose beyond their financial return.

And so it is that all the best businesses start with purpose and values; they stay true to purpose and values in all relationships. They have a clear success model – a constantly evolving hypothesis about how they will earn sufficient income from fulfilling their purpose. Their communications are always a reflection of their purpose and values. Their ethics are founded on their purpose and values. Their interventions in the community, their corporate citizenship is not some self-indulgent leisure activity – it is simply their purpose and values in action, in all, their relationships. Their thinking, which guides their actions, follows the logic of the virtuous circle of governance.

At a global level, the same principles apply, but with even more force. Global companies have become giants, stepping out beyond their own homeland. There is no longer one government which can control what they do. They have to negotiate a tangle of relationships. Before BP started operating in Azerbaijan it had to help the local government to set up the regime by which it would be taxed! Long before national or supranational regulators understood the social implications of their powerful products, Facebook needed to understand and manage their potential misuse by terrorists, manipulators and criminals. The part needed to understand its relationship to the whole, and the constant flow of influence and impact from the business to society and vice versa.[18]

This description of business seems obvious. It has been true for centuries. Yet it has precious little to do with what is taught in our business schools or our management textbooks, let alone the world of corporate finance.

For in all of these sources and places you will enter a parallel universe where money, and not purpose or relationships, is the measure of everything. Where something called corporate social responsibility or sustainability somehow has to be fitted in (and only if we have prepared something called a 'business case'!)

How We Own

Many people find these arguments about the human purposes of business convincing. But then, they say, that's all very well. How do you persuade shareholders to think like that?

That leads us to a different question. Who are the shareholders?

Often, they are presented like alien beings. 'The markets' we are told 'think this'. 'The markets 'have formed a poor impression of this company's latest announcement. That company has had to change course in order to satisfy 'market expectations'.

Here again, we are in a parallel universe and one which denies the importance of harmony. In the real-world, entrepreneurs start businesses; they go to the markets for capital to help them. I've never heard of an entrepreneur starting a business by saying, 'I wonder if there are some shareholders or institutions whom I have never met who might like to own some shares if I started a business?'

The real world of wealth creation is different to the world measured by short-term shareholder value. This is where change is needed if our approach to business is to fulfil the conditions for harmonious business.

The entrepreneurial stage of a business seems better aligned with harmony. The individual is creating their own future, organising work to their own timetable. This is the sowing and planting stage.

But then the scale increases. What happens when the business grows and outside capital is brought in?

The choice of ownership solutions is pivotal. Many businesses start off with good intentions. Yet, as they grow, and bring in outside finance, they often lose the underpinning of their values that is provided by their owners. Listed companies have dispersed, often absentee, owners. The crucial harmony condition of a scale within which relationships can stay meaningful is then lost.

It doesn't have to be this way. A few businesses remain determined to stay faithful to the conditions required by harmony. The answer lies in their approach to ownership. An approach which combines continuity and renewal; inner strength and outer strength.

TATA: A 150-YEAR CASE STUDY IN HARMONIOUS BUSINESS

The history of Tata is a living, enduring and convincing case study in stewardship by owners concerned to achieve harmony between the needs of their business and those of society.

The company started by Jamsetji Tata in 1868 and floated in 1874 with capital subscribed by his friends as The Central India Spinning and Weaving Manufacturing Company was, in his own words, 'Started on sound and straightforward business principles, considering the interests of the shareholders our own, and the health and welfare of the employees the sure foundation of our prosperity'.[19]

Tata Sons has evolved a unique ownership structure with a careful balance between the parts and the whole. The whole is under the influence of Tata and Sons. Two-thirds of the equity of Tata and Sons is held by philanthropic trusts endowed by members of the Tata family. Tata and Sons hold a stake of varying size in each Tata company. Each company is independent, with its own board of directors. Twenty-nine of these companies have their own stock market listing. Thus, while ownership of the different Tata companies is dispersed, the philosophy is shared. The Tata statement of purpose and code of conduct covers all these independent companies. Making money is never seen as an end in itself. JRD Tata, chairman from 1938 to 1991, said, 'I never had any interest in making money. None of my decisions was influenced by whether it would bring me money or wealth'.[20]

Asked how he would define the house of Tata, and what linked the companies together JRD Tata, replied:

First a feeling that they are part of a larger group which carries the name and prestige of the Tatas, and public recognition of honesty and reliability – trustworthiness. Each company enjoys its share of the privilege. They use the Tata emblem. The reason is, you might say, enlightened self-interest.[21]

This then finds its expression in the diverse ways in which Tata subsidiaries engage with the communities and society of which they are a part, and with key stakeholders under the umbrella of the Tata Sustainability Group. Tata recently described its purpose in terms that resonate with the words of earlier generations of its leaders:

At the Tata Group we are committed to improving the quality of life of the communities we serve. We do this by striving for leadership and global competitiveness in the business sectors in which we operate.

Our practice of returning to society what we earn evokes trust among consumers, employees, shareholders and the community. We are committed to protecting this heritage of leadership with trust through the manner in which we conduct our business.[22]

This echoes the words of JRD Tata.

The wealth gathered by Jamsetji Tata and his sons in half a century of industrial pioneering formed but a minute fraction of the amount by which they enriched the nation. The whole of that wealth is held in trust for the people and used exclusively for their benefit. The cycle is thus complete; what came from the people has gone back to the people, many times over.[23]

Influenced by these values, the Taj Hotel group, a Tata company, developed a unique approach to recruiting staff direct from small towns, focusing more on hiring people with integrity and devotion than skills and talent, and training their staff for eighteen months rather than the average of twelve. They believed in teaching people to improvise rather than doing things by the book, and insisted that employees place guests' interests over the company's.

A test of the values of Tata came in November 2008, when terrorists entered the Taj Mahal Hotel in Mumbai, shooting and killing indiscriminately. After a night of terror thirty-one people had died (of whom half were hotel staff) and twenty-eight were hurt. Dozens of guests were saved by the calm and brave actions of the Taj's staff.

Bhisham Mansukhani was a guest who escaped that night. Of the staff who saved his life when they might have run away, he said:

They were just kids. Young boys and girls. Two girls in their early 20s, couple of kitchen staff. Those brave girls had their phone on charging and were guiding the NSG to our location. They were remarkably great. One of them, Rajan Kamble, who was in front of us, was shot in the stomach while helping the guests escape.

Over the period of 11 hours, the staff saved my life several times.

When the National Security Guard knocked on the door at 3:30AM it was with the help of hotel staff that we escaped who had come right back inside the hotel to help. 'They guided us to the lobby and outside. There was such chaos around us, blood and glass was scattered all over. The firing started again. The cops ran away, but the staff formed a human chain around us.[24]

The sacrifice of these staff reflected a pattern of behaviour exemplified by top leaders. Jamsetji Tata, the founder, had started a textile mill in the 1860s, instituted a pension fund for his employees as early as 1886 and began voluntarily to pay accident compensation as early as 1895. After making a success of that first mill, Jamsetji had created a public company to buy and turn around a bigger, older and more inefficient mill in Bombay. The turnaround was slow and the economy unhelpful. When the public company was faced with financial ruin, and the banks would not accept his family trusts as security for a loan, he revoked the family trust into which he had ploughed his personal wealth, sold shares and ploughed the money back into the second mill and made it a success. This is a far cry from many later creators of public companies who were content to see the capital of other shareholders lost while protecting their own fortunes.

Tata has endured for a hundred and fifty years. Its history exemplifies the subtle tension between the role of leaders and the role of owners. Its pioneering founders designed its ownership like a keel that could keep Tata companies on a steady course through times of disruption.

Today there is a cohort of entrepreneurial businesses which have pioneered their own journey in the spirit of balancing the parts and the whole. Some of them have, like Tata, thought carefully about their ownership design. TTP Group is an employee-owned company with more than three decades of achievement in a highly competitive market which exemplifies the same principles. The story of this, and many more, some older, some younger, are described in a recent Tomorrow's Company publication 'The Courage of their Convictions'.[25]

HARMONIOUS BUSINESS: THE ROLE OF OWNERS

Harmonious business can only be achieved in a world in which owners, rather than seeing themselves as traders of shares, see themselves as stewards and wish to pass on the business in a healthier shape than they inherited it.[26]

Such owners give managers the confidence to manage without the arrogance or greed that puts themselves ahead of the organisation. They look outwards. They see the business as a servant of society. Because they are in tune with the changing needs and expectations of society, they adapt and stimulate necessary change while holding on to constancy.

We need such owners precisely because ordinary citizens have become disconnected from the world of business. That disconnection is very risky.

Sometimes the necessary human connection is achieved by family ownership, provided that it is the kind of family ownership in which the family sees itself as a servant of society. As organisations become larger, many family businesses evolve a form of trust ownership of the kind achieved by the Tatas. Sometimes this connectedness may be achieved by mutual ownership. A 2013 study in Canada found that, overall, credit unions outperformed other financial institutions when it came to serving micro, small and medium-sized businesses.[27]

Sometimes it is best achieved by devolved models of social enterprise such as M-Pesa in Kenya.

The story of M-Pesa is a fascinating example. It was a grant from the UK's Department for International Development which set the ball rolling on a powerful piece of innovation that has helped thousands of micro-businesses to start. The grant was given to a not-for-profit knowledge transfer and capacity-building organisation called Gamos, which had been founded by Professor Peter Dunn, an engineering professor at Reading University in the UK. The initial idea of M-Pesa was to create a service which would allow microfinance borrowers to receive and repay loans using the network of Safaricom airtime resellers. As a result of a collaboration between Gamos and the Vodafone Foundation, M-Pesa evolved into a way in which people could send, receive and withdraw money from their mobile devices. M-Pesa spread quickly and by 2010 had become the most successful mobile-phone-based financial service in the developing world. By 2012, a stock of about 17 million M-Pesa accounts had been registered in Kenya. By June 2016, a total of 7 million M-Pesa accounts had been opened in Tanzania by Vodacom. The service has spread to other parts of Africa and beyond. It has given millions of 'unbanked' people access to the formal financial system and the ability to create micro-businesses in areas devoid of conventional finance. In the process the development of this company has also been an exemplar of the second key area for wealth creation – the financing of emerging businesses.

In pre-industrial societies, work is carried out in the family and savings held within the family. The earliest shareholdings would simply have been the wealth that a family built up in a business they had created and passed on.

Now the process of harvest and storage is more complex. People save, their relatives send money back from working overseas and their parents pass on wealth when they die. That money increasingly flows into savings schemes, life insurance and pensions. From there in due course the savings find their way into the world's capital markets. But people aren't connected with them.

We need listed companies and the stock exchanges on which they float to meet the financial needs of companies and of the entrepreneurs who wish to receive back some of the value they have created. We need them, but we need them to operate in a much more inclusive way.

Most of us citizens don't like what capitalism is becoming and what it is doing to our society. The irony is that, apart from the poorest; we are all part-owners of many of the very companies whose behaviour we dislike so much. Citizens are shareholders and shareholders are citizens. Why would they want different things?

For shareholders the quality of the financial harvest cannot be separated from surrounding conditions in which those dividends are received. We all want and need value – a return on our investment consistent with an improving quality of life. What's the point of owning your car, if it can't move for the traffic around it? What's the point of that penthouse suite with the breath-taking views of the city, if you cannot see the city for smog?

How to Square the Circle between Citizens and Shareholders

The question is – how do we square the circle? How can we allow the human being as saver and investor to make shareholder choices in ways that will be welcomed by the human being as citizen? This is where trusteeship and stewardship come in.

Older employees, especially in the public sector, may already be members of an employee pension scheme. Younger savers will have to bear this risk themselves, and their future pension will dependent on the success of the investments made by the fund and the effectiveness of its stewardship of their assets.

In either case, regular payments are deducted from their salary. These are passed on to the trustees of the pension scheme, who in turn pass the payments to the managers of the scheme. The scheme managers are employed to invest the money in line with the trustees' requirements so that the pension promise can be met. The time horizon for these trustees has to be long. The job of those handling their savings is to look several decades ahead. The trustees of a pension scheme, working with the guidance of investment consultants, have the responsibility to define an investment policy and decide on the appropriate asset allocation.

The ordinary saver has no direct idea about the companies into which the money is being invested and no way of assessing the effectiveness of the stewardship of all the different people along the way. There may be financial advisers and investment consultants, stockbrokers and analysts, insurance companies with their own fund managers, third-party fund managers, hedge funds and private equity funds. In many jurisdictions savers will be amazed to discover that the shares that are ultimately owned in their name can be loaned out for a fee so that hedge funds and other owners can go 'short' or 'long' – which essentially means betting on the future price of the shares.

To stay with the Myanmar rice example, by now the saver is a very long way indeed from the monsoon smell of the rice, and some of the grains of rice are seen as gambling chips, not future nourishment.

Common sense – and the principles of harmony – would suggest that all this should happen in a *connected* way under some kind of guidance or criteria decided by the person whose money it is. But that isn't how it works at the moment. Financial services companies occupy a parallel universe. They don't necessarily make the common-sense connection between longer term returns and the enduringly successful companies in which these will be achieved. Most fund managers will be deemed more successful, and so earn better financial rewards, if they back a management that promises to drive up the share price immediately. They will be frightened that if they refuse that take-over bid, or show patience to that management team, the pension trustees will judge them against their rivals who accepted the windfall gain. This puts them at risk of losing their investment mandate.

As savers and investors, we all need to see our investments as Pascal Khoo Thwe saw the harvest – as living processes that we could virtually if not physically smell, taking our values out across the economic system and touching companies in ways which strengthen their ability to become a force for good.

In the decades to come, as savers and investors discover that the risk now lies with them, they will begin to demand that they can see and influence where the money is going and how dividends are achieved. How much more might it mean to the saver if, without abandoning the usual advice about not putting all the eggs in one basket, savers could invest in fixed interest bonds that helped with the improvement of infrastructure in their own region and in social impact investment that was targeting reduced mortality or improved housing stock or local food production in their region, or early-stage finance for young entrepreneurs in the area.

In time, the financial services consumer may eventually be able to use a stewardship label which has the same authority as the Fair-Trade label used in the marketing of tea and coffee. Without this golden thread of effective stewardship, society won't get the robust, principled companies and investment intermediaries that it needs to handle its harvest.[28]

THE CHANGING BUSINESS LANDSCAPE

To summarise, we learn from the past that the five key principles by which business may be conducted harmoniously are

- Health, well-being and wholeness of the individual
- Appropriate and accessible scale of organisation
- Fulfilling work open to all talents
- Science, technology, and the tools and organisations they generate being servants and not masters of human beings
- The right balance between the needs of today's and tomorrow's generations

These principles of harmonious business do not change, even though the business landscape itself is constantly changing. In 'Ill Fares the Land' the historian Tony Judt wrote shortly before his death in 2010 about the breakdown of the post-war system that had combined economic growth with social security. He said that 'We have entered an age of insecurity: economic security, physical insecurity; political insecurity', and added:

Something is profoundly wrong with the way we live today. For thirty years we have made a virtue out of the pursuit of material self-interest: indeed, this very pursuit now constitutes whatever remains of our sense of collective purpose. We know what things cost but have no idea what they are worth. We no longer ask of a judicial ruling or a legislative act: is it good? Is it fair? Is it just? Is it right? Will it help bring about a better society or a better world? Those used to be the political questions, even if they invited no easy answers. We must learn once again to pose them.

The materialistic and selfish quality of contemporary life is not inherent in the human condition. Much of what appears 'natural' today dates from the 1980s: the obsession with wealth creation, the cult of privatization and the private sector, the growing disparities of rich and poor. And above all, the rhetoric which accompanies these: uncritical admiration for unfettered markets, disdain for the public sector, the delusion of endless growth. We cannot go on living like this.[29]

These words seem even more apt today. Take, for example, concerns about the growing dominance of Amazon, Apple, Facebook and Google.

In his critique of these companies and their business model written in 2017, Scott Galloway describes the emergence of the 'winner-takes-all economy'.[30] Four years later the *Financial Times* reported that Google and Facebook between them controlled more than 58% of total US digital advertising spend; Amazon was on course to capture nearly half of America's e-commerce market, and in the same article the FT quoted the IMF as seeing 'evidence of rising market power and declining competition in the US ... This is coupled with signs that the labour share is going down'.[31]

Five years later Harvard Professor Shoshana Zuboff has described Google's 'disregard for the boundaries of private human experience and the moral integrity of the autonomous individual'. She describes how:

> Surveillance capitalists asserted their right to invade at will, usurping individual decision rights in favour of universal surveillance and the self-authorised extraction of human experience for others' profits.[32]

As an example, she describes the subversion of the original intentions of the technologists' vision of the Aware Home – using the Internet of Things to put the householder in better control of the house and all the functions operating there. In its original form the Aware Home was designed on the basis of complete privacy for the householder.

Today that privacy has gone. The major providers of technology to the Aware Home upload all the data for their own use. For example, the Nest thermostat learns all about the behaviours of the householder and family. Its apps gather data from other connected products such as cars, ovens, fitness trackers and beds. Personalised data now flows straight to Google's servers. If the user wishes to deny that data flow, they are no longer supported by the necessary updates.[33]

Health, well-being and wholeness of the individual is the first condition for harmonious business. Human scale is the second. In the twenty-first century this will require firm action by governments and regulators. Decades ago, US regulators forced the breakup of telecommunications giants with positive results. Institutional investors themselves can gain greater value from their investments if the corporate leviathans are broken up. The splitting of ICI into two companies led to the creation of the company that became Astra Zeneca, one of the most productive of major drug companies at the end of the last century. The EU has become increasingly aggressive in dealing with anti-competitive behaviour by Google. This is only the beginning. National and international regulators need to broaden their interest from anti-competitive behaviour, towards positive bias in favour of the principles of harmony. This may well mean the breaking up of companies in the interests of not only of competition and equality but also of human scale and accessibility.

Systemic problems require systemic responses. Neither companies, nor individual savers, nor rich individuals, nor investment institutions nor trade unions or industry associations, nor business educators can on their own shape the conditions for harmonious business. All are part of the answer. If they are to be enabled to contribute to it, then we will require what David Cadman describes as an 'integrative discipline' within which they are understood as part of the whole. With this the assumptions, and framing and teaching and practice of economics, business investment and public policy may gradually evolve into what Schumacher made the subtitle of his book – 'a study of economics as if people mattered'.

NOTES

1.   Frederic Laloux, *Reinventing Organisations: A Guide to Creating Organizations Inspired by the Next Stage of Human Consciousness* (Brussels: Nelson Parker, 2014), p. 195.
2.   David Cadman, 'Harmony and Integration', paper presented by David Cadman, UNESCO Conference, Paris September 2017.
3.   Cadman, 'Harmony and Integration'.
4.   T. B. Bottomore and M. Rubel, *Karl Marx: Selected Writings in Sociology and Social Philosophy* (London: Pelican, 1971), p. 263.
5.   Bottomore and Rubel, *Karl Marx: Selected Writings*, p. 263.
6.   Asa Briggs (ed.), 'How We Live and How We Might Live' in William Morris, *Selected Writings and Designs* (London: Pelican, 1962), pp. 177-78.
7.   John Ruskin, *Unto This Last* (London: Pallas Athene, 2010), p.14.
8.   Ivan Illich, *Tools for Conviviality* (New York: Marion Boyars, 1973), p. x.
9.   Illich, *Tools for Conviviality*, p. xii.
10.   E. F. Schumacher, *Small is Beautiful: A study of economics as if people mattered* (London: Abacus, 1973), p. 27.

11. Schumacher, *Small is Beautiful*, p. 84.

12. Charles Hampden Turner and Fons Trompenaars, *The Seven Cultures of Capitalism* (New York: Doubleday, 1993), p. 32. Also reviewed by Mark Goyder in *The Independent* at https://www.independent.co.uk/news/business/book-review-reliving-goals-over-a-bottle-of-beer-the-seven-cultures-of-capitalism-charles-hampden-5428408.html/

13. Pascal Khoo Thwe, *From the Land of Green Ghosts* (London: Harper Collins, 2002), p. 55.

14. Alan Greenspan, *The Map and the Territory: Risk Human Nature and the Future of Forecasting* (New York: Penguin, 2013) quoted by Bob Garratt, *Stop the Rot: Reframing Governance for Directors and Politicians* (Oxon/New York: Routledge, 2017), p. 65.

15. R. Gopalakrishnan, '150 years on: How the Tatas flew with their dreams', *Rediff Business*, https://www.rediff.com/business/special/-150-years-on-how-the-tatas-flew-with-their-dreams/20180917.htm [accessed 12 April 2019]; See also R. M. Lala, *The Creation of Wealth: The Tatas from the 19th to the 21st Century* (London: Portfolio Penguin, 2017).

16. Andrew Edgecliffe-Johnson, 'Unilever chief admits Kraft Heinz bid forced compromises', *Financial Times*, 18 February 2018, https://www.ft.com/content/ea0218ce-1be0-11e8-aaca-4574d7dabfb6 [accessed 12 April 2019]/

17. See Mark Goyder, 'The Courage of Their Convictions', Tomorrow's Company, 23rd April 2018, https://tomorrowscompany.com/publication/the-courage-of-their-convictions [Accessed 10 May 2019] and Mark Goyder, *Living Tomorrow's Company: Rediscovering the Human Purposes of Business* (Mumbai: Knowledge Partners, 2013).

18. The role of global companies in creating the framework of rules within which they can legitimately operate was described in Luke Robinson, 'Tomorrow's Global Company: Challenges and Choices', *Tomorrow's Company*, 4 May 2007 at https://tomorrowscompany.com/publication/tomorrows-global-company-challenges-and-choices/ [accessed 10 May 2019]

19. Lala, *The Creation of Wealth*, p. 126.

20. Lala, *The Creation of Wealth*, p. 193.

21. Lala, *The Creation of Wealth*, p. 101.

22. Tata, 'Values and Purpose', http://arch.tata.com/aboutus/articlesinside/Values-and-purpose [Accessed 15 May 2019].

23. Lala, *The Creation of Wealth*, p. 184.

24. 'Ratan Tata recalls sacrifice of staff on anniversary of the Taj', *The Hindu*, 17 December 2011 at http://www.thehindu.com/news/national/ratan-tata-recalls-sacrifice-of-staff-on-anniversary-of-the-taj/article2721477.ece [accessed 12 April 2019].

25. 'The Courage of their Convictions', 23 April 2018 at https://tomorrowscompany.com/publication/the-courage-of-their-convictions [accessed 12 April 2019].

26. There has been a concerted move over the last ten years, prompted initially by Tomorrow's Company, to promote Better Stewardship by institutional investors. See Luke Robinson, 'Better Stewardship – an agenda for concerted action', 25 January 2018 at https://tomorrowscompany.com/publication/better-stewardship-an-agenda-for-concerted-action/ [accessed 12 April 2019].

27. Daphne Rixon and Peter Goth, 'Credit Union Commercial Lending: Mitigating Risk through Recording, Monitoring, and Reporting', at https://p.widencdn.net/v4cd7d/432_EXS_CU-Commercial-Lending [accessed 12 April 2019].

28. These ideas are developed in more detail in *Living Tomorrow's Company*, p. 151.

29. Tony Judt, 'Ill Fares the Land', *New York Times*, 16 March 2010 at https://www.nytimes.com/2010/03/17/books/excerpt-ill-fares-the-land.html [accessed 12 April 2019].

30. Scott Galloway, *The Four: the Hidden* DNA *of Amazon, Apple, Facebook and Google* (New York: Penguin, 2017).

31. 'Economists warm on dominance of US corporate giants', *Financial Times*, 16 August 2018 at https://www.ft.com/content/7f88226e-9f0b-11e8-85da-eeb7a9ce36e4 [accessed 12 April 2019].

32. Shoshana Zuboff, *The Age of Surveillance Capitalism: The Fight for a Human Future at the New Frontier of Power* (New York: Hachette Book Group, 2019), p. 19.

33. Zuboff, *The Age of Surveillance Capitalism*, p. 6.

The Circular Economy

Dame Ellen MacArthur

From a talk at the Harmony, Food and Farming Conference, organised by the Sustainable Food Trust, Llandovery College, 10 July 2017

WHAT RESONATES FOR ME PERSONALLY around the idea of Harmony is the fact that it addresses the whole system. As I grew up and went through education, everything I seemed to learn was in silos; it was in columns. It was about one subject there, and one subject here. I think what's so powerful with the Harmony principles is that they reconnect everything. And everything is indeed connected, as it is in nature. Everything in business is connected, every resource that we take from the ground is connected, and every insect is connected with every piece of soil. Yet it's so easy to slip into that mindset that everything is separate, and we tackle things separately.

I believe the only way that we're able to face, and indeed solve, the world's challenges is to look at the world as a system, as a whole, and to look at how those connected elements can really resonate. Nature depends on cycles. It always has. If there's one message that's central to the circular economy, it's that life itself has existed for billions of years. Life itself never created waste; everything was regenerated. But what we've created now is a system which is extractive and consumptive, not restorative and regenerative. The message behind the circular economy is in parallel with Harmony in that you need to build a restorative and regenerative system from the beginning. That involves us all through the education of young people, and right through to the chief executives running the biggest companies in the world.

My journey to this point, right now, to understanding the principles of the circular economy, began with a boat. You may or may not be aware that my history was racing. My goal from the age of four was to sail around the world. I had no idea how I would achieve that, but that was absolutely my goal. I was lucky enough to sail around the world solo twice, once in a race in 2000 called Vendée Globe, and the second time in 2004. And it was quite an interesting insight for me into how systems function. When you set off on one of these boats around the world, you enter a different space, a different mindset. You behave in a different way, you're incredibly stressed, you're full of adrenaline, and you're managing this tiny world, which is your cocoon that keeps you alive. These boats are pretty exciting. It's not a gentle sail in the sunshine, waving at people on the beaches. You generally don't see land when you sail around the world. And boats can go wrong very quickly. I know this very well because I was one of the five crew members on a boat which flipped upside down, literally in five seconds.

Now, that could happen at any time when you're sailing around the world. It could happen when you're in the Southern Ocean, two and a half thousand miles away from the nearest town and if it does then you probably won't make it: if help can get to you, it takes five days, and then five days for that ship to get you back, and to a hospital. You really are isolated, and you really do understand that what you have with you on that boat is all you have, there is no more. It was that understanding that I developed from sailing, that fundamental understanding of what finite truly means. That on that boat, what you have is all you have. There are no more resources, you're two and a half thousand miles to the nearest shop. That led me to look at the economy in a different way.

On the boat you live in a different world, you perceive what's around you in a different way, and I couldn't get that out of my head when I finished that voyage. I couldn't stop thinking about it: it was like a tiny little spark under a rock. You know, a large part of me wanted to put that rock back down and carry on with my dream job of racing around the world, but I couldn't. I couldn't do it. I had to put that rock to one side, and learn more about resources, energy and how the economy functions. We too have finite resources available to us as humanity. About eight years ago I started to learn. I started to study. I went to a coal-fired power station: it's still incredibly important and it was a subject very close to my family because my great grandfather was a coal miner. We were really close. He used to tell me his mining stories from when he worked deep underground with ponies to pull the coal out from the coal face. He was alive until I was eleven years old. This is really not that long ago. Yet, when I went on this journey of learning to try and understand the resources that we have available to us, one of the places I went to was the World Coal Association, and there, in the middle of the homepage, it said, 'We're not about to run out of coal, we've got about 118 years left'.

I did the maths, and I realised that my great grandfather had been born exactly a hundred and eighteen years previously. Then you realise it's no time at all. A hundred and eighteen years is nothing in the broad sweep of history, and it made me make the decision I never, ever thought I would make: to completely leave the sport of sailing and focus on global economics. To understand what success can look like for us, for the global economy, for farming, for children, for everything. Because the way the economy functions at the moment uses up natural resources, as I did on my boat. In contrast to sailing though we cannot re-stock the planet's resources as we can re-stock at the end of a voyage. We know that natural materials are finite – we have them once, and we're using them up at a faster and faster rate. You can define that as a linear economy: one that takes material out of the ground, makes something out of it, and then ultimately, the majority of it gets thrown away. We're able to get some of the material back, but we have a conveyor belt, exactly as farming has moved from being regenerative, even just a hundred years ago, to now, run on the extractive and consumptive principles of a linear economy, and that's worked quite well for a while. Now, we have a growing population, we have more and more pressure on resources, and we know the inputs are finite. As a long-term plan, it simply cannot work. Is there a different way of doing things? To replace the extractive and consumptive principles of a linear economy with a system which is restorative and regenerative?

As I went on my own journey, trying to understand what success could look like, what fascinated me were other ideas: life itself, which has existed for billions of years, has always been regenerative. Is there a way we could shift so that the linear becomes circular, taking ideas such as biomimicry, cradle-to-cradle design, industrial symbiosis, the performance economy, the sharing economy?

When we approach anything that we do within our economy, we do that in a circular way. We look at life itself, the fact that life itself has never had waste of the type that we create now. What if you apply that to the global economy? Not just the materials, such as the biodegradable materials – human waste, farm waste, agricultural waste and food production waste. Not just those materials that cycle, as they have, arguably, for billions of years, but also technical materials, that really will never biodegrade. What if metals, plastics and other polymers were also designed to fit within a cycle? At the beginning, at the design stage, we will design products so that we can get the materials back again. Better still, we will design a product so it can be re-manufactured, so we can collect the materials and re-use them easily, enhancing their value. Actually, the last loop is recycling, feeding materials back into the economy. And we can work with renewable energy, and reduce energy demand by 80%, if we re-manufacture something rather than melt it down or start from scratch: 80% less material, 80% less energy.

When we created the Ellen MacArthur Foundation seven years ago, our goal was to accelerate the transition to this circular economy, working with young people, which is absolutely vital: they see the world, from the beginning, in a circular way. [1] Also, working with businesses that know the linear system can't run in the long-term, working on analysis and insight, looking at the numbers, and asking what does the circular economy mean for the global economy? Does it make more money, or is it expensive? Every single report we've done to date has shown there is a sound economic rationale for shifting from linear to circular. It makes more money. The most hard-nosed businesswomen get this because if they can become circular, they will unlock more value for their company. This is the most important driver, so far, of the circular economy, because although we know the idea makes common sense, if there is economic value to be had, employment to be had, growth of countries to be had, through decoupling growth from resource constraints, then it will happen much more quickly.

There is a great crossover between the circular economy and the Harmony principles which stress that everything is connected: it is about strength and diversity. Diversity comes from the different sizes, types, and locations of companies. And there is strength in that diversity, which we should embrace and understand. In addition, the fact that nature itself depends on cycles, well, that is at the heart of the circular economy.

What are examples of what a circular economy could look like? First, anything biodegradable – that could include a cotton T-shirt, this lectern, agricultural waste, farming waste, food production waste, ourselves even – could re-enter the natural systems that support food and farming, or be used in other industries. A piece of timber might be turned into chipboard, and then into particle board. Then, maybe, at the end of the life of that particle board, if it's designed correctly, it could be bio-digested and turned into fertiliser, heat and biogas. How many times can you use that resource before it gets to the end of its period of use, before it re-enters the biological cycle? Then we have what we call technical materials: metals, plastics, other polymers, and so on. These may not be biodegradable, but we can also look at them through that same lens of circularity, asking 'What if they were designed so that we could keep them at their highest value at all times?' Whether it is a car, a plane, a chair or a phone, we could design it so that at the end of its use, it comes back and the components and materials are reused. Then you've built a restorative, regenerative system.

Even Apple, now, has a tariff in the US where you get a new phone every year. You think, 'Well, what's circular about that?' The moment you go on this tariff, they own the phone, so at the end of the year, they get the phone back. They know everything that's in that phone: they may resell it, they may remanufacture it, or they may recycle the materials and put them in the next phone. That phone no longer stays in our drawer at home when we don't know what to do with it. It feeds back into a system.

To return to biological material – sorry, food waste: in every tonne of food waste, there's $6 of fertiliser, $18 of heat and $26 of electricity.[2] How much of that is fed back into that system from our cities, from our towns? Even from our villages? This is a massive resource: not just food waste, but human waste and agricultural waste. In fact, we did a study asking whether, if we could collect all this together, globally, could we actually replace current chemical fertiliser use? The answer was yes by 2.7 times.[3] Now, we're a long way from that, but it shows the potential of looking at the materials we have available to us, in a regenerative and restorative way.

You look at the aggregate nutrients from cities, materials in landfill sites: if they were all designed correctly, none of them would be waste. We lose between 80 and 120 billion US dollars' worth of plastic packaging every single year because we don't design it, or the systems to recover it, so that it can be valorised.[4] Then we need to consider different ideas for packaging to replace plastic. We now are able to manufacture biodegradable packaging from mycelium, mould and corn husks, on price parity with Styrofoam, which it's designed to replace. And if we can produce alternatives to plastic, then we can do the same for fabrics. One company based in Switzerland makes fabrics that you could actually eat if you wanted, and are used in airplanes, so the air quality is better because they're not full of toxins.

The economic reports that the Foundation issues are absolutely vital in our understanding of a circular economy. We try to understand the numbers in the connected way that both Harmony and the circular economy approach recognise, looking at the entire system. In 2015 we completed a report looking at how we could become more circular from the perspectives of the built environment, mobility and food systems.[5] To include the digital revolution as the Prince of Wales mentioned, we found the figures to be absolutely fascinating. For example, if we harness that digital revolution in Europe, within those three sectors – the built environment, mobility and food systems – we would save 32% of primary materials by 2030 and 53% by 2050. Reduction in CO_2 emissions would be 48% by 2030 and 83% by 2050. And there would be an 80% decrease in chemical fertiliser use by 2050. We really do see significant momentum in putting all this into practice. We have companies all over the world embracing this and understanding how they can become circular. We are working within education, with universities all over the world, with lecturers, understanding their teaching and learning in relation to the circular economy and supporting the research that has to happen to understand it in more detail.

I'd like to finish with cities, which are huge aggregators of both technical and biological materials. We need to look at how cities can be transformative agents to help food and farming systems. More and more people live in cities. Food comes into cities from the countryside, but how do we get the waste to be something of great value and feed it back into that food and farming system, to echo the natural cycles that have existed for billions of years? In the area of food and farming, we can build a system that actually regenerates the land. It doesn't slow down the demise, as we have talked about so often. It actually rebuilds natural capital, faster and faster. When we work with young people, they get this immediately, and they want to build a better system; it's incredibly inspiring.

Just finally, is this possible? Well, to shift the entire global economy from linear to circular is quite a big task. But if we look at what can happen in a lifetime, we can see that anything, really, is possible. When my great grandfather was born in 1894, there were twenty-five cars on the road in the entire world. Twenty-five! That's it. When he was thirteen years old, we built the first aeroplane. Now, three times the population of the world back then fly every single year. When he was forty, we built the first computer, and many said it wouldn't catch on, but we turned that into a microchip within just twenty years.

Ten years before my great grandfather died, the mobile phone arrived. It definitely wasn't a smartphone, but it changed infrastructure across the entire world, including in emerging markets. It showed there was a different way of developing, and we believe the circular economy provides a different model of development for emerging markets. As my great grandfather left this earth, the internet arrived. If ever there was a time when we can change the global economy, it's right now. We can share an idea from Wales to the rest of the world in seconds!

NOTES

1.   Ellen MacArthur Foundation, https://www.ellenmacarthurfoundation.org/ (Accessed 14 September 2019)
2.   Ellen MacArthur Foundation, Towards the Circular Economy Vol. 2: opportunities for the consumer goods sector' (2013). For more information see Ellen MacArthur Foundation, 'Food Initiative', https://www.ellenmacarthurfoundation.org/our-work/activities/food (Accessed 14 September 2019).
3.   Ellen MacArthur Foundation, 'Urban Biocycles' (2017), https://www.ellenmacarthurfoundation.org/publications/urban-biocyles (Accessed 29 October 2019).
4.   Ellen MacArthur Foundation, 'The New Plastics Economy: Rethinking the future of plastics' (2016). For more information see https://www.ellenmacarthurfoundation.org/our-work/activities/new-plastics-economy (Accessed 14 September 2019).
5.   Ellen MacArthur Foundation, Stiftungsfonds für Umweltökonomie und Nachhaltigkeit (SUN) and McKinsey Center for Business and Environment, 'Growth Within: A Circular Economy Vision for a Competitive Europe' (2015).

Implementing Harmony: Wellness Tourism

Leo Downer

As a former public servant, my interest in Harmony is in how its principles can be implemented in practice. My current concern is with the application of Harmony to the travel and tourism sectors, and my initial project involves the creation of a Harmony Quality (or Kite) Mark for wellness tourism. Current estimates indicate that the wellness tourism market was worth $639 billion in 2017, and was projected (pre-pandemic) to reach $919 billion by 2022.[1] Most of this spending occurs in luxury sectors, but wellness tourism can equally flourish in low-cost, local areas. In my view, a Harmony Quality Mark needs to ensure that an organisation's services and delivery are in harmony with nature and active in the repair, restoration and rebalance of its clients and the wider environment. There are already a number of organisations which accredit organisations on the basis of their sustainability policies, mainly relying on environmental impact. These include the ISO, the Global Sustainability Tourism Council, EcoStar and Green Globe Solutions.[2] The factors derived from sustainable tourism accreditation cover a wide range, with Green Globe featuring Sustainable Management (including employee training, sustainable design and construction of buildings and infrastructure, using local principles of sustainable construction and design); Socio/Economic (including community development and support and fair trade); Cultural Heritage (including protection of sites and respect for local culture); Environmental (including environmentally friendly purchasing policies, sustainable energy and water consumption and recycling); and Biodiversity, Ecosystems, and Landscapes (including wildlife and biodiversity conservation).

This is a comprehensive list in which all factors are relevant to the wellness experience. But some may be more immediate. For example, if we consider architecture and design, a harmonious environment with fresh air and light are obviously going to be vital in most circumstances. I consider that a Harmony approach should appreciate the web of interactions, influences and the widest possible chain of consequences that could occur in the management of tourism, avoiding apparently simple solutions that can cause more harm. A Harmony Quality Mark in Wellness Tourism would therefore provide the framework to ensure that the actions taken in the creation, maintenance and delivery of Wellness services consider the client, practitioner, and the wider organisational, social and environmental landscape. It would require a whole system approach occurring throughout every layer and delivery of a service within the organisation or to a client. It would be:

- guided academically by a University standard
- have a robust philosophical and science base
- acknowledge that change alters systems, but this can lead to an alternative balance and more adaptive solutions
- require continual participation, learning and leading in good practice
- be active in the improvement of people, planet and place

While everything that is included in accreditation to sustainability standards would obviously be relevant to Harmony accreditation, I am looking at ways to make Harmony accreditation distinctive. I am looking at the concept of prosperity rather than profit as a key indicator, in that prosperity can be considered in broad terms to include, for example, the health of a community. One phrase which I am working with, drawing on the word Harmony, is off-setting Harm. We would take the widest possible measure of a business or organisation's impact, relying on the kind of analyses familiar to the circular economy and total-cost accounting and, in an analogy with carbon neutrality, aim not just to be harm-neutral, but harm-negative. If we are able to quantify harm and benefit, then in order to achieve a harm-negative status a business or organisation would be required to spend a sum profit to projects which promote sustainability objectives. For example, we could arrive at a financial assessment of harms (harm neutral) plus 5% (of that harm neutral figure), on top of the harm neutral amount, which would make it harm negative. In other words, rather than maintaining balance (which might be a basic definition of maintaining harmony), we would aim to secure a net benefit for the planet, promoting Harmony on a deeper – and broader – level.

NOTES

1. Global Wellness Institute, 'Wellness Industry Statistics and Facts', at https://globalwellnessinstitute.org/press-room/statistics-and-facts/#:~:text=Wellness%20tourism%20is%20a%20%24639,reach%20%24919%20billion%20by%202022.&text=International%20wellness%20tourists%20on%20average,than%20othe%20average%20domestic%20tourist [accessed 7 July 2020].

2. ISO, 'Tourism and related services — Sustainability management system for accommodation establishments — Requirements', at https://www.iso.org/standard/70869.html [accessed 7 July 2020]; Global Sustainability Tourism Council, at https://www.gstcouncil.org/ [accessed 7 July 2020]; Ecostar, at https://www.ecostarhub.com/ [accessed 7 July 2020]; and Green Globe at https://greenglobe.com/ggs/ [accessed 7 July 2020].

A REFLECTION ON HARMONY

Patrick Holden

THE PUBLICATION OF *Harmony: A New Way of Looking at Our World*, has had a profound impact on my life.[1] It has exerted a gentle, but increasingly potent influence in the world, germinating amongst many individuals, including myself, a seed of latent interest, which is, nine years later in my own case, beginning to find expression in my daily work.

Although I read the book shortly after it was published in 2010, it wasn't until several years later that I began to understand its deeper significance. My first impression was that the Prince of Wales was just reminding us of things that I already knew about various laws and principles, examples of which can be found in nature, music and other aspects of the world. My attitude was 'yes, yes, I know all that, it is important. I am a lover of music, I have a spiritual search, I love to visit Gothic Cathedrals, but these are personal and private interests, and, in the meantime, I need to get on with my work in the more material world'. I now see that what I did not fully realise, is that the Prince's calling is to remind us that there is, or should be, no separation between an esoteric understanding of these inner harmonic laws and principles, their expression in the physical universe and, by extension, one's work in the world.

By the time this seed of a potentially much deeper understanding of the Harmony philosophy and principles germinated in my mind, Richard Dunne, who was a primary school head teacher at the time, had also read the book, but unlike me, he immediately understood its deeper significance. He was already a leader in primary school education, with a track record of OFSTED outstanding status and a strong interest in sustainability.[2] Richard realised that despite his achievements, the educational practices in his school were failing to enable his pupils to make sense of the world in which they found themselves or to adequately prepare them to take on the environmental challenges threatening their future well-being. Inspired by the Prince's book, he chose seven key principles that can help us live and work in harmony with Nature and with one another.

In effect he was responding to the real message of the *Harmony* book. The Prince of Wales himself describes his message as revolutionary and a 'call to action', not merely action in thinking, but action in practice, and he is right! The phrase he uses – 'a new way of looking at our world' – needs to be interpreted quite literally. It is a revelation to understand the deeper significance of the Harmony message, and it requires action in the field of work in which one is already engaged. For Richard Dunne this meant embedding the Harmony principles in the educational timetable and curriculum at the school where he was head teacher. Since then, he has produced a guide which enables all teachers with an interest in the Harmony philosophy and principles to apply them on a daily and weekly basis in their particular school, in ways which would be compatible within the existing curriculum.[3]

Inspired by his example, I began to ask myself the same question. If I too had experienced the same revelation, albeit a bit late in the day, how was I going to put this into practice in my field of work, namely food and farming? Because my day job is the Founder and Chief Executive of the Sustainable Food Trust, I thought one of the best ways to explore this question was to organise a conference around the theme of 'Harmony in Food and Farming', engaging with other leaders in other fields of activity which potentially could be revolutionised, to use the Prince's words, by "the application of Harmony principles".

I think it would be fair to say that everyone who attended the Harmony in Food and Farming conference in 2017 – and there were more than 400 of them – was profoundly influenced by the atmosphere that was co-created by the participants. It was wonderful to hear The Prince of Wales speak and to be in the company of people like Dame Ellen MacArthur, Sir John Eliot Gardiner, Rupert Sheldrake, Sir Anthony Seldon, the Bishop of California Marc Andrus, Chef Barny Haughton and Tony Juniper, to name but a few, each of whom in their own way was interested in exploring this question – how could I review the work in which I am already involved through the lens of the Harmony principles and philosophy?

In light of my ongoing observations, the question I would like to pose is this: if we were to review our particular fields of activity through a Harmony lens, how would this inspiration find expression in our future work? The fields of activity where this question is relevant might include Governments, policy makers and NGOs, as well as those working in sectors such as the built environment, music and the arts, sports and tourism. In each one of these fields, the Harmony perspective can provide an antidote to the tendency towards siloed and reductionist thinking.

In relation to my own field of work, the conference marked the beginning of a new chapter. I intend to continue reviewing the work that I am already doing, striving to make sense of it in light of my awareness of the interconnectedness of everything, specifically related to the way in which the principles of Harmony manifest in my farming and in transforming our food and farming systems.[4] It is my hope that this perspective will strengthen the capacity of the Sustainable Food Trust to have a transformative influence on our future food systems.

The progress that I have made to date is only the beginning – it is halting, it is tentative. However, what I have already realised is that it does not require a fundamental change of direction but rather a mindfulness that many of the activities I have already been practising have been informed intuitively by a subliminal understanding of these principles. Having been informed, inspired and empowered by a more conscious understanding of their existence, I can improve and enhance what I am already doing in my work as an individual, in the community and in the wider world.

As mentioned above, might I be bold enough to suggest that Prince Charles is actually calling each of us to undertake a parallel review of our work, against the backdrop of these extraordinary truths, known by the ancients, but perhaps forgotten? This, at least in part, relates to the grammar of Harmony, as expressed most perfectly in geometry but also in the movement of planets, in music, in maths, in nature, in the form of buildings... I could go on. This is, dare I say it, not only my enquiry, but it could form the heart of our future enquiries, if we are to make sense of the world in which we find ourselves and make it a more habitable and, perhaps, a better place for future generations.

NOTES

1.   HRH The Prince of Wales, Tony Juniper, Ian Skelly, *Harmony: A New Way of Looking at Our World* (London: Harper Collins, 2010).
2.   OFSTED is the Office for Standards in Education, Children's Services and Skills, the UK government's regulator of school standards.
3.   Richard Dunne, *Harmony: A new way of looking at and learning about our world. A teacher's guide to purposeful learning* (London and Bristol: The Harmony Project/ Sustainable Food Trust, 2019).
4.   'Harmony in Food and Farming', Llandovery College, 10th-11th July 2017, https:// sustainablefoodtrust.org/events/harmony-in-food-and-farming/ (Accessed 28 October 2019).

CONTRIBUTORS

DR CRYSTAL ADDEY is a Lecturer in Classics at University College Cork, an honorary Research Fellow in Classics at the University of St Andrews, and a Tutor
for the Sophia Centre for the Study of Cosmology in Culture at the University of
Wales Trinity St. David (the latter since 2008). She is the author of *Divination and
Theurgy in Neoplatonism: Oracles of the gods* (Ashgate 2014) and has published
numerous articles and essays on divination, ritual and religion in Neoplatonism
and late antiquity, on the reception of Plato and Socrates in late antiquity, and on
women, gender and ancient philosophy. She is an elected member of the Board of
Directors for the International Society for Neoplatonic Studies, a Research Fellow
of the Foro di Studi Avanzati Gaetano Massa (FSA; Gaetano Massa Forum for the
Advanced Study of the Humanities), Rome, Italy, and an associate member of the
Centre of Late Antique Religion and Culture, Cardiff University.

The RT. REV. DR. MARC HANDLEY ANDRUS is the eighth bishop of the Episcopal
Diocese of California. He was installed as bishop in 2006 – a position of over-
sight for a diocese comprised of 24,000 communicants in Alameda, Contra Costa,
Marin, San Francisco, and San Mateo Counties, and the cities of Los Altos and
part of Palo Alto. Prior to his election as Bishop of California, Marc served as
Bishop Suffragan in the Episcopal Diocese of Alabama.
    Marc received his Bachelor of Science in Plant Science from the University
of Tennessee, Knoxville, in 1979 and a Masters in Social Sciences from Virginia
Polytechnic Institute and State University, Blacksburg, in 1982. After receiving
his master's degree, Andrus went to work as a regional planner for the Acco-
mack-Northampton Planning District Commission on Virginia's Delmarva Pen-
insula. In 1987, he was awarded a Master of Divinity degree from the Virginia
Theological Seminary in Alexandria, Virginia. In 1990, he became Chaplain at
Episcopal High School in Alexandria, Virginia, until 1997 when he became Rec-
tor of Emmanuel Church in Middleburg, Virginia.
    Marc's leadership has focused on key issues related to peace and justice, in-
cluding immigration reform, civil rights for LBGTQ+ persons, health care, and
climate change. Early in his tenure at the diocese, he co-chaired a community
coalition that paved the way for the rebuilding of St. Luke's Hospital in San Francisco. More recently, his climate advocacy work has taken him to the UN Climate
Conferences in Paris (COP21), Marrakesh (COP22), Bonn (COP23), and Kato-
wice, Poland (COP24), as well as the Dakota Access Pipeline demonstrations at
Standing Rock, North Dakota. He is a member of the We Are Still In Leaders'
Circle, a diverse group of ambassadors for American climate action. https://diocal.
org/bishop-marc

EVE ANNECKE is a Bertha Fellow exploring an independent inquiry into the art and power of retreat. She is a teacher and writer, with a special focus on retreat-based work, facilitating transgressive and transformative learning. In South Africa she co-founded Lynedoch Development, Lynedoch EcoVillage, Lynedoch Children's House and the Sustainability Institute. Her Masters level teaching is in leading transitions, ecological ethics and other ways of knowing. Her work at Schumacher College, UK, spans 26 years as participant, facilitator and teacher in story, ecology and spiritualities. She is currently curating an essay series on *Re-imagining Activisms in 2020, when we are sure of nothing*. She holds a teaching series exploring decolonising through *Pedagogies of Sanctuary*. Her intrigue with the feminine as un-gendered ways of being includes co-creating the Aura Fellowship for women activists. She continues her lifelong connection with children through creating a Rights of Nature declaration at Lynedoch EcoVillage with a group of ten year olds. http://www.eveannecke.net/

KAYLEEN ASBO, PHD, is a pianist, composer, lecturer and multimedia event producer. She holds advanced degrees in the three separate fields of music, psychology and mythological studies. She is the resident cultural historian for the Santa Rosa Symphony in California and a preconcert lecturer for the San Francisco Opera and Mendocino Music Festivals, as well as a lecturer on a plethora of topics ranging from Dante and T.S. Eliot to Women's History for the Osher Lifelong Learning Institutes at UC Berkeley, Dominican University and Sonoma State University. A faculty member of the San Francisco Conservatory of Music for nineteen years, she is now the founder and artistic director for Mythica, an organization that focuses on bringing hope and healing to the world through the integration of contemplation, scholarship and the arts through classes, concerts, rituals and pilgrimage to sacred sites. www.kayleenasbo.com

DR LUCI ATTALA is a senior lecturer in Anthropology at UWTSD. Her ethnographic work focuses on how water's physical behaviours in Kenya, Spain and Wales shape people's lives. Adopting a posthuman, New Materialities framework, Luci's work rejects the environmental violence and domination that human exceptionalist perspectives produce and offers an alternative material lens with which to view how the other-than-human agents play a part in determining the ways that lives can be lived. Rather than inert resources, Luci considers materials' shifting patterns of behaviour to be co-generative regulators that provoke and resist human action, as much as enable it. Luci has also written about eating/edibility, plants and the body. She co-edits the *Materialities in Archaeology and Anthropology series* with University of Wales Press; she has been part of the Harmony Initiative from the outset; won the Green Gown Award in 2015; is a member of the Educere Alliance; and sits on the Board of Directors for Campus Global, an alternative education platform that runs out of Oxford University. l.attala@uwtsd.ac.uk

HELEN BROWNING has a very mixed organic farm in Wiltshire, with dairy, beef, pigs, sheep, agro-forestry and arable enterprises. 'Helen Browning's Organic' supplies organic meat to multiple and independent retailers and she also runs a dining pub with rooms, and a restaurant. She is Chief Executive of the Soil Association, a member of the Food Ethics Council, trustee of the RSPB and recently appointed to the RSA's Food Farming and Countryside Commission as well as the BBC's Rural Affairs Advisory Committee. Helen was formerly Director of External Affairs at the National Trust and has had a number of roles in agri-politics over the years, including the 'Curry' commission on the future of food and farming. Helen was awarded an OBE in 1998 for her services to organic farming. http://www.helen-browningsorganic.co.uk/, http://www.soilassociation.org/

WENDY BUONAVENTURA is a dancer, choreographer and writer. Her work explores cultural myths about women, with a focus on the arts. *Serpent of the Nile: Women and Dance in the Arab World* (Saqi) was an *Observer* Book of the Year. She was on the programme of the Bath Literature Festival, talking about her subsequent book *I Put A Spell On You: Dancing Women from Salome to Madonna* (Saqi): 'Buonaventura's theatrical flourish never deserts her in her bravura leaps through history... The breadth of her knowledge is apparent in every gem of an anecdote' (*Daily Telegraph*). Her stage shows combining text, dance and music have been performed at the Paris Institut du Monde Arabe and London's Sadler's Wells Theatre. The TV documentary about her, *Making Mimi*, can be found on her website and on You Tube. Her BBC Radio 4 series *Dances With The Devil* was chosen as Pick of the Day and Choice of the Week. Other radio appearances include 'Woman's Hour' and 'Off The Page'. She has recently created a cabaret using material from her latest book *Dark Venus: Maud Allan and the Myth of the Femme Fatale* (Amberley). www.buonaventura.com

DAVID CADMAN is a Quaker writer and a Harmony Professor of Practice at the University of Wales Trinity Saint. He has held a number of professorial chairs, fellowships and visiting professorships, both in the UK and America. He is a Fellow of the Temenos Academy. He was formerly Chairman of the Prince's Foundation and a Trustee of The Prince's School of Traditional Arts. He is presently Harmony Advisor to The Prince's Foundation. He was the co-editor (and later editor) of Speeches and Articles 1968-2017 (University of Wales Press, 2015 and 2019), a collection of the speeches and writings of HRH The Prince of Wales. He was the co-editor of *Why Love Matters* (Peter Lang, 2016) and of *Peacefulness* (Spirit of Humanity Press, 2017) and is the author *Love and Divine Feminine* (to be published by Panacea Books, 2020). His work can be found at https://davidcadmanatwork.com Under his pen name, William Blyghton, he is the author of The Suffolk Trilogy, which comprises *The House by the Marsh*, *Abraham Soar* and *Noah: An Old Fool* (Panacea Books, 2017, 2018 and 2019) and *Finding Elsewhere*, a collection of stories for our time (Panacea Books, 2018). *Mary's Story: A Kind of Knowing* will be published by Panacea Books in 2020. https://williamblyghton.com

NICHOLAS CAMPION is Director of the Harmony Institute, Associate Professor in Cosmology and Culture, Principal Lecturer in the Institute of Education and Humanities and Director of the Sophia Centre for the Study of Cosmology in Culture at the University of Wales Trinity Saint David. He is Programme Director of the University's MA in Cultural Astronomy and Astrology and the MA in Ecology and Spirituality. His books include the two-volume *History of Western Astrology* (Bloomsbury 2008/9), *Astrology and Cosmology in the World's Religions* (NYU Press 2012), and *The New Age in the Modern West: Counter-Culture, Utopia and Prophecy from the late Eighteenth Century to the Present Day* (Routledge 2015). Current projects include the six-volume *Cultural History of the Universe* (Bloomsbury, forthcoming), for which he is General Editor. His recent publications include 'Einstein's Cosmic Religion', *Intersections of Religion and Astronomy*, ed. Aaron Ricker (Routledge, 2020); 'Adventures in Space: Harmony, Sustainability and Environmental Ethics' in Nicholas Campion and Chris Impey (eds), *Imagining Other Worlds: Explorations in Astronomy and Culture* (Lampeter: Sophia Centre Press, 2018), pp. 69–85; and 'The Importance of Cosmology in Culture: Contexts and Consequences', in Abraao Jesse Capistrano de Souza (ed.), *Cosmology* (Rijeka: InTech Open, 2017), pp. 3–17. https://www.uwtsd.ac.uk/staff/nicholas-campion/

ILARIA CRISTOFARO is a PhD student at Università degli Studi della Campania *Luigi Vanvitelli*, Italy. Her interdisciplinary doctoral project investigates the astronomical orientation of temples and cities in ancient Campania from the Archaic to the Hellenistic period, supervised by archaeologist Prof Carlo Rescigno. Ilaria is an astronomer with a specialisation in cultural astronomy, archaeoastronomy & skyscape archaeology. She holds a B.Sc. in Astronomy from Univerisità *Alma Mater* di Bologna, Italy, and an MA in Cultural Astronomy & Astrology at University of Wales Trinity Saint David, UK. She is president director of Osservatorio Astronomico Sirio at Castellana Grotte (BA, Italy), where she leads outreaching activities to introduce the public to the celestial vault. She is a member of Società Italiana di Archeoastornomia (SIA), the European Society for Astronomy in Culture (SEAC) and the European Association of Archaeologists (EAA), and collaborates with the Sophia Centre for the Study of Cosmology in Culture. https://unina2.academia.edu/IlariaCristofaro , https://ilariacristofaro.wordpress.com/

TANIA DAVIES is the Portfolio Director of Health within the Faculty of Business and Management's School of Sport, Health and Outdoor Education, at the University of Wales Trinity Saint David, with responsibility for a range of health-related programmes. She has been a lecturer for over thirteen years, is an ERSC accredited social researcher and a Senior Fellow of the Higher Education Academy. Her research interests are wellbeing and health, psychopathology and the mental health of those currently and previously serving in the armed forces. She is currently working toward a PhD which examines the links between the Wellbeing of Future Generations (Wales) Act 2015 and wellbeing in the HE (Higher Education) workplace.

LEO DOWNER has over thirty years experience in the management of clinicians and developing health services; he has worked in strategy and delivery towards the improvement of health, wellbeing and safety nationally and internationally. His previous roles have included: microbiologist in Great Ormond Street Hospital for Sick Children; Head of Drug and Crime Reduction in the Government Office for London; Business Manager for Counter-terrorism in the Home Office HQ Strategy and International Directorate; Chief of Staff to Baroness Newlove, House of Lords. Combining an evidenced based approach to policy development, objectives and strategy; transformational team leadership, and a demonstrable communicator with clients, clinicians and senior leaders in government, communities and business, Leo has founded new health approaches, transformed government operations, and led on national delivery.

RICHARD DUNNE is an environmentalist and thought leader in education. He was formerly head-teacher of a primary school in south-east England, where he identified and implemented Harmony principles in educational practice through the school year and across age-groups. He now leads the education work of The Harmony Project. He believes passionately that the best way for children to be motivated in their learning is when it makes sense to them and when they have a key role to play: through combining core skills to purposeful enquiries of learning, the children start to develop their own vision of how they want to see their world and just as importantly what they can do to make it happen. http://www.theharmonyproject.org.uk/

MIKE DURKE is a Senior Lecturer at the University of Wales Trinity Saint David and a Councillor for the City and County of Swansea representing the Cockett ward – the area where he grew up and continues to live. Apart from his family – partner Suzanne, children and grandchildren and his love of cricket – there have been two overriding passions in his working life: child welfare and the potential of communities to shape their own futures. For over thirty-five years, his public service career has involved youth work, social services for adults and children, a stint as a Ministry of Defence Police Constable, management roles in local and national government and twelve years as CEO of a development trust in a much-maligned community full of fantastic people. Following the child care expert Dr Spock, he feels that we should trust ordinary, everyday people to design and deliver the services they need because they know far more than we think they do. http://www.mikedurke.co.uk/

LOUISE EMANUEL is Programme Director for undergraduate programmes at Carmarthen Business School. Her undergraduate teaching includes Global Business Challenges, Technological Change and Innovation and Globally Responsible Business. She is also module tutor for The Challenges of Sustainability module on the school's MBA Sustainability Leadership and supervises a number of PhD and DBA students in the areas of sustainable and social business. Her research interests include sustainable business and place. She is also a Commissioner for the Royal Commission on the Ancient and Historical Monuments of Wales.

ALAN EREIRA was an award-winning BBC documentary producer with BBC radio and television for 30 years from 1965, specialising in history, then setting up his own TV company. He worked for many years with Terry Jones, of Monty Python fame, and they wrote a number of history books together. He became particularly known for his work with the Kogi on two films, in 1990 and 2012, and founded the Tairona Heritage Trust. Married to a distinguished book indexer, he became an Honorary Fellow of UWTSD in 2014 and a Professor of Practice in 2017. He lives in London and has dual British-Portuguese citizenship. http://alanereira.co.uk/

SIR JOHN ELIOT GARDINER stands as an international leader in today's musical life, respected as one of the world's most innovative and dynamic musicians, constantly at the forefront of enlightened interpretation. His work as Artistic Director of his Monteverdi Choir, English Baroque Soloists and Orchestre Révolutionnaire et Romantique has marked him out as a central figure in the early music revival and a pioneer of historically informed performance. As a regular guest of the world's leading symphony orchestras, such as the London Symphony Orchestra, Symphonieorchester des Bayerischen Rundfunks, Royal Concertgebouw Orchestra and Gewandhausorchester Leipzig, Gardiner conducts repertoire from the 17th to the 20th century.

DR SCHERTO GILL is Senior Fellow at the GHFP Research Institute, Visiting Fellow at the University of Sussex, Research Fellow at University of Wales and Fellow of Royal Society of the Arts (FRSA). Through research, grassroots projects and published work, Scherto actively explores ways to implement ideas such as deep dialogue, ethics of caring, holistic well-being, and harmony in social transformation and peace. Scherto writes in the fields of education, ethics, and peace. Her most recent books include: *Happiness, Flourishing and the Good Life: A Transformative Vision of Human Well-Being* (Routledge), *Understanding Peace Holistically* (Peter Lang), *Beyond the Tyranny of Testing* (Oxford University Press), *Ethical Education: Towards an Ecology of Human Development* (Cambridge University Press); and *Being Peace and Making Peace* (Spirit of Humanity Press). http://www.scherto.com/

TOTO GILL is an academic, jazz pianist and composer currently in his second year of undergraduate studies at the University of Oxford, where he was awarded a music scholarship. Within Oxford, Toto has examined neo-Confucian concepts of harmony as a framework for understanding socio-musical practices within jazz from a relational perspective. He is also an associate at the GHFP Research Institute, where he contributes in various ongoing projects within the fields of relational ontology, harmony studies and love within Western philosophy.

MARK GOYDER spent over 15 years working in manufacturing industry. He then joined the Royal Society for the encouragement of Arts, Manufactures and Commerce. While he was there, he organised a range of dialogues with business leaders, and this led to the development of the influential RSA Tomorrow's Company Inquiry – a business led inquiry into the role of business in a changing world. Following the success of this project in 1996 he set up Tomorrow's Company as an independent research organisation to inspire and enable business to be a force for good in society. The organisation laid the foundations for the restatement of directors' duties and has initiated worldwide changes in the approach to the stewardship responsibilities of investors. After over twenty years of challenging and working with business leaders, investors and policymakers, he stepped down as CEO in 2017. http://www.markgoyder.com/

STEPHAN HARDING holds a doctorate in ecology from the University of Oxford and is Deep Ecology Research Fellow and Senior Lecturer in Holistic Science at Schumacher College, where he coordinated the MSc in Holistic Science for about 20 years. In 2007, Stephan was appointed co-holder with James Lovelock of the Arne Naess Chair in Global Justice and the Environment at the University of Oslo, Norway. He has taught Gaia theory, deep ecology, and holistic science all over the world. Stephan is author of *Animate Earth: Science, Intuition and Gaia*. His latest book is *Gaia Alchemy* forthcoming in 2021. https://www.schumachercollege.org.uk/

PATRICK HOLDEN is the founder and chief executive of the Sustainable Food Trust, an organisation founded in 2012 and working internationally to accelerate the transition to more sustainable food systems. Prior to this he was director of the Soil Association (until 2010), during which time his advocacy and campaigning for more sustainable food systems was underpinned through the development of the organic standards and market-place. His farming experience spans nearly 50 years, centred on Holden Farm Dairy, now the longest established organic dairy farm in West Wales, where he produces Hafod cheese from the milk of his 80 Ayrshire cows. He was awarded a CBE for services to organic farming in 2005, is Patron of the UK Biodynamic Farming Association, and was elected an Ashoka Fellow in 2016. Patrick is a regular writer, broadcaster, and speaker at public events. http://www.sustainablefoodtrust.org/, https://www.theharmonyproject.org.uk/

SOPHIE HOWARD teaches sculpture at the Royal West of England Academy in Bristol, UK, where she also runs 'HOURS', a small gallery. She makes permanent sculpture in ceramics, resin and bronze, as well as working in natural materials that disintegrate over time. Her pieces translate visions and thoughts into solid forms, expressing something about life beyond the purely visible, and she works on collections of pieces exploring themes such as the nude, animals and dance. Born in West Dorset in 1950s, she grew up in Yorkshire and trained at Winchester Art School before moving to London and returning to live in the South West of England in the 1980s. http://sophiehoward.co.uk/

JACK HUNTER is an anthropologist exploring the borderlands of consciousness, religion, ecology and the paranormal. He lives in the hills of Mid-Wales with his family. He is a Tutor in the Sophia Centre at the University of Wales Trinity Saint David, as well as an Honorary Research Fellow with the University's Alister Hardy Religious Experience Research Centre. He is also currently an Access to Higher Education lecturer in Humanities and Social Sciences at Newtown College. He completed a Permaculture design course at Chester Cathedral in 2017 and has been working on the One School One Planet project to develop a Permaculture programme for the mainstream curriculum. He is the author of *Why People Believe in Spirits, Gods and Magic* (2012) and *Engaging the Anomalous* (2018), editor of *Strange Dimensions: A Paranthropology Anthology* (2015), *Damned Facts: Fortean Essays on Religion, Folklore and the Paranormal* (2016), and co-editor with Dr David Luke of *Talking With the Spirits: Ethnographies from Between the Worlds* (2014). http://www.jack-hunter.webstarts.com/

TONY JUNIPER CBE is Chair of the official nature conservation agency Natural England and a Professor of Practice at the University of Wales Trinity Saint David. Before joining Natural England in April 2019 he was the Executive Director for Advocacy and Campaigns at WWF-UK, a Fellow with the University of Cambridge Institute for Sustainability Leadership and President of the Wildlife Trusts. From 2008 to 2018 he was a Special Advisor with HRH The Prince of Wales's Rainforests Project and International Sustainability Unit. Juniper speaks and writes widely on conservation and sustainability themes and is the author of many books, including the multi-award-winning bestseller *What has Nature ever done for us?*, published in 2013, and *Harmony*, co-authored with HRH The Prince of Wales and Ian Skelly and published in 2010. Tony began his career as an ornithologist, working with Birdlife International. From 1990 he worked at Friends of the Earth, initially leading the campaign for the tropical rainforests, and from 2003–2008 was the organisation's executive director. From 2000–2008 was Vice Chair of Friends of the Earth International. Tony was the first recipient of the Charles and Miriam Rothschild medal (2009) and was awarded honorary Doctor of Science degrees from the Universities of Bristol and Plymouth (2013). The Ladybird guide to climate change, co-authored with HRH The Prince of Wales and Emily Shuckburgh, was published in January 2017. His latest book, 'Rainforest', was published in April 2018. In 2017 he was appointed a Commander of the British Empire (CBE).

TREVOR LEAMAN is a member of the Australian Indigenous Astronomy Research Group, (www.aboriginalastronomy.com.au). He is also a PhD candidate in the School of Humanities & Languages, the University of New South Wales, researching the astronomical traditions of the Wiradjuri people of central NSW, with support & guidance from Wiradjuri Elders & Communities. In a past life he earned diplomas in civil & mechanical engineering, degrees in biology & forest ecology, and more recently earned his MSc in astronomy, which included a major project examining the astronomy of the Aboriginal people near Ooldea, South Australia. He has also worked as an astronomy educator at Ayres Rock Resort, Launceston Planetarium, and Sydney Observatory. https://unsw.academia.edu/TrevorLeaman , https://www.researchgate.net/profile/Trevor_Leaman

CAROLINE LOHMANN-HANCOCK is a Senior Lecturer/Researcher within Social Justice, equity and diversity at the University of Wales Trinity Saint David's and leads on PhDs in Social Justice. She has also worked in a variety of academic and voluntary settings which focus on equality. Her research interests centre on social justice and exclusion, developing a sustainable and equitable society, power and control in society, identity, learning and teaching and the social reproduction of inequality. She has worked on a range of funded research projects including 'Evaluation of 'Hafal Let's Talk Project'; Raising Awareness and Changing Attitudes with Young People about Violence against Women and Domestic Abuse: Hafan, UNCRC training across Wales: Welsh Government and a Training Needs Assessment of UNCRC for the Jersey Government alongside giving young people a voice within the 'Youth Engagement – Ein Llais' project on behalf of Dyfed-Powys Police Commissioner. www.uwtsd.ac.uk/staff/caroline-lohmann-hancock/

DAME ELLEN MACARTHUR made yachting history in 2005, when she became the fastest solo sailor to circumnavigate the globe. She remains the UK's most successful offshore racer ever, having won the Ostar, the Route du Rhum, and finished second in the Vendée Globe. She received the French Legion of Honour from President Nicolas Sarkozy in 2008, three years after having been knighted by HM Queen Elizabeth II. Having become acutely aware of the finite nature of the resources on which our linear economy relies, she retired from professional sailing to launch the Ellen MacArthur Foundation in 2010. The Foundation works to accelerate the transition to a circular economy, establishing the concept on the agenda of decision makers around the world. Since the publication of its first economic report in 2012, the Foundation has launched global initiatives on plastics, fashion, food and finance, developed innovation networks with educators, businesses, cities and governments, and published more than 20 reports and books. Dame Ellen is a World Economic Forum Global Agenda Trustee for Environment and Natural Resource Security and member of its Platform for Accelerating the Circular Economy and she sat on the European Commission's Resource Efficiency Platform between 2012 and 2014. http://www.ellenmacarthurfoundation.org/

EMILIE MARTIN is a writer and teacher. She has taught in primary schools in and around London, most recently at Ashley C of E Primary School, alongside Richard Dunne. She is a contributor to The Harmony Project, which promotes the application of Harmony principles to education, food and farming, and other fields, and has researched and documented the development of a curriculum based on principles of Harmony in primary schools. https://www.theharmonyproject.org.uk/

Until retiring in 2013 DR JOSEPH MILNE was Honorary Lecturer at the University of Kent where he taught on the MA course in Mysticism and Religious Experience. His interests range from Platonism to medieval mysticism and theology, and in particular the transformations of metaphysical thought that have occurred in different periods of Western civilisation. He is also interested in Patristic scriptural exegesis from Origen to Hugh of St Victor and Meister Eckhart. Currently he is researching classical and medieval Natural Law. He is Editor of *Land & Liberty*, the journal of the Henry George Foundation, and a member of the team editing for *The Annotated Works of Henry George*. Also he is Associate Editor of *Medieval Mystical Theology*, the journal of The Eckhart Society. Publications include *The Ground of Being: Foundations of Christian Mysticism* (2004), *Metaphysics and the Cosmic Order* (2008), *The Mystical Cosmos* (2013), and contributed chapter in *Mystical Theology and Continental Philosophy* (2017). He is a Trustee of The Eckhart Society and a Fellow of the Temenos Academy. http://josephmilne.co.uk/

JEREMY NAYDLER holds a PhD in Theology and Religious Studies, and has written several books on religious life in antiquity and on the history of consciousness. He is a Fellow of the Temenos Academy, a member of its Academic Board, and teaches on the Temenos Academy Foundation Course on the Perennial Philosophy. His most recent book is *In the Shadow of the Machine: The Prehistory of the Computer and the Evolution of Consciousness* (2018).

RACHEL PARKER is a Fellow of the Royal Society of Public Health and a senior mental health consultant. Her current research centres upon adolescent self-harm, exploring what is acceptable and feasible for prevention intervention support in secondary schools in Wales. Her consultancy projects build community sustainability, resilience and well-being, developing quality partnerships to strengthen expertise in community empowerment. By maximising opportunities for knowledge exchange, co-production and research, this generates positive impacts for societal benefit, helping to address the societal challenges we face together. https://thewellbeingdoc.com/

ANGIE POLKEY is a founding member of the Lampeter Permaculture Group and Lampeter Resilience Hub. Her background in ecological research and nature conservation helps her understand and teach the natural principles, patterns and processes that have inspired permaculture. Angie first learnt about and practised permaculture in a tiny urban garden in Sussex and then moved to Wales with her husband in 1997. Their small acreage in west Wales has several forest gardens, vegetable beds, a polytunnel, wildlife habitats, and hard-working chickens, and is designed to minimise inputs and optimise sustainable outputs over time. After a long career in conservation organisations such as the RSPB, Wildlife Trusts, Learning through Landscapes, Denmark Farm Conservation Centre and PONT Cymru, Angie now advises landowners how to improve their sites ecologically, teaches permaculture, and facilitates participatory democracy towards positive sustainable change.

M. A. RASHED earned her master's degree in Cultural Astronomy from the University of Wales Trinity Saint David and is currently pursuing her doctoral studies at the Department of Religion at Rice University in Texas. Her major research interest is pre-Islamic and Islamic cultural astronomy and cosmology in its all manifestations, whether classical or contemporary, 'mathematical' or 'folkloric', 'scientific' or 'mystical', extending in geography from Central Asia to Northern Africa and Andalusia. She is also very interested in studying the corpus of Islamic eschatology and apocalyptica, and Islamic movements which were mobilized by such literature, from the cosmological perspective of cultural astronomy. Her current academic projects include translating to English a seventeenth century Ottoman apocalyptic text, editing and transcribing a sixteenth century Syrian apocalyptic manuscript, and a joint-article on Bedouin poetry of Arabia with Dr. Daniel Varisco, Post-Doctoral Fellow at the Institute for Advanced Studies, School of Historical Studies, Princeton. She had also recently contributed to a forthcoming anthology of apocalyptic authors by Eyüp Öztürk, professor of Islamic History and Arts at Trabzon University in Turkey, and David Cook, professor of Islamic studies at Rice University. To aid her research, Mai is studying Ottoman Turkish, Attic and Byzantine Greek in addition to Classical and Medieval Latin.

SNEHA ROY is a doctoral student at the University of Wales Trinity Saint David, specialising in women's participation in religious movements, leadership, conflict management, and interreligious dialogue. She is an anthropologist by her academic training; a former Commonwealth scholar, a fellow of KAICIID International Dialogue Centre, a fellow of Mindanao Peacebuilding Institute, and a youth advocate with UNESCO.

(AHARON) DAVID CLIVE RUBIN is a ritual scribe, rabbi, astrologer, and clinical hypnotherapist. He has a passion for prophetic meditation, is author of *Eye to the Infinite: A Guide to Jewish Meditation: How to increase Divine awareness*, and gives lectures on Jewish meditation. His interest in astrology has taken him from Ibn Ezra to Vedic, Western Medieval, and, latterly, to the modern humanistic approach. In 2019, he was awarded an MA with Distinction in Cultural Astronomy and Astrology from the University of Wales Trinity Saint David. Daily inspired by his four lovely children, he enjoys writing, gardening, hiking, and meditating. http://jewish-meditation.weebly.com/latest-edition.html

JOHN SAUVEN is the Executive Director of Greenpeace UK and a Harmony Professor of Practice at the University of Wales Trinity Saint David. With a background in forests he was instrumental in getting protection for the Great Bear temperate rainforest in Canada. It was an epic battle between logging companies, timber traders and their retail customers in Europe and North America. He co-ordinated the international campaign to secure a moratoria on further destruction of the Amazon by soya producers and later similar tactics were used to get a cattle moratorium. Similar tactics were used elsewhere to tackle the drivers of deforestation including for paper and palm oil in Indonesia. Ultimately it changed the supply chains of many of the world's biggest corporations. It was one of Greenpeace's most successful campaigns to protect large areas of the world's last intact rainforests providing both climate and biodiversity protection as well as local peoples livelihoods. In 2010 John Sauven started the campaign to protect the Arctic from oil exploration. It turned into a heroic battle first with Russia's Gazprom and then Shell. In 2015 Shell pulled out of the Arctic. www.greenpeace.org.uk

DR RUPERT SHELDRAKE is a biologist and author of more than eighty-five technical papers and nine books, including *Science and Spiritual Practices,* and the co-author of six books. He studied at Cambridge and Harvard Universities. As a Fellow of Clare College, Cambridge, he was Director of Studies in Cell Biology, and was also a Research Fellow of the Royal Society. He worked in Hyderabad, India, as Principal Plant Physiologist at the International Crops Research Institute for the Semi-Arid Tropics (ICRISAT), and also lived for two years in the Benedictine ashram of Father Bede Griffiths on the bank of the Cauvery River in Tamil Nadu. From 2005–2010, he was Director of the Perrott-Warrick Project for the study of unexplained human and animal abilities, funded from Trinity College, Cambridge. He is currently a Fellow of the Institute of Noetic Sciences in Petaluma, California, and of Schumacher College in Dartington, Devon. He lives in London and is married to Jill Purce, with whom he has two sons. His web site is www. sheldrake.org.

DR ANGUS M SLATER is currently a Lecturer in Christian Theology and Interfaith Studies at the University of Wales Trinity St David in Lampeter. His research interests are centred around theological considerations of authority and aesthetics within the Christian and Islamic traditions, including queer approaches to authority, the formation of identity, and the role that aesthetics plays in religious construction. These interests can be seen in the recent publication of *Radical Orthodoxy in a Pluralistic World* with Routledge, as well as further work in the *Journal of Law, Religion, and State, Theology & Sexuality,* and *Interreligious Studies and Intercultural Theology.* Recent activity includes speaking at the UK Department of Education on LGBT+ and Faith Issues, assisting the UNESCO Peace Conference on multifaith initiatives, and advising the Church in Wales Doctrinal Commission regarding religious same-sex marriage.

GUNHILD A. STORDALEN is the founder and executive chair of EAT. She is a driving force in linking climate, health and sustainability issues across sectors to transform the global food system. Together with Petter A. Stordalen, Gunhild established the Stordalen Foundation in 2011, under which she later founded the EAT Initiative together with Professor Johan Rockström and the Stockholm Resilience Center (SRC). In 2014, they gathered 400 leaders from 28 countries for the inaugural EAT Stockholm Food Forum. In 2016, the Wellcome Trust joined the SRC and the Stordalen Foundation as core partner of EAT and helped grow the initiative to become a global multi-stakeholder platform for transforming the world's food system. Gunhild sits on several boards and councils including the United Nation's Scaling Up Nutrition (SUN) Movement Lead Group, the World Economic Forum (WEF) Stewardship Board on Food Systems and the WEF Global Future Council on Food Security and Agriculture, the British Telecom Group's Committee for Sustainable and Responsible Business, the Global Alliance for Improved Nutrition (GAIN) Partnership Council, and the international advisory board of the SRC. Gunhild is a published scientist and a renowned public speaker with a distinct, personal voice and a passionate commitment to her work. WWF Sweden named her Environmental Hero of the Year in 2014 and in 2015 she was appointed a Young Global Leader by WEF. She has been ranked among the 150 most influential business communicators in Sweden by the news magazine Resumé and among the 100 most powerful women in Norway by the financial magazine Kapital. Gunhild is a medical doctor from the University of Oslo and holds a PhD in pathology/orthopedic surgery. She is a Norwegian national.

RHODRI THOMAS is lecturer in Religious Studies at the University of Wales Trinity Saint David, having been appointed in 2016. His research to date includes work on the relationship between religion and ecology. More specifically, he is interested in the function and fate of non-human creation in Jewish and early Christian 'apocalyptic' texts and the extent to which the Bible can be viewed as a useful resource for the formulation of a positive environmental ethic.

GLENDA TINNEY: I started my career as a biologist at Lancaster University. In 2002, I combined my interest in wellbeing, outdoor learning and sustainability when I was appointed a lecturer at the University of Wales, Trinity Saint David, delivering undergraduate and postgraduate programmes linked to environmental awareness, conservation and sustainable development. In 2007 I was appointed within the School of Early Years (now Childhood and Education) where I have been able to link my interest in sustainability and wellbeing with young children's experiences of Education for Sustainable Development and Global Citizenship (ESDGC) in Wales. I have a level 3 Forest School Leader qualification which supports my work with children and adults in the outdoors. I have increasingly been drawn to the Principles of Harmony in Education and the significance of developing early years settings that interconnect children with the nonhuman world. https://www.uwtsd.ac.uk/staff/glenda-tinney/

ANGELA VOSS has been involved in devising and teaching the Masters programmes in Myth, Cosmology and the Sacred in Canterbury UK for the last fifteen years. Her passion for Renaissance music and culture led her to delve into the magical world of the fifteenth century magus Marsilio Ficino, and from there to the Western esoteric traditions and the power of the symbolic to awaken the human soul. She has published extensively on Ficino's astrological music, astrology and divination, Neoplatonism and magic, and, more recently, on transformative learning.

CONTRIBUTORS   587

She has always felt the need to be a bridge builder, bringing imagination, creativity and reflexive writing into academic research. Her most recent publications include *Re-enchanting the Academy* (edited with Simon Wilson), 2017. Most of her work can be found at https://canterbury.academia.edu/AngelaVoss and she can be contacted at imaginalcosmos@gmail.com. Her website is: http://www.mythcosmologysacred.com/

NICHOLA WELTON is a Senior Lecturer at the University of Wales Trinity Saint Davids, Yr Athrofa, Institute of Education and Humanities. Her teaching has focused on Education for Sustainable Development and Sustainable Communities for programmes in Equity and Diversity, and Youth work. Key interests are transformational educational approaches, social-ecology and environmental justice, and addressing social justice issues within the sustainability debate. In addition, as a qualified youth worker, the importance of young people's voices is also central to my work, which has included working with the Welsh Government and Jersey Government on the United Nation's Convention of the Rights of the Child.

INDEX